Translational Bioinformatics

Volume 11

Series editor
Xiangdong Wang, MD, Ph.D.
Professor of Medicine, Zhongshan Hospital, Fudan University Medical School, China
Director of Shanghai Institute of Clinical Bioinformatics, (www.fuccb.org)

Aims and Scope

The Book Series in Translational Bioinformatics is a powerful and integrative resource for understanding and translating discoveries and advances of genomic, transcriptomic, proteomic and bioinformatic technologies into the study of human diseases. The Series represents leading global opinions on the translation of bioinformatics sciences into both the clinical setting and descriptions to medical informatics. It presents the critical evidence to further understand the molecular mechanisms underlying organ or cell dysfunctions in human diseases, the results of genomic, transcriptomic, proteomic and bioinformatic studies from human tissues dedicated to the discovery and validation of diagnostic and prognostic disease biomarkers, essential information on the identification and validation of novel drug targets and the application of tissue genomics, transcriptomics, proteomics and bioinformatics in drug efficacy and toxicity in clinical research.

The Book Series in Translational Bioinformatics focuses on outstanding articles/chapters presenting significant recent works in genomic, transcriptomic, proteomic and bioinformatic profiles related to human organ or cell dysfunctions and clinical findings. The Series includes bioinformatics-driven molecular and cellular disease mechanisms, the understanding of human diseases and the improvement of patient prognoses. Additionally, it provides practical and useful study insights into and protocols of design and methodology.

Series Description

Translational bioinformatics is defined as the development of storage-related, analytic, and interpretive methods to optimize the transformation of increasingly voluminous biomedical data, and genomic data in particular, into proactive, predictive, preventive, and participatory health. Translational bioinformatics includes research on the development of novel techniques for the integration of biological and clinical data and the evolution of clinical informatics methodology to encompass biological observations. The end product of translational bioinformatics is the newly found knowledge from these integrative efforts that can be disseminated to a variety of stakeholders including biomedical scientists, clinicians, and patients. Issues related to database management, administration, or policy will be coordinated through the clinical research informatics domain. Analytic, storage-related, and interpretive methods should be used to improve predictions, early diagnostics, severity monitoring, therapeutic effects, and the prognosis of human diseases.

Recently Published and Forthcoming Volumes

Computational and Statistical Epigenomics
Editor: Andrew E. Teschendorff
Volume 7

Allergy Bioinformatics
Editors: Ailin Tao, Eyal Raz
Volume 8

Transcriptomics and Gene Regulation
Editor: Jiaqian Wu
Volume 9

Pediatric Biomedical Informatics - Computer Applications in Pediatric Research (Edition 2)
Editor: John J. Hutton
Volume 10

More information about this series at http://www.springer.com/series/11057

Xiangdong Wang • Christian Baumgartner
Denis C. Shields • Hong-Wen Deng
Jacques S. Beckmann
Editors

Application of Clinical Bioinformatics

Springer

Editors
Xiangdong Wang
Zhongshan Hospital, Fudan University
Shanghai Institute of Clinical Bioinformatics
Shanghai, China

Denis C. Shields
School of Medicine
University College Dublin
Dublin 4, Ireland

Jacques S. Beckmann
Section of Clinical Bioinformatics
Swiss Institute of Bioinformatics
Switzerland

Christian Baumgartner
Institute of Health Care Engineering with European Notified Body of Medical Devices
Graz University of Technology
Graz, Austria

Hong-Wen Deng
Center for Bioinformatics and Genomics, Department of Biostatistics and Bioinformatics
Tulane University School of Public Health and Tropical Medicine
New Orleans, LA, USA

ISSN 2213-2775 ISSN 2213-2783 (electronic)
ISBN 978-94-017-7541-0 ISBN 978-94-017-7543-4 (eBook)
DOI 10.1007/978-94-017-7543-4

Library of Congress Control Number: 2016933682

Springer Dordrecht Heidelberg New York London

Printed on acid-free paper

Springer Science+Business Media B.V. Dordrecht is part of Springer Science+Business Media (www.springer.com)

Contents

Chapter 1
The Era of Big Data: From Data-Driven Research to Data-Driven Clinical Care

Christian Baumgartner

Abstract When the era of big data arrived in the early nineteen nineties, biomedical research boosted new innovations, procedures and methods aiding in clinical care and patient management. This chapter provides an introduction to the basic concepts and strategies of data-driven biomedical research and application, an area that is explained using terms such as *computational biomedicine* or *clinical/medical bioinformatics*. After a brief motivation it starts with a survey on data sources and bioanalytic technologies for high-throughput data generation, a selection of experimental study designs and their applications, procedures and recommendations on how to handle data quality and privacy, followed by a discussion on basic data warehouse concepts utilized for life science data integration, data mining and knowledge discovery. Finally, five application examples are briefly delineated, emphasizing the benefit and power of computational methods and tools in this field. The author trusts that this chapter will encourage the reader to handle and interpret the huge amount of data usually generated in research projects or clinical routine to exploit mined bioinformation and medical knowledge for individualized health care.

Keywords Computational biomedicine • Data integration and management • Knowledge discovery • Data mining • Clinical applications

1.1 Introduction

In the past two decades, the new era of "big data" in experimental and clinical biomedicine has arrived and grown as a direct consequence of the availability of large reservoirs of data. Data collection in digital form was already underway by the 1960s, allowing for retrospective data management and analysis to be undertaken using computers for the first time. Relational databases arose in the 1980s along with Structured Query Languages (SQL), enabling dynamic, on-demand structural analysis and interpretation of data from complex research designs. The 1990s saw

C. Baumgartner, Ph.D. (✉)
Institute of Health Care Engineering with European Notified Body of Medical Devices, Graz University of Technology, Stremayrgasse 16, A-8010 Graz, Austria
e-mail: Christian.Baumgartner@TUGraz.at

X. Wang et al. (eds.), *Application of Clinical Bioinformatics*,
Translational Bioinformatics 11, DOI 10.1007/978-94-017-7543-4_1

an explosion in the growth of data associated with the emerging use of new high-throughput, lab and imaging technologies in fundamental biomedical research and clinical application. Data warehouses were beginning to be used for storing and integrating various types of data, where different data sources are transformed into a common format and converted to a common vocabulary needed to overcome computational challenges of data-driven research and development. The new era of "computational biomedicine" or "clinical bioinformatics" was born as a multidisciplinary approach that brought together medical, natural and computer sciences, aiming at uncovering unknown and unexpected biomedical knowledge stored in these data sources, which had the potential to transform our current clinical practices (Chang 2005; Wang and Liotta 2011; Coveney et al. 2014). Research areas such as data warehousing and information retrieval, machine learning, data mining, and others thus arose as a response to challenges faced by the computer science and bioinformatics community in dealing with huge amounts of data, enabling a better quality of data-driven decision making. As data are any facts, numbers, images or texts that can be accessed and processed by stand-alone computers or computational networks, the patterns, associations or relationships among available data can provide information about historical patterns and future trends so that undreamt of opportunities emerge for biomedical research and application. This knowledge may help to create a new way of dealing with clinical care and patient management never previously possible. Clinical bioinformatics, which resulted from the big data era, is thus a crucial element of the medical knowledge discovery process where relevant sources of medical information and bioinformation are combined and mined to allow for individualized healthcare.

1.2 The Revolution of High-Throughput and Imaging Technologies and the Flood of Generated Data

In the life sciences, huge amount of data are generated, utilizing the wide spectrum of high throughput and laboratory technologies, and modern health care imaging systems such as MRI or CT. In biomolecular research, microarray based expression profiling and more recently next-generation sequencing (NGS) technologies have become the methodology of choice e.g. for whole transcriptome expression profiling, producing a flood of data that need to be computationally processed and analysed (Worthey 2013; Soon et al. 2013). The most widely used NGS devices, for example, are able to sequence up to 150 bases from both sides of RNA fragments and create a maximum output of up to 1000 GB per run. Most advanced protein profiling technologies are implemented with a broad panel of mass spectrometry-based techniques to separate, characterize and quantify analytes from complex biological samples (Chen and Pramanik 2009; Brewis and Brennan 2010; Woods et al. 2014). Labs are typically equipped with diverse mass spectrometer (MS) systems including TOF-TOF, Quadrupole-TOF, FT-ICR, and LTQ-Orbitrap type analyzers. In this field, shotgun proteomics is a widely used tool for global analysis of protein modifications, where, in a typical LC-MS/MS experiment, hundreds of thousands of tandem mass spectra are typically generated. Sophisticated computational tools for MS sprectra processing and database search strategies are used for the identification of peptide/protein modifications (Baumgartner et al. 2008; Cerqueira et al. 2010; Sjöström et al. 2015). In metabolomics, different fundamental approaches can be distinguished, i.e. untargeted and targeted metabolomics and metabolic fingerprinting (Baumgartner and Graber 2007; Putri et al. 2013; Naz et al. 2014; Zhang et al. 2015). Using targeted metabolomics, quantitiation of a preselected set of known metabolites by determining absolute values of analyte concentrations with the use of internal chemical standards allows for hypothesis-driven research and interpretation of data based on *a-priori* knowledge. To provide a holistic picture of metabolism, untargeted metabolic profiling aims at measuring as many analytes as possible (up to several hundreds) to create a snapshot of the biochemical profile within the analysed sample. The established technologies in metabolomics include – analoguous to proteomics – mass spectrometry based approaches and nuclear magnetic resonance (NMR) spectroscopy, generating thousands to tens of thousands data points per spectrum. Multiple processing steps are required to analyze this huge amount of spectral information, ranging from modalities for denoising, binning, aligning spectra to peak detection and high-level analysis e.g. for biomarker identification and verification (Swan et al. 2013; Netzer et al. 2015).

Nowadays bioimaging devices with increasing resolution are widely used in biological and clinical laboratories, generating imaging data with hundreds of Megabytes or Gigabytes (Eliceiri et al. 2012; Edelstein et al. 2014). Whole-slide bioimaging, for instance, combines light microscopy techniques with electronic scanning of slides and is able to collect quantitative data, currently regarded as one of the most promising avenues for diagnosis or prediction of cancer and other diseases. Traditional health care imaging technologies such as CT, MRI, ultrasound or SPECT and PET make it possible to assess the current status and condition of organs or tissues and to monitor patients over time for diagnostic evaluation or for controlling therapeutic interventions (Smith and Webb 2010; Mikla and Mikla 2013). In particular, CPU-intensive image reconstruction and modeling techniques allow instant processing of 2D signals to create 3D/4D image stacks of enormous amounts of data, typically stored in DICOM file format. This DICOM standard facilitates interoperability of medical imaging instrumentations, providing a standardized medical file format and directory structure, which enables access to the images and patient-related information for further processing, modeling and analysis.

1.3 Study Design and Data Privacy

Different epidemiological study designs such as case-control, (longitudinal) cohort studies or more complex designs such as randomized controlled trials are selected in biomedical research (Dawson and Trapp 2004; Porta 2014). Case-control studies, which are always retrospective, are designed to determine if an exposure is associated or correlated with an outcome, i.e., a disease or biological/physiological condition of interest. This study type is referred to as an observational, non-experimental study where the investigator simply "observes", as the outcome of each subject enrolled in their respective groups is already known by the investigator. The investigator identifies the study groups of interest, i.e. cases (a group known to have the outcome such as patients with a coronary artery disease or patients with prostate cancer classified by the established gleason scoring scheme), and controls, which is a cohort known to be free of the outcome. Note that the same data must be collected in the same way from both groups. As the investigator usually makes use of previously collected data, a major limitation of observational studies is cofounding. By definition, a confounding variable is one which is associated with the exposure and is a cause of the outcome. For example, if researchers investigate whether "smoking" leads to "lung cancer", "smoking" is the independent variable and "lung cancer" is the dependent variable. Confounding variables are any other variables that also have an effect on the dependent variable like "age", which has a hidden effect on the selected dependent variables and may increase the variance in the data by introducing a bias. The big advantage and practical value of a case-control study is that this study design is very efficient, produces rapid results and may also be ideal for preliminary investigations of e.g. a suspected risk factor for a disease. However, results from case-control studies need to be independently verified and confirmed by larger, more accurate prospective cohort studies or randomized controlled clinical trials which are the only way to eliminate all confounding effects in a study. Randomized controlled trials, also known as double blind studies, are the most effective way of determining whether a cause-effect relation exists between a clinical intervention or treatment and outcome. Typically, subjects are allocated at random to receive one of several clinical interventions where one of them may serve as the standard or control investigation. This group usually receives a placebo or no clinical intervention. All intervention groups are treated identically. Although randomized controlled clinical trials are very powerful tools, they are time consuming and costly, and show some limitations by ethical and practical concerns such as recruitment and randomization. Besides the controlled collection of clinical information including biological samples, the acquisition of patient-related information is highly sensitive and data privacy is essential to prevent the unauthorized or unwanted disclosure of information about an individual if this data is not used for individual patient management and care. If this information is needed e.g. for biomedical research, educational purposes for the medical staff, etc., the strong data protection requirements can be overcome by approaches such as anonymization or pseudonymization (Neubauer and Riedl 2008; Elger et al. 2010). Anonymization, a method to disassociate all identifiers from the data, and pseudonymization, which supports an authorized re-identification of personalized data, make it possible to remove information from the data that are not strictly required for the intended purpose of those data and thus guarantees the privacy protection established by law.

1.4 Data Quality and Standard Operating Procedures (SOPs)

A Good Clinical Practice (GCP) is a central principle in biomedical research and clinical patient management, which allows for quality controlled collection and tracking of patient-related records, biosamples and additional study material. In the process of measuring and acquiring bioanalytical information gathered from biological samples such as blood (plasma, serum, dried spots), urine, other body fluids like sputum or lavages, cell cultures or tissue samples, the quality of generated data is crucial for the subsequent steps in data preprocessing and analysis. In general, the entire analytical and computational workflow, ranging from study design to study execution including pre-analytical sample handling, bioanalytical analyses, data aggregation and consolidation, computational tools for data integration, knowledge mining and interpretation, requires controlled procedures and standardized regulatory directives to ensure a high degree of consistency, completeness, and reproducibility of data and results. A guideline is provided by the "Guidance for Industry – Bioanalytical Method Validation" (FDA 2013) for the development of analytical methods and standard operating procedures (SOPs). According to laboratory-specific SOPs, a research lab has to handle a hazardous chemical safely and bring an application within the scope of the special execution procedures. These include the amount and concentration of proposed analytes, technical controls and inspections, as well as personal protective equipment with safety instructions. A controlled collection of samples, for example, is assured by the implementation of specific protocols and standards for sample taking using barcoding, a way to rapidly, accurately, and efficiently gather sample information and transmit it to a central data server for further analysis.

1.5 Life Science Data Warehouse Concepts for Data Integration and Knowledge Retrieval to Support Medical Decision Making

There is a strong need in life science research to integrate and store biomedical information generated in multiple research fields using proper approaches such as data warehouses (Parmanto et al. 2005; Töpel et al. 2008; Kienast and Baumgartner 2011; Hu et al. 2011; Lyne et al. 2013; Galhardas and Rahm 2014; Dander et al. 2014). By definition, data integration is the task of "combining the data residing at different sources, and providing the user with a unified view of the data" (Calì et al. 2001; 2003). These efforts require efficient and feasible IT concepts taking into account quality-assured standards and procedures (Shadbolt et al. 2006). The number, size, and complexity of life science databases continuously grow (Kei-Hoi et al. 2009) meaning that scientists in experimental and clinical research fields demand new concepts to handle (i) the variety and amount of available data,

(ii) data heterogeneity arising from different sources and (iii) a lack of standards for such integration concepts, which is a prominent problem (Kei-Hoi et al. 2009). Generally, heterogeneity in computer science can be divided into four classes, i.e. system heterogeneity (different hardware platforms and operation systems), semantic heterogeneity (differences in the interpretation of different data sources), syntactic heterogeneity (difference of data representation formats) and structural heterogeneity (different data models or structures). As a consequence of these subclasses of heterogeneity, the following challenges in life science data integration need to be taken into account (Kienast and Baumgartner 2011):

(1) The origin of data with different data formats. Basically, three groups of data structures can be defined:

   - structured data, which is organized in a form and structure so that it is identifiable. E.g. databases using Structured Query Language (SQL) for data management and retrieval.
   - semi-structured data which is used to identify certain elements within the data, but lacks a strict data model structure: E.g. metadata in HTML, XML formats and in other mark-up languages.
   - unstructured data that include other data types that are not part of a database: E.g. text (electronic patient records, biomedical literature), biomedical images of diverse formats like DICOM, JPG, TIF.

(2) The identification and interpretation of "synonyms" and "homonyms": In biomedicine scientists often name biological entities and relationships synonymously. For data integration it is crucial to strictly distinguish between synonyms, i.e. words that share meanings with other words, and homonyms i.e. words that sound similar, but have different meanings, and to select the right term in relation to the context. As examples, "entities" include anatomical terms like cells and tissues (cell and tissue types, anatomy, populations, etc.), biomolecules such as genes, proteins, metabolites (amino acids, enzymes, antibodies, protein genes, etc.) and "relationships" that include terms like gene expression, mutation, activation, inhibition, regulation, prognosis, diagnosis or therapy.
(3) The recognition of granularity: Biomedical data sources may provide information at different levels of granularity. For example one data source contains information about different metabolic diseases and their clinical phenotypes, symptoms and therapeutic recommendations while other databases provide detailed information about the same diseases, but characterized by the underlying molecular mechanisms, pathways and networks.
(4) The identification of viable resources: It is crucial to identify relevant and interoperable data sources that use widely accepted, comprehensive standards for the access and exchange of data to avoid unnecessary duplication and incompatibility in the collection, processing, and dissemination of such data.

In general, there are various approaches for integrating different data sources by using warehousing, mediation and Semantic Web technology based approaches

(Töpel et al. 2008; Pasquier 2008; Grethe et al. 2009; Spanos et al. 2012). Warehouse integration consists of cataloguing and accessing data from multiple sources and repositories in a local database, which is called the warehouse and designed with the objective of retrieving information from the data and supporting decision making (Fig. 1.1) (Töpel et al. 2008; Kienast and Baumgartner 2011). Usually relational databases with different database schemata are used such as the Star schema[1] or the Snowflake schema.[2] Basically, a data warehouse consists of two entities, a back room and front room entity, which are mostly separated physically as well as logically. The ETL process (Extract, Transform and Load), the basic concept of a data warehouse back room, is conducted to extract and import the data from the data source (i.e. data from external sources including flat files, XML files or databases) into the needed reference system (repository) of the data warehouse (Hernandez and Kambhampati 2004; Kugler et al. 2008). The back room thus describes the data management component, which permanently stores the data in a physical database and delivers subsets of data retrieved by queries. The front room makes it possible to access the data held in the warehouse by providing tools and methods for intelligent data accessing, mining and information retrieval, which is of great interest for clinical uses e.g. to support medical decision making. Users have access to the data of interest via the front room by two different types of database queries: ad-hoc and intelligent queries. Ad-hoc queries are executed through SQL statements that need to be formulated by a trained user, while intelligent query approaches are extended, more user-friendly queries based on result sets of adhoc queries using a medical knowledge base to process the information requests. Such requests allow for identification of patterns and relationships within the data relying on the concept of different information hierarchies in the data (Lyne et al. 2013).

Mediator based integration focuses more on query translation, where data is not centrally stored, but directly accessed from the distributed sources (Grethe et al. 2009). The data flow between mediators and data sources is provided by specific software components termed Wrappers (Hernandez and Kambhampati 2004). A more targeted approach for the integration of heterogeneous data sources, in particular in biomedical applications, is the use of Semantic Web technologies, whereby existing documents and data are provided with structured meta-information (Cheung et al. 2007; Pasquier 2008; Spanos et al. 2012). A key feature of this approach is the use of semantics by ontologies, which overcomes the problem of interpreting homonyms and synonyms in different sources. It should be noted that ontologies are a type of controlled vocabulary that attempt to capture the knowledge of a specific domain, which is also an important approach for warehousing and mediation based data integration concepts (Bodenreider 2008).

[1] It comprises so-called dimension tables containing data from different data sources and fact tables connecting various dimension tables.

[2] It uses a transformation of the dimension tables to the third normal form with less data space needed, but more complex data queries.

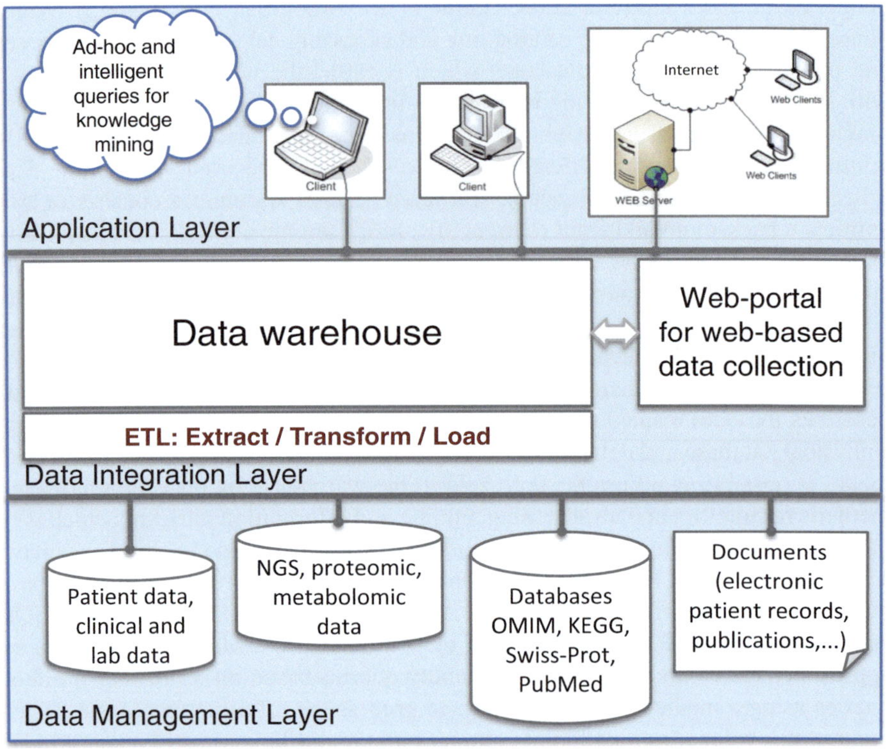

Fig. 1.1 Three-tier architecture of a data warehouse which is composed of a data management layer, data integration layer and application layer

To benefit from ontologies, it is important to annotate instances to metadata ontologies, which should be standardized and machine-readable. The Semantic Web technology provides tools for exchanging metadata information via Extensible Markup Language (XML) to allow for semantic data integration.

1.6 Knowledge Discovery and Data Mining for Clinical Care

According to Fayyad's 1996 definition (Fayyad et al. 1996a, b), knowledge discovery in general is the "nontrivial process of identifying valid, novel, potentially useful and ultimately understandable patterns in data". Data mining (DM) is a step in this process consisting of "particular data mining algorithms that, under some acceptable computationally efficiency limitations, produce a particular enumeration of patterns". The term knowledge discovery thus refers to the process of finding novel knowledge in the data. It does this by using data mining, machine learning or biostatistical methods to extract and identify what is deemed knowledge, according to the specifications of measures and thresholds, using the given database with any required preprocessing, subsampling, and transformations of that data (Fig. 1.2). Note that the terms knowledge discovery and data mining are distinct. So this field is of interest to researchers in machine learning, pattern recognition, databases, statistics, artificial intelligence, knowledge acquisition for expert systems and data visualization, and requires an interdisciplinary view of research, in particular in a biomedical setting (Mitchell 1997; Pardalos et al. 2007; Ting et al. 2009; Dua and Chowriappa 2012; Holzinger and Jurisica 2014).

For the purpose of targeted analyses, data often need to be preprocessed and transformed into a standardized and quality assured format. Data preprocessing is used e.g. to normalize or rescale data (logarithmic scaling), to select data subsets, samples or single features or to remove outliers in the data in order to avoid manipulations of subsequently performed statistical analyses (Kotsiantis et al. 2006; García et al. 2015). Assuming a normal distribution of data, a common model for removing outliers is, for example, the use of the interquartile ranges. This simple statistical approach defines an outlier as observation outside the interquartile range.

Basically, data mining can be distinguished by the forms of supervised and unsupervised learning. In supervised learning or class prediction, knowledge of a particular class, group or study cohort is used to classify the class instances into the correct groups or to select significant features from the data in terms of high discriminatory ability or predictive value using a learning method. In biomarker discovery, the search for biomarker candidates is typically "supervised" because study groups in preclinical experimental studies or controlled clinical trials are typically well-defined and phenotyped. The data are then available in the form of tuples $T = \{(c_j, x) \mid c_j \mid C, x \mid X\}$, where c_j is the class label (e.g. normal, diseased, various stages of a disease, treated, etc.), and $X = \{x \mid x_1, \ldots, x_n\}$ is the set of given data (e.g. *metric* data such as lab measurements, gene expression data or mass spectral data, or *nominal* and *ordina*l data such as medical scores like the Glasgow Coma Scale). In this specific field, basic data mining and computational concepts for the search, prioritization and verification of biomarker candidates constitute filter-based feature selection algorithms or more sophisticated approaches such as embedded or ensemble methods (Lewis et al. 2008; Osl et al. 2008; Netzer et al. 2009; Millonig et al. 2010; Fang et al. 2012a, b; Swan et al. 2013, 2015;

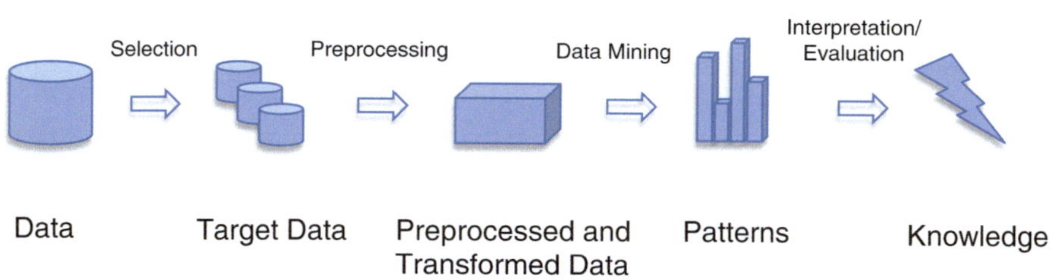

Fig. 1.2 General knowledge discovery process

Assawamakin et al. 2013). A survey of widely-used supervised feature selection techniques considering both independent and dependent samples w.r.t. unpaired and paired test hypotheses can be found in Baumgartner and co-authors (Baumgartner et al 2011). Newer approaches for identifying dynamic metabolic biomarkers using longitudinal or time-series data have been presented in Breit et al. (2015a, b).

The predictive performance and generalization power of validated clinical or biological markers is utilized to build classification models for medical decision making or disease screening. The basic idea of classification is to group or classify the given data $X = \{x \mid x_1, \ldots, x_n\}$ into the correct class $c_j \mid C$. For building classification models, multiple methods are available: logistic regression analysis, a widely-used method in biomedical applications, decision or classification trees, Bayes classifiers, k-nearest neighbor classifiers (k-NN), support vector machines, artificial neural networks or modalities of network-based approaches (Fielding 2006; Swan et al. 2013; Assawamakin et al. 2013). A high predictive value of such models is required to keep the false positive and false negative rate low which is expressed by high values of sensitivity and specificity (typically beyond 95-98 %). Note that such models have to consider the real incidence rate of a disease to correctly estimate the true false-positive rates. Statistical validation of a model is now the process of estimating how well a model, e.g. trained on a single derivation cohort, performs on future as-yet-unseen data by limiting problems like overfitting. Typical validation concepts include "train-and-test" strategies, splitting data into a separate train and test set, stratified cross-validation and permutation modalities, where the given data set is separated into train and test partitions. Multiple rounds of cross-validation are performed using the different partitions, and results are averaged over the rounds. This procedure reduces the variability in the data (Holzinger and Jurisica 2014).

In unsupervised data mining class information is unknown. A data set is typically given as a set of tuples in the form of $T = \{x \mid X\}$, where $X = \{x \mid x_1, \ldots, x_n\}$ is the set of given data (e.g. not annotated clinical scores, gene expression data, mass spectrometry data represented by lists of intensities vs. m/z values, voxels in a biomedical image, etc.). Using cluster analysis, data are grouped into meaningful classes based on similarity distance measures. Well-known methods are partitioning or hierarchical methods (k-means, single or average link) or newer methods that better consider local density structures in data such as DBSCAN (Ester et al. 1996) or Optics (Ankerst et al. 1999) as well as graph-based models (Fielding 2006; Xu and Wunsch 2010; Ye 2011). Association rule mining and regression analysis are complement methodological approaches in knowledge mining (Fig. 1.3).

To improve findings of single experiments, meta analysis as an further layer of integrated data analysis may be used, for example, for the search and verification of clinical and biological markers. This analysis strategy runs through multiple levels: (a) integrated analysis of the different clinical and preclinical experimental data which may arise e.g. from multi center studies and assessment of selected markers

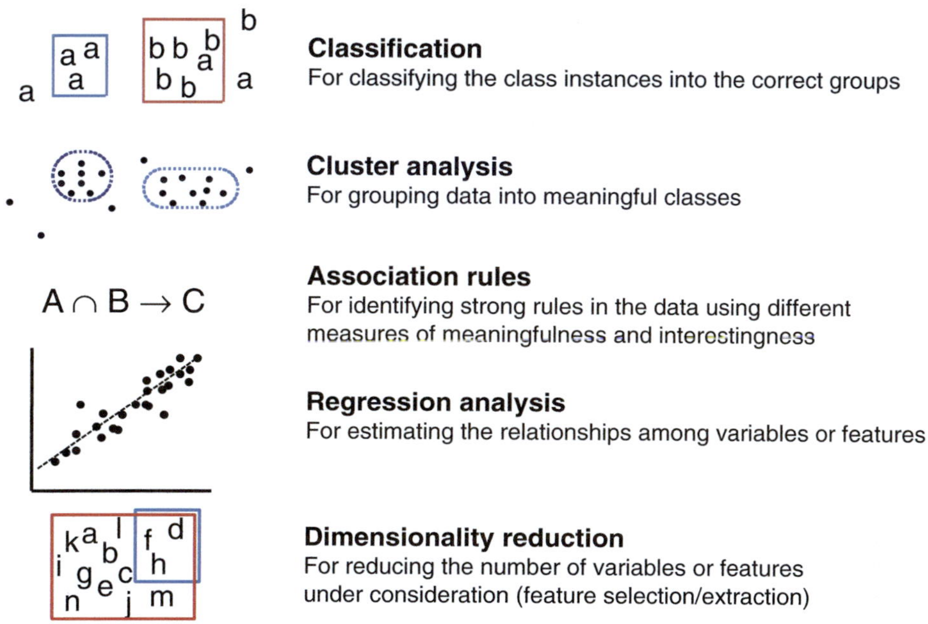

Fig. 1.3 Overview of the basic data mining areas used in biomedical research

with respect to their clinical plausibility and relevance, (b) application of statistical bioinformatics and data mining methods for searching and verifying marker candidates with superiour discriminatory ability with respect to the targeted patient cohorts or populations, (c) advanced correlation analysis by including all patient-relevant antropological, clinical and biomolecular data, (d) evaluation of predictive value of selected marker candidates by decision-outcome analysis and modeling. Tools such as Receiver operating characteristics (ROC) analysis and further approaches of health care technology assessment are used to estimate expected epidemiologic and economic consequences for individuals, the population and the public health (Mak et al. 2012; Tseng et al. 2012; Kaever et al. 2014).

1.7 Bio-Medical Application Examples

In the following section, selected application examples including studies from our research are presented, demonstrating the strength and benefit of computational approaches and concepts for data mining and knowledge discovery used in biomedical research and clinical care.

1.7.1 Example 1

Figure 1.4 shows an example workflow for the discovery of metabolic biomarkers in myocardical injury following the scheme of the general knowledge discovery process (Fayyad et al. 1996a, b, Fig. 1.2). In addition to a case/control study design, a longitudinal design for a biomarker search was selected where each subject serves as his/her own biological control. This design makes it possible to study the kinetic characteristics of circulating metabolites and thus to identify and classify biomarker candidates as early, late or sustained after acute injury (Lewis et al. 2008; Baumgartner et al. 2010).

1.7.2 Example 2

To support medical decision making in patients with Marfan syndrome (MFS), which is an autosomal dominant connective tissue disorder caused by mutations in the gene encoding fibrillin-1(FBN1) with highly variable clinical manifestations in the musculoskeletal, ocular and cardiovascular systems, a multiple logistic regression model was proposed by Baumgartner et al. (Baumgartner et al. 2005a, b). The model includes three cardiovascular parameters, i.e. the normalized diameters of aortic bulb and ascending aorta, and the ascending aortic distensibility. It demonstrated a sensitivity of almost 100 % and a specificity of 95 %, validated in an independent validation cohort. Interestingly, this model allows for the classification of patients with MFS only on three aortic parameters, selected from a pool of more than 30 measured parameters of the musculoskeletal, ocular and cardiovascular systems including genetic information (mutation data) (Baumgartner et al. 2005a, b).

Method Box

Classification model: A logistic regression model of the form $P = 1/(1+e^{-z})$ was selected, where p is the conditional probability $P(z=1 \mid x_1, \ldots, x_n)$ that MFS is present and z is the logit (discriminant function) of the model with three aortic parameters. A cut-off value ($P = 0.5$ by default) classifies controls if $P < 0.5$ and cases of disorder if $P \geq 0.5$. The logit of the MFS regression model is given by the following equation: $z = 4.379 + 2.293 \cdot$ normalized diastolic diameter of aortic bulbus [dimensionless] – $2.449 \cdot$ normalized diastolic diameter of ascending aorta [dimensionless] – $0.247 \cdot$ distensibility of ascending aorta [$kPa^{-1}\ 10^{-3}$] (see Baumgartner et al. 2005a, b).

For phenotype-genotype correlation, hierarchical cluster analysis on a collection of clinical symptoms of the MFS was performed. In this study, four phenotype classes (I, II, III, IV) could be identified, where the presence of missense

General knowledge discovery process (Fayyad et al., 1996a, 1996b, modified by CB)	Example workflow of a biomarker discovery study in myocardial injury (Baumgartner et al., 2010)
Definition of the knowledge discovery goals	Search and verification of metabolic biomarkers in myocardial injury using targeted MS/MS data
Selection of targeted data • Selecting a data set • Focusing on a subset of variables • Data samples, on which discovery is performed	• Longitudinal biomarker cohort study: targeted metabolite profiling of serial blood samples obtained from patients undergoing alcohol septal ablation for HOCM (clinical model of „planned“ MI) • Samples collected before and after intervention (10min,60min, 120min, 240min) • Data of 210 metabolite concentrations per sample
Data cleaning and preprocessing • Removal of noise or outliers • Collecting necessary information to model or account for noise • Strategies for handling missing data fields • Accounting for time sequence information and known changes	• Examination of MS/MS raw data, • Determination if all requested metabolites were accounted for, and targeted measurement levels were met. • Exclusion of metabolites with more than 20% missing values in the dataset • Outlier detection using the interquartile ranges as statistical model
Data mining task 1: Data reduction, projection and feature selection and choosing the data mining algorithm(s) • Finding useful features to represent the data depending on the goal of the task. • Using dimensionality reduction or transformation methods to reduce the effective number of variables or to find invariant representations for the data.	• Data mining task: feature selection for identifying and prioritizing biomarker candidates w.r.t. their discriminatory ability and predictive value • Algorithms: "Biomarker identifier" (Baumgartner et al., 2010) and statistical tests
Data mining task 2: Deciding whether the analysis goal is classification, regression, clustering, etc. and choosing the data mining algorithm(s) • Searching for patterns of interest in a particular representational form or a set of such representations as classification rules or trees, regression, clustering, and so forth	• Data mining task: classification is used (class information is known) to assess the predictive value of selected biomarker candidates • Algorithms: logistic regression analysis and statistical tests • Determination of classification accuracy of single or subsets of metabolites using ROC analysis
Interpretation of mined patterns	• Verification and interpretation of results by metabolic pathway analysis (KEGG database)
Consolidation of discovered knowledge	• Validation of new biomarker candidates by experimental studies and clinical trials before they can go into clinical application

Fig. 1.4 Example workflow for the search for new biomarkers in myocardial injury based on the general scheme of the knowledge discovery process

mutations (substitutions or point mutations) was highly correlated with phenotype classes I and III, while phenotype class II, characterized e.g. by severe cardio-vascular manifestations, was primarily associated with more complex mutations such as stop mutations or deletions with frame shift. These findings were used to specify characteristic clinical phenotypes with respect to different classes of mutations (substitutions vs. stronger forms of mutations such as stop mutations, insertions or deletions with frame shift) (see Fig. 1.5) (Baumgartner et al. 2005a, b, 2006).

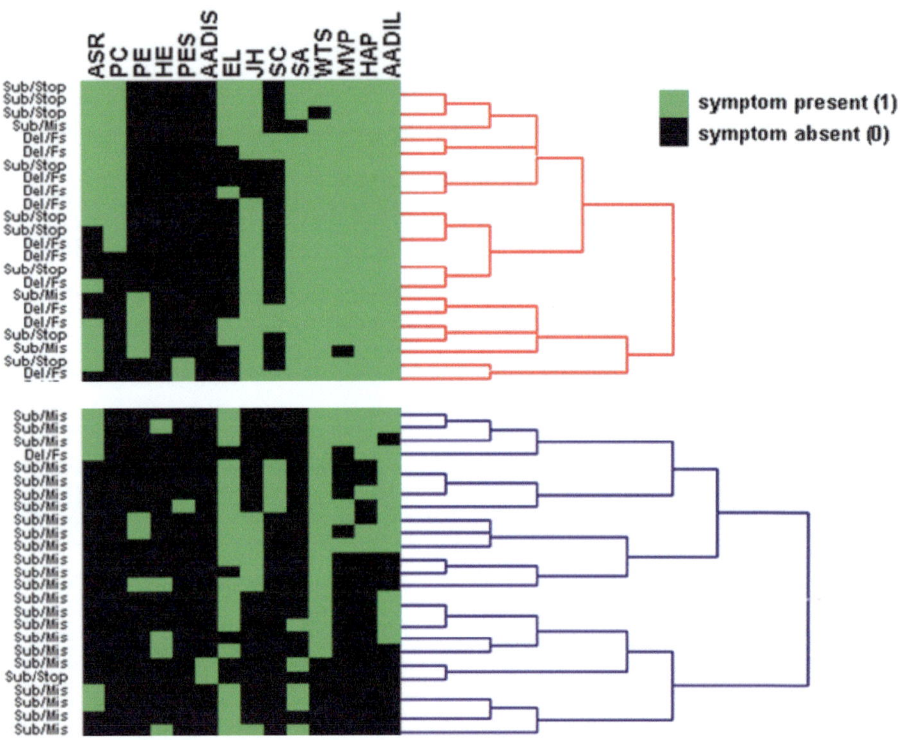

Fig. 1.5 Genotype-phynotype correlation in patients with MFS. Different classes of clinical phenotypes based on 14 examined clinical manifestations in the skeletal, ocular and cardiovascular system were identified using hierarchical cluster analysis. Each cluster was reviewed with respect to different forms of mutations (Sub/Mis ... substitution-missense mutation, Sub/Stop ... substitution with stop codon, Del/Fs ... deletions with frameshift, Ins/Fs ... insertions with frameshift) determined in each patient. Two different phenotype classes (I and II) with weak versus strong forms of mutations (substitutions (Sub/Mis) versus stronger forms of mutations such as stop mutations (Sub/Stop) or deletions with frame shift (Del/Fs) are shown in detail). Clinical manifestations: *EL* ectopia lentis, *AADIL* dilation of ascending aorta, *AADIS* dissection of ascending aorta, *MVP* mitral valve prolapse, *PC* pectus carinatum, *PES* pectus excavatum requiring surgery, *ASR* arm span ratio (ASR), *WTS* wrist or thumb sign, *SC* scoliosis, *PE* moderate pectus excavatum, *JE* joint hypermobility, *HAP* highly arched palate with crowding of teeth, *SA* striae atrophicae, *HE* herniae

Method Box
Cluster analysis: Hierarchical cluster analysis on 14 preselected clinical manifestations was performed to group patients with a similar clinical phenotype (clinical symptom present/absent, nominal measure). The average linkage method was selected, describing the distance between two clusters as the mean distance between an observation in one cluster and an observation in the other cluster. Clustered patient groups were reviewed with regard to their genetic predispositions (identified mutations). In the two-dimensional presentation of the clustered color map, each row represents a single mutation and each column the presence (green)/ absence (black) of examined clinical manifestations (see Fig. 1.5).

1.7.3 Example 3

In the work of Aronica and co-authors (Aronica et al. 2015) whole-genome expression profiles of 41 motor cortex samples of control and sporadic amyotrophic lateral sclerosis (SALS) patients were analyzed. ALS is a rapidly progressive neurodegenerative disease which is characterized by upper and lower motor neuron loss, leading to respiratory insufficiency and death after 3–5 years. Tissue samples were used for RNA extraction. A GenePix microarray scanner was selected to analyze gene expression changes in biological pathways associated with ALS. Although SALS patients could be clearly classified on the basis of their motor cortex gene expression profiles, no significant association between their clinical characteristics and cluster assignment was found.

Method Box
Cluster analysis: Hierarchical cluster analysis (similarity measure: Pearson centered; linkage method: average linkage, similarities were measured over the genes expressed in the motor cortex, 9646 genes in total) was used to separate and group controls from SALS patients. In the two-dimensional presentation of the clustered color map, each row represents a single gene and each column a motor cortex from controls or SALS patients. Colors indicate up-regulation, down-regulation or no change in gene expression.

1.7.4 Example 4

Wang and co-authors (Wang et al. 2014) introduced a comprehensive knowledge base, termed MitProNet, for the mitochondrial proteome, interactome and human disease associated mechanisms with mitochondria. This knowledge base allows for a systematic identification of mitochondrial proteomes and a comprehensive characterization of functional linkages among mitochondrial proteins.

Method Box
Design of the knowledge discovery pipeline: A three step computational pipeline for data integration, modeling, analysis and interpretation was proposed. In step one, an inventory of mammalian mitochondrial proteins is integrated by collecting relevant proteomic datasets, and the proteins are classified using data mining and machine learning methods. A network of functional linkages among mitochondrial proteins is generated in step 2 by integrating 11 genomic features including protein-protein interaction, domain-domain interaction, shared domains, genomic context, genetic interaction, phenotypic semantic similarity, co-expression, GO semantic similarity, protein expression profiles, disease involvement and operon (operon contains a series of genes that are involved in the same biological process) based on the selected Naive bayes model. Step three prioritizes disease candidate genes by utilizing the network of functional linkages and network-based methods such as PageRank with Priors (PRP), Kstep Markov (KSM) or Heat Kernel Diffusion Ranking (HKDR). ROC analysis was selected for evaluating the performances of the various data sources and generated networks.

The system architecture and main contents of MitProNet can be found under doi:10.1371/journal.pone.0111187. The database is freely accessible.

1.7.5 Example 5

A web-based bioinformatics platform for clinical cancer research and routine applications in medical oncology, termed Personalized Oncology Suite (POS), integrating clinical data, NGS data and whole-slide bioimages from tissue sections was introduced by Dander and co-authors (Dander et al. 2014). Interestingly, POS combines biological data (mutations identified via next-generation sequencing and whole-slide bioimaging) and clinical data (information about the cancer patients, TNM staging, and density values of tumor-infiltrating lymphocytes used for immune score estimation) into one platform. As POS contains confidential and patient related data, the platform is secured by an authorization and authentication system (AAS). POS provides a convenient user interface, allowing for data upload, manipulation and visualization of integrated data. In a next release, POS will be extended by knowledge discovery and mining tools to aid in personalized cancer immunotherapy. The platform is open-source and can be downloaded at http://www.icbi.at/POS.

Method Box
Data warehouse application: The software architecture of POS as well as all detailed information on the functionality and provided features of the platform can be found under http://www.biomedcentral.com/content/pdf/1471-2105-15-306.pdf.

1.8 Is There a Need for Biomedical Scientists to Become Data Engineers?

Although not every biomedical scientist or clinician is a mathematician, statistician or computer scientist, it is definitely necessary to have basics skills in these fields. Modern study programs in medicine or biomedical sciences need to offer obligatory modules or tracks to strengthen these skills so that scientists are able to process, review and interpret data gathered in the scientist's field of expertise. Data-driven research approaches require all scientists in their field to develop and apply data analysis capabilities in computational biomedicine and statistics with proper sensitivity and quality assurance.

1.9 Conclusion

This chapter has provided a brief insight into concepts, methods, procedures, recommendations and applications in the field of computational biomedicine and was anticipated to appeal to those who undertake basic biomedical research or are interested in life science applications for clinical care. The presented sections were developed consecutively and cover the entire knowledge discovery process typically found in a biomedical setting. It might assist in the selection of computational methods and strategies for data collection, integration and analysis, which are urgently needed to transform mined knowledge into clinical applications.

References

Ankerst M, Breunig MM, Kriegel HP, Sander J. OPTICS: Ordering points to identify the clustering structure. In: Proceedings of the ACM SIGMOD international conference on management of data (SIGMOD'99), Philadelphia; 1999. p. 49–60.

Aronica E, Baas F, Iyer A, ten Asbroek AL, Morello G, Cavallaro S. Molecular classification of amyotrophic lateral sclerosis by unsupervised clustering of gene expression in motor cortex. Neurobiol Dis. 2015;74:359–76.

Assawamakin A, Prueksaaroon S, Kulawonganunchai S, Shaw PJ, Varavithya V, Ruangrajitpakorn T, Tongsima S. Biomarker selection and classification of "-omics" data using a two-step bayes classification framework. Biomed Res Int. 2013;2013:148014.

Baumgartner C, Graber A. Chapter 7: Data mining and knowledge discovery in metabolomics. In: Masseglia F, Poncelet P, Teisseire M, editors. Successes and new directions in data mining. Hershey, PA: Idea Group Inc; 2007. p. 141–66. ISBN 978-1-59904-639-6.

Baumgartner C, Mátyás G, Steinmann B, Baumgartner D. Marfan syndrome: a diagnostic challenge caused by phenotypic and genetic heterogeneity. Methods Inf Med. 2005a;44:487–97.

Baumgartner D, Baumgartner C, Mátyás G, Steinmann B, Löffler J, Schermer E, Schweigmann U, Baldissera I, Frischhut B, Hess J, Hammerer I. Diagnostic power of aortic elastic properties in young patients with Marfan syndrome. J Thorac Cardiovasc Surg. 2005b;129:730–9.

Baumgartner C, Mátyás G, Steinmann B, Eberle M, Stein JI, Baumgartner D. A bioinformatics framework for genotype-phenotype correlation in humans with Marfan syndrome caused by FBN1 gene mutations. J Biomed Inform. 2006;39:171–83.

Baumgartner C, Rejtar T, Kullolli M, Akella LM, Karger BL. SeMoP: a new computational strategy for the unrestricted search for modified peptides using LC-MS/MS data. J Proteome Res. 2008;7:4199–208.

Baumgartner C, Lewis GD, Netzer M, Pfeifer B, Gerszten RE. A new data mining approach for profiling and categorizing kinetic patterns of metabolic biomarkers after myocardial injury. Bioinformatics. 2010;26(14):1745–51.

Baumgartner C, Osl M, Netzer M, Baumgartner D. Bioinformatic-driven search for metabolic biomarkers in disease. J Clin Bioinform. 2011;1:2.

Bodenreider O. Ontologies and data integration in biomedicine: success stories and challenging issues. In: Bairoch A, Cohen-Boulakia S, Froidevaux C, editors. Data Integration in the life sciences, Lecture notes in computer science, vol. 5109. Berlin/Heidelberg: Springer; 2008. p. 1–4.

Breit M, Baumgartner C, Weinberger KM. Chapter 9: data handling and analysis in metabolomics. In: Current applications of chemometrics. New York: Nova Sciences Publisher; 2015a. p. 181–203. ISBN: 978-1-63463-117-4.

Breit M, Netzer M, Weinberger KM, Baumgartner C. Modeling and classification of kinetic patterns of dynamic metabolic biomarkers in physical activity. PLoS Comput Biol. 2015b; 11(8): e1004454.

Brewis IA, Brennan P. Proteomics technologies for the global identification and quantification of proteins. Adv Protein Chem Struct Biol. 2010;80:1–44.

Calì A, Calvanese D, De Giacomo G, Lenzerini M. Accessing data integration systems through conceptual schemas, conceptual modeling – ER 2001, Lecture notes in computer science, vol. 2224. Berlin/Heidelberg: Springer; 2001. p. 270–84.

Calì A, Calvanese D, De Giacomo G, Lenzerini M. On the expressive power of data integration systems. In: Spaccapietra S, March S, Kambayashi Y, editors. Conceptual modeling – ER 2002, Lecture notes in computer science, vol. 2503. Berlin/Heidelberg: Springer; 2003. p. 338–50.

Cerqueira F, Graber A, Schwikowski B, Baumgartner C. MUDE: a new approach for optimizing sensitivity in the target-decoy search strategy for large-scale peptide/protein identification. J Proteome Res. 2010;9(5):2265–77.

Chang PL. Clinical bioinformatics. Chang Gung Med J. 2005;28(4):201–11.

Chen G, Pramanik BN. Application of LC/MS to proteomics studies: current status and future prospects. Drug Discov Today. 2009;14(9-10):465–71.

Cheung K, Smith A, Yip K, Baker C, Gerstein M. Semantic web approach to database integration in the life sciences. In: Baker CJO, Cheung K-H, editors. Semantic web. New York: Springer; 2007. p. 11–30.

Coveney P, Diaz V, Hunter P, Viceconti M. Computational biomedicine: modelling the human body. Oxford: Oxford University Press; 2014.

Dander A, Baldauf M, Sperk M, Pabinger S, Hiltpolt B, Trajanoski Z. Personalized oncology suite: integrating next-generation sequencing data and whole-slide bioimages. BMC Bioinf. 2014;15:306.

Dawson B, Trapp RG. Basic & clinical biostatistics (LANGE basic science). 4th ed. New York: Lange Medical Books/McGraw-Hill; 2004.

Dua S, Chowriappa P. Data mining for bioinformatics. Boca Raton: CRC Press; 2012.

Edelstein AD, Tsuchida MA, Amodaj N, Pinkard H, Vale RD, Stuurman N. Advanced methods of microscope control using μManager software. J Biol Methods. 2014;1(2):e10.

Elger BS, Iavindrasana J, Lo Iacono L, Müller H, Roduit N, Summers P, Wright J. Strategies for health data exchange for secondary, cross-institutional clinical research. Comput Methods Programs Biomed. 2010;99(3):230–51.

Eliceiri KW, Berthold MR, Goldberg IG, Ibáñez L, Manjunath BS, Martone ME, Murphy RF, Peng H, Plant AL, Roysam B, Stuurman N, Swedlow JR, Tomancak P, Carpenter AE. Biological imaging software tools. Nat Methods. 2012;9(7):697–710.

Ester M, Kriegel HP, Sander J, Xu X. A density-based algorithm for discovering clusters in large spatial databases with noise. In: Proceedings of 2nd International Conference. on Knowledge Discovery and Data Mining (KDD'96). Menlo Park: AAAI Press; 1996. p. 226–231.

Fang W, Chang X, Su X, Jian Xu, Zhang D, Ning K. A machine learning framework of functional biomarker discovery for different microbial communities based on metagenomic data. In: IEEE 6th International Conference on Systems Biology (ISB), Xiang, China, 2012a; p. 106–112.

Fang X, Netzer M, Baumgartner C, Bai C, Wang XD. Genetic network and gene set enrichment analysis to identify biomarkers related to cigarette smoking and lung cancer. Cancer Treat Rev. 2012b;2013(39):77–88.

Fayyad UM, Piatetsky-Shapiro G, Smyth P. Advances in knowledge discovery and data mining, chapter: from data mining to knowledge discovery: an overview. Menlo Park: AAAI Press. 1996a. p. 1–30.

Fayyad UM, Piatetsky-Shapiro G, Smyth P. Knowledge discovery and data mining: towards a unifying framework. In: Simoudis E, Han JW, Fayyad UM (Hrsg.), editors. Proceedings of 2nd international conference on knowledge discovery and data mining, Portland, Oregon, AAAI Press; 1996b. p. 82–88.

FDA. Guidance for industry bioanalytical method validation. 2013. http://www.fda.gov/downloads/drugs/guidancecomplianceregulatoryinformation/guidances/ucm368107.pdf

Fielding AH. Cluster and classification techniques for the biosciences. Cambridge: Cambridge University Press; 2006.

Galhardas H, Rahm E. Data integration in the life sciences, Lecture notes in bioinformatics, vol. 8574. Berlin: Springer; 2014.

García S, Luengo J, Herrera F. Data preprocessing in data, mining, intelligent systems, Lecture notes in bioinformatics, vol. 72. Berlin: Springer; 2015.

Grethe JS, Ross E, Little D, Sanders B, Gupta A, Astakhov V. Mediator infrastructure for information integration and semantic data integration environment for biomedical research. Methods Mol Biol. 2009;569:33–53.

Hernandez T, Kambhampati S. Integration of biological sources: current systems and challenges ahead. SIGMOD Rec. 2004;33(3):51–60.

Holzinger A, Jurisica I. Interactive knowledge discovery and data mining in biomedical informatics, Lecture notes in computer science, vol. 8401. Berlin, Heidelberg: Springer; 2014.

Hu H, Correll M, Kvecher L, Osmond M, Clark J, Bekhash A, Schwab G, Gao D, Gao J, Kubatin V, Shriver CD, Hooke JA, Maxwell LG, Kovatich AJ, Sheldon JG, Liebman MN, Mural RJ. DW4TR: a data warehouse for translational research. J Biomed Inform. 2011;44 (6):1004–19.

Kaever A, Landesfeind M, Feussner K, Morgenstern B, Feussner I, Meinicke P. Meta-analysis of pathway enrichment: combining independent and dependent omics data sets. PLoS ONE. 2014;9(2):e89297.

Kei-Hoi C, Robert F, Scott M, Matthias S, Jun Z, Adrian P. A journey to semantic web query federation in the life sciences. BMC Bioinf. 2009;10 Suppl 10:S10.

Kienast R, Baumgartner C. Chapter 3: data integration on biomedical data using semantic web technologies In: Mahdavi MA, editors. Bioinformatics/Book 1, ISBN 978-953-307-282-1. Rijeka: InTech Open Access Publisher; 2011; p. 57–82.

Kotsiantis SB, Kanellopoulos D, Pintelas PE. Data preprocessing for supervised leaning. Int J Elect Comput Eng. 2006;1:2.

Kugler K, Tejada M, Baumgartner C, Tilg B, Graber A, Pfeifer B. Bridging data management and knowledge discovery in the life sciences. Open Bioinform J. 2008;2:28–36.

Lewis GD, Wei R, Liu E, Yang E, Shi X, Martinovic M, Farrell L, Asnani A, Cyrille M, Ramanathan A, Shaham O, Berriz G, Lowry PA, Palacios I, Tasan M, Roth FP, Min J, Baumgartner C, Keshishian H, Addona T, Mootha VK, Rosenzweig A, Carr SA, Fifer MA, Sabatine MS, Gerszten RE. Metabolite profiling of blood from individuals undergoing planned myocardial infarction reveals early markers of myocardial injury. J Clin Invest. 2008;118:3503–12.

Lyne M, Smith RN, Lyne R, Aleksic J, Hu F, Kalderimis A, Stepan R, Micklem G. metabolicMine: an integrated genomics, genetics and proteomics data warehouse for common metabolic disease research. Database (Oxford). 2013;2013:bat060.

Mak A, Cheung MW, Fu EH, Ho RC. Meta-analysis in medicine: an introduction. Int J Rheum Dis. 2010;13(2):101–4.

Mikla VI, Mikla VV. Medical Imaging Technology, 1st ed. Waltham: Elsevier; 2013. ISBN: 9780124170216.

Millonig G, Praun S, Netzer M, Baumgartner C, Mueller S, Villinger J, Vogel W. Non-invasive diagnosis of liver diseases by breath analysis using an optimized ion-molecule reaction-mass spectrometry approach: a pilot study. Biomarkers. 2010;15(4):297–306.

Mitchell TM. Machine learning. Boston: McGraw-Hill; 1997.

Naz S, Vallejo M, García A, Barbas C. Method validation strategies involved in non-targeted metabolomics. J Chromatogr A. 2014;1353:99–105.

Netzer M, Millonig G, Osl M, Pfeifer B, Praun S, Villinger J, Vogel W, Baumgartner C. A new ensemble based algorithm for identifying breath gas marker candidates in liver disease using ion molecule reaction mass spectrometry. Bioinformatics. 2009;25(7):941–7.

Netzer M, Handler M, Pfeifer B, Dander A, Baumgartner C. Knowledge discovery in proteomic mass spectrometry data. In: Tran QN, Arabnia H, editors. Emerging trends in computational biology, bioinformatics, and systems biology – algorithms and software tools. Waltham: Elsevier/MK; 2015. ISBN 9780128025086.

Neubauer T, Riedl B. Improving patients privacy with pseudonymization. In: Proceedings of the international congress of the European Federation for medical informatics. Studies in health technology and informatics, vol 136. Amsterdam: IOS Press; 2008. ISBN: 978-1-58603-864-9.

Osl M, Dreiseitl S, Pfeifer B, Weinberger K, Klocker H, Bartsch G, Schäfer G, Tilg B, Graber A, Baumgartner C. A new rule-based data mining algorithm for identifying metabolic markers in prostate cancer using tandem mass spectrometry. Bioinformatics. 2008;24:2908–14.

Pardalos PM, Boginski VL, Vazacopoulos A. Data mining in biomedicine. Berlin: Springer; 2007.

Parmanto B, Scotch M, Ahmad S. A framework for designing a healthcare outcome data warehouse. Perspect Health Inf Manag, 2005;2:3.

Pasquier C. Biological data integration using semantic web technologies. Biochimie. 2008;90 (4):584–94.

Porta M. A dictionary of epidemiology. 5th ed. Oxford: Oxford University Press; 2014.

Putri SP, Yamamoto S, Tsugawa H, Fukusaki E. Current metabolomics: technological advances. J Biosci Bioeng. 2013;116(1):9–16.

Shadbolt N, Hall W, Berners-Lee T. The semantic web revisited. IEEE Intell Syst App. 2006;21 (3):96–101.

Sjöström M, Ossola R, Breslin T, Rinner O, Malmström L, Schmidt A, Aebersold R, Malmström J, Niméus E. A combined shotgun and targeted mass spectrometry strategy for breast cancer biomarker discovery. J Proteome Res. 2015;14(7):2807–18.

Smith BS, Webb A. Introduction to medical imaging: physics, engineering and clinical applications (Cambridge texts in biomedical engineering). Cambridge: Cambridge University Press; 2010. ISBN 978-0521190657.

Soon WW, Hariharan M, Snyder MP. High-throughput sequencing for biology and medicine. Mol Syst Biol. 2013;9:640.

Spanos DE, Stavrou P, Mitrou N. Bringing relational databases into the semantic web: a survey. J Sem Web. 2012;3(2):169–209.

Swan AL, Mobasheri A, Allaway D, Liddell S, Bacardit J. Application of machine learning to proteomics data: classification and biomarker identification in postgenomics biology. OMICS. 2013;17(12):595–610.

Swan AL, Stekel DJ, Hodgman C, Allaway D, Alqahtani MH, Mobasheri A, Bacardit J. A machine learning heuristic to identify biologically relevant and minimal biomarker panels from omics data. BMC Genomics. 2015;16(Suppl 1):S2.

Ting SL, Shum CC, Kwok SK, Tsang AHC, Lee WB. Data mining in biomedicine: current applications and further directions for research. J Softw Eng Appl. 2009;2:150–9.

Töpel T, Kormeier B, Klassen A, Hofestädt R. BioDWH: a data warehouse kit for life science data integration. J Integr Bioinform. 2008;5(2):93.

Tseng GC, Ghosh D, Feingold E. Comprehensive literature review and statistical considerations for microarray meta-analysis. Nucleic Acids Res. 2012;40(9):3785–99.

Wang XD, Liotta L. Clinical bioinformatics: a new emerging science. J Clin Bioinform. 2011;1 (1):1.

Wang J, Yang J, Mao S, Chai X, Hu Y, et al. MitProNet: a knowledgebase and analysis platform of proteome, interactome and diseases for mammalian mitochondria. PLoS ONE. 2014;9(10): e111187.

Woods AG, Sokolowska I, Ngounou Wetie AG, Wormwood K, Aslebagh R, Patel S, Darie CC. Mass spectrometry for proteomics-based investigation. Adv Exp Med Biol. 2014;806:1–32.

Worthey EA. Analysis and annotation of whole-genome or whole-exome sequencing-derived variants for clinical diagnosis. Curr Protoc Hum Genet. 2013;79:Unit 9.24.

Xu R, Wunsch 2nd DC. Clustering algorithms in biomedical research: a review. IEEE Rev Biomed Eng. 2010;3:120–54.

Ye Xiao-rong. Analysis on network clustering algorithm of data mining methods based on rough set theory. 2011 fourth international symposium on Knowledge Acquisition and Modeling (KAM), Sanya, 8–9 October. 2011; p. 296–298. ISBN: 978-1-4577-1788-8.

Zhang A, Sun H, Yan G, Wang P, Wang X. Mass spectrometry-based metabolomics: applications to biomarker and metabolic pathway research. Biomed Chromatogr. 2016;30(1):7-12.

Christian Baumgartner, PhD is Professor of Health Care Engineering at Graz University of Technology, Austria. He received his MSc (1994) and PhD degree (1998) in Electrical and Biomedical Engineering from Graz University of Technology, Austria, and his habilitation degree (Assoc.-Prof.) in Biomedical Engineering from the University for Health Sciences, Medical Informatics and Technology (UMIT), Hall in Tyrol, Austria (2006).

From 1998 to 2002, Dr. Baumgartner held an R&D position at Tecan.com, where he developed confocal fluorescence laser scanning systems for micro array applications. From 2007–2008, he was part of the Barnett Institute of Chemical and Biological Analysis, Northeastern University and Harvard Medical School, Boston, MA, where Dr. Baumgartner worked in the field of computational biomarker discovery. In 2009, he was appointed Full Professor, Director of the Institute of Electrical and Biomedical Engineering, and Vice Chair of the Department of Biomedical Informatics and Mechatronics at UMIT. He has been a professor since 2015 and Head of the Institute of Health Care Engineering with European Notified Body of Medical Devices at Graz University of Technology in Austria since 2016.

Dr. Baumgartner is the author of more than 150 publications in refereed journals, books and conference proceedings, and is a reviewer for more than 35 scientific journals. He served as a deputy editor of the "Journal of Clinical Bioinformatics", and is an editorial board member of "Clinical and Translational Medicine", "Methods of Information in Medicine" and "Cell Biology and Toxicology". His main research interests include cellular electrophysiology, biomedical sensors and signal processing, biomedical modeling, simulation, clinical bioinformatics and computational biology.

Chapter 2
Biostatistics, Data Mining and Computational Modeling

Hao He*, Dongdong Lin*, Jigang Zhang, Yuping Wang, and Hong-Wen Deng

Abstract With the rapid development of high-throughput experimental technologies, bioinformatics and computational modeling has been a rapid evolving science field concerned with the development of various analysis methods and tools for investigating these large biological data efficiently and rigorously. There are many methods and tools available for the analysis of single omics dataset. It is a great challenge that biological systems are being further investigated by integrating multiple heterogeneous and large omics data. Many powerful methods and algorithmic techniques have been developed to answer important biomedical questions through integrative analysis. In this chapter, in order to help the bench biologist analyze omics data, we introduced various methods from classical statistical techniques for single marker association and multivariate analysis to more recent advances from gene network analysis and integrative analysis of multi-omics data.

Keywords Multi-omics • Integrative analysis • Gene network analysis • Disease diagnosis • Classification

---

*Author contributed equally with all other contributors

H. He • J. Zhang
Center for Bioinformatics and Genomics, Department of Biostatistics and Bioinformatics, Tulane University, New Orleans, LA 70112, USA

D. Lin • Y. Wang
Center for Bioinformatics and Genomics, Tulane University, New Orleans, LA, USA

Department of Biomedical Engineering, Tulane University, New Orleans, LA, USA

Hong-Wen Deng, Ph.D. (✉)
Center for Bioinformatics and Genomics, Department of Biostatistics and Bioinformatics, Tulane University School of Public Health and Tropical Medicine, New Orleans, LA, USA
e-mail: hdeng2@tulane.edu

X. Wang et al. (eds.), *Application of Clinical Bioinformatics*,
Translational Bioinformatics 11, DOI 10.1007/978-94-017-7543-4_2

2.1 Introduction

In the past decade, with the development of high throughput technologies, massive biological data have been generated from multiple levels of biological systems — including DNA sequence data in genomics, RNA expression levels in transcriptomics, DNA methylation and other epigenetic markers in epigenomics, protein expression in proteomics and metabolic profiling in metabolomics. These omics data are high throughput measurements of the abundance and/or structure features of molecules involved in biological metabolism and regulation. Table 2.1 summarizes the main features of various omics data.

Generally, omics data are high-dimensional data, which means that the number of subject n (e.g., tissue or samples) is much smaller than the number of variables p (e.g., number of SNPs in genome wide association, number of genes in an expression profile). In this setting, we are confronted with thousands of hypothesis testing simultaneously. There is a high risk that statistical models may overfit the omics data. In addition, datasets from diverse genomic levels have unique properties. A better understanding of the data characteristics will help to improve statistical modeling. An increasing number of advanced statistical methods have been developed to address these issues in omics data analysis at different levels.

Table 2.1 Main features of omics data

Omics	Biomarker data	Platforms	Features
Genome	Single nucleotide polymorphism (SNP)	Microarray	Categorical data
	Copy number variation (CNV)	DNA sequencing	Distance-driven correlation
	Loss of heterozygosity (LOH)		Extremely stable over time
	Rare variant		
Transcriptome	Gene expression	Microarray	Continuous data
	Alternative splicing	RNA sequencing	Affected by time and exposures
	Long non-coding RNA		Strong measurement noise
	Small RNA		
Proteome	Protein expression	Microarray	Continuous data
		Mass spectrometry	Affected by time and exposures
Epigenome	DNA methylation	Microarray	Continuous data
	Histone modification	Bisulfite sequencing	Affected by time and exposures
	miRNA		
Metabolome	Metabolite profiling	Mass spectrometry	Continuous data
		Nuclear magnetic resonance (NMR) spectroscopy	Affected by time and exposures
			Structured correlation
			Strongly affected by exposures

Instead of analyzing single omics data, it is interesting to integrate multiple levels of omics data to gain comprehensive insights into biology and disease etiology. It is recognized that multi-scale features do not act in isolation but interact in complex networks (within and across individual omics), e.g., genomics information flow DNA- > RNA- > protein- > traits. Therefore, no single type of omics data can provide a thorough understanding of the complex function/regulatory networks that mediate gene expression/function for disease etiology. Integrative analysis of multiple omics data with the same subjects has the following advantages: 1) multiple omics data can provide diverse information that the identified genetic variants may be consistent in the effects across different omics levels. Consistent results will compensate for unreliable findings in single omics data, which can improve the detection power for those variants with modest effects in individual omics data. Complementary results will confirm the findings to get a more comprehensive understanding of genetic mechanisms of diseases; 2) importantly, integrative analysis of multiple omics data will enable the reconstruction of interplay/regulatory relationship among genetic factors at different levels. The analysis of complex regulatory networks will aid in functional annotation of individual genes/regulatory factors, gaining new insights into the molecular mechanisms underlying disease pathogenesis and generating model hypothesis for further specific testing. Taken together, the integrative trans-omics studies can provide a much more comprehensive view of complex disease etiology than can be achieved by examining individual omics data on their own.

In this chapter, we first briefly review statistical methods for biomarker detection in different omics data. Then we will review integrative statistical analysis involving at least two different types of omics data.

2.2 Statistical Methods for Biomarker Detection in Clinical Bioinformatics

Several types of biological data can be used to identify informative biomarker panels, including SNP data, microarray based gene expression and microRNA. Statistical methods especially predictive models based on these biomarkers are becoming increasingly important in clinical, translational and basic biomedical research. We will first provide illustrations of various statistical methods in the analysis of SNP and gene expression data, attempting to offer practical advice on the appropriate methods to use.

2.2.1 Statistical Analysis for Single Omics Data

2.2.1.1 Single Marker Association

Single SNP Association The objective of genetic association analysis is to establish an association between a phenotype/quantitative trait and a genetic marker.

Usually genetic association tests are performed separately for each individual SNP. A variety of statistical methods could be applied according to the data types of the phenotype/quantitative trait. The phenotype in a study can be case-control (binary), quantitative (continuous), or categorical. First we will discuss analysis for case-control, continuous and categorical disease outcomes and then we will present more advanced statistical methods for multivariate analysis.

Here is the basic problem formulation. Let $\{X_1, \ldots, X_p\}$ be a set of P SNPs for N individuals. Suppose the data with each SNP having minor allele a and major allele A. We use 0, 1, 2 to represent the homozygous major allele, heterozygous allele and homozygous minor allele, respectively. Therefore we have $X_{pn} \in \{0, 1, 2\}, (1 \leq p \leq P, 1 \leq n \leq N)$. Let phenotype be $Y = \{y_1, \ldots, y_n\}$. Depending on the data type, the values of Y can be binary, continuous or categorical.

For case-control phenotype, it can be represented as a binary variable with 0 representing controls and 1 representing cases. The association between a SNP and case-control status is to test the null hypothesis of no association between the marker with disease status in a contingency table, which links disease status by either three genotypes counts (A/A, A/a and a/a) or allele count (A and a). The test of association is given by Pearson χ^2 test for the independence of the rows and columns in the contingency table (Balding 2006). The choice of degrees of freedom is based on recessive, dominant and additive models of inheritance. The contingency table can allow alternative models by summarizing the counts based on the models of inheritance. For instance, to test for a dominant model, the contingency table is summarized as 2×2 table of genotype counts (A/A vs. A/a and a/a). As to a recessive model, the contingency table is summarized as 2×2 table of genotype counts (a/a vs. A/A and A/a). There are two tests commonly used for testing the additive model of inheritance: the allele test and the trend test, also known as the Cochran-Armitage trend test. Both tests have the same null hypothesis: $P_{\text{case}} = P_{\text{control}}$, where P_{case} and P_{control} denote the frequency of A alleles among diseased and non-diseased in a population, respectively. As the underlying genetic model is unknown in most genetic association studies, the test for additive model is most commonly used. However, there is no generally accepted answer to the question about what kind of test to be used. The analyses could be designed optimally according to the information that what proportion of undiscovered disease-predisposing variants function additively and what proportions are dominant and recessive. Table 2.2 summarizes different contingency table methods based on diverse tests of association. Take genotypic association for instance, Table 2.3 is the contingency table. For a SNP and the phenotype Y, we use O_{ij} to denote the number of individuals whose X_p equals i and Y equals j. The Pearson χ^2 statistics is calculated as $\sum_i \sum_j \frac{(O_{ij} - E_{ij})^2}{E_{ij}}$, where $E_{ij} = \frac{O_{i.} O_{.j}}{N}$, $O_{i.} = \sum_j O_{ij}$ and $O_{.j} = \sum_i O_{ij}$. The degree of freedom is 2.

Logistic regression is a statistical method for predicting binary and categorical outcome. It can be applied to both single-locus and multi-locus association studies with covariates in the model. Let $Y \in \{0, 1\}$ be a binary variable and $X \in \{0, 1, 2\}$ be

Table 2.2 Tests of association using contingency table methods

Test	DF	Contingency table description
Genotypic association	2	2 × 3 table of *N* case-control by genotype counts
		(A/A vs. A/a vs. a/a)
Dominant model	1	2 × 2 table of *N* case-control by dominant genotype pattern of inheritance counts
		(a/a vs. not a/a)
Recessive model	1	2 × 2 table of *N* case-control by recessive genotype pattern of inheritance counts
		(not A/A vs. A/A)
Cochran-Armitage trend test	1	2 × 3 table of *N* case-control by genotype counts
		(A/A vs. A/a vs. a/a)
Allelic association	1	2 × 2 table of 2*N* case-control by allele counts
		(A vs. a)

Note: *DF* degrees of freedom

Table 2.3 Contingency table for genotypic association test of a single SNP X_p and a phenotype Y

Count	Genotype aa ($X_p = 0$)	Genotype Aa ($X_p = 1$)	Genotype aa ($X_p = 2$)	Total
$Y=0$ (Control)	O_{00}	O_{01}	O_{02}	$O_{0.}$
$Y=1$ (Case)	O_{10}	O_{11}	O_{12}	$O_{1.}$
Total	$O_{.0}$	$O_{.1}$	$O_{.2}$	N

a SNP. The conditional probability of $Y=1$ given a SNP is $\theta(X) = P(Y = 1|X)$. The logit function is defined as $\mathrm{log}it(X) = \ln\frac{\theta(X)}{1-\theta(X)}$. The *logit* function can be taken as a linear predictor function: $\mathrm{log}it(X) \sim \beta_0 + \beta_1 X$. The model can be modified to incorporate multiple SNP loci and potential covariates. For example, the following model fits two predictor SNPs (X_1 and X_2) and two covariates (Z_1 and Z_2): $\mathrm{log}it(X) \sim \beta_0 + \beta_1 X_1 + \beta_2 X_2 + \beta_3 Z_1 + \beta_4 Z_2$.

For continuous (quantitative) traits, the basic statistical tools are linear regression and analysis of variance (ANOVA).

In **regression models**, there are two types of variables: dependent variable (response variable or outcome variable) and independent variable (explanatory variable or predictor variable). In a regression model, the dependent variable is modeled as a function of one or more independent variables. When this function is a linear combination of one or more model parameters, called regression coefficients, the model is called a linear regression model. A least-squares regression line is often used to find optimal fit between the phenotype and the genotype.

For simplicity, a single SNP genotype is denoted X_i and the phenotype is $Y_i, i = 1, \ldots, n$. For this given data set (X_i, Y_i), we are fitting a simple linear regression model, $Y = \beta_0 + \beta_1 X + \varepsilon$, such that $E(\varepsilon) = 0$ and $Var(\varepsilon) = \sigma^2$, and

ε's are uncorrelated. We can find b_0 and b_1 as least squares estimators for β_0 and β_1, respectively. We have the sums of squares as follows: $S_{XX} = \sum_{i=1}^{n} (X_i - \overline{X})^2$, $S_{YY} = \sum_{i=1}^{n} (Y_i - \overline{Y})^2$, and $S_{XY} = \sum_{i=1}^{n} (X_i - \overline{X})(Y_i - \overline{Y})$, and the following two normal equations, $b_0 + b_1 \sum_{i=1}^{n} X_i = \sum_{i=1}^{n} Y_i$ and $b_0 \sum_{i=1}^{n} X_i + b_1 \sum_{i=1}^{n} X_i^2 = \sum_{i=1}^{n} X_i Y_i$. The estimator of b_1 *is* $\frac{S_{XY}}{S_{XX}}$. Then we can test the null hypothesis against the alternative hypothesis $H_0 : \beta_1 = \beta_{10}$ versus $H_1 : \beta_1 \neq \beta_{10}$, where β_{10} is a specified value that could be zero. The test statistics is calculated as $t = \frac{(b_1 - \beta_{10})}{se(b_1)} = \frac{(b_1 - \beta_{10}) \left\{ \sum (X_i - \overline{X})^2 \right\}^{\frac{1}{2}}}{\sqrt{S^2}}$, where S^2 is the estimate of residual mean square $\sigma_{Y.X}^2$. One can compare |t| with $t(n-2, 1-\frac{\alpha}{2})$ from a t-table with $(n-2)$ degrees of freedom. The test is a two-sided test conducted at the 100α% level.

In **one-way ANOVA** the F-test is used to assess whether the expected values of a quantitative variable within several pre-defined groups differ from each other. For a single SNP, we can divide all the subjects into three groups according to their genotypes. Let $Y_i^{'} (i \in \{0, 1, 2\})$ be the subset of phenotypes for the subjects corresponding to genotype i. The number of subjects with $Y_i^{'}$ is denoted as n_i. Note that $\sum_{i=0}^{2} n_i = N$. The total sum of squares (SST) can be divided into two parts, the between-group sum of squares (SSB) and the within-group sum of squares (SSW).

$SSB = \sum_{i=1}^{2} \left(\overline{Y^{'}}_i - \overline{Y} \right)^2$, $SST = \sum_{i=0}^{2} \sum_{n=1}^{N} \left(Y_{in}^{'} - \overline{Y} \right)^2$, and $SSW = SST - SSB$. The formula of F-test statistic is $F = \frac{SSB}{SSW}$, and F follows the F-distribution with 2 and *N-3* degrees of freedom under the null hypothesis.

Gene Expression Analysis In transcriptomics studies for biomarker discovery among thousands of features, we are interested in which genes/features are differentially expressed under two (or more) conditions. The hypothesis test will be performed individually for each feature. Statistical significance for each hypothesis test is assessed according to its corresponding p-value from a statistical test. Suppose there are K conditions and n_k samples in the kth condition in a total of N samples, where $K \in \{1, 2\}$. Let X_{ijk} be an expression value, where sample $i = 1, 2, \ldots, n_k$, gene features $j = 1, 2, \ldots, m$, and condition $K = 1, 2$. Assume that gene expression values have been background corrected, normalized and transformed by taking the logarithm to base 2. The sample mean and variance of gene feature j in group k are given as $\overline{X}_{jk} = \frac{\sum_{i=1}^{n_k} X_{ijk}}{n_k}$ and $S_{jk}^2 = \frac{\sum_{i=1}^{n} (X_{ijk} - \overline{X}_{jk})^2}{n_k - 1}$, respectively.

Fold change approach is a simple and straightforward way of evaluating the degree of differential expression under two conditions. For a gene feature j, the mean difference is given by $M_j = \overline{X}_{j1} - \overline{X}_{j2}$. Then the fold change is a statistic 2^{M_j}. Gene will be declared as significant if $|M_j|$ is greater than a predefined threshold. Such procedure assumes that the variances are equal across all genes. However, it is not the case for gene expression profile. Therefore, this approach may easily yield many false positive and false negative results in differential expression analyses.

The two-sample t-test is a most used parametric statistical test in differential expression analysis. It compares the means of expression value in two groups taking the variance into consideration. Statistically, we want to test the null hypothesis $H_0 : \mu_{j1} = \mu_{j2}$ against the alternative hypothesis $H_1 : \mu_{j1} \neq \mu_{j2}$ for j = 1,2,. . .m. The test statistic for each j is $t_j = \frac{\sum_{i=1}^{n} \left(\overline{X}_{j1} - \overline{X}_{j2}\right)^2}{S_j}$, *where* $S_j = \sqrt{\left(\frac{1}{n_1} + \frac{1}{n_2}\right) \frac{(n_1-1)S_{j1}^2 + (n_2-1)S_{j2}^2}{n_1+n_2-2}}$, called pooled within-group standard error. Under the null hypothesis, t_j follows Student's t-distribution with $n_1 + n_2 - 2$ degrees of freedom. A p-value can be found using a t-distribution table. By using the pooled within-group standard error estimated from each gene separately, the t-statistic takes into consideration of variance across different genes.

Significance analysis of microarrays (SAM) is a statistical technique for determining whether changes in gene expression are statistically significant (Tusher et al. 2001). In SAM, statistically significant genes will be identified based on gene specific t-tests. A statistic d_j for each gene j measures the strength of the relationship between gene expression and a response variable. Non-parametric statistics is used as the data may not follow a normal distribution. SAM will perform repeated permutations for the data to determine the significance of any gene with the response. The use of permutation-based analysis accounts for correlations in genes and avoids parametric assumptions about the distribution of individual genes. It assumes equal variance and/or independence of genes. This is an advantage over other techniques. Here is the generic procedure for SAM. A statistic d_j is computed as $d_j = \frac{r_j}{s_j + s_0}$, where r_j is a score, s_j is a standard deviation and s_0 is an exchangeability factor. Compared with the standard t-statistic, the SAM's procedure adds a s_0 term to the denominator. The rationale behind it is that the variance s_j tends to be smaller at lower expression levels, making d_j dependent on the expression levels. However, in order to compare d_j across all genes, the distribution of d_j should be independent of the expression levels. Therefore, SAM seeks to find a s_0 such that the dependence of d_j on s_j is as small as possible. An appropriate value of s_0 will be picked such that the coefficient of variation of d_j is approximately constant as a function of s_j. For details of the SAM procedure, please refer to the tutorial document for the software package, SAM, at http://statweb.stanford.edu/~tibs/SAM/sam.pdf.

The **Wilcoxon rank-sum test**, also known as the Mann–Whitney U-test, is a nonparametric test, which can be applied to data with unknown distributions contrary to t-test applied only to normal distributions. It is nearly as efficient as the t-test on normal distributions. The null hypothesis of the test is that two samples come from the same population and an alternative hypothesis is that a particular population tends to have larger values than the other. The Wilcoxon rank-sum test is based on the ranks of the original data values. To perform the Wilcoxon rank-sum test, one first assigns numeric ranks to all the observations, beginning with 1 for the smallest value. Where there are groups of tied values, assigning a rank equal to the midpoint of unadjusted rankings. Second, one adds up the ranks for the observations which came from group 1. The sum of ranks in group 2 is now determinative, since the sum of all the ranks equals $N(N+1)/2$ where N is the total number of observations. Then calculate $U_1 = R_1 - \frac{n_1(n_1+1)}{2}$ and $U_2 = R_2 - \frac{n_2(n_2+1)}{2}$. The smaller value of U_1 and U_2 is the one used when consulting significance tables.

2.2.1.2 Multiple Testing

As mentioned earlier, in omics studies we are confront with a great number of hypotheses to be tested simultaneously. It will result in an inflation of the family wise error rate (FWER) if there is no adjustment for multiple tests. In statistical hypothesis testing, a type I error occurs when the null hypothesis (H_0) is true, but is rejected (a "false positive"). A type II error occurs when the null hypothesis is false, but erroneously fails to be rejected (a "false negative"). A type I error is the incorrect rejection of a true null hypothesis (a "false positive"), while a type II error is the failure to reject a false null hypothesis (a "false negative"). Basically, in hypothesis testing, we want to maximize the power (=1-the type II error) while controlling the type I error less than or equal to a predetermined significance level α. In particular, consider the problem of testing simultaneously m null hypothesis H_j: no differential expression against H_j^a: differential expression, where $j = 1, 2, \ldots,$ m. A gene will be considered as significantly differentially expressed if its p-value is less than the defined significant level α. However, for hypothesis testing, the problem of multiple testing problem results from the increase in type I error that occurs when many statistical tests are used simultaneously. Suppose there are m independent comparisons, the experiment-wide significance level $\overline{\alpha}$, also termed FWER, is given by $\overline{\alpha} = 1 - (1 - \alpha)^m$. $\overline{\alpha}$ increases as the number of comparison increases. Multiple testing correction is to re-calculate the probabilities obtained from a statistical test which was repeated multiple times. In order to retain FWER $\overline{\alpha}$ in an analysis, the error rate for each comparison must be more stringent than α.

A number of procedures for controlling error rates have been developed to solve the multiple-testing problem. One of the most commonly used approaches for multiple comparisons is the Bonferroni procedure for controlling the FWER at level α, which rejects any hypothesis H_j with unadjusted p-value less than or equal to α/m. The Bonferroni procedure is very conservative. A less conservative procedure is the Benjamini–Hochberg procedure (BH step-up procedure), which controls the false discovery rate (at level α). The procedure works as follows: first for a given α, find the largest k such that $P_{(k)} \leq \frac{k}{m}\alpha$. Second, reject all H_j for j = 1, 2, . . ., k. The BH procedure is valid when the m tests are independent and also in various scenarios of dependence.

2.2.1.3 Multivariate Analysis

Although many common genetic variants associated with complex traits have been identified by GWAS, these traits are typically analyzed separately in a univariate manner for association with DNA markers. However, multivariate analysis for correlated traits could be very advantageous in several aspects. First, when there is genetic correlation between different traits, a multivariate analysis can increase power by using the extra information provided by the cross-trait covariance, which is ignored by the univariate analysis. Second, a multivariate analysis of multiple traits can reduce the number of performed tests and alleviate multiple testing burden compared to analyzing all traits separately. Lastly, a multivariate analysis is biologically making more sense as a single genetic marker is associated with multiple traits, compared to the cross-trait comparison in univariate analysis (Galesloot et al. 2014).

A number of multivariate analysis methods in population-based GWAS have been published. Here we briefly introduce six methods including as well as their softwares.

The multivariate test of association MQFAM is implemented in the genetic association analysis software PLINK (MV-PLINK) (Ferreira and Purcell 2009; Purcell et al. 2007). The command used for association testing with MV-PLINK (https://genepi.qimr.edu.au/staff/manuelF/multivariate/main.html) is: *plink.multivariate –noweb –file geno –mqfam –mult-pheno pheno.phen –out output*. For each genetic variant, MV-PLINK produces an F-statistic and a p-value in the additive model. Canonical correlation analysis (CCA), which is a multivariate generalization of the Pearson product-moment correlation, to measure the association between the two sets of variables. Specifically, CCA extracts the linear combination of traits that explain the largest possible amount of the covariation between the marker and all traits. The interpretation of a significant multivariate test is aided by the inspection of the weights attributed by the CCA to each phenotype.

Bayesian multiple phenotype test is implemented in SNPTEST (MV-SNPTEST) (Marchini et al. 2007). The command used to perform additive association testing with MV-SNPTEST is provided in the online tutorial (https://mathgen.stats.ox.ac.uk/genetics_software/snptest/snptest.html#multiple_phenotype_tests). The model is the Bayesian Multivariate Linear model which is specified by $(y_{i1}, \ldots, y_{iq})^T = G_i(\beta_1, \ldots, \beta_q)^T + (e_{i1}, \ldots, e_{iq})^T$,where $(e_{i1}, \ldots, e_{iq})^T \sim N(0, \sum)$ and $(y_{i1}, \ldots, y_{iq})$ is the vector of the q residual phenotypes measured on the ith individual. G_i is the code of the SNP genotype for the *i*th individual. We use the conjugate prior for this model. This is an inverse Wishart prior IW(c,Q) on the error covariance matrix $\sum$ and a matrix normal (N) prior on the vector of parameters $(\beta_1, \ldots, \beta_q) - M{\sim}N(V, \Sigma)$, where M is a mean vector and V is a constant. An inverse Wishart prior [IW(6,4)] was set on the error covariance matrix $\sum$ and a matrix normal prior [N(0.02,$\sum$)] on the vector of parameters, according to recommendations of the authors. Method 'expected' will result in the use of expected genotype counts (~dosages) in the analyses.

MultiPhen is an R package available from CRAN (https://cran.r-project.org/web/packages/MultiPhen/index.html) (O'Reilly et al. 2012). The regression performed at a SNP, g, and a phenotype, k, to test for association between the SNP genotypes and the phenotype is: $Y_{ik} = \alpha_k + \beta_{gk} X_{ig} + \varepsilon_{igk}$, where ε_{igk} is the residual error assumed to be normally distributed. The null hypothesis of no association between SNP and genotype can be tested by performing a t-test on the null hypothesis $\beta_{gk} = 0$. In the MultiPhen approach, the regression is inverted so that the SNP genotype, X, becomes the dependent variable, and K phenotypes under study become the predictor variables. The genotype data is an allele count and is therefore modelled using ordinal regression; we use proportional odds logistic regression. This model defines the class probabilities as follows. $P(X_{ig} \leq m) = \frac{1}{\left(-\alpha_{gm} - \sum_{k=1}^{K} \beta_{gk} Y_{ik}\right)}$. At each SNP g = 1,2,. . .,G, the test for association is a likelihood ratio test (LRT) for model fit, testing the null hypothesis $\beta_{g1} = \ldots = \beta_{gk} = 0$. This results in a p value per trait and a p-value for the LRT.

A Bayesian model comparison and model averaging for multivariate regression is implemented in BIMBAM software (Stephens 2013). The details of statistical method are provided in the reference (Stephens 2013). The BIMBAM software can be run in two different ways. First we test for association between the multivariate traits, all partitioned in the group of directly affected traits, and genotype. Second, we consider all possible partitions of traits into the different categories of traits (directly affected, indirectly affected, and unaffected).

The Principal Component of Heritability Association Test (PCHAT) (Klei et al. 2008) is implemented in the software available at http://www.wpic.pitt.edu/wpiccompgen/PCHAT/PCHAT.htm). First, the sample is split into a training set and a test set. The training set is used to construct the optimal linear combination of traits from a heritability point of view. A test set is used for association testing between genotype and the optimal linear combination of traits. In this way, use of the same data for both estimation of the optimal linear combination of traits and association testing is avoided. In addition, a 'bagging' approach is performed, in which bootstrap samples are drawn from the training sample and the optimal linear combination of traits is averaged across bootstrap samples. The null distribution of the test statistic is obtained in the same way, using permutation of the data.

A Trait-based Association Test (TATES) is based on Extended Simes procedure (van der Sluis et al. 2013). TATES (http://ctglab.nl/software) constitutes a powerful new multivariate strategy that allows researchers to identify novel causal variants. TATES acquire one trait-based p-value by combing p-values in standard univariate GWAS, while correcting for correlations between components. It can detect both genetic variants which are common to multiple phenotypes and those which are specific to a single phenotype. It requires a correlation matrix of the traits and univariate association results as input. The *corr* function in R can be used to generate the full and symmetrical correlation matrices. TATES was run in R and the output contains the TATES trait-based p-value corrected for the correlations between the traits.

2.2.1.4 Gene Set Analysis

In transcriptomics study, massive throughput techniques, such as microarray and RNA sequencing, allow to identify differentially expressed genes (DEGs) associated with diseases or phenotypes from genome-wide gene expression profile. The challenge in expression data analysis in recent years has shifted from single DEG analysis to gene set analysis (GSA), as biologically many complex diseases may be modestly regulated by a set of related genes rather than a single gene. The gene sets are defined based on prior biological knowledge, e.g., biochemical pathways or coexpression in previous experiments. GSA can alleviate the difficulty in interpretation of multiple testing lists of DEGs and provide insights into biological mechanisms for complex diseases. The first and most popular GSA is gene set enrichment analysis (GSEA) (Subramanian et al. 2005), which is a computational method that determines whether an a priori defined set of genes shows statistically significant, concordant differences between two biological states (e.g. phenotypes). The GSEA method is implemented in a freely available software package at http://www.broadinstitute.org/gsea/index.jsp. The basic idea for this method is presented as follow (Subramanian et al. 2005):

Step 1: Calculate an Enrichment Score. Rank genes by their expression difference in two biological states and then compute cumulative sum over ranked genes. The magnitude of increment depends on correlation of gene with phenotype. Record the maximum deviation from zero as the enrichment score.

Step 2: Estimate significance. Permute phenotype labels 1000 times and compute ES score for each permutation. Then compare ES score for actual data to distribution of ES scores from permuted data.

Step 3: Multiple Hypothesis Testing. Normalize the ES accounting for size of each gene set to obtain the normalized enrichment score (NES). Calculate FDR for each NES to control proportion of false positives by comparing tails of the observed and null distributions for the NES.

Another interesting GSA method proposed by Efron and Tibshirani attempts to combine gene and sample randomization in one procedure (Efron and Tibshirani

2007). It shows that it is more powerful based on the "maxmean" statistic than the modified Kolmogorov-Smirnov statistic used in GSEA. This method can be implemented by the R package "GSA". The basic procedures are summarized here:

1. Compute a summary statistic z_i for each gene, for example the two sample t-statistic for two-class data. Let z_s be the vector of z_i values for genes in a gene-set S.
2. For each gene-set S, choose a summary statistic S = s(z): the maxmean statistic $\left\{\left|\frac{\sum_{i=1}^{m} I(z_i > 0)z_i}{m}\right|, \left|\frac{\sum_{i=1}^{m} I(z_i < 0)z_i}{m}\right|\right\}$
3. Standardize S by its randomization mean and standard deviation as $S' = \frac{(S-mean(s))}{std(s)}$. For summary statistics such as the mean, mean absolute value or maxmean, this can be computed from the genewise means and standard deviations, without having to draw random sets of genes.
4. Compute permutations of the outcome values (e.g., the class labels in the two-class case) and re-compute S' on each permuted dataset, yielding permutation values. Use these permutation values to estimate p-values for each gene-set score S' and false discovery rates applied to these p-values for the collection of gene-set scores.

In 2007, Wang et al. extended the GSEA to GWAS of complex diseases (Wang et al. 2007), where multiple genes in the same GS/pathway contribute to disease etiology but where common variations in each of those genes make modest contributions to disease risk. Gene set analysis tests disease association with genetic variants in a group of functionally related genes, such as those belonging to the same biological pathway. It can potentially improve the power to detect causal GS/pathways and disease mechanisms by considering multiple contribution factors together, rather than focusing on the top SNPs associated with disease. Individual SNPs in univariate analysis only account for a small proportion of the heritability of complex diseases. The method assesses the enrichment of significant associations for genes in the GS/pathway (as compared with those outside the GS/pathway) using a weighted Kolmogorov–Smirnov running-sum statistic. The GSEA method is modified to fit GWAS data. For each SNP V_i ($i = 1, \ldots, L$, where L is the total number of SNPs in a GWA study), its test statistic value is calculated, r_i (e.g., a χ^2 statistic for a case-control association test). We next associated SNP V_i with gene G_j ($j = 1, \ldots, N$, where N is the total number of genes represented by all SNPs) if the SNP is located within or <500 kb away from the gene. The highest statistic value among all SNPs mapped to the gene, is assigned as the statistic value of the gene. For all N genes that are represented by SNPs in the GWA study, their statistic values are sorted from largest to smallest, denoted by $r_{(1)}, \ldots, r_{(N)}$. For any given gene set S, composed of N_H genes, a weighted Kolmogorov-Smirnov–like running-sum statistic is calculated which reflects the overrepresentation of genes within the set S at the top of the entire ranked list of genes in the genome.

Over recent years, various methods have been published for gene-set or pathway-based association analysis for GWAS. Basically, these statistical methods can be classified into two categories based on whether the required input data sets are a collection of SNP p-values or individual-level SNP genotypes. Additionally, the null hypothesis can also be categorized as 'self-contained' versus 'competitive' based on whether comparisons were made between genes in a specific pathway and non-associated genes or other genes in the genome. Some of these published algorithms as well as software implementations or web servers are summarized in the review (Wang et al. 2010).

2.2.1.5 Gene Network Analysis

Recent years many network theories have been applied to gene coexpression network analysis. As gene expression microarrays measure the transcription levels of thousands of genes simultaneously, it provides great opportunities to explore large scale gene regulatory networks. Genes with similar expression patterns may participate in pathways and in regulatory and signaling circuits and their products may form complexes. Gene networks provide a systematic understanding of molecular mechanisms underlying biological processes, and the visualization of direct dependencies facilitates systematic interpretation and comprehension of the relationships among genes. Most complex human diseases are arising not from a single gene but from interactions with many other genes, especially in a gene network. The hub genes, which interact with many other genes, are likely to be drivers of the disease status. The analysis on the hub genes has become a promising approach for identifying the key candidate genes for complex diseases.

A great number of statistical methods for gene network reconstruction from gene expression microarray data have been proposed in recent years. There are four main categories of statistical methods: (1) Probabilistic networks-based approaches, mainly Bayesian networks (BN), (2) correlation-based methods, (3) partial-correlation-based methods, and (4) information-theory-based methods (Allen et al. 2012). The representative method in each category and the implementation software are summarized below.

Probabilistic networks, mainly Bayesian networks, are based on a probabilistic graphical model that represents a set of variables and their probabilistic independencies. The Bayesian networks expand the joint probability in terms of simpler conditional probabilities, which allow them to handle noise inherent in both biological processes and microarray experiments. Generally, the joint likelihood function of nodes $X_1,\ldots,X_p$ in a Bayesian network can be expressed as $P(X_1,\ldots,X_p) = \prod_{i=1}^{p} P(X_i|\prod_i^G)$, where graph $G=(V,E)$ represents the topological structure of the Bayesian network, in which $V=\{X_1,\ldots,X_p\}$ denotes the set of nodes and $E=\left\{X_j \to X_i, X_j \in \prod_i^G\right\}$ denotes the set of edges. Werhli's implementation for Bayesian network construction method is most used and outperforms other implementations (Werhli et al. 2006). A Bayesian network models the distribution of observations and a causal network models the distributions of observations and effects of interventions. A causal network can be interpreted as a Bayesian network, when we are willing to make the Causal Markov Assumptions: given the values of a variable's immediate causes, it is independent of its earlier causes (Friedman et al. 2000).

Correlation-based methods are the most straightforward and popular way to explore the gene co-expression network. They have been successfully applied in many studies and have shown their usefulness in identifying important gene modules and in interpreting biological results. Basically a gene co-expression similarity matrix is defined as $S = [S_{i,j}]$, where $S_{i,j}$ is the pair-wise transcription correlation coefficients between gene *i* and *j*. *S* is the correlation matrix (Zhang and Horvath 2005). Particularly, Weighted Correlation Network Analysis (WGCNA) is a representative method for the correlation-based approach (Langfelder and Horvath 2008). The implementation of WGCNA is in R package, which is used for identifying modules/subnetworks using hierarchical clustering approaches. The WGCNA R package includes interfaces with Cytoscape (Shannon et al. 2003) for network visualization and The database for annotation, visualization and integrated discovery (DAVID) (Dennis et al. 2003) for enrichment analysis. The comprehensive set of online tutorials that guide users through the major steps for gene network analysis by WGCNA are provided in the website http://labs.genetics.ucla.edu/horvath/htdocs/CoexpressionNetwork/Rpackages/WGCNA/Tutorials/index.html. In the tutorials, R code in each step is provided so that the user can copy and paste into an R session. The tutorials cover the following major topics: correlation network construction, step-by-step and automatic module identification, consensus module detection, eigengene network analysis and differential network analysis.

Here we briefly review the key concepts of the WGCNA framework. The nodes in a gene coexpression network correspond to genes, labeled by indices *i*, $j = 1,2,\ldots,n$. The edge between two nodes is determined by the pairwise correlation. The network can be specified by its *adjacency matrix* **A**, a symmetric matrix with entries a_{ij} in [0,1] that encode the strength of the link between genes *i* and *j*. An unsigned network is defined by the adjacency **A** in terms of *coexpression similarity* $S_{ij} = \left|cor\left(x_i, x_j\right)\right|$, in which positive and negative correlations are treated equally. Also if we want to preserve the sign of the correlation, we can use a *signed* similarity defined as $S_{ij} = \frac{\left(1+cor\left(x_i,x_j\right)\right)}{2}$. The main difference between signed and unsigned similarities is that genes with a high negative correlation (close to -1) will have a low similarity in a signed network but a high similarity in an unsigned network. A weighted network can preserve the continuous nature of the co-expression information by using a soft thresholding parameter, $\beta \geq 1$. By using a power function, the connection strength can be assessed, $a_{ij} = S_{ij}^{\beta}$. The default values $\beta = 6$ and $\beta = 12$ are used for unsigned and signed networks.

In WGCNA, genes are clustered into network modules based on their coexpression. Highly coexpressed genes have a small dissimilarity. For example, the adjacency-based dissimilarity measure is $dissAdj_{ij} = 1 - a_{ij}$. The dissimilarity measure can be used as input in average linkage hierarchical clustering. Then, modules are defined as branches of the resulting cluster tree. If larger and more robust modules are desired, one can use a dissimilarity measure based on the topological overlap matrix (TOM): $dissTOM_{ij} = 1 - TOM_{ij} = 1 - \frac{\sum_{u \neq i} a_{iu} a_{uj} + a_{ij}}{\min(k_i, k_j) + 1 - a_{ij}}$, where $k_i = \sum_{u \neq i} a_{ui}$ denotes the network connectivity. TOM combines the connection strength between a pair of genes with their connections to other 'third party' genes, which has been shown to be a highly robust measure of network interconnectedness (proximity). In order to summarize the module genes by a single representative expression profile, module eigengene is defined as the first principal component of the standardized expression profiles of a given module, which is considered as the weighted average of the module gene expressions. We can correlate the module eigengenes with the trait of interest *y*. The correlation coefficient or its corresponding p-value is referred to as the eigengene significance. For each module, the module significance is defined as the average absolute gene significance for all genes in the module. WGCNA can alleviate the multiple testing problem in DEG analysis, as it focuses on a few modules with the trait rather than thousands of genes and these modules may be included into some important biological pathways.

Partial-correlation-based methods are based on Gaussian graphic model. These methods infer the conditional dependency by the non-zero entries in the precision matrix, $C = [C_{i,j}] = S^{-1}$, which is the inverse of covariance matrix (Allen et al. 2012). The zero entries in the precision matrix imply conditional independency between the expression levels of gene *i* and *j* given the expression of all other genes, which means two genes do not interact directly with each other. The sparse partial correlation estimation (SPACE) algorithm is a representative partial-correlation-based method (Peng et al. 2009). It converts the concentration matrix estimation problem to a regression problem and optimizes the results with a symmetric constraint and an L_1 penalization.

Information-theory-based methods use mutual information (MI) to determine how similar the joint distribution P(X, Y) is to the products of factored marginal distribution P(X)P(Y). It can determine the dependency among the genes and then remove indirect interactions. Algorithm for the Reconstruction of Accurate Cellular Networks (ARACNE) is a successful and popular information-theory-based method, which has been successfully applied to construct gene regulatory networks in the context of specific cellular types (Margolin et al. 2006). The calculation of MI does not assume a monotonic relationship; therefore it is able to identify the non-linear or irregular dependencies, which will be missed by Pearson correlation. If the gene network contains non-monotonic dependencies the ARACNE could outperform correlation-based methods.

2.2.2 *Computational Methods for Integrating Multi-Omics Data*

A variety of statistical methods and tools have been proposed for integrating two or more omics data. These methods aim to help understand molecular mechanism or biological pathways underlying variation of different types of clinical traits. Also they explore the relationship or interactions among diverse omics data for complex network structure reconstruction and thereby identifying risk modules associated with clinical outcomes. Integrated information is finally used for subtyping clinical diseases or predicting the outcome for prospective patients. These computational methods can be broadly categorized into four types in terms of the objective of analysis and the way of integrating omics data.

2.2.2.1 Multi-Stage Method: Analyzing Multi-Omics Data Sequentially

Multi-stage method is a way to divide multi-omics analysis into multiple stages, where each stage only incorporates two levels of omics and subsequently relates biomarkers to the trait or phenotype of interest. For example, a three stage strategy is commonly applied for identifying genetic variants associated with the phenotype and relating the other levels of omics, e.g., gene expression (Holzinger and Ritchie 2012).

Step1. Identifying those significant genetic variants (e.g., SNPs) associated with phenotype by genome-wide association test with multiple testing corrected.

Step2. Testing those identified SNPs for association with the other omics data, such as gene expression, DNA methylation, protein expression and other functional profiling. The corresponding associated SNPs are called expression quantitative loci (eQTLs (Jansen and Nap 2001)), methylation QTL (meQTLs (Kerkel et al. 2008)), protein QTL(pQTLs (Melzer et al. 2008)) respectively.

Step3. Those omics features having at least one QTL are further tested for the association with phenotype. Subsequently, biological pathways can be derived; some SNPs associate with phenotype through other omics data while some SNPs can affect phenotype independent of the other omics data. One benefit of multi-stage method is that each single stage analysis is performed independently with a variety of statistical methods (Cantor et al. 2010). For example, to identify significant biomarkers at the first and third stage, both univariate test (e.g., linear regression or logistic regression) and multivariate methods (e.g., region or pathway based test (Khatri et al. 2012)) can be applied for genome-wide detection. At the second stage, many approaches proposed for identifying eQTLs can also be applied for the analysis of meQTLs, or pQTLs, such as single-trait QTL tests, multi-trait QTL methods, and QTL test with pedigree or error correction (Kendziorski et al. 2006).

Some multi-stage methods have been proposed for sequential analysis of multi-omics data. For instance, Schadt et al. applies multistep method to analyze DNA methylation, gene expression and other complex traits to determine if the variation of DNA methylation that leads to the change of gene expression traits statistically supports an independent, causative or reactive function relative to the complex traits (Schadt et al. 2005). Hao et al. performed a systematic analysis and identified two modules underlying BMD by incorporating GWASs, human PPI network, and gene expression (He et al. 2014). The tool, Multiple Concerted Disruption (MCD) is proposed to sequentially search for a set of genes which exhibit concerted disruption through multiple genomic dimension (DNA methylation, copy number and allelic status) and consequential change in gene expression (Chari et al. 2010). The procedure involves four sequential steps with increasing number of genomic data incorporated to filter out those genes lacking concerted disruption. Similar method for exploring the relationship between copy number alternation and methylation (CNAmet) is also proposed (Louhimo and Hautaniemi 2011). In addition, prior knowledge such as KEGG pathway, gene ontology or functional annotation of the region (e.g., transcription factor binding, methylated or regulatory motifs) could also be incorporated into the analysis to refine the specific regions of interest for the subsequent multi-stage analysis.

Although it is easy to model the relationship among multi-omics data by exploring their pair-wise relationship sequentially, there is a limitation for the stepwise hypothesis. If different omics interplay to have joint effect, for example, miRNA and DNA methylation may simultaneously affect the gene expression, the multi-stage methods may lose their efficiency.

2.2.2.2 Parallel Analysis: Combining Individual Omics Analysis Results

Parallel analysis combines multi-omics data into the analysis simultaneously. It can be generally divided into two categories: concatenation-based integration and model-based integration.

Concatenation-Based Integration This method is to straightforwardly concatenate all of omics data from the same subjects, resulting in a large combined matrix. One advantage of this integration is the applicability of many single omics analysis methods if combing features appropriately. For example, a variety of univariate and multivariate association tests could be applied for biomarker detection from different levels of features, especially the penalized likelihood methods which can handle high dimensionality of data. Lasso is a very useful penalized method and has been widely used for feature selection (Tibshirani 1996). Recently significant test based on lasso is also proposed to control the type I error (Lockhart et al. 2014). Other penalized methods such as sparse logistic regression (Shevade and Keerthi 2003), cox lasso (Wang et al. 2009), and sparse multinomial regression (Krishnapuram et al. 2005) have also been used for genetic biomarker identification corresponding to different types of phenotypes (e.g., categorical or survival traits). These methods can be extended to the analysis of concatenated matrix consisting multi-omics data.

Another advantage of concatenating datasets is that they can account for relationship among features from different levels of omics data. For example, SNP and DNA methylation measure the effect of genetic mutation and environmental factors on complex traits respectively. They may interact with each other to deregulate gene expression, leading to the variation of traits. Fridley et al. used Bayesian modeling to incorporate the relationship between SNPs and mRNA gene expression into the concatenation-based association model for the prediction of drug cytotoxicity (Fridley et al. 2012). In penalized likelihood methods, elastic net is used to simultaneously select features and account for the correlation among features (Ogutu et al. 2012). Group based penalties (e.g., group lasso, sparse group lasso, group Bridge, and overlapping group lasso) were proposed to group different levels of features based on their genomic annotation (e.g., gene or pathway) to increase the detection power on group level (Huang et al. 2012). In addition, Lando et al. used the correlation between copy number and phenotype to weight the penalty of gene expression in a penalized regression model. Genes corresponding to important CNVs were less penalized in expression regression model (Lando et al. 2009).

In spite of the advantages of concatenating multi-omics data, it is still a challenge to find an appropriate way to combine these data matrices collected from different platforms with different scales into one model. In addition, the combination of these high-dimensional matrices will largely expand the dimension of the model, which could increase computational burden. Therefore, the concatenation of multiple datasets is more applicable for omics data integration if there exists an appropriate way of concatenating matrix and the dimension of data is moderate.

Model-Based Integration To avoid the issues of combing data directly, some studies try to build a model for each data separately and then transform each model into an intermediate form, and finally integrate transformed outputs for multi-omics analysis. Tyekucheva et al. performed gene-level and gene set-level tests on gene expression and copy number data separately and combined the gene set scores by meta-analytical approaches (e.g., geometrically averaged P-values and minimum P-values) to derive the combined gene-set score (Tyekucheva et al. 2011). The integrative approach identified more reliable glioblastoma multiforme tumor related gene sets than individual data analysis. Similarly, Poisson et al. proposed the sum of square statistics to combine gene set score from gene expression and metabolites to test integrative set enrichment (Soneson et al. 2010). Xiong et al. developed a tool, Gene Set Association Analysis (GSAA), to test gene-set enrichment by combing SNP-set and gene expression using different score based combination methods (e.g., z-score sum, rank sum and fisher's test) (Xiong et al. 2012). Analysis Tool for Heritable and Environmental network Association (ATHENA) is another model-based analysis tool for performing integrative analysis of different omics data as well as their association with clinical outcomes (Holzinger et al. 2013).

Besides the statistical model or score integration, multi-task learning is another powerful strategy to jointly model different but related tasks simultaneously. Biomarker identification in each single omics is treated as a task and then multiple tasks are combined by multitask learning. Bennett et al. used multi-task learning to consider enrichment analysis scores from both SNP and gene expression to identify several pathways with both genetic and expression differences related to the phenotype (Bennett et al. 2012). Lin et al. adopted two bi-level penalties in multitask regression model to integrate multiple diverse genomics datasets under different level and/or platform for identifying common biomarkers (e.g., genes or gene-set) (Lin et al. 2014a). They assumed a regression model for each dataset as a task, and then considered multiple regression models as multiple tasks. Variables from all datasets were grouped by specific units (e.g., genes) and penalized by sparse group penalties. The integration shows higher power of detecting risk genes than single omics data analysis and meta-analysis under the scenarios of both fixed effect and random effect.

It is noted that model-based integration methods need to build a model for each data set and then combine the models or their intermediate outputs. The scale of model errors or the intermediate outputs needs to be comparable for integration. If each omics data is extremely heterogeneous, this integration method may yield little improvement over separated analysis.

2.2.2.3 Latent Variable Models: Transform Variables into New Feature Space for Integration

The high dimensionality of diverse genomic data is a challenge. One commonly used strategy is to project high dimensional genomic data into low dimensional space before an integrative analysis is performed. Principle component analysis (PCA) is popularly used to explain the variance–covariance structure in a single data. It is widely used for handling pleiotropy with multiple correlated traits (e.g., eQTL) with the assumption that multiple correlated traits are able to reveal stronger signals than are obtained from univariate analysis of each trait separately. PCA based method collapses a number of correlated variables into a smaller number of uncorrelated variables as new phenotypes, which captures most variability and then test association for each new phenotype separately. Christine et al. used PCA to detect pleiotropic QTLs for boar taint and paternal fertility traits (Große-Brinkhaus et al. 2015). Jane et al. applied PCA on 70 skeletal traits to explore pleiotropy pattern through skeleton as well as genetic mechanism of each pattern (Kenney-Hunt et al. 2008).

Some latent variable models work in two- or multi-block way such as canonical correlation analysis (CCA) and partial least squares (PLS) with the aim to estimate latent variate from each dataset respectively (a linear combination of variables) by maximizing the correlation (CCA) or covariance (PLS) between them. Soneson et al. applied CCA to explore two pairs of highly correlated features from the gene expression and copy number variable sets, which represent different characteristic in leukemia. Tang et al. proposed a gene-based association test using CCA to detect QTLs associated with multiple quantitative traits (Tang and Ferreira 2012). Boulesteix et al. used PLS to predict transcription factor activities from combined analysis of gene expression and chromatin immunoprecipitation (ChIP) data (Boulesteix and Strimmer 2007). To integrate multiple datasets or clinical traits, some multi-block approaches such as multi-set CCA and multi-block PLS-correlation have also been proposed by summarizing pairwise correlations (or covariances) among different data sources (Lin et al. 2014b). In addition, parallel independent component analysis (pICA) and joint ICA are also two block methods widely used in genetic, imaging and clinical integration to explore independent components from each modality respectively while maximizing the correlation of the components simultaneously (Sui et al. 2012). Shen et al. show the robustness of joint ICA in integrating multi-omics data for biomarker detection and combined gene expression and copy number variation to identify significant genes associated with breast cancer (Sheng et al. 2011).

The above latent variables models mainly focus on the linear relationship among omics data. It may be interesting to consider non-linear relationship to explore more complicated genetic regulatory mechanism. 'Kernel trick' is a popular strategy which maps omics data into feature space by kernel matrix (e.g., Gaussian kernel matrix). Reverter et al. used kernel PCA to reduce dimension of metabolomics and genomics data and combined them for better representation of samples (Reverter et al. 2014). Yamannishi et al. proposed two types of kernel CCA to measure the correlation between several heterogeneous datasets, and to extract sets of genes which share similarities with respect to multiple biological attributes (Yamanishi et al. 2003).

Due to high dimensionality and small sample size of multi-omics data, there are usually issues of multi-collinearity (linear dependence) in the data and overfitting of the model. To address these issues, one way is to introduce the sparse regularizations into the conventional latent model to perform feature selection and correlative analysis simultaneously. Several types of regularized latent variable models have been proposed by enforcing different sparse penalties (e.g., lasso, elastic net and sparse group lasso penalty) on the loading vectors in the model. Waaijenborg et al. (2008) introduced the L-1 norm and elastic net penalties to the CCA model to analyze the correlation between gene expression and DNA-markers. Parkhomenko et al. (2009) proposed a CCA method with lasso penalty based on SVD (Singular value decomposition). Le Cao et al. (2009) used the penalized CCA with the elastic net to identify sets of co-expressed genes from two different microarray platforms. Witten et al. (2009) developed penalized matrix decomposition (PMD) method and applied it to solve CCA with lasso and fused lasso penalties. Lin et al. presented a unified framework of formulating these sparse CCA models as in (2.1):

$$\min_{\boldsymbol{u},\boldsymbol{v}} -\boldsymbol{u}^t\Sigma_{XY}\boldsymbol{v} + \lambda_1\|\boldsymbol{u}\|_G + \tau_1\|u\|_1 + \lambda_2\|v\|_G + \tau_2\|\boldsymbol{v}\|_1 \quad s.t.\, \boldsymbol{u}^t\Sigma_{XX}\boldsymbol{u} \leq 1, \boldsymbol{v}^t\Sigma_{YY}\boldsymbol{v} \leq 1 \tag{2.1}$$

where $\boldsymbol{X}$,$\boldsymbol{Y}$ are the two data matrices; $\boldsymbol{u}$ and $\boldsymbol{v}$ are the loading vectors constrained by sparse terms;$\|\boldsymbol{u}\|_1$ and $\|\boldsymbol{v}\|_1$ are $l-1$ norm lasso penalty for performing the selection of individual variable/feature, and $\|\boldsymbol{u}\|_G = \sum_{l=1}^{L} \omega_l \|\boldsymbol{u}_l\|_2$, $\|\boldsymbol{v}\|_G = \sum_{h=1}^{H} \mu_h \|\boldsymbol{v}_h\|_2$ are the group penalties to account for joint effects of features within the same group. The group penalty uses the non-diffentialbility of $\|\boldsymbol{u}_l\|_2$ (or $\|\boldsymbol{v}_h\|_2$) at $\boldsymbol{u}_l = 0$ ($\boldsymbol{v}_h = 0$) to set the coefficients of the group to 0 so the entire group of features will be removed to achieve the group sparsity.

Figure 2.1a shows the results of recovered loading vectors u and v by CCA-l1, CCA-group and CCA-sparse group methods respectively. It can be seen that the CCA-sparse group method can better estimate true u and v than CCA-l1, CCA-group method. Figure 2.1b compares the accuracy of recovering loading vectors from three methods with respect to different noise levels (standard deviation changes from 0.1 to 1 with interval 0.1), corresponding to different degrees of correlations between the two data sets. The result shows that the CCA-group model can recover the most correlated variables but gives the highest total discordance. CCA-sparse group has a comparable recovering accuracy as CCA-group model but much less total discordance especially when noise level decreases. These methods were also applied to fMRI data and SNP data and other omics data to identify significant correlated features.

Several other latent variable models were also proposed. Chun et al. proposed sparse PLS for simultaneous dimension reduction and feature selection in gene expression and transcriptional factor data. sPLS discriminant analysis (sPLS-DA), included in mixomics packages (Lê Cao et al. 2011), incorporated disease phenotype to extract those latent variables from gene expression or SNPs which are discriminative in multiclass disease, e.g., Leukemia. Li et al. introduced a sparse Multi-Block Partial Least Squares (sMBPLS) regression method to identify multidimensional regulatory modules from copy number variation, DNA methylation, gene expression and microRNA expression (Li et al. 2012).

2.2.2.4 Integrative Network Analysis

Networks represent the interactions of features within or across different levels of omics. The methods for reconstructing genetic network in single omics data have been well studied, as introduced in Sect. 2.2.1.4. However, they are limited to understand complex biological networks underlying cell and organ functions by single level of omic data. Integration of different levels of omics data to reconstruct comprehensive network is able to enrich our understanding of biological processes and improve the discovery of disease biomarkers. There are mainly two categories of integrative network reconstruction algorithms: single-stage reconstruction and multi-stage reconstruction.

Single-Stage Integrative Network Reconstruction This type of method tends to incorporate multi-omics data directly into the model for network construction. A simple way is using correlation based measurement to weight the interactions among omics features. WGCNA was used to construct network between metabolomics and transcriptomics data to identify clusters of metabolites and transcriptional factors associated with body weight change. A correlation derived

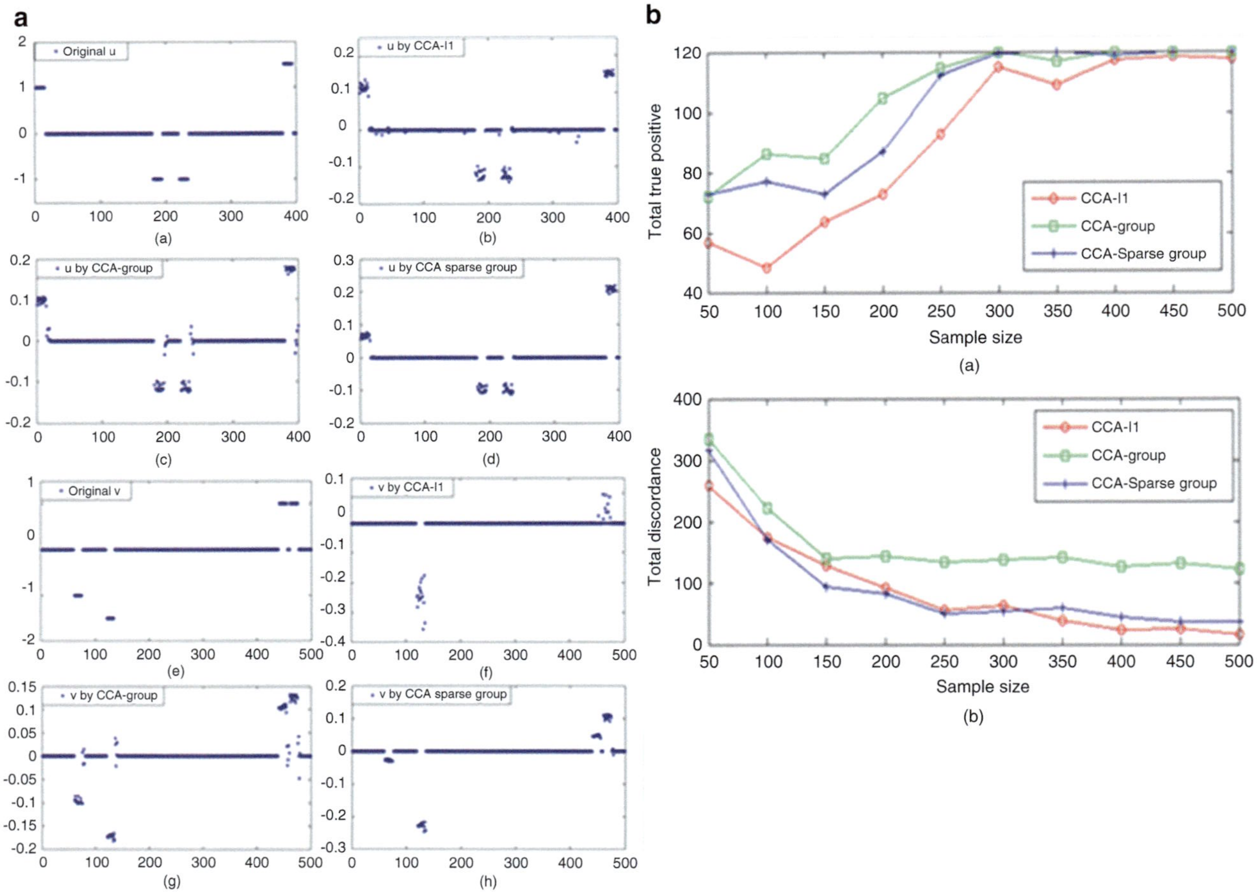

Fig. 2.1 A comparison of group sparse CCA with the other sparse CCA methods(e.g., CCA-lasso, CCA-group lasso). (**a**) The comparison of accuracy of recovering the loading vectors (u and v) by three methods. (**b**) The comparison of accuracy of recovering u and v and total discordance with respect to different correlation values between omics datasets

topological matrix was used for clustering correlated features and cutting into different modules for association analysis (Wahl et al. 2015). Kayano et al. developed a statistical method based on low-order partial correlations with a robust correlation coefficient for estimating metabolic networks from metabolome, proteome, and transcriptome data (Kayano et al. 2013).

Another way is Bayesian network, which is a directed probabilistic graphical model with each edge representing the dependence between nodes (e.g., genes). Bayesian network is based on both prior distribution assumptions and observed data to design a model which can be mostly trusted. Prior distributions could be informative, such as conjugate prior, or mostly be non-informative. Some prior knowledge such as protein-protein interaction database could be incorporated to improve the accuracy and efficiency of network reconstruction. Conditional independence facilitates the integration of diverse data in a coherent way. Zhu et al. combined genotypic, expression, transcription factor binding site (TFBS), and protein–protein interaction (PPI) data to reconstruct causal gene networks. Three levels of Bayesian networks (BN_raw, BN_eQTL and BN_full) incorporating different prior knowledge (e.g., eQTL) were reconstructed and compared in terms of their power to infer causal regulators for validated signature gene sets (Zhu et al. 2008). Some Bayesian clustering models were designed to cluster genes from multiple omics data based on their interactions. Multiple dataset integration (MDI) was developed to identify groups of genes that are co-regulated and additionally their protein products appearing in the same complex (Kirk et al. 2012). To constrain the consistency of identified clusters across multiple omics sources, Bayesian consensus clustering was built to find consensus genetic clusters shared in different omics levels (Lock and Dunson 2013). Instead of finding clusters of genetic markers, Pathway recognition algorithm using data integration on genomic models (PARADIGM) was used to infer the molecular pathways altered in a patient sample by integrating genomic and functional genomic datasets (Vaske et al. 2010). Pathways were constructed based on prior knowledge database following CNV->gene expression->protein activity assumption and all measurements were categorized into three discrete states (inhibited, normal and activated). Joint posterior distribution was then computed based on observed data. The difference between pre- and post-activity levels indicated the quantitative alternation induced by the disease. Similarly, Multi-level Ontology Analysis (MONA) was a computationally efficient method to approximate the marginal posteriors of ontology terms based on three basis model assumptions (base, cooperative, and inhibitory models), given lists of genes responding to experimental conditions (Sass et al. 2013). iNET takes a "feature-specific" approach to model eight underlying biological basis models for constructing Bayesian network (Wang et al. 2013).

Multi-Stage Integrative Network Reconstruction There are generally two major steps: constructing network in each single level of omics data; and fusing multiple networks to an integrated network. The first step could be achieved by using various single omics network reconstruction algorithms. Network alignment and fusion methods are usually needed for the second step. Network alignment is the algorithm to map the nodes from two or multiple types of networks in such a way that maximizes the topological and biological similarity between pairs of aligned nodes (Mitra et al. 2013). This technique is helpful in identifying previously undiscovered conserved modules that have been maintained across different species and revealing functionally similar subnetworks. Computational methods for network alignment consist of pair-wise alignment for aligning two networks only and multiple alignment to find transitive alignments among multiple networks. Some alignment algorithms, e.g., local alignment, aim to identify conserved regions between the input networks, which is particularly useful in finding known functional components (e.g., pathways) in a new species. For instance, PathBLAST allows the comparison of simple pathways (e.g., linear pathways) or subnetworks (e.g., modules) based on homology and interaction confidence (Kelley et al. 2004). NetworkBLAST finds highly conserved local regions greedily using inferred phylogeny (Kalaev et al. 2008). Some algorithms, e.g., global alignment, align every node in the smaller network to the larger network to find an overall network which enables species-level comparisons and discovery of functional orthologs. For instance, IsoRank and IsoRankN identify a stationary random walk distribution to perform global network alignment (Singh et al. 2008; Liao et al. 2009).

Network fusion is a technique to fuse multiple distinct but complementary biological networks to gain comprehensive insights of cellular structure and function. One of these approaches is integrating biological networks across different types of molecular interactions to identify composite modules. A cytoscape-based tool, PanGIA is designed to detect composite modules by identifying overlapping clusters of physical and genetic networks (Srivas et al. 2011). Physical interactions are mainly represented by protein–protein and protein–DNA interactions. Genetic interactions represent functional relationships between genes, in which the phenotypic effect of one gene is modified by another. Composite modules are extracted based on the physical interactions while cluster of genetic interactions between two different composite modules reflect inter-modular dependencies. Integrative analysis of both physical and genetic networks can reveal physical mechanism of phenotype associated with genes in the composite module and also predict the genetic dependence between composite modules mapped in physical binding assays. Another Cytoscape tool, GeneMANIA builds a composite functional association network by taking a weighted average of individual functional association networks (Mostafavi et al. 2008). It first assigns weights to each of interaction networks. The composite network is then set to be the weighted average of the individual networks. Each network weights are calculated on demand and are tailored to the query list.

2.2.3 Statistics for Clinical Disease Diagnosis and Classification

The above has discussed the analysis of single omics or multi-omics data for biomarker detection, genetic regulatory network inferring as well as the exploration of genetic pathways underlying complex diseases. The next step is translating this knowledge into clinical diagnosis or prediction. Predictive modeling, particularly classification, is critical in clinic research where risk biomarkers may vary largely with different diseases and even the subjects from one group may have subject-specific genetic variations. An effective method for classification of complex disease is demanded. We generally categorize them into two types: supervised learning method and unsupervised learning method. The former usually needs labelled training dataset for searching the optimal values of model parameters, which helps to build an accurate model and is more applicable for disease classification. The latter is data-driven method without knowing the class label from training, which is more likely to be used for subtyping to explore new subclass of diseases.

2.2.3.1 Supervised Learning in Omics Data

We will introduce several commonly used supervised classifiers in genetic data for classification of complex diseases. Assume there are m types of omics dataset, denoted by $\boldsymbol{X} = [\boldsymbol{X}_1, \boldsymbol{X}_2, \ldots, \boldsymbol{X}_m]$, where $\boldsymbol{X}_i \in R^{N \times P_i}, i = 1, 2, \ldots, m$, P_i is the dimension of features in the i-th omics data. $\boldsymbol{Y} \in R^{N \times c}$, c is the number of classes, and the subjects belonged to the j-th class are denoted by $\{w_j\}, j = 1, 2, \ldots, c$. The object is to predict the class of a new sample $\boldsymbol{y}$ given the observed omic feature matrices $\boldsymbol{X}$.

Discriminant Analysis Linear discrinant analysis (LDA) and quadratic discriminant analysis (QDA) are popularly used methods in clinical genomic analysis for risk feature identification and classification. LDA is a latent variable model which projects original high dimensional variables (e.g., gene expression measurements) into a new feature space by linear combinations $\boldsymbol{X\alpha}$ with large ratios of between-group to within-group sums of squares, that is, maximizing the ratio $\boldsymbol{\alpha}^T\boldsymbol{B\alpha}/\boldsymbol{\alpha}^T\boldsymbol{W\alpha}$, where $\boldsymbol{B}$ denotes the between-classes covariance matrix, and $\boldsymbol{W}$ denotes the within-class covariance matrix. The calculation of $\boldsymbol{B}$ and $\boldsymbol{W}$ are given by

$$\boldsymbol{B} = \sum_{i=1}^{C} N(\boldsymbol{\mu}_i - \boldsymbol{\mu})(\boldsymbol{\mu}_i - \boldsymbol{\mu})^T; \boldsymbol{W} = \sum_{i=1}^{C} \sum_{\boldsymbol{x} \in w_j} (\boldsymbol{x} - \boldsymbol{\mu}_i)(\boldsymbol{x} - \boldsymbol{\mu}_i)^T$$

where $\boldsymbol{\mu}_i = \frac{1}{N} \sum_{\boldsymbol{x} \in w_j} \boldsymbol{x}$, $\boldsymbol{\mu} = \frac{1}{N} \sum_{\forall \boldsymbol{x}} \boldsymbol{x}$. For a new subject $\boldsymbol{x}$, it can be projected to new feature space by the estimated $\boldsymbol{\alpha}$ and classified to the class which has the minimum distance by the classification rule:

$$C(\boldsymbol{x}, L) = argmin_k \, D_k(\boldsymbol{x})$$

where L is the training dataset to estimate LDA model and D(.) is the function to measure the distance between new subject with each class. LDA is a non-parametric method that is also a special form of a maximum likelihood discriminant rule for multivariate normal class densities with the same covariance matrix. QDA is similar to LDA with the slight difference that QDA needs to estimate the covariance for each class separately. Zhang compared the two methods in recognition of two splice sites (acceptor site and donor site) in exons (Zhang 1997). The features from internal exons and their flanking regions (e.g., in-frame hexamer frequency bias) were adopted in LDA to distinguish acceptor site from donor site. To further consider the complex correlation structure among various acceptor sites or donor sites among exons, the covariance matrix may not be same between two sites. QDA was applied and shown better identification accuracy than LDA. There are also some other modifications of LDA to account for the specific characteristics in the omics data. For example, sparse LDA is combined with sparse regularizations to perform feature selection in discriminant analysis with high dimensional dataset, e.g., gene expression data (Clemmensen et al. 2011). Ye et al. also proposed unrelated LDA to handle the under-sampled data in genetic analysis and used generalized singular value decomposition method to make the features in transformed space be uncorrelated (Ye 2005). The method shows effectiveness in classification of tumors by gene expression data. Huang et al. compared LDA with other four modified methods on tumor classification by gene expression and showed the advantage of LDA modification methods over traditional LDA in terms of the average error and found no significant difference (Huang et al. 2009).

Decision Tree

Decision tree is one of most widely used machine learning methods. A decision tree model is built by a tree-like structure, where each internal node represents a specific test of an attribute, each branch represents one of the possible test results, and each leaf node represents an outcome. There are mainly two types of decision tree: decision tree classification and decision tree regression. The former aims to output the classifications labels (e.g., class) while the latter can output any real number of measurement. Decision tree can be learned by splitting the node into subsets according to the attribute value test. The splitting process is repeated in a recursive manner until the subsets of a node have all the same value of target variable or no more information could be added after splitting. Several algorithms have been developed to determine if splitting the node at each step, such as Gini impurity, information gain or variance reduction, leads to several types of decision trees, e.g., C4.5, C5, IDE, GINI, Codrington's and CART (classification and regression tree). Chen et al. used CART tree to select important genes for improving cancer classification (Chen et al. 2014). CART was also applied to explore the influence of the interactions among those genes that influence androgen in prostate cancer and if these interactions are able to improve the cancer prediction (Barnholtz-Sloan et al. 2011). There are also many other successful biological applications of decision tree based classification, including coding and noncoding DNA classification (Langfelder and Horvath 2008), protein secondary structure prediction (Shannon et al. 2003), and operon structure classification (Dennis et al. 2003).

Support Vector Machine

Support vector machines (SVM) are a family of classifiers which transform the input samples into a high dimensional space by a linear or kernel function, named feature space. Then a linear hyperplane could be drawn to separate two classes mapped in the feature space. To avoid overfitting, SVMs choose a specific hyperplane that maximizes the minimum distance from the hyperplane to the closest training point which is called support vectors. The optimal hyperplane is defined by the pair (w, b) by solving the following problem:

$$\min \|\boldsymbol{w}\|^2$$

$$s.t.\ y_i(\boldsymbol{w} \cdot x_i + b) - 1 \geq 0,\ \forall i = 1, 2, \ldots, N$$

where $\|\boldsymbol{w}\|^2$ measures the inverse of distance between two boundaries to obtain the maximum margin. $\boldsymbol{w} \cdot \mathrm{x_i} + \mathrm{b} = \pm 1$ indicates two boundary hyperplanes separating subjects from two different classes ($\mathrm{y} = 1$ or -1). Boundary hyperplanes are built on the support vectors. It is efficient for SVM to classify new examples since the majority of the training examples can be safely ignored. In order to transform original variables into high dimensional feature space and measure the non-linear correlation in feature space, a kernel function $K(x_i, x_j)$ is usually applied such as polynomial kernel, Gaussian radial basis function and hyperbolic function.

Support vector machines have drawn a lot of research efforts from diverse fields (Noble 2004). In bioinformatics, it is widely used for cancer diagnosis and classification, protein structure and function prediction and gene expression pattern recognition. An early application example of SVM is to identify important genes and further improve the classification on leukemia and colon cancers (Guyon et al. 2002). Ferry et al. used SVM to not only classify cancer tissue samples based on microarray data but also identify those samples wrongly classified by experts. Hua and Sun used SVMs to perform protein classification with respect to subcellular localization (Hua and Sun 2001). A 20-feature composition kernel function is applied and shown to produce more accurate classifications than other competing methods, including a neural network, a Markov Distinguishing model and the covariant discriminant algorithm. Yeang et al. extended SVM to multi-class SVM which can address the multiple classes issue. The method was applied for multi-class tumor classification on a data set of 190 samples from 14 tumor classes (Yeang et al. 2001). Nguyue et al. compared several multi-lass SVM algorithms on protein secondary structure prediction including: one-against-all, one-against-one, and directed acyclic graph, and two approaches for multi-class problem by solving one single optimization problem (Nguyen and Rajapakse 2003). The results demonstrated better recovery accuracy of multi-class SVMs proposed by Vapnik and Weston than the other multi-class SVMs, including binary SVMs.

Ensemble Learning

Ensemble learning is an effective technique that constructs a set of classifiers and combines them to improve overall prediction accuracy (Dietterich 2000). There are a lot of ensemble methods that have been applied to biological data analysis in addressing small sample size but high dimensional data sets and reducing the overfitting risk. The classification accuracy is also improved by generating multiple prediction models and aggregating these multiple models (called basis classifiers) to make the final prediction in a consensus way. There are several types of ensemble learning algorithms including bagging (Breiman 1996), boosting (Freund and Schapire 1996) and random forests (Breiman 2001). Being the principle ensemble learning methods, they are usually combined with the other classifiers such as decision trees.

There are several applications of ensemble learning methods such as sample/tissue classification and gene-gene interaction prediction. Ben-Dor et al. (2000) and Dudoit et al. (2002) applied bagging and boosting methods to classify tumors using gene expression profiles. Both studies compared the ensemble methods with other individual classifiers such as k-nearest neighbors (kNN), clustering based classifiers, SVM, LDA, and classification trees. The conclusion was that ensemble methods (e.g., bagging and boosting) performed similarly to other single classification algorithms. Wu et al. (2003), compared several methods for the classification of ovarian cancer based on MS spectra including the ensemble methods of bagging, boosting, and random forests to individual classifiers, e.g., LDA, QDA, kNN, and SVM. The study found that among all methods random forests outperforms the others with the lowest error rate. Moon et al. developed a new ensemble-based classification algorithm, Classification by Ensembles from Random Partitions (CERP) combined with classification and decision tree (CART) and applied it to genomic data on leukemia patients and on breast cancer patients (Moon et al. 2006). The performance was compared with other classifiers such as single decision tree (e.g., CART), SVM, diagonal LDA and other ensemble learning methods (e.g., RF and boosting). The results demonstrate that CERP is a consistently better algorithm and maintains a good balance between sensitivity and specificity even in case of unbalanced sample size.

2.2.3.2 Unsupervised Learning in Omics Data

Clustering is a popular unsupervised learning method and commonly applied in omics data analysis such as clustering genes based on their expression, or clustering samples based on their omics features to identify subgroups or subtypes of diseases. There are several clustering methods proposed including partition clustering and hierarchical clustering.

Partition Clustering

This type of clustering methods mainly partition objects and change the clusters based on the dissimilarity or distance between objects with clusters. The fixed number of clusters could be specified before the clustering.

K-means clustering is a popular method for clustering genes or subjects. The general procedure is as follows:

(1) Randomly generate k clusters and calculate the centroid of each cluster;
(2) Calculate the distance of each point with each cluster centroid and assign each point to the cluster with shortest distance.
(3) Update the centroid of each new cluster;
(4) Repeat until certain convergence is met, e.g., no changes of assignment of each point.

There are some applications of k-means in bioinformatics, such as gene clustering or subtyping. Lehmann et al. used k-means to analyze gene expression profiles of 587 TNBC cases from 21 breast cancer to subtype TNBC. Each TNBC case contained 13,060 genes after normalization for clustering analysis by K-means. The optimal number of clusters was determined by the change of proportion of area under empirical cumulative distribution curve and consequently, 6 Triple-negative breast cancer subtypes were identified with unique gene expression and ontologies (Lehmann et al. 2011). Further they predicted "driver" signaling pathways of each subtypes to show that analysis of distinct GE signatures can inform therapy selection.

Fuzzy C-means (FCM) clustering is another clustering method using the 'soft' clustering instead of 'hard' clustering in k-means. For each subject, FCM assigns a degree of membership in each cluster, which can account for the uncertainty of some subjects. It has been widely used in imaging analysis (Li et al. 2013) since it is more suitable for the scenario that there is overlapping among clusters, which is also common in clinical analysis such as tumor classification where unlabeled tumor samples may not necessarily be clear members of one class or another. Wang et al. applied FCM clustering on gene expression data for tumor classification and gene prediction (Wang et al. 2003). Given a dataset $\boldsymbol{X} = [\boldsymbol{X}_1, \boldsymbol{X}_2, \ldots, \boldsymbol{X}_N] \in R^{N \times p}$ from N tumor subjects measured on p gene expression levels. We assume the existence of Nc tumor classes, whose centers are denoted by $\boldsymbol{C} = [\boldsymbol{C}_1, \boldsymbol{C}_2, \ldots, \boldsymbol{C}_{Nc}]$ which are unknown and to be estimated. $\boldsymbol{U} = [\boldsymbol{U}_{i,1}, \boldsymbol{U}_{i,2}, \ldots, \boldsymbol{U}_{i,Nc}]$ is fuzzy membership matrix for the i-th subject on all of tumor classes, whose value between zero and one. FCM clustering can be obtained by solving the optimization issue:

$$\min_{U,C} \sum_{k=1}^{Nc} \sum_{i=1}^{N} u_{k,i}^{q} \|\boldsymbol{X}_i - \boldsymbol{C}_k\|^2 \text{ , } \mathit{subject\ to} \sum_{k=1}^{Nc} u_{k,i}^{q} = 1$$

where q is a weight on each fuzzy membership and determines the degree of fuzziness. Each tumor subject will have a membership in every class; membership close to one indicates a high degree of similarity between the subject and a tumor class while membership close to zero implies little similarity. The subject is assigned to the class with the highest membership values. The second term is used to constrain that the summation of membership of different classes equals one to make sure the value of membership is between zero and one. The tests on four different tumor datasets show the efficiency of FCM clustering in terms of reduced error rates and the importance of selected features for medical diagnostics and cancer classification.

Hierarchical Clustering

Hierarchical clustering is a clustering method to represent the objects in a tree-like structure, where each node has zero or more child nodes below it. There are mainly two types of strategies to generate the hierarchical tree: agglomerative, a 'bottom up' approach which takes each object as its own cluster and merge clusters as one moves up the hierarchy; divisive, a 'top down' approach which takes all objects as one cluster and split it recursively as one moves down the hierarchy. Here shows the procedure of agglomerative as an example:

(1) Start with n clusters with each contains one object;
(2) Merge the most similar pair of clusters from the proximity matrix which can be built based on different distance measurements, e.g., single linkage, complete linkage and average linkage, which take the minimum, maximum and average of pairwise distance between two clusters, respectively.
(3) Update the proximity matrix by replacing the individual clusters with merged cluster;
(4) Repeat until only one cluster is left.

Hierarchical clustering is also applied for clinical classification and gene clustering. Makretsov et al. used hierarchical clustering to determine the efficiency in improving prognostication in patients with invasive breast cancer by multiple immunomarkers (protein expression profiles) (Makretsov et al. 2004). They identified three cluster groups with significant differences in clinical outcome and demonstrated that hierarchical clustering by using multiple markers can group breast cancers into classes with clinical relevance and outperform individual prognostic markers. Furlan et al. applied unsupervised hierarchical clustering analysis to 126 colorectal carcinomas to combine 13 routinely assessed clinicopathologic features and all five molecular markers to distinguish four molecular subtypes of sporadic colorectal carcinomas (Furlan et al. 2011). The results demonstrate the superiority of classification based on the combination of clinicopathologic and molecular features of colorectal cancers over single features, and also indicate that hierarchical clustering is a useful tool to define a diagnostic and prognostic signature for different carcinomas.

References

Allen JD, Xie Y, Chen M, Girard L, Xiao G. Comparing statistical methods for constructing large scale gene networks. PLoS ONE. 2012;7:e29348.

Balding DJ. A tutorial on statistical methods for population association studies. Nat Rev Genet. 2006;7:781–91.

Barnholtz-Sloan JS, Guan X, Zeigler-Johnson C, Meropol NJ, Rebbeck TR. Decision tree–based modeling of androgen pathway genes and prostate cancer risk. Cancer Epidemiol Biomark Prev. 2011;20:1146–55.

Ben-Dor A, Bruhn L, Friedman N, Nachman I, Schummer M, et al. Tissue classification with gene expression profiles. J Comput Biol. 2000;7:559–83.

Bennett BD, Xiong Q, Mukherjee S, Furey TS. A predictive framework for integrating disparate genomic data types using sample-specific gene set enrichment analysis and multi-task learning. PLoS ONE. 2012;7:e44635.

Boulesteix A-L, Strimmer K. Partial least squares: a versatile tool for the analysis of high-dimensional genomic data. Brief Bioinform. 2007;8:32–44.

Breiman L. Bagging predictors. Mach Learn. 1996;24:123–40.

Breiman L. Random forests. Mach Learn. 2001;45:5–32.

Cantor RM, Lange K, Sinsheimer JS. Prioritizing GWAS results: a review of statistical methods and recommendations for their application. Am J Hum Genet. 2010;86:6–22.

Chari R, Coe BP, Vucic EA, Lockwood WW, Lam WL. An integrative multi-dimensional genetic and epigenetic strategy to identify aberrant genes and pathways in cancer. BMC Syst Biol. 2010;4:67.

Chen K-H, Wang K-J, Tsai M-L, Wang K-M, Adrian AM, et al. Gene selection for cancer identification: a decision tree model empowered by particle swarm optimization algorithm. BMC Bioinf. 2014;15:49.

Clemmensen L, Hastie T, Witten D, Ersbøll B. Sparse discriminant analysis. Technometrics. 2011;53:406–13.

Dennis Jr G, Sherman BT, Hosack DA, Yang J, Gao W, et al. DAVID: database for annotation, visualization, and integrated discovery. Genome Biol. 2003;4:P3.

Dietterich TG. Ensemble methods in machine learning. In: Multiple classifier systems. Berlin/ Heidelberg: Springer; 2000. p. 1–15.

Dudoit S, Fridlyand J, Speed TP. Comparison of discrimination methods for the classification of tumors using gene expression data. J Am Stat Assoc. 2002;97:77–87.

Efron B, Tibshirani R. On testing the significance of sets of genes. The Annals of Applied Statistics 2007;1:107–29.

Ferreira MA, Purcell SM. A multivariate test of association. Bioinformatics. 2009;25:132–3.

Freund Y, Schapire RE. Experiments with a new boosting algorithm. 1996:148–56.

Fridley BL, Lund S, Jenkins GD, Wang L. A Bayesian integrative genomic model for pathway analysis of complex traits. Genet Epidemiol. 2012;36:352–9.

Friedman N, Linial M, Nachman I, Pe'er D. Using Bayesian networks to analyze expression data. J Comput Biol. 2000;7:601–20.

Furlan D, Carnevali IW, Bernasconi B, Sahnane N, Milani K, et al. Hierarchical clustering analysis of pathologic and molecular data identifies prognostically and biologically distinct groups of colorectal carcinomas. Mod Pathol. 2011;24:126–37.

Galesloot TE, van Steen K, Kiemeney LA, Janss LL, Vermeulen SH. A comparison of multivariate genome-wide association methods. PLoS ONE. 2014;9:e95923.

Große-Brinkhaus C, Storck LC, Frieden L, Neuhoff C, Schellander K, et al. Genome-wide association analyses for boar taint components and testicular traits revealed regions having pleiotropic effects. BMC Genet. 2015;16:36.

Guyon I, Weston J, Barnhill S, Vapnik V. Gene selection for cancer classification using support vector machines. Mach Learn. 2002;46:389–422.

He H, Zhang L, Li J, Wang YP, Zhang JG, et al. Integrative analysis of GWASs, human protein interaction, and gene expression identified gene modules associated with BMDs. J Clin Endocrinol Metab. 2014;99:E2392–9.

Holzinger ER, Ritchie MD. Integrating heterogeneous high-throughput data for meta-dimensional pharmacogenomics and disease-related studies. Pharmacogenomics. 2012;13:213–22.

Holzinger ER, Dudek SM, Frase AT, Pendergrass SA, Ritchie MD. ATHENA: the analysis tool for heritable and environmental network associations. Bioinformatics; 2013;30(5):698–705.

Hua S, Sun Z. Support vector machine approach for protein subcellular localization prediction. Bioinformatics. 2001;17:721–8.

Huang D, Quan Y, He M, Zhou B. Comparison of linear discriminant analysis methods for the classification of cancer based on gene expression data. J Exp Clin Cancer Res. 2009;28:149.

Huang J, Breheny P, Ma S. A selective review of group selection in high-dimensional models. Stat Sci Rev J Instit Math Stat. 2012;27:481–99.

Jansen RC, Nap J-P. Genetical genomics: the added value from segregation. TRENDS Genet. 2001;17:388–91.

Kalaev M, Smoot M, Ideker T, Sharan R. NetworkBLAST: comparative analysis of protein networks. Bioinformatics. 2008;24:594–6.

Kayano M, Imoto S, Yamaguchi R, Miyano S. Multi-omics approach for estimating metabolic networks using low-order partial correlations. J Comput Biol. 2013;20:571–82.

Kelley BP, Yuan B, Lewitter F, Sharan R, Stockwell BR, et al. PathBLAST: a tool for alignment of protein interaction networks. Nucleic Acids Res. 2004;32:W83–8.

Kendziorski C, Chen M, Yuan M, Lan H, Attie A. Statistical methods for expression quantitative trait loci (eQTL) mapping. Biometrics. 2006;62:19–27.

Kenney-Hunt JP, Wang B, Norgard EA, Fawcett G, Falk D, et al. Pleiotropic patterns of quantitative trait loci for 70 murine skeletal traits. Genetics. 2008;178:2275–88.

Kerkel K, Spadola A, Yuan E, Kosek J, Jiang L, et al. Genomic surveys by methylation-sensitive SNP analysis identify sequence-dependent allele-specific DNA methylation. Nat Genet. 2008;40:904–8.

Khatri P, Sirota M, Butte AJ. Ten years of pathway analysis: current approaches and outstanding challenges. PLoS Comput Biol. 2012;8:e1002375.

Kirk P, Griffin JE, Savage RS, Ghahramani Z, Wild DL. Bayesian correlated clustering to integrate multiple datasets. Bioinformatics. 2012;28:3290–7.

Klei L, Luca D, Devlin B, Roeder K. Pleiotropy and principal components of heritability combine to increase power for association analysis. Genet Epidemiol. 2008;32:9–19.

Krishnapuram B, Carin L, Figueiredo MA, Hartemink AJ. Sparse multinomial logistic regression: fast algorithms and generalization bounds. IEEE Transac Pattern Anal Mach Intell. 2005;27:957–68.

Lando M, Holden M, Bergersen LC, Svendsrud DH, Stokke T, et al. Gene dosage, expression, and ontology analysis identifies driver genes in the carcinogenesis and chemoradioresistance of cervical cancer. PLoS Genet. 2009;5:e1000719.

Langfelder P, Horvath S. WGCNA: an R package for weighted correlation network analysis. BMC Bioinformatics. 2008;9:559.

Le Cao KA, Martin PGP, Robert-Granie C, Besse P. Sparse canonical methods for biological data integration: application to a cross-platform study. BMC Bioinf. 2009;10:34.

Lê Cao K-A, Boitard S, Besse P. Sparse PLS discriminant analysis: biologically relevant feature selection and graphical displays for multiclass problems. BMC Bioinf. 2011;12:253.

Lehmann BD, Bauer JA, Chen X, Sanders ME, Chakravarthy AB, et al. Identification of human triple-negative breast cancer subtypes and preclinical models for selection of targeted therapies. J Clin Invest. 2011;121:2750.

Li W, Zhang S, Liu C-C, Zhou XJ. Identifying multi-layer gene regulatory modules from multi-dimensional genomic data. Bioinformatics. 2012;28:2458–66.

Li J, Lin D, Cao H, Wang Y-P. An improved sparse representation model with structural information for Multicolour Fluorescence In-Situ Hybridization (M-FISH) image classification. BMC Syst Biol. 2013;7:S5.

Liao C-S, Lu K, Baym M, Singh R, Berger B. IsoRankN: spectral methods for global alignment of multiple protein networks. Bioinformatics. 2009;25:i253–8.
Lin D, Zhang J, Li J, He H, Deng H-W, et al. Integrative analysis of multiple diverse omics datasets by sparse group multitask regression. Frontiers in cell and developmental biology. 2014a;2:62.
Lin D, Cao H, Calhoun VD, Wang Y-P. Sparse models for correlative and integrative analysis of imaging and genetic data. J Neurosci Methods. 2014b;237:69–78.
Lock EF, Dunson DB. Bayesian consensus clustering. Bioinformatics. 2013;29(20):2610–6.
Lockhart R, Taylor J, Tibshirani RJ, Tibshirani R. A significance test for the lasso. Ann Stat. 2014;42:413.
Louhimo R, Hautaniemi S. CNAmet: an R package for integrating copy number, methylation and expression data. Bioinformatics. 2011;27:887–8.
Makretsov NA, Huntsman DG, Nielsen TO, Yorida E, Peacock M, et al. Hierarchical clustering analysis of tissue microarray immunostaining data identifies prognostically significant groups of breast carcinoma. Clin Cancer Res. 2004;10:6143–51.
Marchini J, Howie B, Myers S, McVean G, Donnelly P. A new multipoint method for genome-wide association studies by imputation of genotypes. Nat Genet. 2007;39:906–13.
Margolin AA, Nemenman I, Basso K, Wiggins C, Stolovitzky G, et al. ARACNE: an algorithm for the reconstruction of gene regulatory networks in a mammalian cellular context. BMC Bioinf. 2006;7 Suppl 1:S7.
Melzer D, Perry JR, Hernandez D, Corsi A-M, Stevens K, et al. A genome-wide association study identifies protein quantitative trait loci (pQTLs). PLoS Genet. 2008;4:e1000072.
Mitra K, Carvunis A-R, Ramesh SK, Ideker T. Integrative approaches for finding modular structure in biological networks. Nat Rev Genet. 2013;14:719–32.
Moon H, Ahn H, Kodell RL, Lin C-J, Baek S, et al. Classification methods for the development of genomic signatures from high-dimensional data. Genome Biol. 2006;7:R121.
Mostafavi S, Ray D, Warde-Farley D, Grouios C, Morris Q. GeneMANIA: a real-time multiple association network integration algorithm for predicting gene function. Genome Biol. 2008;9:S4.
Nguyen MN, Rajapakse JC. Multi-class support vector machines for protein secondary structure prediction. Genome Inform. 2003;14:218–27.
Noble WS. Support vector machine applications in computational biology. In: Kernel methods in computational biology. The MIT Press; 2014. p. 71–92.
Ogutu JO, Schulz-Streeck T, Piepho H-P. Genomic selection using regularized linear regression models: ridge regression, lasso, elastic net and their extensions. BioMed Cent Ltd. 2012;6(2):1–6.
O'Reilly PF, Hoggart CJ, Pomyen Y, Calboli FC, Elliott P, et al. MultiPhen: joint model of multiple phenotypes can increase discovery in GWAS. PLoS ONE. 2012;7:e34861.
Parkhomenko E, Tritchler D, Beyene J. Sparse canonical correlation analysis with application to genomic data integration. Stat Appl Genet Mol Biol. 2009;8:1–34
Peng J, Wang P, Zhou N, Zhu J. Partial correlation estimation by joint sparse regression models. J Am Stat Assoc. 2009;104:735–46.
Purcell S, Neale B, Todd-Brown K, Thomas L, Ferreira MA, et al. PLINK: a tool set for whole-genome association and population-based linkage analyses. Am J Hum Genet. 2007;81:559–75.
Reverter F, Vegas E, Oller JM. Kernel-PCA data integration with enhanced interpretability. BMC Syst Biol. 2014;8:S6.
Sass S, Buettner F, Mueller NS, Theis FJ. A modular framework for gene set analysis integrating multilevel omics data. Nucleic Acids Res. 2013;41:9622–33.
Schadt EE, Lamb J, Yang X, Zhu J, Edwards S, et al. An integrative genomics approach to infer causal associations between gene expression and disease. Nat Genet. 2005;37:710–17.
Shannon P, Markiel A, Ozier O, Baliga NS, Wang JT, et al. Cytoscape: a software environment for integrated models of biomolecular interaction networks. Genome Res. 2003;13:2498–504.

Sheng J, Deng H-W, Calhoun V, Wang Y-P. Integrated analysis of gene expression and copy number data on gene shaving using independent component analysis. IEEE/ACM Transac Comput Biol Bioinform (TCBB). 2011;8:1568–79.

Shevade SK, Keerthi SS. A simple and efficient algorithm for gene selection using sparse logistic regression. Bioinformatics. 2003;19:2246–53.

Singh R, Xu J, Berger B. Global alignment of multiple protein interaction networks with application to functional orthology detection. Proc Natl Acad Sci. 2008;105:12763–8.

Soneson C, Lilljebjörn H, Fioretos T, Fontes M. Integrative analysis of gene expression and copy number alterations using canonical correlation analysis. BMC Bioinf. 2010;11:191.

Srivas R, Hannum G, Ruscheinski J, Ono K, Wang P-L, et al. Assembling global maps of cellular function through integrative analysis of physical and genetic networks. Nat Protoc. 2011;6:1308–23.

Stephens M. A unified framework for association analysis with multiple related phenotypes. PLoS ONE. 2013;8:e65245.

Subramanian A, Tamayo P, Mootha VK, Mukherjee S, Ebert BL, et al. Gene set enrichment analysis: a knowledge-based approach for interpreting genome-wide expression profiles. Proc Natl Acad Sci U S A. 2005;102:15545–50.

Sui J, Adali T, Yu Q, Chen J, Calhoun VD. A review of multivariate methods for multimodal fusion of brain imaging data. J Neurosci Methods. 2012;204:68–81.

Tang CS, Ferreira MA. A gene-based test of association using canonical correlation analysis. Bioinformatics. 2012;28:845–50.

Tibshirani R. Regression shrinkage and selection via the lasso. J Royal Stat Soc Ser B (Methodol). 1996;58(1):267–88.

Tusher VG, Tibshirani R, Chu G. Significance analysis of microarrays applied to the ionizing radiation response. Proc Natl Acad Sci U S A. 2001;98:5116–21.

Tyekucheva S, Marchionni L, Karchin R, Parmigiani G. Integrating diverse genomic data using gene sets. Genome Biol. 2011;12:R105.

van der Sluis S, Posthuma D, Dolan CV. TATES: efficient multivariate genotype-phenotype analysis for genome-wide association studies. PLoS Genet. 2013;9:e1003235.

Vaske CJ, Benz SC, Sanborn JZ, Earl D, Szeto C, et al. Inference of patient-specific pathway activities from multi-dimensional cancer genomics data using PARADIGM. Bioinformatics. 2010;26:i237–45.

Waaijenborg S, Hamer PCVDW, Zwinderman AH. Quantifying the association between gene expressions and DNA-Markers by penalized canonical correlation analysis. Stat Appl Genet Mol Biol. 2008; 7

Wahl S, Vogt S, Stückler F, Krumsiek J, Bartel J, et al. Multi-omic signature of body weight change: results from a population-based cohort study. BMC Med. 2015;13:48.

Wang J, Bø TH, Jonassen I, Myklebost O, Hovig E. Tumor classification and marker gene prediction by feature selection and fuzzy c-means clustering using microarray data. BMC Bioinf. 2003;4:60.

Wang K, Li M, Bucan M. Pathway-based approaches for analysis of genomewide association studies. Am J Hum Genet. 2007;81:1278–83.

Wang S, Nan B, Zhu N, Zhu J. Hierarchically penalized Cox regression with grouped variables. Biometrika. 2009;96:307–22.

Wang K, Li M, Hakonarson H. Analysing biological pathways in genome-wide association studies. Nat Rev Genet. 2010;11:843–54.

Wang W, Baladandayuthapani V, Holmes CC, Do K-A. Integrative network-based Bayesian analysis of diverse genomics data. BMC Bioinf. 2013;14:S8.

Werhli AV, Grzegorczyk M, Husmeier D. Comparative evaluation of reverse engineering gene regulatory networks with relevance networks, graphical gaussian models and bayesian networks. Bioinformatics. 2006;22:2523–31.

Witten DM, Tibshirani R, Hastie T. A penalized matrix decomposition, with applications to sparse principal components and canonical correlation analysis. Biostatistics. 2009;10:515–34.

Wu B, Abbott T, Fishman D, McMurray W, Mor G, et al. Comparison of statistical methods for classification of ovarian cancer using mass spectrometry data. Bioinformatics. 2003;19:1636–43.

Xiong Q, Ancona N, Hauser ER, Mukherjee S, Furey TS. Integrating genetic and gene expression evidence into genome-wide association analysis of gene sets. Genome Res. 2012;22:386–97.

Yamanishi Y, Vert J-P, Nakaya A, Kanehisa M. Extraction of correlated gene clusters from multiple genomic data by generalized kernel canonical correlation analysis. Bioinformatics. 2003;19:i323–30.

Ye J. Characterization of a family of algorithms for generalized discriminant analysis on undersampled problems. J Mach Learn Res JMLR. 2005;6:483–502.

Yeang C-H, Ramaswamy S, Tamayo P, Mukherjee S, Rifkin RM, et al. Molecular classification of multiple tumor types. Bioinformatics. 2001;17:S316–22.

Zhang MQ. Identification of protein coding regions in the human genome by quadratic discriminant analysis. Proc Natl Acad Sci. 1997;94:565–8.

Zhang B, Horvath S. A general framework for weighted gene co-expression network analysis. Stat Appl Genet Mol Biol. 2005;4:Article17.

Zhu J, Zhang B, Smith EN, Drees B, Brem RB, et al. Integrating large-scale functional genomic data to dissect the complexity of yeast regulatory networks. Nat Genet. 2008;40:854–61.

Chapter 3
Gene Expression and Profiling

Yu Zhou, Chao Xu, Jigang Zhang, and Hong-Wen Deng

Abstract Transcriptomics analysis has been widely applied to explore the unknown genes or factors in various biological processes, traits, diseases, and drug treatment. By screening entire RNAs, this technique implicates novel genes and pathways related to a particular condition. Two major types of transcriptomic profiling methods are microarray and RNA sequencing. The two approaches generates and preprocess their respective raw data quite differently, but further analysis is nearly identical. In this chapter, we briefly describe the principles of both methods and compare the differences between them. Since RNA quality is a key practical factor in transcriptomic study, we also introduce isolation and quality control methods and popular software packages for each step in the data analysis process. Finally, we provide an example of a clinical project which used transcriptomics approach to study a disease etiology.

Keywords Transcriptomics • Microarray • RNA sequencing (RNA-Seq) • Data analysis

Abbreviations

BH	Benjamini-hochberg
CLL	Chronic lymphocytic leukemia
DEG	Differentially expressed gene
EBI	European bioinformatics institute

Y. Zhou
Center for Bioinformatics and Genomics, Tulane University, New Orleans, LA, USA

Department of Cell and Molecular Biology, Tulane University, New Orleans, LA, USA
e-mail: yzhou3@tulane.edu

C. Xu • J. Zhang
Center for Bioinformatics and Genomics, Department of Biostatistics and Bioinformatics, Tulane University, New Orleans, LA 70112, USA
e-mail: cxu2@tulane.edu

H.-W. Deng (✉)
Center for Bioinformatics and Genomics, Department of Biostatistics and Bioinformatics, Tulane University School of Public Health and Tropical Medicine, New Orleans, LA, USA
e-mail: hdeng2@tulane.edu

X. Wang et al. (eds.), *Application of Clinical Bioinformatics*,
Translational Bioinformatics 11, DOI 10.1007/978-94-017-7543-4_3

eQTL	Expression quantitative trait loci
FDR	False discovery rate
FPKM	Fragments per kilobase of transcript per million mapped reads
GAPDH	Glyceraldehyde-3-phosphate dehydrogenase
GO	Gene ontology
GSEA	Gene set enrichment analysis
GWAS	Genome-wide association studies
ICGC	International cancer genome consortium
LOD	Logarithm of the odds
MM	Mismatch probes
NCBI	National Center for Biotechnology Information
PBMC	Peripheral blood mononuclear cells
PM	Perfect match probes
RIN	Rna integrity number
RMA	Robust microarray average
RNA-Seq	RNA sequencing
RPKM	Reads per kilobase of transcript per million mapped reads
SAM	Significance analysis of microarrays

3.1 Introduction

Transcriptomics is the study of the whole set of RNA transcripts expressed in the specific tissues or cells via high-throughput techniques. In molecular biology, the central dogma could be roughly described as "DNA is transcribed into RNA, and RNA is translated into protein." Using DNA as a template, the synthesis of RNA is the first step of biological sequential information transfer and the key stage of gene expression regulation. If mRNA is identified in a biological sample, it signifies that the corresponding gene was expressed, and the protein product of this gene may be translated. Via transcriptomics, researchers are trying to figure out

- Which genes are expressed differentially and what are their functional roles in studied various conditions?
- What are the interaction networks of these genes and how are these genes regulated?
- Is there any gene or gene expression pattern that could be used as biomarker for the studied conditions?

As opposed to genomics research, genome-wide gene expression studies include the information of tissues/cells at specific condition and specific time point. All the genes in the genome are screened simultaneously in transcriptomic studies, allowing researchers to identify the genes involved in the specific biological processes and discover novel candidate genes which may contribute to the process. With this comprehensive transcriptome-wide approach, it is not necessary to specify target genes for study during experimental design. This means that more and novel genes may be discovered, giving us a better and comprehensive understanding of the gene patterns related to biological processes and/or conditions. Diseases in particular are normally affected and regulated by series of genes rather than individual genes. Transcriptomics is a powerful tool for identifying these functionally related genes and improving our fundamental understanding of disease etiology.

Transcriptomics will also identify the expression levels of genes, allowing us to understand the activity patterns of all the genes under various studied conditions. Gene expression may be compared in several experimental conditions (such as normal vs. diseased tissue, or cell lines with or without drug treatment). With the aid of bioinformatics and biostatistics tools, we can estimate the correlation effects (interactions) among the genes and genetic effects functioning in particular conditions. Moreover, understanding gene-gene interactions and the regulatory networks can provide insight into the specifics of various biological processes. The details of these methods will be introduced later in this chapter.

Currently, microarray and RNA sequencing (RNA-Seq) are the most widely used technologies for transcriptome studies. DNA microarray is based on DNA hybridization rules (adenine binds to thymine, cytosine binds to guanine) and is the first high-throughput method to perform expression profiling of transcriptome. Millions of "reporter probes", which are composed of parts of the genes' sequences, are fixed on a solid-based platform, like a glass slide. Each probe may contain a few million copies of the DNA sequences and can hybridize with fluorescent-labeled cDNA synthesized from the mRNA or total RNA extracted from cells or tissues of interest. By measuring the label intensities of each probe on the array, the relative expression level of the corresponding gene is quantitated. Microarray technology has been steadily improving for the last 20 years.

Currently, it is relatively inexpensive for many laboratories to acquire the tools and software necessary for microarray data analyses. Many specific microarray platforms have been designed, including probes for the non-coding, non-translated and non-transcribed chromosomal regions (Hey and Pepper 2009). Since the principles of the analysis of microarray data are nearly identical for both non-coding RNAs and mRNAs, we will take mRNA analysis as an example.

With the development of next-generation sequencing technology, RNA-Seq is becoming more and more popular for transcriptomic studies. In this method, after isolation, RNA is converted to a library of cDNA fragments, which are sequenced by high-throughput technologies. The sequence reads are then mapped to the reference transcriptome or assembled *de novo* into a transcriptome. The copies of sequences mapping to each gene represent the gene expression level in the sample.

Compared to microarray technology, RNA-Seq has several advantages. First, the microarray is limited to detecting existing genomic sequencing information, while the pre-designed known species- or transcript-specific probes are not required in RNA-Seq. So it is possible for RNA-Seq to discover novel transcripts, single nucleotide variants and/or alternative splicing. Second, with microarray, hybridization errors can occur, like cross-hybridization or non-ideal hybridization; while, the reads in RNA-Seq will be mapped to a unique sequence of transcriptome. This feature offers RNA-Seq higher specificity and sensitivity than microarray approach. Third, RNA-Seq offers a broader dynamic range. In cells, gene expression level has a large dynamic range. However, because it is limited to hybridization technology, microarray does not perform well in detecting low abundance RNAs (Wang et al. 2009).

Overall, in most cases, microarray is still a suitable choice for quick and inexpensive experiments. With accurate probe annotations and well-developed analysis tools, microarray approach still offers results as reliable as RNA-Seq (Zhao et al. 2014). However, RNA-Seq offers advantages of better sensitivity and specificity, novel discovery, and larger range of gene expression. As the cost continues to decrease, RNA-Seq is expected to become the main stream tool for transcriptomic studies.

3.2 RNA Sample Preparation

3.2.1 Tissue Selection

The rapid advancement of high-throughput technologies for acquisition of genomic, transcriptomic, epigenomic, and proteomic data has led to an explosion of transcriptomic and epigenomic studies for a variety of complex human disorders. Findings from these studies may assist in identification of new biomarkers for prognosis and/or diagnosis of diseases.

In transcriptomic studies (as opposed to genomics), the ideal RNA materials should be isolated from homogeneous samples, like a cultured cell line or a specific type of cells *in vivo*. However, the majority of past and current studies are performed on heterogeneous tissue samples related to a given disease, peripheral blood mononuclear cells (PBMC), or tissue/blood-cell-derived cell lines that may not have definitive or direct relationship with the disorders in question (Johannes et al. 2008). On the one hand, it is understandable that studies for many complex diseases may be hindered by the inaccessibility and/or impracticability of obtaining an adequate quantity of cells directly related to the disease. On the other hand, such a study design is potentially problematic because transcriptomic, epigenomic, and proteomic profiles are generally presented in a cell-, tissue- and organ- specific manner (Johannes et al. 2008; Reinius et al. 2012; Velculescu et al. 1999; Xu et al. 2002; Bossi and Lehner 2009) and disease-associated functional genomic and epigenetic variations can be cell-specific.

Tissue/mixed cell samples (e.g., PBMC) are generally composed of multiple distinct cell types. The proportion (in general) and even the cell type composition (in particular) of such multiple distinct cell types may vary in different biosamples. Failure to account for such cellular heterogeneity can easily yield false positive and false negative results (Johannes et al. 2008; Michels et al. 2013). In addition, the disease state itself may alter the proportional distribution of different cell types in the tissue/mixed cell samples, and thus measured functional genomic/epigenetic differences between cases and controls may only reflect differences in cell-type composition rather than true functional/epigenetic differences. Furthermore, if the disease-associated functional genomic/epigenomic variations are restricted to a certain cell type that represents only a small proportion of the tissue sampled, the disease-related functional variations may not be readily detected in the presence of cell heterogeneity, particularly when their effects on disease susceptibility are modest and the sample size is limited.

A few statistical deconvolution methods have been developed to correct for cell mixture proportions in functional genomic and epigenomic studies (Houseman et al. 2012; Abbas et al. 2009). However, these methods require the availability of reference transcriptomes/epigenomes for various cell types in population sub-groups defined by age, sex and ethnicity, which may be laborious or expensive to collect. In addition, for some tissues such as placenta, saliva, adipose or tumor tissue, the relevant component cell types may not be known (Houseman et al. 2014). To overcome this limitation, Houseman et al. (Houseman et al. 2014) recently proposed a novel method for conducting epigenome-wide association studies when a reference dataset is unavailable. Based on the simulation and empirical data analyses, it was suggested that this reference-free method can perform similar or better than methods that use reference datasets (Houseman et al. 2014). Another limitation of these statistical deconvolution methods is that the measured transcriptomic/epigenomic profile is assumed to be a linear mixture of the distinct cell-specific profiles, which is biologically motivated but sometimes may not be valid.

Finally, cells expanded *in vitro* generally have distorted gene expression and/or epigenetic profiles (Saferali et al. 2010; Caliskan et al. 2011) and thus are not solid for use in the study looking for the *in vivo* functional genomic and epigenomic mechanisms. Therefore, it is critical to perform functional genomics and epigenomics studies in a *single* type of relatively homogeneous cells that have a direct relationship to the diseases of interest, in order to unravel genuine functional genomic and epigenomic mechanisms underlying disease etiology.

3.2.2 RNA Sample Preparation

The most important step for transcriptomic experiments is the isolation of high-quality RNA.

Firstly, it is important to find the most appropriate method for RNA isolation. For different tissues or cells, a suitable lytic agent and RNA isolation kit should be chosen from large bio-reagent companies like Qiagen (https://www.qiagen.com/us/) or ThermoFisher (https://www.thermofisher.com/us/en/home.html).

If RNA isolation is difficult to process immediately after sample collection (for example, if samples are large or numerous, or need transportation), because RNA can be easily degraded, RNA stabilization reagents should be added to postpone RNA isolation for a few days without sacrificing the integrity of the RNA (Ohmomo et al. 2014).

After extraction from tissues or cells, RNA quality should be tested before any transcriptomic experiments. UV spectroscopy is a traditional method for measuring RNA concentration and purity. The ratio of absorbance value of a diluted RNA sample at 260 and 280 nm, A260/A280, is used to assess RNA purity. The ratio is affected by pH and ionic strength; a value of 1.8–2.1 is indicative of good RNA purity. Another important RNA quality feature is the integrity. The Agilent® 2100 Bioanalyzer™ instrument is commonly used for this measurement. Agilent® RNA 6000 Nano/Pico System is the kit compatible with this instrument. After the test, the RNA Integrity Number (RIN) is calculated by the instrument software to determine the integrity of the RNA sample. The normal requirement for RNA-Seq or microarray is RIN>7. Before the downstream experiment, RNA should be stored at −80 °C and suspended in RNase-free solution.

3.3 Profiling Methods

3.3.1 Microarray

3.3.1.1 Platform Choice

Various platforms and technical improvements have been developed for microarray since its invention in early 1990s. There are two major types of microarray platforms: cDNA microarray and oligonucleotide microarrays.

In cDNA microarray, cDNA libraries, which are reversely transcribed from RNA and amplified, are built. The cDNAs are then spotted on a nylon membrane or a glass slide. This microarray may be prepared in the laboratory, so it is sometimes called a "home-made" microarray.

Oligonucleotide microarray is the more conventional and popular type of microarray. The technology was first developed by Affymetrix (Santa Clara, CA) (http://www.affymetrix.com/) (Fodor et al. 1993), but some other major microarray commercial vendors have also developed their own, such as Agilent Technologies (Palo Alto, CA) (http://www.chem.agilent.com/), Illumina, Inc. (San Diego, CA) (http://www.illumina.com/) and NimbleGen Systems Inc. (Madison, WI) (http://www.nimblegen.com/). The probe in oligonucleotide microarray is designed to represent a gene or a specific RNA fragment, so it is normally short: 25-mer in the Affymetrix microarray or 60-mer in Agilent's. Because the oligonucleotide microarray is manufactured industrially, the quality control is better than cDNA microarray and suitable for integrated data analysis between different projects. Here, we will discuss oligonucleotide microarray, using the Affymetrix microarray as an example.

3.3.1.2 Data Processing and Introduction of Related Software

The raw microarray data are the image of the whole microarray with a fluorescence value in each spot representing the relative expression level of a given gene. Microarray manufacturers usually provide the corresponding software along with the scanner instruments, and the process of extracting expression data from the image can be completed by either the instrument or the facility where the microarray experiment is performed. Here, we introduce the subsequent steps: background correction and normalization.

Background correction is necessary because there will be non-specific hybridization spots in the microarray. Background noise can be estimated by the intensity of empty spots on the arrays or some more complex design. For example, on Affymetrix arrays, there are two types of probes, "mismatch probes (MM)" and "perfect match probes (PM)". The MM is almost identical to the PM, except for one base in the middle of probe, so the MM intensity is used as background intensity. After subtracting for background intensity, the spot intensity will represent relative gene expression. The data are then log-transformed. Logarithm base 2 value of the expression ratio is taken to treat upregulation and downregulation at the same magnitude. For example, fourfold upregulation and fourfold downregulation will be changed to 2 vs −2 instead of 4 vs 0.25.

In multiple microarray experiments, there are many sources of systematic variation, like experimental error and biological variation (Lee et al. 2000). Normalization is essential to allow the comparison across two or more microarray data sets and transform the data for traditional statistical methods. There are several methods for normalizing microarray data (Quackenbush 2001). The common normalization methods, such as ratio-based decisions and the quantitative analysis of cDNA microarray images, simply force the data from arrays to have the same mean (Irizarry et al. 2003). The basic assumption is that the average expression level of genes is same among all the tested samples. It means that expression of most genes do not change in different conditions and amount of upregulated genes and downregulated genes are equal. Another option is to normalize all expression values to a set of "housekeeping genes", which are expressed stably in all cells of an organism under normal and pathological conditions, including beta-actin, glyceraldehyde-3-phosphate dehydrogenase (GAPDH) and so on. This approach is also adopted by Affymetrix software. The gene expression value is divided by the mean expression value of these housekeeping genes on the same array. The microarray manufacturers normally offer the commercial normalization software for their chips, like MAS 5.0 for Affymetrix microarray. In Bioconductor software, there are also various free packages and tools for microarray data preprocessing (Gentleman et al. 2004), including the well-known Robust Microarray Average (RMA) (Irizarry et al. 2003) and GC-RMA (Zhijin Wu et al. 2004).

3.3.2 *RNA-Seq*

3.3.2.1 Parameter Choice

RNA-Seq is a transcriptome profiling method that uses developed deep-sequencing technologies (Wang et al. 2009). Various sequencing platforms can be applied to RNA-Seq, such as Illumina HiSeq, Roche 454 GS FLX+, Life Technologies Ion sequencing, and the Pacific Biosciences RS series sequencer (PacBio) (Liu et al. 2012; Quail et al. 2012; Glenn 2011). By virtue of the capability of deep-sequencing, it can not only quantify the RNA expression level but also look into the structure of the RNA, such as the boundaries of isoform and locations of the alternative splicing.

On the other hand, the performance of RNA-Seq strongly depends on the settings of the deep-sequencing. Some critical settings include the single/paired-end sequencing, sequence read depth (coverage) and read length. Compared with single-end sequencing, paired-end sequencing reads the nucleotides from both ends of the insert rather than a single end. The paired-end reads provide more information, which increases mapping accuracy and is especially useful for the RNA-Seq in isoform detection. Sequence read depth is the average number of reads representing a given nucleotide in the reconstructed sequence. Intuitively, the higher the depth, the better the performance in all aspects of RNA-Seq. Specifically, there is an asymptotic log-linear relationship observed between sequence read depth and gene detection, showing that increasing sequencing depth leads to the discovery of more genes (Li et al. 2014b). Read length is another important factor for an effective design of RNA-Seq. Longer read length gives more accurate information on the relative positions of the bases in a genome. It is critical for the detection of splice junction, even more important than library preparation or sequencing chemistry (Li et al. 2014b). As has been noted, appropriate settings of the RNA-Seq, especially the read depth and read length, are vital for the study design and following analyses of the transcriptome profiling.

3.3.2.2 Data Processing and Software Introduction

The profiling of RNA-Seq data analysis usually consists of three steps: sequence alignment, transcriptome reconstruction, and expression level quantification.

Beginning with the raw reads generated by the sequencing facility, the first step is aligning reads, typically with a reference. The standard reads files produced by the sequencer are FASTQ files, which contain the nucleotide and quality score for each position on each read. The alignment here is similar to the alignment of DNA-Seq. Many aligners for DNA-Seq can also be used for RNA-Seq alignment. Some commonly used aligners include Bowtie (Langmead et al. 2009), BWA (Li and Durbin 2009), SOAP (Li et al. 2008), and Samtools (Li et al. 2009). Meanwhile, a new generation of alignment software has been developed specifically for RNA-Seq, and has several advancements over previous DNA-Seq alignment programs. The new tools first discover exon junctions through initial read alignments, then the junctions are used to guide final alignment. These two-stage aligners for RNA-Seq include GEM (Marco-Sola et al. 2012), MapSplice (Wang et al. 2010a), RUM (Grant et al. 2011) and TopHat (Trapnell et al. 2009; Kim et al. 2013) among many others.

Through sequencing, RNA-Seq is capable of recovering multiple isoforms and alternative splicing sites within genes. The packages assembling exons and reporting all the isoforms can be divided into two classes, genome-guided and genome-independent (Garber et al. 2011). The former connects the exons under the framework of existing gene definition such as Cufflinks (Trapnell et al. 2010) or Scripture (Guttman et al. 2010). Conversely, the latter method makes *de-novo* transcripts, such as Velvet (Zerbino and Birney 2008), TransABySS (Robertson et al. 2010), making the genome-independent approach useful when there is no reference genome or for *de-novo* analysis. Both classes take the alignment files as the input and output the identified transcripts. In practice, combining results from both strategies may be the best way to fully utilize known information and discover novel isoforms (Garber et al. 2011).

Given the current RNA-Seq protocols based on mRNA fragmentation, the relative expression level of a gene is proportional to the number of reads mapped onto the gene. Thus, the read counts are used to estimate the gene expression level. However, there are two main biases caused by the RNA-Seq protocols which need to be normalized to acquire an accurate estimate from the read counts. First, longer genes tend to generate more reads than shorter genes at the same abundance level (Oshlack and Wakefield 2009). Second, each sequencing run produces a different number of reads, which leads to variability in the estimation across different runs. To address these biases, the reads per kilobase of transcript per million mapped reads (RPKM) (Mortazavi et al. 2008) and the fragments per kilobase of transcript per million mapped reads (FPKM) (Trapnell et al. 2010) are used to normalize the gene read counts by the gene length and the mapped reads in the sample. Unlike RPKM, FPKM takes into consideration the dependency between the paired-end reads. Thus, FPKM is mainly used in processing reads from paired-end sequencing.

Once the raw reads are normalized, the gene expression level can be quantified using union of exons or reads. Alexa-seq (Griffith et al. 2010), Cufflinks (Trapnell et al. 2010), ERANGE (Johnson et al. 2007) and many other options may be applied. Another interesting result would be the quantification of specific isoform. Unlike the direct counting method for gene quantification, isoform expression is usually estimated by a likelihood-based approach, such as Cufflinks (Trapnell et al. 2010), MISO (Katz et al. 2010), or RSEM (Wang et al. 2010b). Essentially, it uses the common, inclusive and exclusive reads among exons to recover the expression level of isoforms.

Before the final quantification of gene expression, other systematic variations may also exist under certain conditions which require further normalization. For example the GC content, gene body coverage evenness, nucleotide composition biases caused by library preparation, base error rate caused by sequencing, and batch effects caused by analyzing data from different batches or platforms (Li et al. 2014a). Accordingly, several tools have been proposed to correct them, including PEER (Stegle et al. 2010), sva (Leek et al. 2012; Leek and Storey 2007), and cqn (Hansen et al. 2012) among others. Those normalizations are usually included as an important step in the RNA-Seq analysis pipeline.

To summarize, many well-designed tools are available for the profiling of RNA-Seq. Furthermore, a number of pipelines have been developed to wrap up the RNA-Seq profiling and extend to further analysis. Here we introduce two typical pipelines. One is TopHat-Cufflinks pipeline implemented in a Linux environment. The other is an end-to-end workflow in R using DESeq2 (Love et al. 2014) along with other Bioconductor packages.

The TopHat-Cufflinks pipeline consists of several different programs that work together to perform a number of different analyses for RNA-Seq experiments. The complete pipeline and all the types of analyses it can conduct is summarized in Fig. 3.1, which refers to the online manual of Cufflinks (http://cole-trapnell-lab.github.io/cufflinks/manual/). The TopHat-Cufflinks workflow includes read mapping with TopHat, assembly and quantification with Cufflinks. The protocol available from the Nature Protocols (Trapnell et al. 2012) illustrates the pipeline in more detail. A newer, more advanced pipeline was introduced with Cufflinks version 2.2.0. In the new version, to deal with the gene expression quantification and normalization of large numbers of samples, the Cuffquant and Cuffnorm were developed to enhance the functionality of the TopHat-Cufflinks pipeline.

For those who are unfamiliar with the Linux or Mac OS environment which is required for the previous pipeline, an end-to-end workflow is also available in R (downloadable at http://www.bioconductor.org/help/workflows/rnaseqGene/). As shown in Fig. 3.2, it starts with aligned bam files, then performs reads count, normalization, differential expression analysis, visualization and other analysis using a set of available R packages. Some packages for specific functions are listed in Fig. 3.2. A great advantage of this workflow is that it is fully embedded in the R environment. Thus the analysts are free to choose their preferred packages to fulfill the workflow and further data analysis, like meta-analysis or network analysis.

3.4 Further Analysis

After the basic results of transcriptomics analysis are obtained (for example, a matrix or a table with each gene expression level in each sample), further analysis is often performed, including the following (network analysis is introduced in Chap. 14).

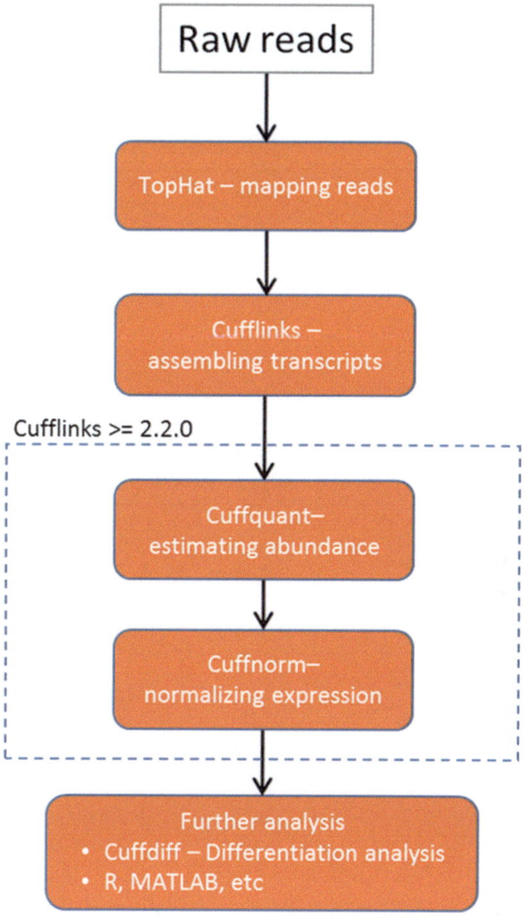

Fig. 3.1 The "classic" TopHat-Cufflinks workflow

3.4.1 Differential Expression Analysis

Identification of differentially expressed genes (DEGs) is the most common goal of transcriptomic studies. The methods for differential expression analysis are essentially the same for both RNA-Seq and microarray data.

The classical approach is the hypothesis testing in statistics. The "null hypothesis" is that there is no difference in gene expression levels among the conditions in study. The "alternative hypothesis" is that the genes are expressed differentially. The commonly used tests include t-tests, moderated t-tests, Mann–Whitney test and ANOVAs. The choice of statistic methods depends on the project aims, the transcriptomics platform used, the experimental design, the sample size, the number of tested groups and the number of replicates in each condition. For example, paired t-test is suitable for comparison between two paired groups, while ANOVA is used to compare three or more groups.

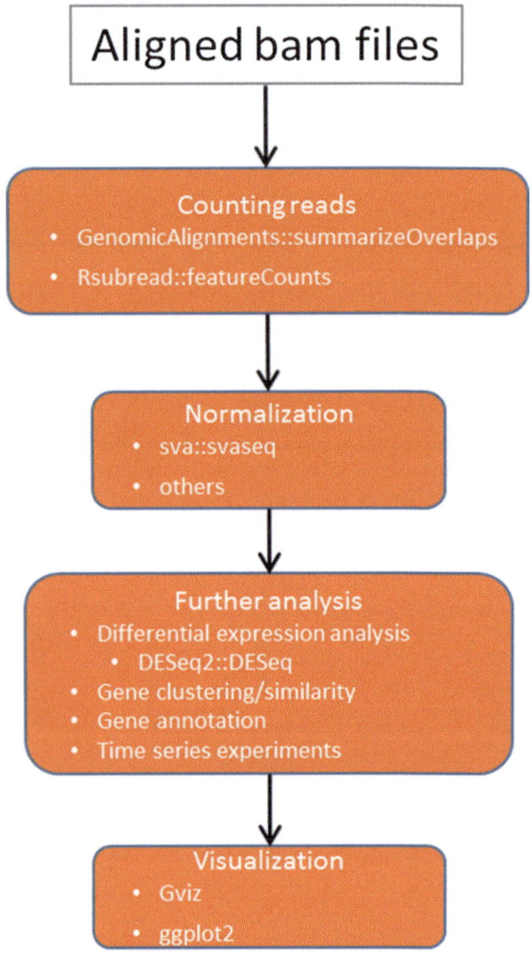

Fig. 3.2 Software and packages for RNA-Seq analysis workflow

However, because of the large number of simultaneous tests in transcriptomics analyses, the balance between false positives and false negatives must be considered for the significance threshold choice. Even at a significance level of 1 % (raw p-value <0.01), 200 false positives will be created in a 20,000 genes comparison study. On the other hand, Bonferroni correction, a typical statistical correction for multiple testing, is so stringent or conservative that a large number of false negatives will arise. Therefore, a False Discovery Rate (FDR) correction is widely used to take place of raw p-value. FDR represents the expected proportion of false positives among all the gene identified as differentially expressed. It means that less than 5 genes would be the false positives in 100 DEGs with FDR (p-adjusted or q-value) <0.05. The empirical Bayes methods such as linear models for microarray and RNA-seq data (limma) in Bioconductor (Smyth 2004) and permutation approaches like significance analysis of microarrays (SAM) (http://www-stat.

stanford.edu/~tibs/SAM/) (Tusher et al. 2001) will both test for significantly DEGs and offer FDR values.

The existing extensive statistical methods for differential expression analysis using microarray are directly applicable for RNA-Seq. Moreover, RNA-Seq technically provides more helpful information for the identification of DEGs. For example, EdgeR (Robinson et al. 2010), DESeq (Anders and Huber 2010) and Cuffdiff (Trapnell et al. 2010) assess the significance of DEGs by accounting for the variance in read counts across samples. While those methods are able to test statistical significance, caution should be used when interpreting these findings, as they do not necessarily draw relevant conclusions.

3.4.2 Class Discovery Analysis

The class discovery analysis is also called unsupervised classification or knowledge discovery. The goal of this approach is to discover or identify the subgroups sharing common features in objects, patients, or tested genes. Identifying naturally existing subgroups will be helpful to understand the pathological mechanism of specific disease or the interaction network of the genes. Many unsupervised classification techniques can be used with transcriptomics data to identify novel clusters (classes) (Peterson 2013). Without hypothesis-driven, the cluster analysis is based on iterative pattern recognition or statistical learning methods to identify the clusters in the data. The common cluster analysis methods include hierarchical cluster analysis, k-means cluster analyses (de Souto et al. 2008), self-organizing maps, neural gas and Genomic Signal Processing based clustering (Istepanian et al. 2011). There are many software packages available in R, Matlab, and many other analysis software packages. Two widely used clustering software are Gene Cluster and TreeView (Eisen et al. 1998), which contain several kinds of algorithms.

3.4.3 Class Prediction Analysis

The class prediction analysis approach is called supervised classification. This approach tries to find a rule to assign objects, patients or genes into a specific group based on the prior knowledge. It will be useful for disease diagnosis or function prediction of individual gene. The commonly applied supervised analysis algorithms (Peterson 2013) are linear regression, k-nearest neighbor, learning vector quantization, decision tree analysis, random forests, naive Bayes, logistic regression, kernel regression, artificial neural networks, support vector machines, mixture of experts, and supervised neural gas. Due to the features of class prediction analysis, an adequate samples size is necessary for both training and test dataset.

3.4.4 Pathway and Gene Enrichment Analysis

An important goal of the biological sciences is to improve the understanding of the mechanisms of biological processes. After obtaining a list of DEGs, the further step is to interpret the biological mechanism or processes which they are involved in.

Gene Ontology (GO) categories (molecular function, biological process or cellular component of the gene products) were proposed to annotate the genes (Draghici et al. 2003b; Khatri et al. 2002). Using Fisher's exact test or a Chi-square test, the GO category could be identified as over-represented in the condition being studied if the proportion of DEGs in this category is significantly larger than the same proportion in the whole gene set. The p-value represents hypergeometric probability if the list of DEGs is associated the GO category (Draghici et al. 2003a; Khatri et al. 2004). There are dozens of software packages for GO term analysis (Khatri and Draghici 2005).

An alternative approach for enrichment analysis is the Gene Set Enrichment Analysis (GSEA) (Tian et al. 2005). In GO term analysis, only the DEGs are selected into further analysis. Meanwhile, GSEA ranks all genes based on the correlation between their expression and the given phenotypes and performs a functional class scoring (Pavlidis et al. 2004; Goeman et al. 2004).

There are also many tools to analyze metabolic or regulatory pathways to understand biological meaning behind large list of genes, including DAVID (Huang da et al. 2009), Kegg database (Kanehisa and Goto 2000), MAPPFinder (Doniger et al. 2003), Pathway-Express (Khatri et al. 2005) and Cytoscape (Shannon et al. 2003).

3.4.5 Meta-Analysis

With the rapid development of high-throughput genomic measurement technology, hundreds of gene expression studies have been conducted generating a tremendous amount of experimental data. In the past decade, several large public transcriptomic databases have been established, such as Gene Expression Omnibus from the National Center for Biotechnology Information (NCBI) and ArrayExpress from the European Bioinformatics Institute (EBI). With this externally available information, researchers are able to increase the study power for findings and validate them through meta-analysis.

Meta-analysis is a statistical procedure that integrates related results of several independent studies (Egger et al. 1997). It is widely used in many areas and also applicable to gene expression analysis. While meta-analysis can be employed to address several research questions, like pathway detection and co-expression analysis, the most common application is DEG detection.

Generally, there are four sorts of methods to integrate information for DEG identification: combine p-values, combine effect size, combine ranks and merge data after normalization (Tseng et al. 2012). Combining p-values is most flexible and is used most frequently, as p-value is all it needs for information integration. Usually, the p-value from different studies is transformed and summed to fit a known distribution, through which an integrated p-value can be calculated; for instance, the Fisher's method that adds up minus log-transformed p-values, and Stouffer's method which adopts the inverse normal transformation. In light of the fact that the effect sizes across studies are essentially combinable, fixed and random effects models, as well as the Bayesian model, have been applied to DEG meta-analysis to combine effect size (Tseng et al. 2012). One concern for effect size models is that they are vulnerable to outliers. Combining ranks is an alternative approach. Gene ranks from different data sets are summed, multiplied or processed by other manipulations to build a test statistic. The integrated p-value can then be calculated through permutation. Another method often used in analysis is directly merging the raw data after normalization to remove the cross-study discrepancy. The combinable studies in the direct merging approach are mostly restricted to studies from the same or similar platforms. Furthermore, its performance heavily depends on the selected normalization methods (Tseng et al. 2012) because the normalizations are not guaranteed to thoroughly eliminate the cross-study discrepancy.

Currently, a number of online databases are available for transcriptomic meta-analysis. Two main data sources are GEO from NCBI and ArrayExpress from EBI. GEO (http://www.ncbi.nlm.nih.gov/geo/) is an international public open repository for microarray, next-generation sequencing, and other forms of high-throughput functional genomics data submitted by the scientific community. As of July 31, 2015, GEO contained 41,024 expression profiles by array, 5124 expression profiles by high throughput sequencing and 1,545,524 samples. ArrayExpress (http://www.ebi.ac.uk/arrayexpress/) is a similar multi-species functional genomics data repository containing 1,730,383 assays (as of 7/31/15) from microarray, RNA-Seq and other technologies. Other databases are also available online, such as ExPASy (http://www.expasy.org/transcriptomics), The Cancer Genome Atlas (https://tcga-data.nci.nih.gov/tcga/) and GENEVESTIGATOR (https://genevestigator.com/gv/index.jsp).

In addition to the many online databases for data support, many tools can be used to implement transcriptomic meta-analysis. Most are free to use and packed in a R/Bioconductor environment, including GeneMeta (https://www.bioconductor.org/packages/release/bioc/html/GeneMeta.html) on fixed and random effects models, metaMA (Marot et al. 2009) on random effects models and Stouffer's method, OrderedList (http://compdiag.molgen.mpg.de/software/index.shtml) on gene rank, and many others.

While meta-analysis is a powerful tool to increase DEG detection power by collecting as many samples as possible, there are some key requirements for a successful meta-analysis. One of the biggest potential obstacles is dataset quality (Eysenck 1994). The inclusion of a poor quality or outlying study in the information integration can greatly dilute information, weaken statistical power and even distort final biological conclusions (Kang et al. 2012). To mitigate such potential pitfalls in meta-analysis, researchers are highly encouraged to use some objective criteria or tools, such as MetaQC, for quality control of the selected data set (Kang et al. 2012).

In summary, meta-analysis is a great approach for gene expression analysis which can increase the power to detect DEG when it is used appropriately. Many tools and online databases are available for transcriptome meta-analysis. At the same time, the inclusion data quality is another important issue for a successful meta-analysis. Researchers should pay special attention to the study design and subsequent interpretation of the meta-analysis.

3.4.6 eQTL Integrated Analysis

The expression quantitative trait loci (eQTL) are genes of which the abundance are directly modified by polymorphism in regulatory elements or other gene regions (Cookson et al. 2009). Similar to QTL mapping, the aim of eQTL mapping is to find the association of genetic markers with gene expression. Further, study designs and statistical methods that are traditionally used to map QTLs can be successfully applied to the identification of eQTLs. A typical classification for the identified regulatory variants in eQTL mapping is either cis or trans-acting, depending on the physical distance from the gene it regulates. The selection of the threshold to define the cis and trans-acting varies, like upstream and downstream 100 kb or 250 kb (Franke and Jansen 2009).

Genome-wide association studies (GWAS) have been widely conducted during the past decade. Thousands of GWAS identified a great number of disease associated single nucleotide polymorphisms (SNPs). However, how those SNPs work on multifactorial disease remains unclear. The regulation of gene expression due to SNPs may have important effects, and eQTL mapping is a way to characterize the biological basis underlying the disease associations. It is helpful to identify gene networks involved in disease pathogenesis by integrating gene expression and genomic variant information.

With the intention of finding association between gene abundance and SNP, eQTL analysis started from gene expression profiling and SNP genotyping. SNP genotyping can be performed by either microarray or RNA-Seq, after which the effect of a given SNP on the expression of a given gene can be tested.

The logarithm of the odds (LOD) score is a frequently used statistical measure of association strength. Essentially, the LOD score is equivalent to the F statistic from ANOVA, which is the ratio of the log10 likelihood under the alternative hypothesis over the log10 likelihood under the null hypothesis (Broman and Sen 2009). A large LOD score favors eQTL signal, while the threshold for claiming significance depends on the multiple testing needs to be adjusted. For testing potential cis-regulatory variants, only SNPs located in the vicinity of the given gene are involved, whereas a genome-wide scan is conducted to test potential trans-regulatory SNPs. The association power of cis-regulatory effects is usually higher than the power of trans-regulatory effects.

Although it is straightforward that the QTL mapping method are introduced in eQTL analysis, eQTL mapping is different from QTL in thousands of phenotypes (Kendziorski et al. 2006). The repeated application of LOD score requires multiple testing adjustments which may result in inflated FDR. To circumvent this, advanced approaches were developed for eQTL mapping, including Q-ALL (Storey et al. 2004), MOM (Kendziorski et al. 2006) and others. Equally important, several R packages are available to implement the traditional and latest eQTL mapping methods, including eqtl developed from qtl (Broman et al. 2003), eqtlM for MOM (Kendziorski et al. 2006), and MatrixEQTL for fast computation (Shabalin 2012), among others.

As RNA-Seq is replacing microarrays as the default technique for the profiling of gene expression, a few pioneer studies of RNA-Seq based eQTL mapping have emerged (Pickrell et al. 2010). Unlike microarray based eQTL mapping, RNA-Seq provides rich information of allele-specific gene expression and makes it possible for isoform-specific eQTL mapping to be performed (Sun and Hu 2013). Recently, a statistical framework for RNA-Seq based eQTL mapping was proposed by Sun (Sun 2012). The R package asSeq can be used to implement this approach. Considering the biological probability that gene expression may vary in an allele-specific manner or at the isoform level, eQTL analysis using RNA-Seq data is a promising way to understand the pathogenesis of the complex diseases.

3.5 Example

Chronic lymphocytic leukemia (CLL) is one of the most common leukemias among adults in the Western world (Zenz et al. 2010). Recently, a comprehensive CLL transcriptome profile with unprecedented resolution was characterized by performing RNA-Seq on a large cohort of CLL samples (Ferreira et al. 2014). Thousands of differentially expressed transcriptional elements between the CLL and normal B cells were identified, including not only protein-coding genes but also noncoding RNAs and pseudogenes. CLL-specific splicing patterns were observed in about 2000 genes, while most of them were not differentially expressed. Pathway, gene set enrichment, and network analysis were performed in this study which provides a global view of the CLL transcriptional landscape.

In this study, there were 98 patients with CLL and 9 healthy subjects. Tumor CLL cells from the patients and normal B-cells from healthy samples were used for RNA-seq. The RNA-Seq libraries were prepared following the standard Illumina protocol with the mRNA-Seq TruSeq. Those libraries were sequenced by the Illumina HiSeq 2000 sequencer with 76-bp paired-end reads. In total, nearly six billion paired-end reads were generated with a median of 45M reads per sample.

They employed GEM mapper (Marco-Sola et al. 2012) to align the paired-end reads to the human genome version hg19 and an exon-junction annotation database from GENCODE (Harrow et al. 2006). To identify as many novel splice junctions as possible, the unmapped paired-end reads were split-mapped (split the paired reads to single reads and mapped independently) again. After alignment, the exon and gene expression values were computed by all reads mapping to the exon and gene. The transcript abundance levels were deconvoluted from the gene expression by the Flux Capacitor program (Montgomery et al. 2010). The RPKM values were reported as the expression value for the downstream analysis. Interestingly, the overall coverage of the human genome was significantly different (p-value 1.8×10^{-8}) between CLL (13.6 %) and normal samples (10.5 %).

Ninety-five study patients together with another 124 CLL patients from the International Cancer Genome Consortium (ICGC) CLL project were also studied by microarray expression profile constituting a validation data set. Microarray profiling was collected by the Affymetrix Human Genome Array U219 array. The mRNA expression values normalized by RMA were generated by the Expression Console software from Affymetrix. The gene expression value quantified by microarray was highly correlated with the data from RNA-Seq with the range between 0.81 and 0.88.

Based on the RNA-Seq profiling, the non-parametric Wilcoxon rank sum test with Benjamini-Hochberg (BH) adjustment was used to find the potential DEGs. In addition, a fold change difference in the median between groups was calculated. The genes with multiple testing adjusted p-value <0.01 and fold change >3 were claimed to be significantly differentially expressed. Based on the stringent criteria, 1089 differential expressed gene were identified, including but not limited to 814 - protein-coding genes, 127 long noncoding RNAs, and 47 lincRNAs. Further, about 2000 genes were found to be significantly different in the relative ratios of alternative splice isoforms between CLL and normal cells. Among them, several genes contained well-known alternative isoforms as cancer biomarkers, such as BCL2L1, CD44, and RAC1 (Pajares et al. 2007).

Moreover, they identified a previously unreported CLL subdivision C1/C2, which was validated by the microarray expression data of the 95 cases. Through the hierarchical clustering of the gene expression of protein-coding and long noncoding genes, not only the normal lymphocytes and tumor samples were clearly separated, but also two subgroups (C1 and C2) within CLL samples were strongly defined. Using the microarray data from independent and published datasets, the gene set enrichment analyses for C1 and C2 subgroups strongly supported the subdivision of C1 and C2. Between C1 and C2, there were 128 differentially expressed genes, which were preferentially related to a few pathways, such as the MAPK/ERK signaling pathway. Differences in splicing patterns between the C1 and C2 subgroups were also observed. For instance, there were a larger number of splicing alterations in C2 samples than C1 (317 vs. 204). Additionally, C1/C2 classification were related to some clinical impacts, which were also confirmed by the microarray profiling. For instance, C1 patients had a less frequency (9 %) of mutations in genes related to adverse outcome compared with 27 % in C2.

In summary, this study represented the CLL transcriptome with unprecedented resolution through RNA-Seq and validated the findings with microarray analysis.

The uncovered CLL transcriptome characterization refines the more traditional etiology of the disease and sheds new insights into the pathogenesis of CLL.

References

Abbas AR, Wolslegel K, Seshasayee D, Modrusan Z, Clark HF. Deconvolution of blood microarray data identifies cellular activation patterns in systemic lupus erythematosus. PLoS One. 2009;4(7):e6098. doi:10.1371/journal.pone.0006098.

Anders S, Huber W. Differential expression analysis for sequence count data. Genome Biol. 2010;11(10):R106. doi:10.1186/gb-2010-11-10-r106.

Bossi A, Lehner B. Tissue specificity and the human protein interaction network. Mol Syst Biol. 2009;5:260. doi:10.1038/msb.2009.17.

Broman KW, Sen Ś. A guide to QTL mapping with R/qtl. New York: Springer; 2009.

Broman KW, Wu H, Sen S, Churchill GA. R/qtl: QTL mapping in experimental crosses. Bioinformatics. 2003;19(7):889–90.

Caliskan M, Cusanovich DA, Ober C, Gilad Y. The effects of EBV transformation on gene expression levels and methylation profiles. Hum Mol Genet. 2011;20(8):1643–52. doi:10.1093/hmg/ddr041.

Cookson W, Liang L, Abecasis G, Moffatt M, Lathrop M. Mapping complex disease traits with global gene expression. Nat Rev Genet. 2009;10(3):184–94. doi:10.1038/nrg2537.

da Huang W, Sherman BT, Lempicki RA. Systematic and integrative analysis of large gene lists using DAVID bioinformatics resources. Nat Protoc. 2009;4(1):44–57. doi:10.1038/nprot.2008.211.

de Souto MC, Costa IG, de Araujo DS, Ludermir TB, Schliep A. Clustering cancer gene expression data: a comparative study. BMC Bioinf. 2008;9:497. doi:10.1186/1471-2105-9-497.

Doniger SW, Salomonis N, Dahlquist KD, Vranizan K, Lawlor SC, Conklin BR. MAPPFinder: using Gene Ontology and GenMAPP to create a global gene-expression profile from microarray data. Genome Biol. 2003;4(1):R7.

Draghici S, Khatri P, Bhavsar P, Shah A, Krawetz SA, Tainsky MA. Onto-Tools, the toolkit of the modern biologist: Onto-Express, Onto-Compare, Onto-Design and Onto-Translate. Nucleic Acids Res. 2003a;31(13):3775–81.

Draghici S, Khatri P, Martins RP, Ostermeier GC, Krawetz SA. Global functional profiling of gene expression. Genomics. 2003b;81(2):98–104.

Egger M, Smith GD, Phillips AN. Meta-analysis: principles and procedures. BMJ. 1997;315 (7121):1533–7.

Eisen MB, Spellman PT, Brown PO, Botstein D. Cluster analysis and display of genome-wide expression patterns. Proc Natl Acad Sci U S A. 1998;95(25):14863–8.

Eysenck HJ. Meta-analysis and its problems. BMJ. 1994;309(6957):789–92.

Ferreira PG, Jares P, Rico D, Gomez-Lopez G, Martinez-Trillos A, Villamor N, Ecker S, Gonzalez-Perez A, Knowles DG, Monlong J, Johnson R, Quesada V, Djebali S, Papasaikas P, Lopez-Guerra M, Colomer D, Royo C, Cazorla M, Pinyol M, Clot G, Aymerich M, Rozman M, Kulis M, Tamborero D, Gouin A, Blanc J, Gut M, Gut I, Puente XS, Pisano DG, Martin-Subero JI, Lopez-Bigas N, Lopez-Guillermo A, Valencia A, Lopez-Otin C, Campo E, Guigo R. Transcriptome characterization by RNA sequencing identifies a major molecular and clinical subdivision in chronic lymphocytic leukemia. Genome Res. 2014;24(2):212–26. doi:10.1101/gr.152132.112.

Fodor SP, Rava RP, Huang XC, Pease AC, Holmes CP, Adams CL. Multiplexed biochemical assays with biological chips. Nature. 1993;364(6437):555–6. doi:10.1038/364555a0.

Franke L, Jansen RC. eQTL analysis in humans. Methods Mol Biol. 2009;573:311–28. doi:10.1007/978-1-60761-247-6_17.
Garber M, Grabherr MG, Guttman M, Trapnell C. Computational methods for transcriptome annotation and quantification using RNA-seq. Nat Methods. 2011;8(6):469–77. doi:10.1038/nmeth.1613.
Gentleman RC, Carey VJ, Bates DM, Bolstad B, Dettling M, Dudoit S, Ellis B, Gautier L, Ge Y, Gentry J, Hornik K, Hothorn T, Huber W, Iacus S, Irizarry R, Leisch F, Li C, Maechler M, Rossini AJ, Sawitzki G, Smith C, Smyth G, Tierney L, Yang JY, Zhang J. Bioconductor: open software development for computational biology and bioinformatics. Genome Biol. 2004;5(10):R80. doi:10.1186/gb-2004-5-10-r80.
Glenn TC. Field guide to next-generation DNA sequencers. Mol Ecol Resour. 2011;11(5):759–69. doi:10.1111/j.1755-0998.2011.03024.x.
Goeman JJ, van de Geer SA, de Kort F, van Houwelingen HC. A global test for groups of genes: testing association with a clinical outcome. Bioinformatics. 2004;20(1):93–9.
Grant GR, Farkas MH, Pizarro AD, Lahens NF, Schug J, Brunk BP, Stoeckert CJ, Hogenesch JB, Pierce EA. Comparative analysis of RNA-Seq alignment algorithms and the RNA-Seq unified mapper (RUM). Bioinformatics. 2011;27(18):2518–28. doi:10.1093/bioinformatics/btr427.
Griffith M, Griffith OL, Mwenifumbo J, Goya R, Morrissy AS, Morin RD, Corbett R, Tang MJ, Hou YC, Pugh TJ, Robertson G, Chittaranjan S, Ally A, Asano JK, Chan SY, Li HI, McDonald H, Teague K, Zhao Y, Zeng T, Delaney A, Hirst M, Morin GB, Jones SJ, Tai IT, Marra MA. Alternative expression analysis by RNA sequencing. Nat Methods. 2010;7(10):843–7. doi:10.1038/nmeth.1503.
Guttman M, Garber M, Levin JZ, Donaghey J, Robinson J, Adiconis X, Fan L, Koziol MJ, Gnirke A, Nusbaum C, Rinn JL, Lander ES, Regev A. Ab initio reconstruction of cell type-specific transcriptomes in mouse reveals the conserved multi-exonic structure of lincRNAs. Nat Biotechnol. 2010;28(5):503–10. doi:10.1038/nbt.1633.
Hansen KD, Irizarry RA, Wu Z. Removing technical variability in RNA-seq data using conditional quantile normalization. Biostatistics. 2012;13(2):204–16. doi:10.1093/biostatistics/kxr054.
Harrow J, Denoeud F, Frankish A, Reymond A, Chen CK, Chrast J, Lagarde J, Gilbert JG, Storey R, Swarbreck D, Rossier C, Ucla C, Hubbard T, Antonarakis SE, Guigo R. GENCODE: producing a reference annotation for ENCODE. Genome Biol. 2006;7 Suppl 1:S4.1–9. doi:10.1186/gb-2006-7-s1-s4.
Hey Y, Pepper SD. Interesting times for microarray expression profiling. Brief Funct Genomic Proteomic. 2009;8(3):170–3. doi:10.1093/bfgp/elp012.
Houseman EA, Accomando WP, Koestler DC, Christensen BC, Marsit CJ, Nelson HH, Wiencke JK, Kelsey KT. DNA methylation arrays as surrogate measures of cell mixture distribution. BMC Bioinf. 2012;13:86. doi:10.1186/1471-2105-13-86.
Houseman EA, Molitor J, Marsit CJ. Reference-free cell mixture adjustments in analysis of DNA methylation data. Bioinformatics. 2014;30(10):1431–9. doi:10.1093/bioinformatics/btu029.
Irizarry RA, Hobbs B, Collin F, Beazer-Barclay YD, Antonellis KJ, Scherf U, Speed TP. Exploration, normalization, and summaries of high density oligonucleotide array probe level data. Biostatistics. 2003;4(2):249–64. doi:10.1093/biostatistics/4.2.249.
Istepanian RS, Sungoor A, Nebel JC. Comparative analysis of genomic signal processing for microarray data clustering. IEEE Trans Nanobioscience. 2011;10(4):225–38. doi:10.1109/TNB.2011.2178262.
Johannes F, Colot V, Jansen RC. Epigenome dynamics: a quantitative genetics perspective. Nat Rev Genet. 2008;9(11):883–90. doi:10.1038/nrg2467.
Johnson DS, Mortazavi A, Myers RM, Wold B. Genome-wide mapping of in vivo protein-DNA interactions. Science. 2007;316(5830):1497–502. doi:10.1126/science.1141319.
Kanehisa M, Goto S. KEGG: kyoto encyclopedia of genes and genomes. Nucleic Acids Res. 2000;28(1):27–30.

Kang DD, Sibille E, Kaminski N, Tseng GC. MetaQC: objective quality control and inclusion/exclusion criteria for genomic meta-analysis. Nucleic Acids Res. 2012;40(2):e15. doi:10.1093/nar/gkr1071.

Katz Y, Wang ET, Airoldi EM, Burge CB. Analysis and design of RNA sequencing experiments for identifying isoform regulation. Nat Methods. 2010;7(12):1009–15. doi:10.1038/nmeth.1528.

Kendziorski CM, Chen M, Yuan M, Lan H, Attie AD. Statistical methods for expression quantitative trait loci (eQTL) mapping. Biometrics. 2006;62(1):19–27. doi:10.1111/j.1541-0420.2005.00437.x.

Khatri P, Draghici S. Ontological analysis of gene expression data: current tools, limitations, and open problems. Bioinformatics. 2005;21(18):3587–95. doi:10.1093/bioinformatics/bti565.

Khatri P, Draghici S, Ostermeier GC, Krawetz SA. Profiling gene expression using onto-express. Genomics. 2002;79(2):266–70. doi:10.1006/geno.2002.6698.

Khatri P, Bhavsar P, Bawa G, Draghici S. Onto-Tools: an ensemble of web-accessible, ontology-based tools for the functional design and interpretation of high-throughput gene expression experiments. Nucleic Acids Res. 2004;32(Web Server issue):W449–56. doi:10.1093/nar/gkh409.

Khatri P, Sellamuthu S, Malhotra P, Amin K, Done A, Draghici S. Recent additions and improvements to the Onto-Tools. Nucleic Acids Res. 2005;33(Web Server issue):W762–5. doi:10.1093/nar/gki472.

Kim D, Pertea G, Trapnell C, Pimentel H, Kelley R, Salzberg SL. TopHat2: accurate alignment of transcriptomes in the presence of insertions, deletions and gene fusions. Genome Biol. 2013;14(4):R36. doi:10.1186/gb-2013-14-4-r36.

Langmead B, Trapnell C, Pop M, Salzberg SL. Ultrafast and memory-efficient alignment of short DNA sequences to the human genome. Genome Biol. 2009;10(3):R25. doi:10.1186/gb-2009-10-3-r25.

Lee ML, Kuo FC, Whitmore GA, Sklar J. Importance of replication in microarray gene expression studies: statistical methods and evidence from repetitive cDNA hybridizations. Proc Natl Acad Sci U S A. 2000;97(18):9834–9.

Leek JT, Storey JD. Capturing heterogeneity in gene expression studies by surrogate variable analysis. PLoS Genet. 2007;3(9):1724–35. doi:10.1371/journal.pgen.0030161.

Leek JT, Johnson WE, Parker HS, Jaffe AE, Storey JD. The sva package for removing batch effects and other unwanted variation in high-throughput experiments. Bioinformatics. 2012;28(6):882–3. doi:10.1093/bioinformatics/bts034.

Li H, Durbin R. Fast and accurate short read alignment with Burrows-Wheeler transform. Bioinformatics. 2009;25(14):1754–60. doi:10.1093/bioinformatics/btp324.

Li R, Li Y, Kristiansen K, Wang J. SOAP: short oligonucleotide alignment program. Bioinformatics. 2008;24(5):713–14. doi:10.1093/bioinformatics/btn025.

Li H, Handsaker B, Wysoker A, Fennell T, Ruan J, Homer N, Marth G, Abecasis G, Durbin R, Genome Project Data Processing Subgroup. The sequence alignment/map format and SAMtools. Bioinformatics. 2009;25(16):2078–9. doi:10.1093/bioinformatics/btp352.

Li S, Labaj PP, Zumbo P, Sykacek P, Shi W, Shi L, Phan J, Wu PY, Wang M, Wang C, Thierry-Mieg D, Thierry-Mieg J, Kreil DP, Mason CE. Detecting and correcting systematic variation in large-scale RNA sequencing data. Nat Biotechnol. 2014a;32(9):888–95. doi:10.1038/nbt.3000.

Li S, Tighe SW, Nicolet CM, Grove D, Levy S, Farmerie W, Viale A, Wright C, Schweitzer PA, Gao Y, Kim D, Boland J, Hicks B, Kim R, Chhangawala S, Jafari N, Raghavachari N, Gandara J, Garcia-Reyero N, Hendrickson C, Roberson D, Rosenfeld J, Smith T, Underwood JG, Wang M, Zumbo P, Baldwin DA, Grills GS, Mason CE. Multi-platform assessment of transcriptome profiling using RNA-seq in the ABRF next-generation sequencing study. Nat Biotechnol. 2014b;32(9):915–25. doi:10.1038/nbt.2972.

Liu L, Li Y, Li S, Hu N, He Y, Pong R, Lin D, Lu L, Law M. Comparison of next-generation sequencing systems. J Biomed Biotechnol. 2012;2012:11. doi:10.1155/2012/251364.

Love MI, Huber W, Anders S. Moderated estimation of fold change and dispersion for RNA-seq data with DESeq2. Genome Biol. 2014;15(12):550. doi:10.1186/s13059-014-0550-8.

Marco-Sola S, Sammeth M, Guigo R, Ribeca P. The GEM mapper: fast, accurate and versatile alignment by filtration. Nat Methods. 2012;9(12):1185–8. doi:10.1038/nmeth.2221.

Marot G, Foulley JL, Mayer CD, Jaffrezic F. Moderated effect size and P-value combinations for microarray meta-analyses. Bioinformatics. 2009;25(20):2692–9. doi:10.1093/bioinformatics/btp444.

Michels KB, Binder AM, Dedeurwaerder S, Epstein CB, Greally JM, Gut I, Houseman EA, Izzi B, Kelsey KT, Meissner A, Milosavljevic A, Siegmund KD, Bock C, Irizarry RA. Recommendations for the design and analysis of epigenome-wide association studies. Nat Methods. 2013;10(10):949–55. doi:10.1038/nmeth.2632.

Montgomery SB, Sammeth M, Gutierrez-Arcelus M, Lach RP, Ingle C, Nisbett J, Guigo R, Dermitzakis ET. Transcriptome genetics using second generation sequencing in a Caucasian population. Nature. 2010;464(7289):773–7. doi:10.1038/nature08903.

Mortazavi A, Williams BA, McCue K, Schaeffer L, Wold B. Mapping and quantifying mammalian transcriptomes by RNA-Seq. Nat Methods. 2008;5(7):621–8. doi:10.1038/nmeth.1226.

Ohmomo H, Hachiya T, Shiwa Y, Furukawa R, Ono K, Ito S, Ishida Y, Satoh M, Hitomi J, Sobue K, Shimizu A. Reduction of systematic bias in transcriptome data from human peripheral blood mononuclear cells for transportation and biobanking. PLoS One. 2014;9(8): e104283. doi:10.1371/journal.pone.0104283.

Oshlack A, Wakefield MJ. Transcript length bias in RNA-seq data confounds systems biology. Biol Direct. 2009;4:14. doi:10.1186/1745-6150-4-14.

Pajares MJ, Ezponda T, Catena R, Calvo A, Pio R, Montuenga LM. Alternative splicing: an emerging topic in molecular and clinical oncology. Lancet Oncol. 2007;8(4):349–57. doi:10.1016/S1470-2045(07)70104-3.

Pavlidis P, Qin J, Arango V, Mann JJ, Sibille E. Using the gene ontology for microarray data mining: a comparison of methods and application to age effects in human prefrontal cortex. Neurochem Res. 2004;29(6):1213–22.

Peterson LE. Classification analysis of DNA microarrays. Hoboken: Wiley; 2013.

Pickrell JK, Marioni JC, Pai AA, Degner JF, Engelhardt BE, Nkadori E, Veyrieras J-B, Stephens M, Gilad Y, Pritchard JK. Understanding mechanisms underlying human gene expression variation with RNA sequencing. Nature. 2010;464(7289):768–72. doi:http://www.nature.com/nature/journal/v464/n7289/suppinfo/nature08872_S1.html.

Quackenbush J. Computational analysis of microarray data. Nat Rev Genet. 2001;2(6):418–27. doi:10.1038/35076576.

Quail MA, Smith M, Coupland P, Otto TD, Harris SR, Connor TR, Bertoni A, Swerdlow HP, Gu Y. A tale of three next generation sequencing platforms: comparison of Ion Torrent, Pacific Biosciences and Illumina MiSeq sequencers. BMC Genomics. 2012;13:341. doi:10.1186/1471-2164-13-341.

Reinius LE, Acevedo N, Joerink M, Pershagen G, Dahlen SE, Greco D, Soderhall C, Scheynius A, Kere J. Differential DNA methylation in purified human blood cells: implications for cell lineage and studies on disease susceptibility. PLoS One. 2012;7(7):e41361. doi:10.1371/journal.pone.0041361.

Robertson G, Schein J, Chiu R, Corbett R, Field M, Jackman SD, Mungall K, Lee S, Okada HM, Qian JQ, Griffith M, Raymond A, Thiessen N, Cezard T, Butterfield YS, Newsome R, Chan SK, She R, Varhol R, Kamoh B, Prabhu AL, Tam A, Zhao Y, Moore RA, Hirst M, Marra MA, Jones SJ, Hoodless PA, Birol I. De novo assembly and analysis of RNA-seq data. Nat Methods. 2010;7(11):909–12. doi:10.1038/nmeth.1517.

Robinson MD, McCarthy DJ, Smyth GK. edgeR: a Bioconductor package for differential expression analysis of digital gene expression data. Bioinformatics. 2010;26(1):139–40. doi:10.1093/bioinformatics/btp616.

Saferali A, Grundberg E, Berlivet S, Beauchemin H, Morcos L, Polychronakos C, Pastinen T, Graham J, McNeney B, Naumova AK. Cell culture-induced aberrant methylation of the imprinted IG DMR in human lymphoblastoid cell lines. Epigenetics Off J DNA Methyl Soc. 2010;5(1):50–60.

Shabalin AA. Matrix eQTL: ultra fast eQTL analysis via large matrix operations. Bioinformatics. 2012;28(10):1353–8. doi:10.1093/bioinformatics/bts163.

Shannon P, Markiel A, Ozier O, Baliga NS, Wang JT, Ramage D, Amin N, Schwikowski B, Ideker T. Cytoscape: a software environment for integrated models of biomolecular interaction networks. Genome Res. 2003;13(11):2498–504. doi:10.1101/gr.1239303.

Smyth GK. Linear models and empirical bayes methods for assessing differential expression in microarray experiments. Stat Appl Genet Mol Biol. 2004;3:Article3. doi:10.2202/1544-6115.1027.

Stegle O, Parts L, Durbin R, Winn J. A Bayesian framework to account for complex non-genetic factors in gene expression levels greatly increases power in eQTL studies. PLoS Comput Biol. 2010;6(5):e1000770. doi:10.1371/journal.pcbi.1000770.

Storey JD, Taylor JE, Siegmund D. Strong control, conservative point estimation and simultaneous conservative consistency of false discovery rates: a unified approach. J R Stat Soc Ser B Stat Methodol. 2004;66(1):187–205. doi:10.1111/j.1467-9868.2004.00439.x.

Sun W. A statistical framework for eQTL mapping using RNA-seq data. Biometrics. 2012;68 (1):1–11. doi:10.1111/j.1541-0420.2011.01654.x.

Sun W, Hu Y. eQTL mapping using RNA-seq data. Stat Biosci. 2013;5(1):198–219. doi:10.1007/s12561-012-9068-3.

Tian L, Greenberg SA, Kong SW, Altschuler J, Kohane IS, Park PJ. Discovering statistically significant pathways in expression profiling studies. Proc Natl Acad Sci U S A. 2005;102 (38):13544–9. doi:10.1073/pnas.0506577102.

Trapnell C, Pachter L, Salzberg SL. TopHat: discovering splice junctions with RNA-Seq. Bioinformatics. 2009;25(9):1105–11. doi:10.1093/bioinformatics/btp120.

Trapnell C, Williams BA, Pertea G, Mortazavi A, Kwan G, van Baren MJ, Salzberg SL, Wold BJ, Pachter L. Transcript assembly and quantification by RNA-Seq reveals unannotated transcripts and isoform switching during cell differentiation. Nat Biotechnol. 2010;28(5):511–15. doi:10.1038/nbt.1621.

Trapnell C, Roberts A, Goff L, Pertea G, Kim D, Kelley DR, Pimentel H, Salzberg SL, Rinn JL, Pachter L. Differential gene and transcript expression analysis of RNA-seq experiments with TopHat and Cufflinks. Nat Protoc. 2012;7(3):562–78.

Tseng GC, Ghosh D, Feingold E. Comprehensive literature review and statistical considerations for microarray meta-analysis. Nucleic Acids Res. 2012;40(9):3785–99. doi:10.1093/nar/gkr1265.

Tusher VG, Tibshirani R, Chu G. Significance analysis of microarrays applied to the ionizing radiation response. Proc Natl Acad Sci U S A. 2001;98(9):5116–21. doi:10.1073/pnas.091062498.

Velculescu VE, Madden SL, Zhang L, Lash AE, Yu J, Rago C, Lal A, Wang CJ, Beaudry GA, Ciriello KM, Cook BP, Dufault MR, Ferguson AT, Gao Y, He TC, Hermeking H, Hiraldo SK, Hwang PM, Lopez MA, Luderer HF, Mathews B, Petroziello JM, Polyak K, Zawel L, Kinzler KW, et al. Analysis of human transcriptomes. Nat Genet. 1999;23(4):387–8. doi:10.1038/70487.

Wang Z, Gerstein M, Snyder M. RNA-Seq: a revolutionary tool for transcriptomics. Nat Rev Genet. 2009;10(1):57–63. doi:10.1038/nrg2484.

Wang K, Singh D, Zeng Z, Coleman SJ, Huang Y, Savich GL, He X, Mieczkowski P, Grimm SA, Perou CM, MacLeod JN, Chiang DY, Prins JF, Liu J. MapSplice: accurate mapping of RNA-seq reads for splice junction discovery. Nucleic Acids Res. 2010a;38(18):e178. doi:10.1093/nar/gkq622.

Wang X, Wu Z, Zhang X. Isoform abundance inference provides a more accurate estimation of gene expression levels in RNA-seq. J Bioinform Comput Biol. 2010b;8 Suppl 1:177–92.

Xu Q, Modrek B, Lee C. Genome-wide detection of tissue-specific alternative splicing in the human transcriptome. Nucleic Acids Res. 2002;30(17):3754–66.

Zenz T, Mertens D, Kuppers R, Dohner H, Stilgenbauer S. From pathogenesis to treatment of chronic lymphocytic leukaemia. Nat Rev Cancer. 2010;10(1):37–50. doi:10.1038/nrc2764.

Zerbino DR, Birney E. Velvet: algorithms for de novo short read assembly using de Bruijn graphs. Genome Res. 2008;18(5):821–9. doi:10.1101/gr.074492.107.

Zhao S, Fung-Leung WP, Bittner A, Ngo K, Liu X. Comparison of RNA-Seq and microarray in transcriptome profiling of activated T cells. PLoS One. 2014;9(1):e78644. doi:10.1371/journal.pone.0078644.

Zhijin Wu RAI, Gentleman R, Martinez-Murillo F, Spencer F. A model-based background adjustment for oligonucleotide expression arrays. J Am Stat Assoc. 2004;99(469):9.

Chapter 4
The Next Generation Sequencing and Applications in Clinical Research

Junbo Duan*, **Xiaoying Fu***, **Jigang Zhang**, **Yu-Ping Wang**, **and Hong-Wen Deng**

Abstract This chapter provides a survey about the next-generation sequencing technologies, as well as four selective applications in clinical researches. After reading Sect. 4.2, we hope the readers to have a general view of the current sequencing technologies in terms of main stream platforms, experimental protocols, data analysis working flow, international projects and databases, state-of-art techniques; by reading the last section, we hope that the readers have a specific view of the clinical applications of next generation sequencing to mutation detection, targeted sequencing, cell free circulating DNA sequencing, and single cell sequencing.

Keywords Next generation sequencing • Mutation detection • Targeted sequencing • Cell free circulating DNA sequencing • Single cell sequencing

---

*Author contributed equally with all other contributors

J. Duan, Ph.D. (✉)
Department of Biomedical Engineering, Xi'an Jiaotong University, Xi'an, China
e-mail: junbo.duan@mail.xjtu.edu.cn

X. Fu • J. Zhang, Ph.D.
Center for Bioinformatics and Genomics, Department of Biostatistics and Bioinformatics, Tulane University, New Orleans, LA 70112, USA
e-mail: xfu@tulane.edu; jzhang9@tulane.edu

Y.-P. Wang, Ph.D.
Center for Bioinformatics and Genomics, Department of Biostatistics and Bioinformatics & Department of Biomedical Engineering, Tulane University, New Orleans, LA, USA
e-mail: wyp@tulane.edu

H.-W. Deng, Ph.D.
Center for Bioinformatics and Genomics, Department of Biostatistics and Bioinformatics, Tulane University School of Public Health and Tropical Medicine, New Orleans, LA, USA
e-mail: hdeng2@tulane.edu

X. Wang et al. (eds.), *Application of Clinical Bioinformatics*, Translational Bioinformatics 11, DOI 10.1007/978-94-017-7543-4_4

4.1 Introduction

The next-generation sequencing (NGS) technology has developed rapidly and spread widely over the past decade. Featured by its high throughput and high genomic resolution, NGS enables us to study the whole genome of human, animal, plant, *etc.* in a much faster and more informative way. Accompanied with rapid development of sequencing technology and platform, demands for more powerful and efficient bioinformatics tools grow in clinical studies.

This chapter provides a survey of NGS technologies and four clinical applications. The first part starts with classical genomic technologies, then introduces mainstream NGS platforms, and experimental protocols of NGS. Afterwards, the analysis working flow of NGS data is presented. Finally, related contents such as the international researches, public repositories, as well as the latest third generation sequencing technologies are introduced briefly.

The second part dedicates to the applications of NGS in clinical researches. Mutation or variation is a very important subject in medical and biological science, so the first application of NGS focuses on the detection of various forms of mutations, including the single nucleotide polymorphism (SNP), insertion and deletion (indel), structural variation (SV) and copy number variation (CNV). To study the function of specific regions of genomes, targeted sequencing is applied in many clinical studies, *e.g.* the exome sequencing, which is subject of the second application. The third application of NGS is the cell free circulating DNA sequencing. In the last application, we show that the NGS can be applied to sequencing the DNA in a single cell.

4.2 The Next Generation Sequencing Technologies

4.2.1 Traditional Genomic Studies

Following Mendel's discovery of the principles of genetics, pioneer geneticists dedicated to figure out the substance basis of heredity. Boveri and Sutton discovered that the chromosomes are those vectors, and afterwards chromosomes came into the view of genetic studies. Early cytogenetic studies employed microscope to observe chromosomes from morphological point of view, namely the karyotype, which includes the number of chromosomes, the length of each chromosome, banding patterns, *etc*. Later on array was employed to detect SNPs, much smaller variants that cannot be observed with cytogenetic methods. But variant detection is not the goal, so genome wide linkage study (GWLS) and genome wide association study (GWAS) were developed to associate genetic variants with phenotypical traits. Nowadays, representative technologies such as fluorescence in situ hybridization (FISH) and comparative genomic hybridization (CGH) are still widely used.

FISH was developed in the 1980s by Langer-Safer et al. to detect and localize the presence or absence of specific DNA segments on chromosomes (Langer-Safer et al. 1982). FISH uses fluorescent probe that highly complementary to the target region of a chromosome, and therefore can bind to the region of interest. Under fluorescence microscope, one can find out where the fluorescent probes are bound to the target region. Furthermore, when multiple color fluorescent probes are used, more regions can be dyed at the same time. And by analyzing the combination of color channels, one can study several target regions. This technique is called multicolor FISH (M-FISH) (Speicher et al. 1996), which is often used to detect translocation, a form of structural variation caused by rearrangement of segments between chromosomes. However, since it is limited by the resolution of the microscope used, the genomic resolution of FISH is usually rather low, which limits the utility of FISH in state-of-art scientific research.

CGH is another cytogenetic method. In a CGH experiment, the DNA of both test and control samples are differentially labeled, and simultaneously hybridized to a reference. If the test sample harbors an unbalanced structural variation, *e.g*. CNV, the comparative hybridization will be differential, causing the ratio between the densities of the two fluorescence deviated from one, which can be detected with statistic tools. CGH was originally developed to study the variations of tumor verse normal control tissue, where CNVs almost always involve. Compared to G-binding and FISH, CGH has an improved resolution of 5–10 mega bases. Combined with microarray techniques, the array CGH (aCGH) was developed to increase the genomic resolution of variation detection as well as the throughput of whole genome analysis. It was reported that the resolution of aCGH can reach up to 200 bp (Urban et al. 2006). To reach the ultimate resolution, *i.e*. base-pair level, sequencing is needed.

4.2.2 The Emergence of Next Generation Sequencing

After Watson and Crick first discovered the double helix structure of DNA molecular, many biological and medical researchers want to know the precise physical order of nucleobases in a DNA molecule, whose determination is called DNA sequencing.

In 1970s Frederick Sanger and colleagues developed the well-known Sanger sequencing, which is characterized by the chain-termination method. By mixing normal dNTPs with modified ddNTPs, DNA strand elongation terminates randomly when a ddNTP is incorporated. The ddNTPs are labeled with radiation or fluorescence and can be detected by sensors. After combining all random termination loci, the order of nucleobases of a DNA segment can be induced. Later on automation and capillary electrophoresis greatly improved the efficiency and reduced the cost, known as the first generation of DNA sequencing.

The Sanger sequencing is widely used in scientific studies for almost 30 years because of its reliability. But high price and low throughput (see Table 4.1) hamper its further application. Promoted by the rapid evolution of modern techniques, NGS emerged and enabled large scale studies in the last decade. However, nowadays the Sanger method is still used in small scale studies and in the situation when reliable sequencing and long contiguous reads are needed.

The NGS is characterized by its low cost and high throughput (see Table 4.1), which is advanced greatly by outstanding sequencing companies such as Illumina, Roche, Life Technologies, *etc*.

Table 4.1 Comparison of sequencing platforms

Company	Illumina	Roche	Life technologies	Life technologies
Platform	HiSeq	454	SOLiD	Sanger
Read length	50 bp	700 bp	50 bp	400–900 bp
Output per run	600 Gbp	1 Gb	100 Gbp	1–100 Kbp
Time per run	3 days to 1 week	1 day	1 week	1–3 h
Price per Mbp (USD)	0.07	10	0.13	2400

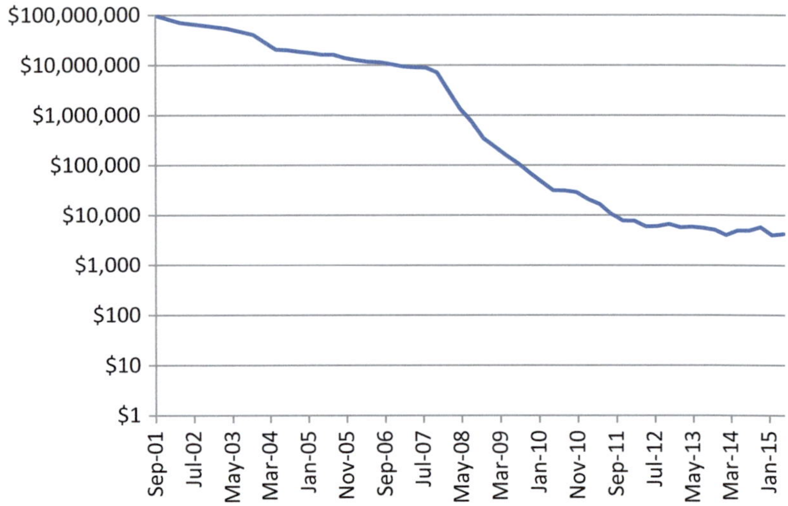

Fig. 4.1 Total cost of sequencing a human genome estimated by the NHGRI (http://www.genome.gov/sequencingcosts/)

The Illumina and Life Technologies platforms have high throughput, but short read length degenerates assembly accuracy. The Roche platform has long read length and fast sequencing, but the sequencing price is relatively high (Liu et al. 2012). Figure 4.1 displays the cost of a human genome in the last decade, which shows that the cost decreases dramatically from 100 million dollars per genome in 2001 to less than ten thousands nowadays.

The main features of mainstream NGS sequencing platforms are summarized in Table 4.1. Even though these platforms utilize different sequencing method, *e.g.* Pyrosequencing by Roche 454, Sequencing by Oligonucleotide Ligation and Detection (SOLiD) by Life Technologies, they follow similar protocol presented in the next subsection.

4.2.3 Protocol of Next Generation Sequencing

The quality of sample preparation influents the quality of sequencing significantly. NGS library preparation is the procedure to generate the library for sequencing from the DNA sample. According to different study designs, sequencing platforms, and/or preparation tool kits, this procedure may be different, but in general it follows similar procedure, that is: (1) DNA fragmentation: shearing DNA either mechanically or by enzymatic digestion. (2) Gel-base selection: filter the target fragment by size. (3) DNA fragment end-repair: filling in or removing the protruding $3'$ and $5'$ ends. (4) $3'$ ends a-tailing: adding 'A' base to facilitate ligation. (5) Ligation with platform-specific adapters. Since upstream adapters (called A) and downstream adapters (called B) are required for each fragment, unwanted fragments with pattern A-A and B-B are filtered out in the final library.

In the case of targeted sequencing, library preparation requires target enrichment. Target enrichment is classified into hybrid capture method and PCR-based amplicon enrichment. The step of hybridization targeted regions or genes is mostly followed by step (5) in the previous passage, while PCR amplicon enrichment is usually performed before DNA is sheared (Linnarsson 2010). The application of these two approaches will be addressed in more details in Sect. 4.3.2.

PCR library amplification, with the purpose of enhancing detectable sequencing signal, is the last step before sequencing. This stage has already being integrated into multiple sequencing platforms. Emulsion PCR (emPCR) and solid-phase PCR (also called Bridge PCR) are commonly used. In emPCR, DNA fragments with adaptors are amplified within the water-in-oil emulsion soon after each droplet encapsulating primer-attached bead along with a single DNA fragment. EmPCR is widely used in Roche 454, SOLiD 3, Polonator and Ion Torrent platforms. In solid-phase PCR, fragments are amplified upon $3'$ and $5'$ primers that are coated on a flat surface, forming DNA clusters. This method is adapted in Illumina platforms (Metzker 2010). After library amplification, the platforms will pool samples for multiplexing sequencing. Here, multiplexing sequencing refers to sequencing multiple sample libraries simultaneously during a single run. Unique molecular tags consisting of three or more base pairs are used as barcodes to distinguish samples. These barcodes can be attached to the fragments either as part of the adapters in ligation or as part of primers in PCR amplification.

Due to the fact that PCR is sensitive with GC content in the genome, bias is usually raised when PCR is largely involved in the above process (Head et al. 2014). Lots efforts have been made to minimize these bias, such that applying novel DNA polymerases that can reduce PCR error and developing PCR-free techniques. What's more, the fully automated library preparation devices with reliable performance have been marketed to reduce the burden of labor work to the greatest extent (*e.g.* Illumina's NeoPrep Library Prep System).

When library is ready, sequencer is used to determine the physical order of nucleobases ('A', 'T', 'C' or 'G') in one end or both ends of DNA fragments, which is called *single-end (SE)* or *paired-end* (*PE,* or mate-pair) sequencing, respectively. Here determine is also called *base-calling*, which assigns a score to each nucleobase type based on the measured signal, and chooses the nucleobase type with the largest score. For example, *Phred* is a mature software that performs base-calling and assigns an error probability to each called base.

Phred quality score was firstly proposed by the software *Phred* (Ewing et al. 1998; Ewing and Green 1998), and now is widely used to characterize the sequence quality. A Phred quality score Q is defined as

$$Q = -10log_{10}P$$

where P is the probability that base-calling is wrong. So the higher the score Q is, the less likely the base-calling is incorrect.

The string of called nucleobases of one end of a DNA segment is call a read, whose length is an important indicator for a sequencing platforms. Longer read length is always favorable, but at the cost of increased price and decreased accuracy. As is shown in Table 4.1, Sanger sequencing has longer read length and higher accuracy, but the cost is much more expensive.

Researchers use coverage C to describe the sequencing amount, which is defined as

$$C = \frac{LN}{G},$$

where L is the average read length, N is the total number of reads, and G is the genome length. From the definition, one can see that coverage is actually the average number of times the sequencing reads can "cover" each base of the genome. A higher sequencing coverage indicates each base is covered by more reads, therefore have higher degree of confidence. As a result, sequencing coverage varies by application, depending on the trade-off between budget and required degree of confidence.

Compared with traditional sequencing method, one advantage of NGS is the high throughput. The output data set of tens or hundreds of gigabytes brings huge challenges to statistics and computer science. In order to store such high volume data set, *FASTA* format was proposed and now widely used as the standard file format in bioinformatics. *FASTA* is a plain text file in which each nucleotide is represented with an English letter ('A', 'T', 'C', 'G', see (Tao) for the meaning of each letter). A FASTA file consists of one or more sequences, each includes two parts. The first part is called the header, which occupies one line and starts with a greater-than sign (">"). The word following ">" is the identifier of the sequence, and the rest of the line is the descriptor which is optional. The second part is the data body, which usually consists of several lines, and each line should not exceed 80 characters. The data body continues until the next ">" appears, indicating another sequence. A toy example of a FASTA format file containing three sequences looks like:

```
>r45640315
CAGAAAGCTCATGTGACTTCTAACTAGAATTTTCAA
>r45640316
ACCCTTCCAGACATACTTTTAAGAGAACTGACAGTT
```

>r45640317
ACTGGTTGAGCTAGATTACAGGTCTGGGTGGTGCCA

4.2.4 Sequencing Data Analysis Working Flow

There are a series of steps to analyze NGS data. The first step of most NGS projects is the quality control, which is very important before further analysis. The second step is read alignment or *de novo* assembly, which is followed by normalization, variant calling, and finally variant annotation.

Quality control aims to preprocess the data, check the integrity of the data file, and sequencing quality, and library preparation problems such as potential artifact, contamination or overrepresentation. Common used software are Trimmomatic (Bolger et al. 2014) and FastQC (Andrews 2015).

Trimmomatic provides several useful processing for both paired-end and single-end data. The processing steps of Trimmomatic includes: (1) cut adapter and other sequences from the read; (2) cut sliding window if the average quality within the window falls below a threshold; (3) cut the specified number of bases from the start of the read, or cut bases off the start or end of a read if below a threshold quality; (4) cut reads to a specified length; (5) convert quality scores to Phred-33 or Phred-64.

FastQC provides a simple way to check the quality of the sequencing data. Without much expertise, one can have an overview of the quality of the data, and decide whether go to further analysis, or pay attention to the quality of the data. With FastQC, one can import data directly from SAM or BAM file, have a quick and automated analysis showing which region maybe problematic, summarize and visualize the statistics of the data, and export reports to an HTML file.

De novo assembly reconstructs a new genome from short reads by using the overlaps between reads. When we study the genome of human or model organisms, the short reads can be aligned to the reference genome which is already known after predecessors' efforts. However, when we investigate a new specie, or a specific individual, the reference genome is unknown, so *de novo* assembly is needed. *De novo* is the Latin expression meaning “from the beginning”.

Many *de novo* software have been developed, such as Velvet, ABySS, SOAPdenovo, *etc*. Table 4.2 gives a list of *de novo* tools. Systematical analysis of their relative performance under various conditions can be found in (Lin et al. 2011; Earl et al. 2011).

Alignment is also called mapping, which aligns sequencing reads back to a given reference genome. For short reads aligners, speed and accuracy are two main concerns for users. As is shown in Table 4.1, NGS platforms output gigabytes data per day, so the alignment speed of such huge volume data is challenging.

Since alignment impacts the downstream analysis, the accuracy of alignment is essential to NGS projects. Due to the short read length and repetitive regions in the reference genome, there are reads that cannot be aligned uniquely, *i.e.* aligned to multiple loci. Furthermore, sequencing error, genetic variations such as SNPs,

Table 4.2 A selection of *de novo* software

Software	Reference and URL
Velvet	Zerbino and Birney (2008), http://www.ebi.ac.uk/~zerbino/velvet/
ABySS	Simpson et al. (2009), http://www.bcgsc.ca/platform/bioinfo/software/abyss
SOAPdenovo	Li et al. (2009), http://soap.genomics.org.cn/soapdenovo.html
SSAKE	Warren et al. (2007), http://www.bcgsc.ca/platform/bioinfo/software/ssake
Edena	Hernandez et al. (2008), http://www.genomic.ch/edena.php
SHARCGS	Dohm et al. (2007), http://sharcgs.molgen.mpg.de/
Euler-sr	Chaisson and Pevzner (2008), http://euler-assembler.ucsd.edu
Celera WGA Assembler	Miller et al. (2008), http://wgs-assembler.sourceforge.net/wiki/index.php?title=Main_Page

indels, SVs and CNVs complicate the alignment, yielding some reads cannot be aligned to any locus of a reference genome. Compared with single-end reads, alignment of paired-end reads is more complicated: two ends may be aligned to two loci but with significantly differential span length than the average of library length, or one of the pair cannot be aligned, so the alignment accuracy is also challenging.

Traditional alignment software such as BLAST (Ye et al. 2006) were designed for long reads, therefore several alignment tools have been proposed recently (Li and Homer 2010; Magi et al. 2010) for short reads. Since different characteristics (Magi et al. 2010) are utilized, those tools have their own advantages and disadvantages. In general, most tools need to build an index dictionary from the reference genome in advance for fast alignment. According to the property of the index, these tools can be grouped into three classes: hash table based tools, suffix tree based tools, and merge sorting based tools (Li and Homer 2010). Table 4.3 summarizes a selection of short reads alignment software. A more complete list can be found at http://www.ebi.ac.uk/~nf/hts_mappers/ and (Magi et al. 2010).

After short reads having been aligned to a reference genome, signature for different applications can be extracted. In the following, we briefly introduce the read depth signature as an example for copy number variation (CNV) detection, which will be expanded in Sect. 4.3.1.

Read depth is a well-known signature proposed to detect copy number variation (Chiang et al. 2009). The read depth is the count of aligned reads (normally consider the first base of $3'$ end) or bases in a genomic region (usually called a window). Since the loci of shot-gun sequencing distribute randomly and evenly along the genome, so ideally the read depth signal is overall a horizontal line but with local fluctuation. However, if a genomic segment harbors a CNV gain/loss, the count of aligned reads within the CNV region shall significantly increase/decrease, yielding a plateau/valley in the corresponding read depth signal. Therefore, read depth signal can be used as the signature to detect CNV.

Table 4.3 A selection of short read alignment software

Software	Platform	Speed (per day)	Reference and URL
SSAHA2	Illumina, SOLiD, 454	0.5 Gbp	Ning et al. (2001), http://www.sanger.ac.uk/resources/software/ssaha2/
MAQ	Illumina and SOLiD	0.2 Gbp	Li et al. (2008), http://maq.sourceforge.net/
SOAP2	Illumina	7 Gbp	Li et al. (2009), http://www.sanger.ac.uk/resources/software/ssaha2/
Bowtie	Illumina	7 Gbp	Langmead et al. (2009), http://bowtiebio.sourceforge.net/index.shtml
BWA	Illumina, SOLiD, 454	7 Gbp	Li et al. (2010), http://biobwa.sourceforge.net/bwa.shtml

Normalization is important before variant calling. For read depth signal, mappability and GC-content are two main biases that affects CNV detection, and therefore corresponding normalizations are needed.

Mappability is measured as a score that quantify the uniqueness of the reference genome. This score is defined as the inverse of frequency $f_k(x)$, which is the number of times the k-mer starting at locus x appears in the genome and its reverse complement (Liu et al. 2012; Derrien et al. 2012). As a result, mappability ranges between zero and one. *e.g.* One represents a unique alignment and 0.5 represents that the k-mer occurs twice, *etc*. To correct the mappability bias, the read depth is divided by the score to compensate the nonuniqueness (Miller et al. 2011).

GC-content is the percentage of bases G and C in the underlying sequence. Bentley et al. (2008) first observed that the read depth is locally correlated with GC-content. Therefore, this bias should be removed to increase the homogeneity of the read depth. Yoon et al. (2009) and Abyzov et al. (2011) proposed to utilize following method

$$RD^{\mathrm{i}}_{\mathrm{corrected}} = \frac{\overline{RD}_{global}}{\overline{RD}_{gc}} RD^{i}_{raw}$$

where i is the index of windows, $RD^{\mathrm{i}}_{\mathrm{corrected}}$ and RD^{i}_{raw} are corrected and raw read depth of the ith window respectively, $\overline{RD}_{global}$ is the global average read depth of all windows, and $\overline{RD}_{gc}$ is the average read depth of all windows that with the same GC-content as in the *i*th window.

Afterwards, segmentation tools can be used to cut the read depth signal into pieces, and CNV regions can be called as the pieces with significantly high or low read depth compared with normal one. CNVs that do not have sufficient confidence are filtered out, and the rest will be annotated. In Sect. 4.3.1, we will give a detailed survey of CNV detection methods.

After variants being detected, attaching annotation information such as gene symbol, exonic function and base pair change to the variant would be helpful for further studies. There are lots of information that could be annotated, such as CpG islands, MAF, enhancers, transcription factor binding sites, *etc*. Functional score is an efficient way in mutation function prediction. *e.g*. the Combined Annotation Dependent Depletion (CADD) database uses score to measure the potential deleteriousness mutations by comparing variants survived through natural selection. The software ANNOVAR, together with annotation function integrated in GATK and SnpEff, are able to carry out an integrated annotation.

If a NGS project is performed with very high fold coverage, a large amount of single nucleotide variants (SNV) would be detected in initial process. Therefore it is necessary to reduce candidate variants by filtering. The steps of variant filtering usually contains: (1) remove variant calls with low quality; (2) remove common polymorphisms; (3) prioritize variants with high functional impact; (4) compare with known disease genes; (5) consider mode of inheritance; and (6) consider the segregation in family (Bao et al. 2014).

4.2.5 International Researches and Public Repositories

There are several international research projects and public repositories that aim to share the data and discoveries about the next generation sequencing. In the follows we select a few among them.

The 1000 Genomes Project is a well-known NGS project (http://www.1000genomes.org/home), as declared by their website, "The 1000 Genomes Project is an international collaboration to produce an extensive public catalog of human genetic variation, including SNPs and structural variants, and their haplotype contexts. This resource will support genome-wide association studies and other medical research studies." This project aims to sequence the genomes of about 2500 people from about 25 populations around the world, using multiple mainstream sequencing platforms from multiple sequencing centers, and make the data of the project freely and publicly accessible to researchers worldwide.

The project is divided into the initial phase (or the pilot project), phase I, II, and III. The pilot project consists of three subprojects (The 1000 Genomes Project Consortium 2010 and 2012): the high-coverage trio subproject, the low-coverage subproject and the exome subproject. The high-coverage trio subproject sequenced the whole genome of a Yoruba Ibadan (YRI) trio and a European ancestry in Utah (CEU) trio with high coverage (42×) and multiple platforms. Each trio consists father, mother and their daughter. The low-coverage subproject sequenced the whole genomes of 59 YRI subjects, 60 CEU subjects, 30 Han Chinese subjects in Beijing (CHB) and 30 Japanese subjects in Tokyo (JPT), with low coverage (2–6×). The exome subproject sequenced the 8140 exonic regions from 906 randomly selected genes of 697 subjects from seven populations with high coverage (50×). This project was completed in 2009. Phase I sequenced and analyzed the low-coverage and exome data from the first

1092 subjects, phase II increased the data set to around 1700 subjects, and phase III even increased this number to 2500.

Following the 1000 Genomes Project, there are also the 1000 Plant Genomes Project (http://sites.google.com/a/ualberta.ca/onekp/), which aims to generate the sequencing data of over 1000 species of plants; the Genome 10 K Project (http://genome10k.soe.ucsc.edu/), which aims to generate the whole genome sequence of 10,000 vertebrate species; the 1001 Genomes Project (http://1001genomes.org/index.html), which aims to sequence the whole-genome in 1001 strains of the model plant Arabidopsis thaliana.

The dbGaP is another remarkable project. As their website (http://www.ncbi.nlm.nih.gov/gap) declares, "the database of Genotypes and Phenotypes (dbGaP) was developed to archive and distribute the results of studies that have investigated the interaction of genotype and phenotype."

The dbGaP includes four major data types (Mailman et al. 2007). First, the study documents including study design, experimental protocol, platform specification, and/or questionnaires; second, the genotype data such as SNP array data, gene expression data, NGS data, *etc.*; third, the phenotype data such as tables of individual trait data, pedigree information; and forth, the analysis results and conclusions, such as the phenotype genotype association results of GWAS, linkage analyses, or meta analysis.

In order to protect the confidentiality of study subjects, the dbGaP offers two ways to access the data: open access and controlled access. Without any required permission, anyone can access and download summarized phenotype and genotype data, study documents and results. To access individual-level phenotype and genotype, researchers are required to apply for approval.

The International HapMap Project (http://hapmap.ncbi.nlm.nih.gov/) aims to generate a haplotype map (HapMap) of the human genome. By sharing the generated data freely with researchers worldwide, HapMap hopes to find information of genetic variations across human population that may associated with diseases, including the genotypes, haplotypes, SNPs, CNVs, and their frequencies. This project is an international collaboration among scientists from Canada, China, Japan, Nigeria, United Kingdom, and the United States. There are 270 human subjects were involved in phase I and II of this project, including 30 YRI trios, 30 CEU trios, 45 JPY, and 45 CHB. In the phase III, more than 1000 samples from 11 population were analyzed.

Nowadays, cloud-based NGS data storage and computation is becoming popular. In fact, about 200 TB of data from 1000 genome is currently stored in Amazon Simple Storage Service (Amazon S3) which is a cloud-based file system and charged by usage. Synapse from Amazon and Galaxy are two cloud-based workflow tools that can handle raw NGS data all the way through annotation and filtering variants, with personal optimization of algorithms available in each step. Overall, the cloud-based storage and computation can greatly reduce the computational burden when dealing with the big data.

4.2.6 Latest Sequencing Techniques

While NGS continuously dominates the sequencing market with its affordable price and feasible application, the third generation sequencing (TGS) is under rapid evolution with the promise to sequencing longer reads, shorter time and lower cost. Single molecular sequencing (SMS) is the technique adapted in TGS. SMS doesn't require PCR library amplification, which taking great advantage over NGS by avoiding PCR-induced bias. Moreover, DNA modification signals such as DNA methylation, would not be diluted out during PCR and thus could be captured if the detection technology available. Nevertheless, an inevitable disadvantage of SMS is high raw read error rate, induced by weak signal generated from single molecular. Detection these signal from background noise is the biggest challenge in SMS technology development.

The first commercially available SMS is Helicos Genetic Analysis Platform developed by Helicos BioSciences, which profiled the first human genome on 2009 (Pushkarev et al. 2009). However, this pioneer SMS platform can only generate similar read length compared with NGS. Suffering from higher error rate introduced by SMS and expensive sequencing cost (Schadt et al. 2010), the company announced 50 % layoff on 2010 and went bankrupt on 2012. Learning from Helicos' failure, the nowadays TGS are better tailored for the market. For the time being, the most representative TGS platforms are PacBio RS II and MinION, utilizing single-molecule real-time (SMRT) approach and nanopore sequencing technology respectively. The two platforms both perform the real-time sequencing and store their time-series raw imaging data. Adding the time axis makes it possible to predict DNA structure variance from the data, since the speed of sequencing is impacted by variety DNA structures. On the contrary, data analysis would surely be more complicated and time consuming when new dimensions are added (Liu et al. 2012).

SMRT, the sequencing system developed by Pacific Biosciences (PacBio), become commercially available since 2011. Zero-mode waveguides (ZMWs) is the special fluorescence-based sequencing technique innovated for this system. As a tiny well with a 10 nm diameter hole in the bottom, ZMW allows light to illuminate only the bottom of the well where DNA polymerase and the template complex is immobilized. At the same time, ZMW only recruits the light emitted from fluorescence when sequencing, to the sequencing signal can be observed and recorded. Tens of thousands of ZMWs are contained in a SMRT cell. Furthermore, instead of labeling the base with fluorescence, the dye in SMRT system is attached to the phosphate and thus would be naturally cleaved off, which avoids the problem that large size of dye can hinder activity of the DNA polymerase. These two key innovations enable SMRT to generate long read lengths in hours, however, with a high raw read error rate (around 15 %). Fortunately, compared with NGS whose error is produced from the system, SMRT generates error stochastically. It means that the error from SMRT can be remarkably reduced by averaging the reads if the same DNA is sequenced multiple times (Roberts et al. 2013). PacBio RS II is the

2nd generation of the PacBio sequencing adapted SMRT technology. Upon improvement, continuous long reads (CLRs) and circular consensus (CCS) are supplied as two additional modes in the system. CLR, with its application focusing on *de novo* assembly, is able to achieve read lengths averaged 10–15 kb; while in CCS, focusing on targeted sequencing, is capable to reduce error rate down to 2 % with short read lengths maximized at six kb (Travers et al. 2010). Besides the merit of long read lengths, PacBio is capable to predict nucleotide modification with direct measurement of kinetic, detect minor variants with high sensitivity, finish a run as little as 30 min and not be biased by GC contents or amplification. However, as a cost of these benefits, the yielding from PacBio RS II is only 1 Gb per run.

The applications of PacBio focus on genome assembly improvement, annotation gap filling, difficult regions (*e.g.* GC rich and high repetitive regions) profiling and specified SV detection. It was reported that, besides extend 55 % of the interstitial gaps in human reference genome, specified SV, such as inversions complex insertions and long tracts of tandem repeats, were detected in base-pair resolution through PacBio sequencing (Chaisson et al. 2015). The first comprehensive analysis of diploid human genome through PacBio sequencing, combined with the technology of single-molecule mapping, had gained markedly improve contiguity and completeness, compared with traditional shotgun methods. This analysis also identified complex SVs ignored by NGS and generated high quality haplotypes when integrated PacBio sequencing with NGS (Pendleton et al. 2015). Moreover, in another biomedical study, known low copy repeats (LCR) that causing Potocki–Lupski syndrome (PTLS) were confirmed with PacBio, while several novel repetitive sequences junctions in related to PTLS were discovered (Wang et al. 2015). As a matter of fact, Pacific Biosciences is already in cooperation with Roche Diagnostics to develop diagnostic products for clinical based on SMRT technology (Shen et al. 2015).

Library preparation in PacBio sequencing is similar to NGS. Although PacBio sequencing is new to the industry, multiple kits and protocols for template preparation are already available, together with globally located vendors. Meanwhile, PacBio provides an open source software SMRT Analysis for their sequencing data, which integrated multiple algorithms and pipelines towards different study applications. An alternative way, running SMRT analysis on Amazon can be accessed through SMRT Portal. SMRT View, serving as a whole genome browser to visualize secondary analysis data, is also available both in standalone version and Amazon version. The documentation for sample preparation in details can be found through their page of support, while open source software along with sample data and tutorials are available from DevNet (http://www.pacb.com/devnet). Although the price for PacBio sequencing is high, it can be used as the supplemental tool in NGS sequencing at this moment. PacBio is still working on improving sample preparation and sequencing chemistry, most importantly, increasing sequencing throughout. Whereas PacBio sequencing won't dominate the sequencing market as Illumina in the few years, its application is expected to have large impact in biomedical research and clinical diagnosis.

MinION, a pocket-size sequencing device, was released by Oxford Nanopore Technologies on November, 2013 based on the nanopore sequencing technology (Feng et al. 2015). The MinION Access Programme (MAP) was started from spring, 2014 to allow researchers testing the device in advance. Nanopore is a nano-scale hole which can be fabricated from protein or synthetic materials. Immersing nanopore into a conducting fluid, when fluid is applied with voltage, electric current will be generated along the nanopore. At this point, passing a single base of DNA through the nanopore would disrupt the current, leaving the signal of reads. Nanopore sequencing is the technology of recording signals generated from single strand DNA when passing through a nanopore. MinION attracted attention quickly with the advantages of small size, low cost, simple library preparation and acquisition data in real time (Madoui et al. 2015). The average read lengths from MinION is 5.4 kb on average, which is longer than regular NGS sequencing but not comparable with PacBio sequencing (Feng et al. 2015). However, MinION suffered from severe error rate and was difficult to be mapped with reference genome. To solve this problem, a two-dimension read (2D read) was later introduced into the device. In the 2D read mode, the two strands of a DNA are linked by a hairpin and sequenced consecutively. Nevertheless, even with 2D read, the accuracy reported from a bacterial genome study didn't exceed 72 % (Ashton et al. 2015). With the fact that the signal from nanopore sequencing is complicated to be interpreted, novel read correction and new alignment algorithms are continuously being proposed (Madoui et al. 2015; Jain et al. 2015; Karamitros and 2015). Among them, similar with the case in PacBio sequencing, integration sequencing data from MinION and Illumina Mi-Seq was proposed, with a study achieving high accuracy in bacterial genome (Madoui et al. 2015). On the other hand, some studies proved that with new algorithms, signals generated from SNV can be efficiently identified from MinION data (Jain et al. 2015). Together with low throughput, Oxford nanopore has made great progress by bringing novel technology into sequencing, but still there's a long way to go towards clinical application.

Beyond the two representative technologies in the industry, there are several other technologies in TGS, for example, real-time DNA sequencing using fluorescence resonance energy transfer, tunneling and transmission-electron-microscopy-based approaches for DNA sequencing, direct imaging of DNA sequence using scanning tunneling microscope tips and transistor-mediated DNA sequencing (Schadt et al. 2010). These on-going development of sequencing systems provide a high potential for us to better understand DNA sequence with novel signals. Nevertheless, one should keep in mind that, sequencing for clinical purpose should be fast and accurate. Thus, at present, PacBio sequencing is leading TGS market with achievement of long reads, fast sequencing and correctable error. Concurrently, MinION aims gaining market by lower price and minimize size. To sum up, the latest sequencing techniques, especially the PacBio sequencing, have been applied in different files of DNA sequencing research and yield significant improvement in *de novo* assembly, SNV detection and deep sequencing. For now, combing TGS in NGS analysis maybe the best solution to produce DNA sequencing with high resolution, high accuracy and reasonable price.

4.3 The Applications in Clinical Researches

4.3.1 Mutation Detection

A very important application of NGS is to detect genomic mutations or variations. There are several types of genomic mutation, and we will cover the following ones: single nucleotide polymorphism (SNP), insertion and deletion (indel), structural variation (SV) and copy number variation (CNV), which are sorted according to their size from small to large.

1. Single nucleotide polymorphism (SNP)

SNP is the smallest genomic mutation, which consists of only one nucleotide. For example, if a test DNA segment is ATCCGCTA, and the corresponding reference segment is ATCGGCTA, then there is a SNP. The SNP database (dbSNP, http://www.ncbi.nlm.nih.gov/projects/SNP/) of NCBI archives about 150 million, and among which, hundreds have been associated with Mendelian or complex diseases.

The detection of SNPs from NGS data is not complicated. During the alignment step of the NGS data analysis working flow, *i.e.* aligning short reads of a test genome to a reference, if most reads that align at nearby region share a mismatch at the same locus (see following demonstration, A with underlines), then it is likely that there is a SNP at this locus of the test genome. The following shows a SNP with "A" in the test genome compared with "C" in the reference genome.

```
REFERENCE   ......ATCGAATTCCGGAAATTTCCCGGGAAAATTTTCCCCGGGG......
READ1        ......ATTCCGGAAATTTCCAGGGAAAATTTTCCCCGGGG......
READ2           ......GGAAATTTCCAGGGAAAATTTTCCCCGGGG......
READ3              ......TTTCCAGGGAAAATTTTCCCCGGGG......
```

2. Insertion and deletion (indel)

The word "indel" is coined by combining both *in*sertion and *del*etion, so an indel is the genomic mutation in the forms of insertion, deletion, or combination of them. The size of indels varies from one base to several hundreds. Indels with size larger than 1 kbp are often considered as structure variation, which will be introduced later. Here we only refer to these with small size. Following shows a deletion of "CCG" in the test genome.

```
REFERENCE   ......ATCGAATTCCGGAAATTTCCCGGGAAAATTTTCCCCGGGG......
READ1        ......ATTCCGGAAATTTCGGAAAATTTTCCCCGGGG......
READ2           ......GGAAATTTCGGAAAATTTTCCCCGGGG......
READ3              ......TTTCGGAAAATTTTCCCCGGGG......
```

The detection of indels is more complicated compared with that of SNPs. During the alignment step of the NGS data analysis working flow, indels can be detected by those short reads that cover the region with indel. Since those short reads that cover an indel cannot be aligned to the reference genome, they will be labeled with unmapped reads. If we split these unmapped read into two or three parts with all possibility, and align both ends to maximize the similarity measure, then indels with small size can be captured. Algorithms such as Smith–Waterman algorithm (Smith and Waterman 1981) can be employed to analyze unmapped reads, and if two or more unmapped reads confirm an indel uniformly, then it is likely that there is an indel in the test genome.

3. Structural variation (SV)

SV is the variation in structure of an organism's genome, which has been associated with several human diseases (Stankiewicz and Lupski 2010). SV consists of the following kinds of variations:

(i) Deletion: a segment is removed from a chromosome.
(ii) Duplication: a segment is duplicated to a chromosome.
(iii) Insertion: a segment is added into a chromosome.
(iv) Inversion: a segment is broken off from a chromosome, inverted, and reattached back to the break points.
(v) Translocation: two segments from two chromosomes are exchanged.

Differ from indel, SV usually is referred with more than one kbp insertion or deletion. SV that does not change the size of a genome is called balanced SV, such as an inversion, and that does change is called unbalanced SV, such as a deletion, duplication, or insertion.

Since the size of a SV is larger than the size of sequencing reads, no single read can cover the whole range of a SV (unless a deletion type), therefore the methodology of detecting SV is different from that of detecting indels.

There are several signatures to detect SVs (Zhao et al. 2013; Korbel et al. 2007): paired-end mapping (PEM), read depth (RD), split read (SR), *de novo* assembly of a genome (AS). Table 4.4 gives a list of PEM-based, SR-based, and AS-based SV detection tools. Since each single signature has its own advantage and disadvantage, combination of multiple signatures yields better detection performance. In the follows, we focus on AS, PEM and SR. RD is mainly used for detecting CNV, which is an important subtype of SV, and will be introduced in a separate part.

The idea of AS-based methods is quite simple: first assemble contigs of the test genome with *de novo* assembly tools, or with the guidance of the reference genome. When the test genome is known, it's straightforward to detect variations by comparing the test genome with the reference genome. For AS-based methods, high sequencing coverage is needed to assemble the contigs with high confidence.

Figure 4.2 shows the PEM signature of insertion, deletion and inversion. When a pair of reads is aligned back to the reference genome, if the span of the pair from the test genome is larger/shorter than a specified cutoff, a deletion/insertion could be identified. For inversion, from Fig. 4.2 we can see that the orientation of one end

Table 4.4 A selection of SV detection software

Software	Language	Input format	Reference and URL
BreakDancer	Perl, C++	Alignment files	Chen et al. (2009), http://breakdancer.sourceforge.net/
PEMer	Perl, Python	FASTA	Korbel et al. (2009), http://sv.gersteinlab.org/pemer/
GASV	Java	BAM	Sindi et al. (2009), http://code.google.com/p/gasv/
Pindel	C++	BAM / FASTQ	Ye et al. (2009), http://www.ebi.ac.uk/~kye/pindel/
SLOPE	C++	SAM/ FASTQ/MAQ	Abel et al. (2010), http://www-genepi.med.utah.edu/suppl/SLOPE
VariationHunter	C	DIVET	Hormozdiari et al. (2010), http://compbio.cs.sfu.ca/strvar.htm
commonLAW	C++	Alignment files	Hormozdiari et al. (2011), http://compbio.cs.sfu.ca/strvar.htm
AGE	C++	FASTA	Abyzov et al. (2011), http://sv.gersteinlab.org/age
Magnolya	Python	FASTA	Nijkamp et al. (2012), http://sourceforge.net/projects/magnolya/
Cortex assembler	C	FASTQ/ FASTA	Iqbal et al. (2012), http://cortexassembler.sourceforge.net/

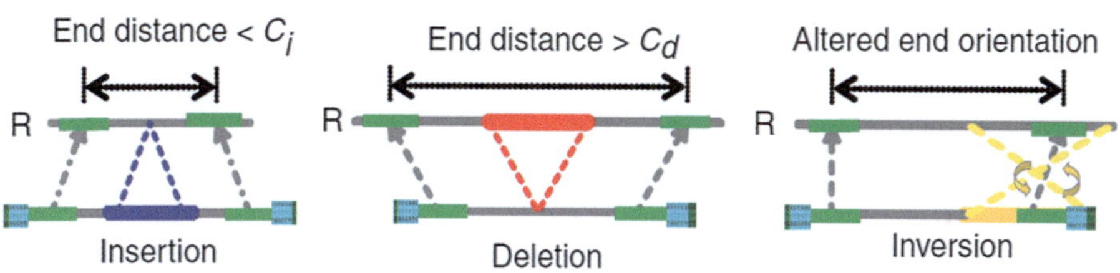

Fig. 4.2 Signatures for insertion, deletion and inversion

inverts, and the other does not. These are the basic signatures to identify SVs. To detect complicated SVs such as translocations, combination of these basic signatures and more sophisticated signatures are needed.

SR signature provides precise break point information (up to base pair level). If one read of the pair aligned uniquely to the reference genome, while the other fails to align, this pair may be informative. By splitting the unmapped read into two fragments, and either fragment can be aligned to nearby regions, then the precise break-point of a SV can be detected. The effectiveness of SR heavily relies on the read length, so is preferable for 454 platform.

4. Copy number variation (CNV)

CNV is commonly referred to as a subtype of SV, and involves a duplication or deletion of DNA segment of size more than 1 kbp (Freeman et al. 2006). CNV was reported to be discovered frequently in both human and other mammal genomes. Several diseases have been associated with CNVs, such as autism (Sebat et al. 2007), schizophrenia (Stefansson et al. 2008), cancer (Campbell et al. 2008), Alzheimer disease (Rovelet-Lecrux et al. 2006), osteoporosis (Yang et al. 2008) and *etc*.

The majority of CNV detection tools are RD-based. The methodology is based on the assumption that copy number is locally proportional to the read depth signal, which is the count of aligned reads in a non-overlapping fix-sized window or a sliding window (also see Sect. 4.2.4). Therefore, a significant increase or decrease of the read depth signal indicates a duplication or deletion event.

Compared with PEM and SR signatures, which are good at break-point localization and detection of small size events, RD signature is more appropriate to estimate the copy numbers, and events with large size.

Since CNVs have been associated with several diseases, the detection of CNVs from NGS is hot topic, and several methods have been proposed. Table 4.5 shows a list of CNV detection software.

According to the experimental design, these methods cluster into three categories: single sample, case-control pair samples, and random samples. For the single sample, since most loci are diploid (*e.g.* for human genome), and therefore absolute number of copies can be estimated by comparing the RD distribution of the local region with the global one. The case-control pair samples are widely used in oncology. Similar to CGH, by calculating the point wise ratio between the test RD and matched control RD, the relative number of copies can be estimated. Since the matched control helps to reduce experimental perturbations, case-control methods yield better CNV detection. To detect both common and rare CNVs from population, random samples is needed for analysis, and loci with differential RD among samples are reported.

RD-based methods usually include these steps for CNV detection: (1) Map sequencing reads to the reference genome (*e.g.*,NCBI37/hg19). (2) Extract the RD signal with fix-sized bins, or a sliding window. (3) Normalize RD signals. (4) Divide normalized RD signal, or ratio of RD signals into segments according to the depths. Classical segmentation algorithms such as circular binary segmentation (CBS) (Olshen et al. 2004), and hidden Markov model (HMM) can be applied. (5) Calculate the copy number status of each segment by statistical hypothesis testing, *e.g.*, Poisson or negative-binomial distribution. (6) Combine consecutive segments that have the same copy number status. (7) Output and display CNV calls, including CNV class (gain or loss), break-point loci and size, number of copies, *etc*. Some of these steps are optional. *e.g.*, step (3) is necessary for single sample, since GC-content correction and mappability correction are needed in order to reduce these biases (see Sect. 4.2.4), while this normalization step is not necessary for case-control pair of samples.

Table 4.5 A selection of CNV detection software (Duan et al. 2013)

Method	Language	Control required?	Input format	Methodology	Reference
CNV-seq	R, perl	Yes	Hits	Statistical testing	Xie and Tammi (2009)
cn.MOPS	R, C++	No	BAM or data matrix	Mixture of Poissons, MAP, EM, CBS	Klambauer et al. (2012)
CNAnorm	R	Yes	SAM, BAM	Linear regression or CBS	Gusnanto et al. (2012)
cnvHiTSeq	Java	No	BAM	HMM	Bellos et al. (2012)
CNAseg	R	Yes	BAM	Wavelet transform and HMM	Ivakhno et al. (2010)
cnD	D	No	SAM, BAM	HMM, Viterbi algorithm	Simpson et al. (2010)
CNVer	C	No	BAM	Maximum-likelihood, graphic flow	Medvedev et al. (2010)
CNVnator	C	No	BAM	Mean shift algorithm	Abyzov et al. (2011)
CopySeq	Java	No	BAM	MAP estimator	Waszak et al. (2010)
EWT (RDXplorer)	R, python	No	BAM	Statistical testing	Yoon et al. (2009)
FREEC	C	Optional	SAM, BAM, bed, etc.	LASSO regression	Boeva et al. (2011)
JointSLM	R, Fortran	No	Data matrix	HMM, ML estimator, Viterbi algorithm	Magi et al. (2011)
readDepth	R	No	Bed	CBS, LOESS regression	Miller et al. (2011)
rSW-seq	NA	Yes	NA	Smith-Waterman algorithm	Kim et al. (2010)
SegSeq	Matlab	Yes	Bed	Statistical testing, CBS	Chiang et al. (2009)
SeqCBS	R	Yes	Bed	Poisson Processes, CBS	Shen and Zhang (2012)
WaveCNV	Matlab, Perl	Yes	BAM	Wavelet transform	Carson et al. (2013)

4.3.2 *Targeted Sequencing*

The feature of high-throughput has allowed NGS to identify the pathological mutations and variants in numerous clinical studies. Whole genome sequencing (WGS) is considered as a comprehensive way to identify different kinds of variation in genome, including gene fusion. However, even when few samples are involved in WGS study, the demand of data storage and computational complexity remains high. To identify the causal variants in WGS, several variants filters should be applied since the incidental findings are fairly high when large sequencing data exists. Although the price to WGS has dropped dramatically, in the context of clinical application, targeted sequencing of human genome remains as the most cost-effective and powerful method in NGS applications. Moreover, when comparing with WGS, targeted sequencing also reduces the incidental findings and increases coverage over regions of interest. For the studies of microbial sequencing, please refer to Chap. 8 in this book.

As described previously, selection or isolation regions of interest from DNA samples is a key step in targeted sequencing. This step is called target enrichment. PCR-amplicon and hybridization capture are the two major technologies currently used in enrichment. To achieve high sensitivity and specificity in targeting, several improvements had been made in both approaches. The RainStorm platform is one of the most commonly used PCR-based methods, which performs independent PCR in micro droplets by holding single primer in each droplet. This technology effectively prevents the interactions of different primers and competition of multiplex PCRs for the same reagent pool. Fluidigm solved the problem very much alike by using multilayer soft lithography. In terms of hybridization, when traditional technology hybridizes library DNA to the probes immobilized on a microarray, solution-based capture enables a further completion reaction by making probes exceeding over template (Mamanova et al. 2010). In general, PCR-amplicon is capable for flexibility targeting in large samples, while hybridization is perfect for capture large regions. A comparison of these methods is as Table 4.6. However in NGS, there still exists some regions that are particular difficult to be sequence, including: (1) GC rich regions; (2) regions with pseudogenes that making it hard to target uniquely; (3) regions with repetitive elements. In particular, to avoid allelic drop out, it is necessary to check for primer binding site SNPs in PCR-based enrichment (Abbs et al. 2014).

Targeted sequencing is already widely used in clinical research in finding mutations for Mendelian disorder and common complex disease. Beyond nuclear genome, furthermore targeted sequencing on mitochondrial DNA is now applicable to study respiratory chain disorders. The comprehensive profiles genetic architecture would guide us in developing tools for genetic screen, complex disease diagnosis, as well as personalized therapy. Usually, targeted sequencing is categorized into whole exome sequencing (WES) and targeted deep sequencing, the latter can be further divided into sequencing with multi-gene panel and designed regions.

WES investigates all of protein-coding DNA, that is, less than 2 % of the human genome but around 85 % known disease-causing variants. As an alternative strategy of WGS, WES is able to detect previous unobserved variants (even with low-frequency) in exons by achieving an average of 100-fold coverage. In particular, for monogenic rare diseases with unknown causes, WES can identify the causal variants with very small sample size. For example, Ng et al. discovered causal gene *OHODH* for Miller syndrome when performed WES in four patients (Ng et al. 2010). With larger sample size, WES also shows its strength in identifying causal variants for both rare and common diseases with genetic heterogeneity.

Table 4.6 Comparison of different target enrichment methods

Method	Target size	Throughput	Advantage	Disadvantage
Multiplex PCR	Small	Low	Simple for designed target	Target size is limited and cost inefficient
Array PCR	Small to medium	Low	Cost efficient PCR reaction	Pricy in synthesizing primers and amplification efficiency varies
Microdroplet PCR	Medium	Low	Ten thousands of amplicons can be applied, highly cost efficiency	Pricy in synthesizing primers and amplification efficiency varies, special technique needed to form microdroplet
Array-based capture	Medium to large	Low	Large target size, low cost	Special technique needed to elude captured DNA and low coverage for GC rich regions
Solution-based capture	Medium to large	High	Large target size, low cost and high throughput	Cost inefficient for small target region and low coverage for GC rich regions

As an example of investigating causal variants for rare diseases which are defined by symptoms but induced via different mechanisms, WES on 110 unrelated patients with progressive myoclonus epilepsy (PME), a rare inherited disorders characterized with action myoclonus, tonic-clonic seizures and ataxia, identified a major causal *de novo* mutation in *KCNC1* gene when pathogenic mutations in known PME-associated genes existed (Muona et al. 2015). In cancer genetics research, with similar sample size, WES was performed to investigate melanoma, especially the mutations induced by ultraviolet light exposure. Mutation in *RAC1* was identified in 9.2 % of sun-exposed melanomas while mutations in *PPP6C* were found in 12 % patients who also had mutations in *BRAF* or *NRAS* (Krauthammer et al. 2012). Meanwhile, applying WES to a large population would gain a profound understanding of the genetic effects on disease. A WES study with 2500 simplex families of autistic spectrum disorder concluded that 12 % of autism diagnoses can be explained by 13 % of *de novo* missense mutations while 9 % of autism by 43 % of *de novo* likely gene-disrupting mutations, by comparing affected to unaffected siblings (2500 pairs). Besides, this study also found that the regions of gene-disrupting mutations in female significantly overlapped with the known regions in relate with lower intelligence quotient in male (Iossifov et al. 2014).

Nevertheless, the rationale of this strategy is that pathological variants are more likely in exons, thus WES can't be used to explore the regulatory mechanisms induced by variants located in introns or intergenic regions. Likewise, sequencing every exon restricts the read depth and discovery power. Targeted deep sequencing, along with prior knowledge for diseases, can remedy some of these drawbacks. Multi-gene panel sequencing, namely sequencing bulks of candidate genes via array hybridization, is an efficient and timely approach in clinical research. Moreover, most of the panels cover introns for investigation. Various gene panels met for distinct clinical demands are now available through vendors, with the number of genes ranging from 70 to 377 in the panel. Gene panel sequencing is more appropriate to use in disease whose genetic locus heterogeneity are significant (*e.g.* muscular dystrophies panel), in disorders with overlapping phenotype (*e.g.* cardiomyopathy panel), in disorders share same manifestation but different overall presentation (*e.g.* epilepsy panel) and in diseases induced by genes from a common pathway or structure (*e.g.* RASopathies panel) (Xue et al. 2015). Similar to the gene panel, we can further capture desired genes or regions in a custom-designed manner, either through PCR amplicons or hybridization, to explore the causality with deeper reads. A classic example is sequence 21 genes known to be in association with breast and ovarian cancer. Walsh et al. captured coding and intronic regions as well as 10 kb upstream of each genic region by in solution hybridization. Greater than 1200-fold average coverage was achieved in this study. Every known pathogenic changes including single-nucleotide substitutions, small insertion and deletion mutations, and large genomic duplications and deletions was identified with this read depth. This study suggests an efficient and accurate NGS design for personalized risk assessment (Walsh et al. 2010). In some studies, targeted deep sequencing was used as a validation step following WES.

As mentioned in previous examples, NGS sampling could vary from numbers of particular cases to large population, depending on study design, type of disease and budget. While small sample size is enough for investigation on common variants, it is still difficult to gain adequate power to discover rare variants in complex diseases. A comparison of study designs with optional sample size for rare variants is listed in Table 4.7. Besides, group-based analysis and meta-analysis are two powerful statistical approaches for rare variants analysis.

4.3.3 Cell Free Circulating DNA Sequencing

Cell-free circulating DNA (cfDNA) is the DNA segments in the circulation released from cell apoptosis and necrosis. Studies of cfDNA mainly focus on prenatal diagnosis and cancer monitoring because that fetal-derived and tumor-derived cfDNA are largely different from the main background of cfDNA. Profiling cfDNA takes the advantage of being a noninvasive tool while cfDNA can be isolated from blood plasma using commercially available kits. Both cell-free fetal DNA (cffDNA) and circulating tumor DNA (ctDNA) exhibit rapid clearance within hours. This means that after delivery or tumor removal operation, these circulating DNA would soon become undetectable.

To date, three types of analyses on NGS sequencing data had been applied in cfDNA studies, including: (1) allelic count-based methods, that is, counting the tagged fragments after mapping them to chromosomes, to estimate cfDNA concentration and detect aneuploidy; (2) regional genomic representation-based methods (same as targeted sequencing) to detect aneuploidy, copy number changes and genomic rearrangement; and (3) size-based analysis for estimation of fetal DNA concentration and detection of aneuploidy. If an estimation algorism is not

Table 4.7 Study design for rare variant discovery in predisposition genes

Study design	Sample size in previous studies	Advantages	Disadvantages
Family-based WGS/WES	Less or equal to 200 pedigrees (Thompson et al. 2012; Park et al. 2012; Neale et al. 2012; Roberts et al. 2012; Palles et al. 2013)	High detection power, efficiency for Mendelian diseases	Less efficiency for complex disease and sporadic disease
Case–control deep sequencing	125–500 pairs of case vs. control (Beaudoin et al. 2012; Hoehe et al. 2000; Chien et al. 2013; Li et al. 2013; Lin et al. 2014)	Power can be attained with feasible sample size, fairly affordable	Limited ability in novel gene discovery
Case–control WGS/WES	500–1000 pairs of case vs. control (Tang et al. 2014; Ellinghaus et al. 2013; Liu et al. 2013; Siemiatkowska et al. 2013)	Capable in novel gene discovery	Pricy and poor power when sample size is limited
Case-only exome sequencing	100–1000 subjects	Cost efficiency	Difficulty in new susceptibility loci identification

available, parental genotyping is needed in the allelic count-based approaches (Chan and Jiang 2015).

The concentration of cffDNA increases with gestational age, while the fragment size of cffDNA is shorter than maternal-derived cfDNA. Currently, the clinical application of NGS on cffDNA is prenatal screening for aneuploidy, especially trisomy on chromosome 13, 18 and 21. Compared with conventional screening tools, cffDNA presented a higher sensitivity and specificity upon performance (Cuckle et al. 2015). The detection of aneuploidy can be achieved by all of the above methods (Yu et al. 2014; Sparks et al. 2012; Chen et al. 2011), moreover, assessment of twin zygosity and aneuploidy in dizygotic twins can also be estimated through similar approaches (Leung et al. 2013). It is possible to construct the entire fetal genome from cffDNA when parental sequencing available, thus detect the mutation in the fetal genome (Fan et al. 2012). However, it is still too expensive and elaborate to carry out this detection in clinical screening.

Similar to cffDNA, concentration of ctDNA is positively correlates with tumor size and stage, but the size of ctDNA could be either shorter or longer than background cfDNA. SNV, CNV and rearrangements in ctDNA have been identified by WGS, WES and targeted deep sequencing in studies of cancer patients in different stages. Some specific techniques were already developed to improve detection sensitivity for targeted deep sequencing in ctDNA, including tagged amplicon deep sequencing (TAm-Seq) (Dawson et al. 2013; Forshew et al. 2012), safe-sequencing system (Safe-Seq) (Bettegowda et al. 2014), cancer personalized profiling by deep sequencing (CAPP-Seq) (Newman et al. 2014). Since ctDNA can widely representative the underlying tumor genome, NGS ctDNA can identify tumor type through mutation and guide the targeted therapy. Besides, while ctDNA is a very sensitive biomarker, monitoring ctDNA can lead an early detection of cancer, as well as assessment of therapeutic response and treatment resistance. The noninvasive and timely manner of ctDNA sequencing, makes it one of the most promising tools in future cancer research (Ignatiadis and Dawson 2014).

4.3.4 Single Cell Sequencing

Somatic variation, known as genomic heterogeneity, exists after sufficient acquirement of genetic mutations during every cell division. It is considered as the cause of many disorders, such as cancer. Single cell sequencing (SCS) is the technology that can provide a view from single cell to observe the genomic change during cell developmental processes (Macaulay and Voet 2014). Current SCS research focus on tissue mosaicism, germline transmission and cancer. The first stage of SCS is the isolation of single cells. The traditional single cell isolation approaches require abundant cells in suspension, for example, flow-assisted cell sorting (FACS), mouth pipetting, serial dilution, robotic micromanipulation, and microfluidic platforms. In cancer SCS studies, aside from isolation single cells from tissues, circulating tumor cell (CTC) also serves as a type of target single cell. Similar to ctDNA, CTC is highly sensitive with the presence of tumor. Nevertheless, in blood, the frequency of CTC is only one in one million, which leads to development of commercial platforms for these rare cells (<1 %) isolation. A comparison of rare cell isolation is listed in Table 4.8. Followed single cell isolation and DNA extraction, an amplification stage is necessary since the average concentration of DNA in a single cell is around six pg. The old fashion PCR, which is mentioned in the earlier section, will surely introduce artificial biases in this process. To better handle DNA amplification from single cell, several new methods had been developed to acquire low bias and high coverage. For example, multiple displacement amplification (MDA) has been widely used currently by using ϕ29 DNA polymerase to achieve relatively low amplification bias and high genome coverage.

An alternative method, multiple annealing, looping-based amplification cycle (MALBAC) which combines features of linear amplification with PCR is later promoted. MALBAC has advantages of low amplification error rate, extremely low amplification bias and high genome coverage, thus better performance in SNV and CNV detection in SCS studies. Except for the cell isolation and DNA amplification, the workflow for SCS is similar to NGS studies. Algorithms for SNP calling and SNV detection can be adapted in SCS analysis (Ning et al. 2014).

In previous tissue mosaic studies, SCS on human frontal cortex neurons detected 13–41 *% de novo* CNVs in neurons, as well as that deletions are twice as common as duplications, suggesting CNV mosaic events are abundant in cortical neurons (McConnell et al. 2013). In germline transmission research, average of 22.8 and 26 recombination events per cell were observed in two studies of single sperm cells, using MDA and MALBAC for amplification respectively (Lu et al. 2012; Wang et al. 2012), while 43 recombination events per single cell were revealed in research of single oocytes (Hou et al. 2013). Meanwhile in cancer research, most SCS

Table 4.8 Methods for isolating single cells from rare populations

Methods	Principle	Company	Advantages	Disadvantages
Laser-capture microdissection	Capture cells via laser under microscope	–	Present spatial context	Potential UV damage
Nano-fabricated filters	Selection based on size	Creatv MicroTech	Cost efficiency and straight forward	Cells may adhere to filters
MagSweeper	Selection based on EpCAM antibodies with magnet	Illumina	High enrichment and cost efficiency	Bias may induced by markers
CellSearch	Selection based on EpCAM and CD45 antibodies with magnet	Johnson & Johnson	High throughput, approved by FDA	Bias may induced by markers
CellCelector	Robotic capillary micromanipulator	Automated Lab Solutions	High throughput	Expensive
DEP-Array	Capture charged cells with dielectrophoretic cages in microchip	Silicon Biosciences	High sensitivity	Time consuming, low throughput, expensive

studies utilized single-cell exome sequencing to examine intratumor heterogeneity and clonal evolution. A study identified a monoclonal population of cells which shared a common genetic lineage in related to *JAK2*-positive myeloproliferative neoplasm (Hou et al. 2012), while another study concluded that clear cell renal cell carcinoma was more genetic complex than expected since no significant clonal subpopulation was found (Xu et al. 2012). More details of clonal evolution was observed in a breast cancer study. It showed that in breast cancer, aneuploidy rearrangements occurred early in tumor evolution but remained stable during clonally expanded while point mutations evolved gradually over time generating extensive clonal diversity (Wang et al. 2014). Beyond the research of single cells from solid tissue, profiling single CTCs through exome sequencing in lung cancer suggested CNV patterns were consistent through cells in the same subtype of cancer while mutation appeared heterogeneous among cancer cells (Ni et al. 2013). Overall, SCS technology provides a chance to explore single cell genomic architecture and addresses important implications for clinical diagnosis as well as therapeutic treatment in personalized care.

4.4 Conclusion

In this chapter we addressed the fundamental concepts in NGS, from principle of both wet and dry laboratory technologies to rational behind detection of different types of variants that are related to diseases. In addition, in order to keep the pace with the fast-moving sequencing technologies, we also cover the contents of most novel NGS applications, public repositories and the third generation sequencing. However, there are still a lot challenges exist in NGS which will be solved by methodologies and technologies in the future. The assumptions and hypothesis of NGS study designs may also be modified with the new findings from human genome. Even so, we hope that the comprehensive knowledge we elucidated in this chapter can hereafter provide a guideline for readers to investigate their own clinical research problems.

Bibliography

Abbs CM, et al. Practice guidelines for targeted next generation sequencing analysis and interpretation. Assoc Clin Genet Sci. 2014.

Abel HJ, et al. SLOPE: a quick and accurate method for locating non-SNP structural variation from targeted next-generation sequence data. Bioinformatics. 2010;26(21):2684–8.

Abyzov A, et al. CNVnator: an approach to discover, genotype, and characterize typical and atypical CNVs from family and population genome sequencing. Genome Res. 2011;21 (6):974–84.

Andrews S. A quality control tool for high throughput sequence data. 2015. Available from: http://www.bioinformatics.babraham.ac.uk/projects/fastqc/.

Ashton PM, et al. MinION nanopore sequencing identifies the position and structure of a bacterial antibiotic resistance island. Nat Biotechnol. 2015;33(3):296–300.

Bao R, et al. Review of current methods, applications, and data management for the bioinformatics analysis of whole exome sequencing. Cancer Inform. 2014;13 Suppl 2:67–82.

Beaudoin M, Lo KS, N'Diaye A, Rivas MA, Dubé MP, Laplante N, Phillips MS, Rioux JD TJ, Lettre G. Pooled DNA resequencing of 68 myocardial infarction candidate genes in French Canadians. Circ Cardiovasc Genet. 2012;5:547–54.

Bellos E, et al. cnvHiTSeq: integrative models for high-resolution copy number variation detection and genotyping using population sequencing data. Genome Biol. 2012;13:R120.

Bentley DR, et al. Accurate whole human genome sequencing using reversible terminator chemistry. Nature. 2008;456(7218):53–9.

Bettegowda C, Sausen M, Leary RJ, et al. Detection of circulating tumor DNA in early- and late-stage human malignancies. Sci Transl Med. 2014;6:224ra24.

Boeva V, et al. Control-free calling of copy number alterations in deep-sequencing data using GC-content normalization. Bioinformatics. 2011;27:268–9.

Bolger AM, Lohse M, Usadel B. Trimmomatic: a flexible trimmer for Illumina sequence data. Bioinformatics. 2014;30(15):2114–20.

Campbell PJ, et al. Identification of somatically acquired rearrangements in cancer using genome-wide massively parallel paired-end sequencing. Nat Genet. 2008;40:722–9.

Chaisson MJ, Pevzner PA. Short read fragment assembly of bacterial genomes. Genome Res. 2008;18:324–30.

Chaisson MJ, et al. Resolving the complexity of the human genome using single-molecule sequencing. Nature. 2015;517(7536):608–11.

Chan LL, Jiang P. Bioinformatics analysis of circulating cell-free DNA sequencing data. Clin Biochem. 2015;48(15):962–75.

Chen K, et al. BreakDancer: an algorithm for high-resolution mapping of genomic structural variation. Nat Methods. 2009;6:677–81.

Chen EZ, Chiu RWK, Sun H, Akolekar R, Chan KCA, et al. Noninvasive prenatal diagnosis of fetal trisomy 18 and trisomy 13 by maternal plasma DNA sequencing. PLoS One. 2011;6(7): e21791.

Chiang DY, et al. High-resolution mapping of copy-number alterations with massively parallel sequencing. Nat Methods. 2009;6(1):99–103.

Chien WH, Gau SS-F, Liao HM, Chiu YN, Wu YY, Huang YS, Tsai WC, Tsai HM, Chen C-H. Deep exon resequencing of DLGAP2 as a candidate gene of autism spectrum disorders. Mol Autism. 2013;4:26.

Cuckle H, Benn P, Pergament E. Cell-free DNA screening for fetal aneuploidy as a clinical service. Clin Biochem. 2015;48(15):932–41.

Dawson SJ, et al. Analysis of circulating tumor DNA to monitor metastatic breast cancer. N Engl J Med. 2013;368(13):1199–209.

Derrien T, et al. Fast computation and applications of genome mappability. PLoS One. 2012;7(1): e30377.

Dohm JC, et al. SHARCGS, a fast and highly accurate short-read assembly algorithm for de novo genomic sequencing. Genome Res. 2007;17:1697–706.

Duan J, et al. Comparative studies of copy number variation detection methods for next-generation sequencing technologies. PLoS One. 2013;8(3):e59128.

Earl D, et al. Assemblathon 1: a competitive assessment of de novo short read assembly methods. Genome Res. 2011;21(12):2224–41.

Ellinghaus D, et al. Association between variants of PRDM1 and NDP52 and Crohn's disease, based on exome sequencing and functional studies. Gastroenterology. 2013;145(2):339–47.

Ewing B, Green P. Base-calling of automated sequencer traces using phred. II. Error probabilities. Genome Res. 1998;8(3):186–94.

Ewing B, et al. Base-calling of automated sequencer traces using phred. I. Accuracy assessment. Genome Res. 1998;8(3):175–85.

Fan HC, et al. Non-invasive prenatal measurement of the fetal genome. Nature. 2012;487 (7407):320–4.

Feng Y, et al. Nanopore-based fourth-generation DNA sequencing technology. Genomics Proteomics Bioinformatics. 2015;13(1):4–16.

Forshew T, Murtaza M, Parkinson C, et al. Noninvasive identification and monitoring of cancer mutations by targeted deep sequencing of plasma DNA. Sci Transl Med. 2012;4:136ra68.

Freeman JL, et al. Copy number variation: new insights in genome diversity. Genome Res. 2006;16(8):949–61.

Gusnanto A, et al. Correcting for cancer genome size and tumour cell content enables better estimation of copy number alterations from nextgeneration sequence data. Bioinformatics. 2012;28(1):40–7.

Head SR, et al. Library construction for next-generation sequencing: overviews and challenges. Biotechniques. 2014;56(2):61–4, 66, 68, passim.

Gusnanto A, et al. Correcting for cancer genome size and tumour cell content enables better estimation of copy number alterations from nextgeneration sequence data. Bioinformatics. 2012;28(1):40–7.

Hernandez D, et al. De novo bacterial genome sequencing: millions of very short reads assembled on a desktop computer. Genome Res. 2008;18:802–9.

Hoehe MR, Köpke K, Wendel B, Rohde K, Flachmeier C, Kidd KK, Berrettini WH, Church GM. Sequence variability and candidate gene analysis in complex disease association of μ opioid receptor gene variation with substance dependence. Hum Mol Genet. 2000;9:2895–908.

Holt C, et al. WaveCNV: allele-specific copy number alterations in primary tumors and xenograft models from next-generation sequencing. Bioinformatics. 2014;30(6):768–74.

Hormozdiari F, et al. Next generation VariationHunter: combinatorial algorithms for transposon insertion discovery. Bioinformatics. 2010;26(12):i350–7.

Hormozdiari F, et al. Simultaneous structural variation discovery among multiple paired-end sequenced genomes. Genome Res. 2011;21:2203–12.

Hou Y, et al. Single-cell exome sequencing and monoclonal evolution of a JAK2-negative myeloproliferative neoplasm. Cell. 2012;148(5):873–85.

Hou Y, et al. Genome analyses of single human oocytes. Cell. 2013;155(7):1492–506.

Ignatiadis M, Dawson SJ. Circulating tumor cells and circulating tumor DNA for precision medicine: dream or reality? Ann Oncol. 2014;25(12):2304–13.

Iossifov I, et al. The contribution of de novo coding mutations to autism spectrum disorder. Nature. 2014;515(7526):216–21.

Iqbal Z, et al. De novo assembly and genotyping of variants using colored de Bruijn graphs. Nat Genet. 2012:226–32.

Ivakhno S, et al. CNAseg–a novel framework for identification of copy number changes in cancer from second-generation sequencing data. Bioinformatics. 2010;26:3051–8.

Jain M, et al. Improved data analysis for the MinION nanopore sequencer. Nat Methods. 2015;12 (4):351–6.

Karamitros T, Magiorkinis G. A novel method for the multiplexed target enrichment of MinION next generation sequencing libraries using PCR-generated baits. Nucleic Acids Res. 2015;43(22):e152.

Kent WJ. BLAT–the BLAST-like alignment tool. Genome Res. 2002;12(4):656–64.

Kim TM, et al. rSW-seq: algorithm for detection of copy number alterations in deep sequencing data. BMC Bioinform. 2010;11:432.

Klambauer G, et al. cn.MOPS: mixture of poissons for discovering copy number variations in next-generation sequencing data with a low false discovery rate. Nucleic Acids Res. 2012;40(9):e69.

Korbel JO, et al. Paired-end mapping reveals extensive structural variation in the human genome. Science. 2007;318(5849):420–6.

Korbel JO, et al. PEMer: a computational framework with simulation-based error models for inferring genomic structural variants from massive paired-end sequencing data. Genome Biol. 2009;10:R23.

Krauthammer M, et al. Exome sequencing identifies recurrent somatic RAC1 mutations in melanoma. Nat Genet. 2012;44(9):1006–14.

Langer-Safer PR, Levine M, Ward DC. Immunological method for mapping genes on Drosophila polytene chromosomes. Proc Natl Acad Sci U S A. 1982;79(14):4381–5.

Langmead B, et al. Ultrafast and memory-efficient alignment of short DNA sequences to the human genome. Genome Biol. 2009;3:R25.

Leung TY, et al. Noninvasive twin zygosity assessment and aneuploidy detection by maternal plasma DNA sequencing. Prenat Diagn. 2013;33(7):675–81.

Li H, Durbin R. Fast and accurate long-read alignment with Burrows-Wheeler transform. Bioinformatics. 2010;5:589–95.

Li H, Homer N. A survey of sequence alignment algorithms for next-generation sequencing. Brief Bioinform. 2010;11(5):473–83.

Li H, et al. Mapping short DNA sequencing reads and calling variants using mapping quality scores. Genome Res. 2008;11:1851–8.

Li R, et al. SOAP2: an improved ultrafast tool for short read alignment. Bioinformatics. 2009;15:1966–7.

Li JM, et al. Exonic resequencing of the DLGAP3 gene as a candidate gene for schizophrenia. Psychiatry Res. 2013;208(1):84–7.

Lin Y, et al. Comparative studies of de novo assembly tools for next-generation sequencing technologies. Bioinformatics. 2011;27(15):2031–7.

Lin H, et al. Targeted sequencing in candidate genes for atrial fibrillation: the Cohorts for Heart and Aging Research in Genomic Epidemiology (CHARGE) Targeted Sequencing Study. Heart Rhythm. 2014;11(3):452–7.

Linnarsson S. Recent advances in DNA sequencing methods – general principles of sample preparation. Exp Cell Res. 2010;316(8):1339–43.

Liu L, et al. Comparison of next-generation sequencing systems. J Biomed Biotechnol. 2012;2012:251364.
Liu L, et al. Analysis of rare, exonic variation amongst subjects with autism spectrum disorders and population controls. PLoS Genet. 2013;9(4):e1003443.
Lu S, Zong C, Fan W, Yang M, Li J, Chapman AR, Zhu P, Hu X, Xu L, Yan L, Bai F, Qiao J, Tang F, Li R, Xie XS. Probing meiotic recombination and aneuploidy of single sperm cells by whole-genome sequencing. Science. 2012;338:1627–30.
Macaulay IC, Voet T. Single cell genomics: advances and future perspectives. PLoS Genet. 2014;10(1):e1004126.
Madoui MA, et al. Genome assembly using Nanopore-guided long and error-free DNA reads. BMC Genomics. 2015;16:327.
Magi A, et al. Bioinformatics for next generation sequencing data. Genes. 2010;1(2):294–307.
Magi A, et al. Detecting common copy number variants in high-throughput sequencing data by using JointSLM algorithm. Nucleic Acids Res. 2011;39:e65.
Mailman MD, et al. The NCBI dbGaP database of genotypes and phenotypes. Nat Genet. 2007;39(10):1181–6.
Mamanova L, et al. Target-enrichment strategies for next-generation sequencing. Nat Methods. 2010;7(2):111–18.
McConnell MJ, Lindberg MR, Brennand KJ, Piper JC, Voet T, Cowing-Zitron C, Shumilina S, Lasken RS, Vermeesch JR, Hall IM, Gage FH. Mosaic copy number variation in human neurons. Science. 2013;342:632–7.
Medvedev P, et al. Detecting copy number variation with mated short reads. Genome Res. 2010;20:1613–22.
Metzker ML. Sequencing technologies – the next generation. Nat Rev Genet. 2010;11(1):31–46.
Miller CA, et al. ReadDepth: a parallel R package for detecting copy number alterations from short sequencing reads. PLoS One. 2011;6(1):e16327.
Miller JR, et al. Aggressive assembly of pyrosequencing reads with mates. Bioinformatics. 2008;24:2818–24.
Muona M, et al. A recurrent de novo mutation in KCNC1 causes progressive myoclonus epilepsy. Nat Genet. 2015;47(1):39–46.
Neale BM, et al. Patterns and rates of exonic de novo mutations in autism spectrum disorders. Nature. 2012;485(7397):242–5.
Newman AM, et al. An ultrasensitive method for quantitating circulating tumor DNA with broad patient coverage. Nat Med. 2014;20(5):548–54.
Ng SB, et al. Exome sequencing identifies the cause of a Mendelian disorder. Nat Genet. 2010;42(1):30–5.
Ni X, et al. Reproducible copy number variation patterns among single circulating tumor cells of lung cancer patients. Proc Natl Acad Sci U S A. 2013;110(52):21083–8.
Nijkamp JF, et al. De novo detection of copy number variation by co-assembly. Bioinformatics. 2012;28(24):3195–202.
Ning L, et al. Current challenges in the bioinformatics of single cell genomics. Front Oncol. 2014;4:7.
Ning Z, et al. SSAHA: a fast search method for large DNA databases. Genome Res. 2001;11:1725–9.
Olshen AB, et al. Circular binary segmentation for the analysis of array-based DNA copy number data. Biostatistics. 2004;5(Oct):557–72.
Palles C, et al. Germline mutations affecting the proofreading domains of POLE and POLD1 predispose to colorectal adenomas and carcinomas. Nat Genet. 2013;45(2):136–44.
Park DJ, et al. Rare mutations in XRCC2 increase the risk of breast cancer. Am J Hum Genet. 2012;90(4):734–9.
Pendleton M, et al. Assembly and diploid architecture of an individual human genome via single-molecule technologies. Nat Methods. 2015;12(8):780–6.

Pushkarev D, Neff NF, Quake SR. Single-molecule sequencing of an individual human genome. Nat Biotechnol. 2009;27(9):847–50.

Roberts NJ, et al. ATM mutations in patients with hereditary pancreatic cancer. Cancer Discov. 2012;2(1):41–6.

Roberts RJ, Carneiro MO, Schatz MC. The advantages of SMRT sequencing. Genome Biol. 2013;14(7):405.

Rovelet-Lecrux A, et al. APP locus duplication causes autosomal dominant early-onset Alzheimer disease with cerebral amyloid angiopathy. Nat Genet. 2006;38(1):24–6.

Schadt EE, Turner S, Kasarskis A. A window into third-generation sequencing. Hum Mol Genet. 2010;19(R2):R227–40.

Sebat J, et al. Strong association of de novo copy number mutations with autism. Science. 2007;316:445–9.

Shen JJ, Zhang NR. Change-point model on nonhomogeneous poisson processes with application in copy number profiling by next-generation DNA sequencing. Ann Appl Stat. 2012;6(2):476–96.

Shen T, et al. Clinical applications of next generation sequencing in cancer: from panels, to exomes, to genomes. Front Genet. 2015;6:215.

Siemiatkowska AM, et al. Mutations in the mevalonate kinase (MVK) gene cause nonsyndromic retinitis pigmentosa. Ophthalmology. 2013;120(12):2697–705.

Simpson JT, et al. ABySS: a parallel assembler for short read sequence data. Genome Res. 2009;19:1117–23.

Simpson JT, et al. Copy number variant detection in inbred strains from short read sequence data. Bioinformatics. 2010;26:565–7.

Sindi S, et al. A geometric approach for classification and comparison of structural variants. Bioinformatics. 2009;25(12):i222–30.

Smith TF, Waterman MS. Identification of common molecular subsequences. J Mol Biol. 1981;147(1):195–7.

Sparks AB, et al. Selective analysis of cell-free DNA in maternal blood for evaluation of fetal trisomy. Prenat Diagn. 2012;32(1):3–9.

Speicher MR, Gwyn Ballard S, Ward DC. Karyotyping human chromosomes by combinatorial multi-fluor FISH. Nat Genet. 1996;12(4):368–75.

Stankiewicz P, Lupski JR. Structural variation in the human genome and its role in disease. Annu Rev Med. 2010;61:437–55.

Stefansson H, et al. Large recurrent microdeletions associated with schizophrenia. Nature. 2008;455(7210):232–6.

Tang H, et al. A large-scale screen for coding variants predisposing to psoriasis. Nat Genet. 2014;46(1):45–50.

Tao T. Available from: http://www.ncbi.nlm.nih.gov/staff/tao/tools/tool_lettercode.html.

The 1000 Genomes Project Consortium. A map of human genome variation from population-scale sequencing. Nature. 2010;467(7319):1061–73.

The 1000 Genomes Project Consortium. An integrated map of genetic variation from 1,092 human genomes. Nature. 2012;491(7422):56–65.

Thompson ER, et al. Exome sequencing identifies rare deleterious mutations in DNA repair genes FANCC and BLM as potential breast cancer susceptibility alleles. PLoS Genet. 2012;8(9): e1002894.

Travers KJ, et al. A flexible and efficient template format for circular consensus sequencing and SNP detection. Nucleic Acids Res. 2010;38(15):e159.

Urban AE, et al. High-resolution mapping of DNA copy alterations in human chromosome 22 using high-density tiling oligonucleotide arrays. Proc Natl Acad Sci USA. 2006;103 (12):4534–9.

Walsh T, et al. Detection of inherited mutations for breast and ovarian cancer using genomic capture and massively parallel sequencing. Proc Natl Acad Sci USA. 2010;107(28):12629–33.

Wang J, et al. Genome-wide single-cell analysis of recombination activity and de novo mutation rates in human sperm. Cell. 2012;150(2):402–12.
Wang Y, et al. Clonal evolution in breast cancer revealed by single nucleus genome sequencing. Nature. 2014;512(7513):155–60.
Wang M, et al. PacBio-LITS: a large-insert targeted sequencing method for characterization of human disease-associated chromosomal structural variations. BMC Genomics. 2015;16 (1):214.
Warren RL, et al. Assembling millions of short DNA sequences using SSAKE. Bioinformatics. 2007;23:500–1.
Waszak SM, et al. Systematic inference of copy number genotypes from personal genome sequencing data reveals extensive olfactory receptor gene content diversity. PLoS Comput Biol. 2010;6:e1000988.
Xie C, Tammi MT. CNV-seq, a new method to detect copy number variation using high-throughput sequencing. BMC Bioinf. 2009;10(80):1–9.
Xu X, et al. Single-cell exome sequencing reveals single-nucleotide mutation characteristics of a kidney tumor. Cell. 2012;148(5):886–95.
Xue Y, et al. Solving the molecular diagnostic testing conundrum for Mendelian disorders in the era of next-generation sequencing: single-gene, gene panel, or exome/genome sequencing. Genet Med. 2015;17(6):444–51.
Yang TL, et al. Genome-wide copy-number-variation study identified a susceptibility gene, UGT2B17, for osteoporosis. Am J Hum Genet. 2008;83(6):663–74.
Ye J, et al. BLAST: improvements for better sequence analysis. Nucleic Acids Res. 2006;34:W6–9.
Ye K, et al. Pindel: a pattern growth approach to detect break points of large deletions and medium sized insertions from paired-end short reads. Bioinformatics. 2009;25(21):2865–71.
Yoon S, et al. Sensitive and accurate detection of copy number variants using read depth of coverage. Genome Res. 2009;19:1586–92.
Yu SC, et al. Size-based molecular diagnostics using plasma DNA for noninvasive prenatal testing. Proc Natl Acad Sci USA. 2014;111(23):8583–8.
Zerbino DR, Birney E. Velvet: algorithms for de novo short read assembly using de Bruijn graphs. Genome Res. 2008;18:821–9.
Zhao M, et al. Computational tools for copy number variation (CNV) detection using next-generation sequencing data: features and perspectives. BMC Bioinformatics. 2013;14 Suppl 11:S1.

Chapter 5
Clinical Epigenetics and Epigenomics

Nian Dong, Lin Shi, Chengshui Chen, Wenhuan Ma, and Xiangdong Wang

Abstract Epigenetics is a molecular phenomenon that pertains to heritable changes in gene expression that do not involve changes in the DNA sequence. Epigenome, epigenetic modifications in a whole genome, play an essential role in the regulation of gene expression in both normal development and disease. DNA methylation, histone modification, and RNA-mediated targeting regulate many biological processes that are fundamental to the genesis of a spectrum of diseases. Here, we give a historical overview of the epigenomics field and focus on the recent progress that has been made in understanding the pathogenic role of cancerous disease, autoimmute disease, and metabolic disorder. We also discuss available traditional epigenetic therapies, epigenetic therapies currently in development, and the potential future use of epigenetic therapeutics in a clinical setting.

Keywords Clinical epigenetics • Epigenetic modification • Epigenomics • Epigenetic therapies

5.1 Introduction

Epigenetics is typically defined as the study of heritable changes in gene expression that are not due to changs in DNA sequence (Feinberg and Tycko 2004; Pogribny and Beland 2009; Wolffe and Matzke 1999; Riddihough and Zahn 2010). Through its

N. Dong • C. Chen (✉)
Department of Pulmonary Medicine, The First Affiliated Hospital, Wenzhou Medical University, Wenzhou, China
e-mail: wzchencs@163.com

L. Shi
Zhongshan Hospital, Shanghai Institute of Clinical Bioinformatics, Fudan University Center for Clinical Bioinformatics, Biomedical Research Center, Fudan University, Shanghai, China

W. Ma
Zhabei District Hospital of Traditional Chinese Medicine, Yanchang Middle Road No. 288, Jingan District, Shanghai, China
e-mail: mawenhuan@126.com

X. Wang
Zhongshan Hospital, Fudan University, Shanghai Institute of Clinical Bioinformatics, Shanghai, China

X. Wang et al. (eds.), *Application of Clinical Bioinformatics*,
Translational Bioinformatics 11, DOI 10.1007/978-94-017-7543-4_5

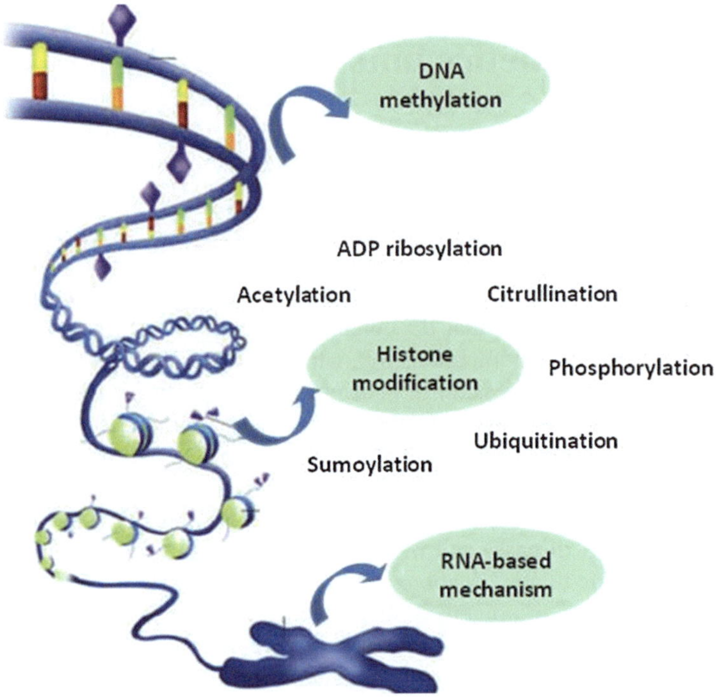

Fig. 5.1 Three common types of epigenetic modifiers: DNA methylation, histone modifications and ncRNAs

ability to mediate gene silencing or gene activation, epigenetic mechanism have a profound effect on diverse biological properties from the morphology of flowers and eye colour in fruitflies. Compared to epigenetic modifications which occur at distinct regions throughout the genome, epigenomics is the study of epigenetic modifications throughout the entire genome with the advent of whole – genome approaches. Epigenetics and epigenomics are the study of chromatin: the complex of DNA, proteins (histone proteins and non-histone proteins), and non-coding RNAs (ncRNAs) that form the structural matrix of a chromosome (Hamm and Costa 2015). As chromatin is essential for nuclear packing and gene regulation, the way how the chromatin structure is maintained and organized is key to a better understanding of the origins of epigenetic alterations in embryonic development, cell differentiation and disease.

The cytosines in DNA methylation and post-translational modifications of histone proteins are the most well documented mechanistic layers in the field of epigenetics, other epigenetic mechanisms include the emerging ncRNA (Costa 2008) (Fig. 5.1). The first type of modification, DNA methylation, involves the covalent attachment of a methyl group onto the C5 position of a cytosine residue on CpG (cytosine-phosphate-guanine) islands, considered one of the epigenetic processes that lead mainly to gene silencing and subsequently to inhibition of the gene transcription (Chiang et al. 1996; Holliday 1990). In general, DNA methylation occurs on the different CpGs clustered as islands on the majority of the genes. Millions of potentially methylated CpG islands cluster through the gene body and play a critical role within the promoter regions. Once methylated, gene transcription might not occur. By contrast, an active promoter may allow interaction with the various transcription factors controlling gene activation (Fujita et al. 2003; Perera et al. 2009).

In comparsion with DNA methylation directly affecting the genomic DNA, post-translational histoine modification act differently. The ability of the nucleosome to condense and organize the genome is related to both the molecular characteristics of the DNA, as well as, the molecular characteristics of the proteins that form the nucleosome. The core of the nucleosome is made up of eight proteins, consisting of two of each of the following histone proteins: H2A, H2B, H3, and H4. Histone proteins are subject to a myriad of post-translational modifications, including acetylation, ADP ribosylation, citrullination, clipping, methylation, phosphorylation, sumoylation, ubiquitination and others (Lister et al. 2009; Ray Gallet and Almouzni 2010; Hassan et al. 2002; Sanchez and Zhou 2009). Of the mentioned modifications, acetylation is the most extensively studied. It is a reversible process catalyzed by two enzymes, the histone acetylase (HAT) and HDAC. The presence of an acetyl group decreases interaction between negatively charged DNA and the positively charged histone tail which results in a less compact nucleosome, enables easier access for transcription factor complexes (Feng and Fan 2009; Mau and Yung 2014). Therefore, removal of the acetyl group by HDAC leads to gene transcription repression. The effects of such modifications is alteration in the electrostatic interaction between the histones and nearby DNA and accessibility to the transcriptional machinery.

The third modification involves different classes of ncRNAs, which can physically bind to the DNA, alter its conformation and, in the case of microRNAs, silence genes by post-translational control. The high-throughput genomic platforms have established that virtually the entire genome is transcribed; however, only 2 % is subsequently translated and the remaining "noncoding" RNAs (ncRNAs) would be roughly categorized into small (under 200 nucleotides) and large ncRNAs (Amaral et al. 2008). The small ncRNA, including small nucleolar RNAs (snoRNAs), PIWIinteracting RNAs (piRNAs), small interfering RNAs (siRNAs), and microRNAs (miRNAs), show high degree of sequence conservation across species and are involved in transcriptional and posttranscriptional gene silencing through specific base pairing with their targets. In contrast, the long ncRNAs (lncRNAs) demonstrate poor cross-species sequence conservation, and the mechanism of action in transcriptional regulation is more diverse. Notably, the lncRNAs appear to have a critical function at chromatin, acting as molecular chaperones or scaffolds for various chromatin regulators (Wang and Chang 2011).

Taking together, different types of epigenetic modification would solely or multiply have an effect on the epigenetic status of a particular locus in the genome. Given the role of epigenetic modifications in organismal development where stable and distinct cellular functions must be established froman identical genotype, it is no surprising that epigenetic deregulation is involved in the pathogenesis of a growing amount of diseases. From the first description of abberant epigenetic alteration in colon cancer (Feinberg and Vogelstein 1983), epigenetic pathway have been recognized as a hallmark of human cancer. Similarly, epigenetic abnormalities have been detected in several non-cancerous diseases. As opposed to genetic alterations, epigenetic alterations are reversible. The involvement of epigenetic abnormities in various human pathologies indicates that specific diseases might benefit from epigenetic targeted therapies. Inspiringly, durg therapy aimed at targeting epigenetic defects such as Dacogen (Decitabine; Eisai Inc) and Zolinza (Vorinostat; Merck) is coming into reality in clinical settings (Giacinti et al. 2008; Ptak and Petronis 2008). Furthermore, the development of Genome Wide Association Study (GWAS) to characterize and functionally analyze the variety of epigenetic modifications through the entire genome (epigenomics) will provide insight into the function of epigenetic modifications not only in normal development and but also in the subsequent transition to disease states, ultimately leading to the future development of more effective epigenetic-based therapies (Hamm and Costa 2011).

5.2 Epigenetics and the Emergence of Epigenomics—A Burgeoning Field

The term epigenetics was first introduced by Conrad H. Waddington as early as the mid-twentieth century to describe the variety of developmental phenomena above the level of genomes that connected genotype to phenotype. The originally definition of epigenetics was completely based on observations at the level of organism development depending on the interactions with the environment, which is currently known as the epigenotype (Jamniczky et al. 2010). Subjected to the limited knowledge and research technique, the definition of epigenetics remanins both contentious and ambiguous (Berger et al. 2009). Oster and Albert once proposed that physical interactions of tissues and extracellular matrices are crucial for the developmental process and epigenetic components might play an important role, while Newman and Muller refined epigenetics as the interactions fo cells with each other and with the surrounding microenvironment (Newman and Muller 2000).

Currently, it is clear that Waddington's observation at the level of the organism was a consequence of a series of molecular changes that occur in the DNA of the cells after interactions with the microenvironment and epigenetics act as a bridge between genetic and environmental factors. Decades after the word epigenetics coined by Waddington to link the fields of developmental biology and genetics, 5-methyl cytosine was justified to have a role in silencing gene expression at the molecular level and that the patterns of DNA methylation are somehow heritable (Holliday 2006). Griffith and Mahler made the first suggestion that the gain or loss of DNA methylation has an important biological role when studying brain memory and Holiday and Pugh proposed a molecular model for turning genes on and off based on changes in DNA methylation (Griffith and Mahler 1969; Holliday and Pugh 1975). At last, through summarization of the previous research, Holliday revisited Waddington's ideas and published a crucial article connecting the molecular and phenotypic aspects of epigenetics (Holliday 1987). In view of the development of DNA methylation, the epigenetics have experienced substantial growth over the past decades.

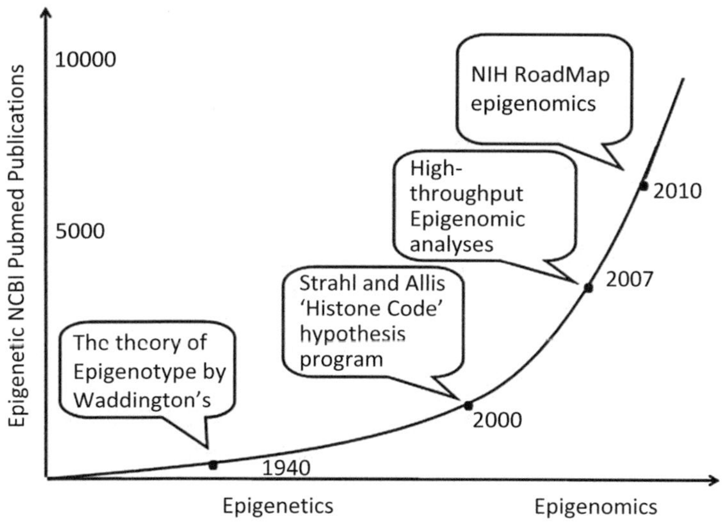

Fig. 5.2 Historical perspective of epigenetics and epigenomics, and the growing number of publications in this field

Epigenetic modifications in a whole genome, known as the epigenome, has emerged with the advances in technologies of molecular biology and next-generation DNA sequencing. Epigenome was defined as a new discipline that studies epigenetic modifications at the molecular level in an entire genome instead of single gene or a smaller number of genes (Callinan and Feinberg 2006). Compared to epigenetics, epigenomics aim at interrogating entire genomes for methylation changes or histone modifications by combining traditional epigenetic analyses with next-generation DNA sequencing. Genome-wide association studies (GWAS) are moving toward including epigenetic analysis (epigenomics) alongside the analysis of single nucleotide polymorphisms (SNPs) in individuals to build a more complete hereditary profile for complex diseases. Examples include epigenomic DNA methylation analyses in complex diseases, such as cancer and psychiatric disorders (Xie et al. 2010; Kato 2009). Epigenetics has seen substantial growth over the past 20 years, and the number of epigenetic research articles published in a given year reflects the growth of the Field. A summary of the research articles, theories, hypothesis and main discoveries in the epigenetics and epigenomics are shown in Fig. 5.2. Despite significant advances in the field, epigenetics and epigenomics are still in infancy. Along with the advances in molecular biology and next-generation DNA sequencing in defining and unraveling the myriad of epigenetic alterations in particular disease states, it will provide opportunity for the development of epigenetic therapeutics across a spectrum of human diseases.

5.3 Clinical Epigenetics and Epigenomics in Complex Diseases

Epigenetics is rising to prominence in biology as a mechanism by which environmental factors have intermediate-term effects on gene expression without changing the underlying genetic sequence. Epigenetics is not just participating in organism development where stable and distinct cellular functions must be established from an identical genotype (Handel et al. 2010). Taking mammalian embryogenesis for example, mammalian genome is demethylated by the time the morula stage embryo has developed; however, de novo DNA methylation is detected in the blastocyst stage embryo (Santos et al. 2002; Mayer et al. 2000). Given the precise regulatory mechanism of epigenetics in regulating appropriate differentiation of embryonic stem cells, it is no doubt that the dyregulated epigenetics participate in the pathogenesis of a variety of diseases. The ability to dissect the epigenomic landscape is not only essential for a more complete understanding of normal development, but is also necessary to gain insight into the etiology of complex diseases. It is gradually recognised that multiple genetic and epigenetic factors contribute to complex diseases and might change throughout the course of the disease. As listed by the World Health Organization, the leading causes of death such as coronary artery disease (referred to as cardiovascular disease), diabetes mellitus (type I and type II) and cancer are diseases that might have an epigenetic and epigenomic component (Kargul et al. 2015). Besides, Epigenetic abnormalities might have a role in infertility, as the impairment of embryo implantation in endometriosis resulting from abnormal DNA hypermethylation and the subsequent silencing of HOX genes (Cakmak and Taylor 2011; Endoh et al. 2012). While epigenetic deregulation occours in a wide range of dieseases, it is important to emphasize that epigenetic changes are reversible. The reversible nature of epigenetics provides plausible treatment or prevention prospects for diseases previously thought hard-coded into the genome.

5.3.1 *Epigenetics and Epigenomics in Cancerous Disease*

Cancer represents a group of over 300 specific diseases that share a number of genetic, epigenetic, and pathological features and tumorigenesis is a multistep process, including initiation, promotion and progression (Sandoval and Esteller 2012; Kinzler and Vogelstein 1996). In contrast to genetic defects, epigenetic deregulation has been increasingly recognized as a hallmark of cancer for the last decade with the advent of whole-genome approaches known as epigenomics. Epigenetic modifications precede genetic changes, and usually occur at an early stage in development of a neoplasm, but may be involved in its invasion and spread as well (Fig. 5.3). DNA methylation, histone modification, nucleosome remodeling, and RNA-mediated targeting regulate many biological processes that are fundamental to the genesis of cancer (Valdespino and Valdespino 2015).

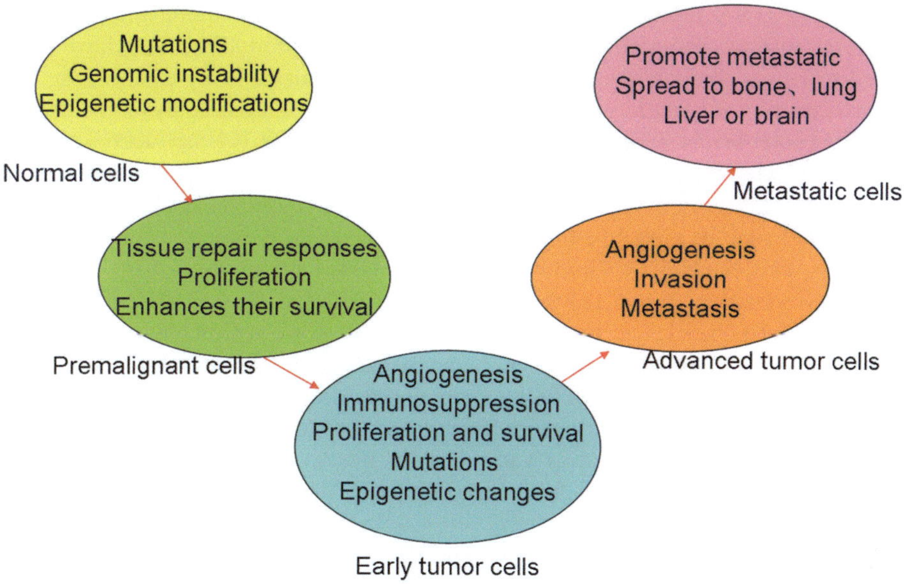

Fig. 5.3 The multifaceted role of epigenetic modifications in each stages of tumorigenesis. The mutations, genomic instability, and epigenetic modifications can lead to tumor initiation. Then, Inflammation activates tissue repair responses, induces proliferation of premalignant cells. Further to stimulate angiogenesis, cause localized immunosuppression, and form a inflammatory microenvironment in which pre-malignant cells can survive, expand, and in turn accumulate additional mutations and epigenetic changes to promote tumor cells spread to different organs, such as lung, bone, brain and liver

The best-studied epigenetic alterations in cancer are the methylation changes that occur within CpG islands, which are present in 80 % of all mammalian promoters. It is confirmed that in various cancer genomes up to 5–10 % of normally unmethylated CpG promoter islands are abnormally methylated. Meanwhile, CpG hypermethylation of promoters not only influence the expression of protein coding genes but also the expression of various noncoding RNAs, which have a role in malignant transformation (Baylin and Jones 2011). In the process of DNA methylation, three active DNA methyltransferases (DNMTs) have been identified. DNMT1 is a maintenance methyltransferase that recognizes hemimethylated DNA generated during DNA replication and then methylates newly synthesized CpG dinucleotides, conversely DNMT3a and DNMT3b function primarily as de novo methyltransferases to establish DNA methylation during embryogenesis (Klose and Bird 2006). Various studies designed at sequencing of cancer genomes have identified recurrent mutations in DNMT cancer, such as DNMT3A in acute myeloid leukemia (AML) (Ley et al. 2010). Understanding the cellular consequences of normal and aberrant DNA methylation remains a key area of interest, especially because hypomethylating agents are one of the few epigenetic therapies that have gained FDA approval for routine clinical use.

Besides DNA methylation, abberant patterns of histone modifications are a characteristic of cancer. Genome-wide studies of histone modifications have been done to better characterize the chromatin of malignant cells by establishing the overall profile of histone modifications in cancer. The global analysis of histone modifications levels reveals a pattern of altered dimethyl-K4 and acetyl-K18 of histone H3 in prostate cancer, which are proposed as being markers of high risk of recurrence (Seligson et al. 2005; Dawson Mark and Kouzarides 2012). The prognostic relevance of global histone modification levels was reported for non small cell lung cancer (NSCLC) and high dimethyl-H3K4 or low acetyl-H3K9 levels were corrletated with a better survival as well (Barlesi et al. 2007). Moreover, global acetyl-H3K9 levels were used to screen out these patients with low-grade bladder cancer who experienced disease recurrence after transurethral resection of the bladder (Barbisan et al. 2008). As genetic lesions in histone-modifying complexes are associated with an aberrant histone modifications in cancer and HDACs are frequently overexpressed in various types of cancer, tumor types would be distinguished on the basis of the expression patterns of their histone-modifying enzymes (Ozdag et al. 2006).

DNA methylation and histone modifications are the most well studied epigenetic modifications in cancer, and the role of miRNAs in the etiology, progression and prognosis of cancer is emerging. Several studies have proposed the profiles of miRNA expression differ between normal and tumor tissues and between tumor types (He and Hannon 2004). Alongside the genome-wide approaches, it is capable of production of miRNA fingerprints in a range of and the identification of new potential biomarkers to distinguish tumor tissue from its normal counterpart (Budhu et al. 2008; Nam et al. 2008).

In summary, the advent of microarray-based technologies and the more recently developed next-generation sequencing technology has enormously increased the data available for assessing epigenetic features of the various human cancer genomes. The combination and integration of epigenomics, genomics and all the other 'omics', including transcriptomic, proteomic, will be definitely accelerating the progress towards a full understanding of the underlying molecular mechanisms that govern the initiation and development of cancer processes. Altogether, these cancer signatures will help identify new potential prognostic and detection tools and, eventually, to develop effective clinical epigenetic therapies.

5.3.2 Epigenetics and Epigenomics in Autoimmune Disease

Autoimmune diseases are complex multifactorial diseases characterized by impaired immunological response against healthy cells and tissues due to the lack of recognition and the loss of immunological tolerance versus self (Hewagama and Richardson 2009; Zhernakova et al. 2013). Although autoimmune diseases possess diverse epidemiology or symptoms, it is clear that genetic predisposition is involved in the etiopathology of autoimmune disorders. However, the incomplete concordance rate of the autoimmune disease in monozygotic twins ranging between 12 and 67 % and the presence of a strong genetic association only in a proportion patients strongly supports the involvement of non-genetic mechanisms in these pathologies (Dang et al. 2013; Renz et al. 2011). Accumulating evidence indicates a role for environmental factors, one of the challenging questions is how such environmental components mechanistically influence autoimmune diseases. In the regard the concept of epigenetic modifications have gained great interest. The development and differentiation of immune cells, as well as innate and adaptive responses, are precisely regulated by dynamic epigenetic modifications (Harb and Renz 2015; Zhao et al. 2015). It has been disclosed that identified epigenetic alterations give rise to several typical human autoimmune diseases such as systemic lupus erythematosus (SLE), rheumatoid arthritis (RA) and multiple sclerosis (MS).

SLE is a chronic multiorgan disease characterized by acute and chronic inflammation with different clinical manifestations due to autoantibodies against nuclear and cytoplasmic antigens. Over the past decade, various studies have shown that gene-specific hypomethylation, particularly of several autoreactivity-related genes, plays an essential role in the pathogenesis of SLE. Clearly, CD11a (ITGAL), perforin (PRF1), CD70 (TNFSF7), and CD40LG (TNFSF5) have all been shown to be overexpressed in lupus CD4+ T cells and contribute to the over-production of auto-antibodies by B cells upon activation by autoreactive T cells (Deng and Tsao 2014; Lu et al. 2002, 2003, 2005). Moreover, aberrant DNA hypomethylation was reported in over-expressed subtype-specific genes in T cell as well, such as interferon related genes including ifng, IFI44 and OSA1, Th2 cytokines IL-4 and IL-6 and the Th17 cytokine IL-17a, leading to dysregulated immune homeostasis in SLE (Absher et al. 2013; Talaat et al. 2015). As for the mechanism leading to aberrant epigenetic modifications in lupus T cells, it was proposed that decreased DNMT1 expression contributes to passive demethyla-tion and increased Gadd45α causes active demethylation in CD4+T cells of SLE patients (Zhang et al. 2013). On the other side, researchers have indentified that defective ERK signaling and up-regulation of microRNA-126 and microRNA-29b leads to the inhibition of DNMT1 expression and contributes to DNA hypomethylation and lupus T cell autoreactivity (Qin et al. 2013). In addition to DNA methylation, global histone H3 and H4 hypoacetylation of CD4 T cells of SLE was observed and the degree of histone H3 acetylation (Hu et al. 2008). Dai et al. showed a significant alteration of H3K4me3 in many key relevant candidate genes in PBMCs of SLE, the alterations of which were associated with the pathogenesis of SLE and could represent a potential biomarker for epigenetic-based lupus therapies (Dai et al. 2010).

The role of epigenetic in the RA onset has been investigated showing a wide DNA methylation status in RA synovial fibroblasts or peripheral blood mononuclear cells (PBMCs). A genome wide evaluation of RA synovial fibroblasts has shown a differential methylation status of several genes, which appear to be involved in immune response, inflammation and leukocyte recruitment (Nakano et al. 2013). In RA synovial cells, Death receptor 3, a factor of apoptosis-inducing Fas gene family was hypermethylated, leading to downregulation of proteins involved in the resistance to apoptosis. The methylation of a single CpG in the

IL-6 promoter region which affects the regulation of IL-6 gene in PBMCs of RA was indentified, suggesting a role in the pathogenesis of RA with a local hyperactivation of inflammatory pathways (Nile et al. 2008). Different from involvement of DNA methylation, histone modifications are so far limited in RA. Reduced HDAC activity plays an important role in the transcriptional regulation of proinflammatory genes in RA resulting in histones hyperacetylation. As a result, the balance of histone acetylase/HDAC activityis strongly shifted to histone hyperacetylation in RA patients and HDAC inhibitors represent a promising treatment approach.

Taking SLE and RA for example, ranging from candidate-gene to genome-wide association studies, a large number of susceptibility genes have been identified in the pathogenesis of autoimmune diseases. Autoimmune diseases recognize multiple genetic and environmental determinants and epigenetic changes are emerging as related factors. The endeavor to completely understand molecular mechanisms governing epigenetic alterations will provide more research orientations and it is noteworthy that epigenetic alterations vary among different tissue and cell types (Glossop et al. 2013; Zhang and Zhang 2015; Picascia et al. 2015). The discovery of shared aspects of disease pathogenesis could offer the opportunity to use known drugs or therapeutic strategies across multiple diseases.

5.3.3 Epigenetics and Epigenomics in Metabolic Disease

The most prevalent metabolic disorders are diabetes mellitus, obesity, dyslipidemia, osteoporosis and metabolic syndrome, the pathophysiologies of which are oxidative stress, Nrf2 pathways, epigenetic modifications (Tabatabaei-Malazy et al. 2015). Clinical and experimental studies indicate that life experiences, perhaps spanning multiple generations, influence lifelong risk of metabolic dysfunction through epigenetic mechanisms (Gluckman 2012). Epigenetic inheritance does not involve alterations in gene sequencing, gene addition, or gene deletion. Rather, epigenetic dysregulation involved in metabolic disease through modifications of: (1) cell differentiation, (2) dosage compensation, (3) genome structure maintenance, (4) genomic/parental imprinting, and (5) repetitive element repression (Bays and Scinta 2015).

Obesity is closely associated with the development of both type diabetes (T2D) and metabolic syndrome and increasing evidence implies that other than individual lifestyle choices, developing obesity is in part due to genetic disposition, especially epigenetic processes as well. A GWAS on 459 individuals of European origin was performed to explore the relationship between DNA methylation and BMI, and the analysis highlighted that increased BMI in these individuals is linked to an increase in methylation at the HIF3A locus in blood cells and in adipose tissue (Pokrywka et al. 2014; Heijmans et al. 2008). Genes that regulate adipogenesis, glucose homeostasis, inflammation and insulin signaling are regulated by epigenetic mechanisms, including genes encoding hormones (for example, leptin), nuclear receptors (adipogenic and lipogenic transcription factors PPARγ and PPARα, respectively), gluconeogenic enzymes (phosphoenolpyruvate carboxykinase (PEPCK)) and transmembrane proteins (such as uncoupling protein 1) (Stepanow et al. 2011; Noer et al. 2007; Yang et al. 2011; Milagro et al. 2009). It was shown that consumption of a high-fat diet resulted in elevated level of methylation of leptin gene promoter in retroperitoneal adipocytes, which was associated with lower circulating leptin levels, suggesting leptin methylation affects leptin gene expression. Similarly, adipogenesis is driven by adipocyte differentiation and induction of adipogenic transcription factors (PPARγ, C/EBPα) via epigenetic mechanisms (Desai et al. 2015).

Other than Type I diabetes mellitus (T1DM) categorized as autoimmune disease, Type II diabetes mellitus (T2DM) is a widespread metabolic disease characterized by insulin resistance pancreatic beta-cell failure. As pancreatic beta-cell failure is core at the pathogenesis of T2DM, various mechanisms have been reported, including decreased insulin signaling, endoplasmic reticulum stress, oxidative stress, and inflammation. One of the underlying mechanisms is epigenetic modification, such as the reduction of histone acetylation and increase of methylation in the promoter region of the Pdx1 gene, which encodes an important transcription factor for pancreatic beta-cell function, leading to the reduction of Pdx1 expression levels (Kido 2013). Meanwhile, numerous susceptibility genes for type 2 diabetes, including KCNQ1, have been gradually identified in humans using genome-wide analyses and other related studies. Insulin resistance is a the other common feature of T2DM and impaired response to insulin results in reduced capacity to clear the glucose from blood stream. Skeletal muscle and adipose tissue play a central role in glucose homeostasis. Genome-wide DNA methylation in human skeletal muscle and adipose tissue from individuals with or without a family history of T2DM indentified the differential methylation baseline of T2DM candidate genes PPARGC1A, TFAM, PPARD, PDK4, MEF2A, THADA, NDUFC2, and IL-7 (Barres et al. 2009, 2012). Compared to the DNA hypermethylation, the diabetes mellitus drugs glucagon-like peptide 1 (GLP-1) and glucose-dependent insulinotropic-peptide 1 (GIP), the mechanisms of which are mediated through the ability of these compounds to increase histone H3 acetyltransferase activity and decrease HDAC activity have come into clinic (Kim et al. 2009). Acumulating evidence highlight a key role for epigenetics in the growing incidence of metabolic disease. Epigenetics holds promises for therapeutic advances, with numerous studies of, for instance, HDAC inhibitors clearly demonstrating beneficial effects on diabetic phenotypes (Ronn and Ling 2015).

5.4 Traditional Epigenetic Therapies

Given the prevalence of epigenetic abnormalities in various types of disease, epigenetic therapy holds great promise for clinical treatment. Two main classes of epigenetic therapies are DNA methyltransferase inhibitors and histone deacetylase (HDAC) inhibitors, which act globally by promoting a more-open chromatin structure and subsequently they promote gene expression. Inhibition of DNA methylation and histone deacetylation has shown promise in clinical trials in myelodysplastic syndrome, acute myeloid leukaemia and T cell lymphoma with still more promising treatment candidates on the horizon (Ptak and Petronis 2008; Wijermans et al. 2000; Garcia-Manero et al. 2006; Duvic et al. 2007).

However, there are unavoidable limitations of the traditional epigenetic treatments. The most serious is a lack of specificity: though epigenetic silencing of tumour suppressor genes inevitably leads to carcinogenesis, while simply inducing global DNA demethylation would result in chromosomal instability. Indeed, DNA hypomethylation promotes tumor formation for activation of genes that are normally epigenetically silenced (Gaudet et al. 2003). Taken together, current DNA methyltransferase inhibitors and histone deacetylase (HDAC) inhibitors are proved to have clinical benefits in diseases that arise from repressive chromatin-mediated gene silencing, while these non-specific drugs should be carefully examined to determine whether the therapeutic benefits outweigh the potential adverse effects.

5.5 New Avenues for Epigenomic Therapy

Epigenetic therapy is emerging as a potentially effective therapy for a variety of diseases. The combination of additional epigenomic (DNAmethylation and chromatin immunoprecipitation or (ChiP)) analysis with next-generation DNA sequencing will obtain a complete molecular profile of the epigenomic landscape in both normal and disease states. Only when the epigenomic profiling truly be global, and focus on regions containing protein-coding sequences as well as other regions of the genome (such as regulatory sequences, ncRNAs, and repetitive elements) will new disease-associated epigenomic changes and new therapeutic targets be identified. For example, PAD4 which catalyzes post-translational modification of arginine to citrulline in histone proteins (H2A, H3 and H4), and is associated with transcriptional repression is a newly identified epigenetic target. Moreover, F-amidine and Cl-amidine, the selective PAD4 inhibitors, show promise for rheumatoid arthritis and multiple sclerosis with abnormal PAD4 activity (Denis et al. 2009).

The discovery of PAD4 selective inhibitor signifies that the future epigenetic therapy would aim at abberant epigenetic molecules rather than blindly introduction of global DNA methyltransferase inhibitors and histone deacetylase (HDAC) inhibitors. Except for selective DNA methyltransferase inhibitors and histone deacetylase (HDAC) inhibitors, RNA molecules that would specifically interfere with abberant epigenetic changes. Given that miRNA-221 is up-regulated in highly aggressive tumors, systemic administration of anti-miRNA-221 (2-O-methylphosphorothioate- modified- anti-miRNA-221) was shown to exert an anti-tumor effect (Park et al. 2011). Moreover, systemic administration of miRNA-26a, which is frequently down-regulated in HCC, showed anti-tumor efficacy in vivo by inhibiting cancer progression via apoptosis (Kota et al. 2009).

Considering the adverse events of the traditional epigenetic therapy, great progress has been made in developing drugs capable of targeting abberant epigenetic alterations. That is to say, epigenomic profiling is fostering in the era of pharmacoepigenetics and phamacoepigenomics, fields that involve the study of the relationship between the epigenome and optimal drug dosage and/or response, with a goal of optimizing individualized treatment and discovering new drug targets (Anestopoulos et al. 2015).

5.6 Conclusions and Future Prospects

The importance of epigenetics as well as epigenomics is highlighted by, but not limited to, its function in normal development and physiology. Although still in its infancy, epigenomics holds substantial promise in helping to explain many previously intractable conundrums in human genetics. Epigenetic modifications, including DNA methylation, histone modification, and ncRNAs, regulate gene expression and in turn determine the dynamic molecular and cellular events during disease initiation and progression. For example, hyper-methylation (silencing of tumor-suppressor genes) and hypo-methylation (i.e. activation of proto-oncogenes) events have been considered as critical ones for various diseases. Moreover, aberrant expression of histone modifying enzymes and non-coding RNAs (micro-RNAs and lnc-RNAs) completes the epigenetic landscape. However, the mentioned each epigenetic type should be considered not as an isolated event but rather as a network of crosstalk and cooperation that contributes to the disease pathophysiology. Fortunately, the continued development of new molecular technologies will aid in high resolution mapping of the epigenomic landscape. In the meantime, it will provide opportunities to identify novel target and signaling pathways that might act as modifiers of the epigenome. Thus, a greater understanding of epigenomics and the factors that mediate changes to the epigenome will lead to a better knowledge of gene regulation and will also translate into more effective disease treatments. Three HDAC inhibitors (HDACi) are currently approved by the FDA and several HDACi are in clinical trial. Other include non-coding RNAs such as microRNAs and long RNA molecules that are emerging as important epigenomic modifiers. In the future, epigenetic modifying agents may provide a means to improve the effectiveness of existing drugs and be a promising field for clinical interventions not just for the mentioned cancerous disease, autoimmune disease and metabolic disease, but for other complex diseases. That is just the beginning of the epigenomics era.

References

Absher DM, Li X, Waite LL, et al. Genome-wide DNA methylation analysis of systemic lupus erythematosus reveals persistent hypomethylation of interferon genes and compositional changes to CD4+ T-cell populations. PLoS Genet. 2013;9:e1003678.

Amaral PP, Dinger ME, Mercer TR, et al. The eukaryotic genome as an RNA machine. Science. 2008;319:1787–9.

Anestopoulos I, Voulgaridou GP, Georgakilas AG, et al. Epigenetic therapy as a novel approach in hepatocellular carcinoma. Pharmacol Ther. 2015;145:103–19.

Barbisan F, Mazzucchelli R, Santinelli A, et al. Immunohistochemical evaluation of global DNA methylation and histone acetylation in papillary urothelial neoplasm of low malignant potential. Int J Immunopathol Pharmacol. 2008;21:615–23.
Barlesi F, Giaccone G, Gallegos-Ruiz MI, et al. Global histone modifications predict prognosis of resected non small-cell lung cancer. J Clin Oncol. 2007;25:4358–64.
Barres R, Osler ME, Yan J, et al. Non-CpG methylation of the PGC-1alpha promoter through DNMT3B controls mitochondrial density. Cell Metab. 2009;10:189–98.
Barres R, Yan J, Egan B, et al. Acute exercise remodels promoter methylation in human skeletal muscle. Cell Metab. 2012;15:405–11.
Baylin SB, Jones PA. A decade of exploring the cancer epigenome – biological and translational implications. Nat Rev Cancer. 2011;11:726–34.
Bays H, Scinta W. Adiposopathy and epigenetics: an introduction to obesity as a transgenerational disease. Curr Med Res Opin. 2015;31(11):2059–69.
Berger SL, Kouzarides T, Shiekhattar R, et al. An operational definition of epigenetics. Genes Dev. 2009;23:781–3.
Budhu A, Jia HL, Forgues M, et al. Identification of metastasis-related microRNAs in hepatocellular carcinoma. Hepatology. 2008;47:897–907.
Cakmak H, Taylor HS. Implantation failure: molecular mechanisms and clinical treatment. Hum Reprod Update. 2011;17:242–53.
Callinan PA, Feinberg AP. The emerging science of epigenomics. Hum Mol Genet. 2006;15(Spec No 1):R95–101.
Chiang PK, Gordon RK, Tal J, et al. S-Adenosylmethionine and methylation. FASEB J. 1996;10:471–80.
Costa FF. Non-coding RNAs, epigenetics and complexity. Gene. 2008;410:9–17.
Dai Y, Zhang L, Hu C, et al. Genome-wide analysis of histone H3 lysine 4 trimethylation by ChIP-chip in peripheral blood mononuclear cells of systemic lupus erythematosus patients. Clin Exp Rheumatol. 2010;28:158–68.
Dang MN, Buzzetti R, Pozzilli P. Epigenetics in autoimmune diseases with focus on type 1 diabetes. Diabetes Metab Res Rev. 2013;29:8–18.
Dawson Mark A, Kouzarides T. Cancer epigenetics: from mechanism to therapy. Cell. 2012;150:12–27.
Deng Y, Tsao BP. Advances in lupus genetics and epigenetics. Curr Opin Rheumatol. 2014;26:482–92.
Denis H, Deplus R, Putmans P, et al. Functional connection between deimination and deacetylation of histones. Mol Cell Biol. 2009;29:4982–93.
Desai M, Jellyman JK, Ross MG. Epigenomics, gestational programming and risk of metabolic syndrome. Int J Obes (Lond). 2015;39:633–41.
Duvic M, Talpur R, Ni X, et al. Phase 2 trial of oral vorinostat (suberoylanilide hydroxamic acid, SAHA) for refractory cutaneous T-cell lymphoma (CTCL). Blood. 2007;109:31–9.
Endoh M, Endo TA, Endoh T, et al. Histone H2A mono-ubiquitination is a crucial step to mediate PRC1-dependent repression of developmental genes to maintain ES cell identity. PLoS Genet. 2012;8:e1002774.
Feinberg AP, Tycko B. The history of cancer epigenetics. Nat Rev Cancer. 2004;4:143–53.
Feinberg AP, Vogelstein B. Hypomethylation distinguishes genes of some human cancers from their normal counterparts. Nature. 1983;301:89–92.
Feng J, Fan G. The role of DNA methylation in the central nervous system and neuropsychiatric disorders. Int Rev Neurobiol. 2009;89:67–84.
Fujita N, Watanabe S, Ichimura T, et al. Methyl-CpG binding domain 1 (MBD1) interacts with the Suv39h1-HP1 heterochromatic complex for DNA methylation-based transcriptional repression. J Biol Chem. 2003;278:24132–8.
Garcia-Manero G, Kantarjian HM, Sanchez-Gonzalez B, et al. Phase 1/2 study of the combination of 5-aza-2′-deoxycytidine with valproic acid in patients with leukemia. Blood. 2006;108:3271–9.
Gaudet F, Hodgson JG, Eden A, et al. Induction of tumors in mice by genomic hypomethylation. Science. 2003;300:489–92.

Giacinti L, Vici P, Lopez M. Epigenome: a new target in cancer therapy. Clin Ter. 2008;159:347–60.
Glossop JR, Nixon NB, Emes RD, et al. Epigenome-wide profiling identifies significant differences in DNA methylation between matched-pairs of T- and B-lymphocytes from healthy individuals. Epigenetics. 2013;8:1188–97.
Gluckman PD. Epigenetics and metabolism in 2011: epigenetics, the life-course and metabolic disease. Nat Rev Endocrinol. 2012;8:74–6.
Griffith JS, Mahler HR. DNA ticketing theory of memory. Nature. 1969;223:580–2.
Hamm CA, Costa FF. The impact of epigenomics on future drug design and new therapies. Drug Discov Today. 2011;16:626–35.
Hamm CA, Costa FF. Epigenomes as therapeutic targets. Pharmacol Ther. 2015;151:72–86.
Handel AE, Ebers GC, Ramagopalan SV. Epigenetics: molecular mechanisms and implications for disease. Trends Mol Med. 2010;16:7–16.
Harb H, Renz H. Update on epigenetics in allergic disease. J Allergy Clin Immunol. 2015;135:15–24.
Hassan AH, Prochasson P, Neely KE, et al. Function and selectivity of bromodomains in anchoring chromatin-modifying complexes to promoter nucleosomes. Cell. 2002;111:369–79.
He L, Hannon GJ. MicroRNAs: small RNAs with a big role in gene regulation. Nat Rev Genet. 2004;5:522–31.
Heijmans BT, Tobi EW, Stein AD, et al. Persistent epigenetic differences associated with prenatal exposure to famine in humans. Proc Natl Acad Sci U S A. 2008;105:17046–9.
Hewagama A, Richardson B. The genetics and epigenetics of autoimmune diseases. J Autoimmun. 2009;33:3–11.
Holliday R. The inheritance of epigenetic defects. Science. 1987;238:163–70.
Holliday R. DNA methylation and epigenetic inheritance. Philos Trans R Soc Lond B Biol Sci. 1990;326:329–38.
Holliday R. Epigenetics: a historical overview. Epigenetics. 2006;1:76–80.
Holliday R, Pugh JE. DNA modification mechanisms and gene activity during development. Science. 1975;187:226–32.
Hu N, Qiu X, Luo Y, et al. Abnormal histone modification patterns in lupus CD4+ T cells. J Rheumatol. 2008;35:804–10.
Jamniczky HA, Boughner JC, Rolian C, et al. Rediscovering Waddington in the post-genomic age: operationalising Waddington's epigenetics reveals new ways to investigate the generation and modulation of phenotypic variation. Bioessays. 2010;32:553–8.
Kargul J, Irminger-Finger I, Laurent GJ. Epigenetics regulation of disease: there is more to a gene than its sequence. Int J Biochem Cell Biol. 2015;67:43.
Kato T. Epigenomics in psychiatry. Neuropsychobiology. 2009;60:2–4.
Kido Y. Progress in diabetes. Rinsho Byori. 2013;61:941–7.
Kim SJ, Nian C, McIntosh CH. Glucose-dependent insulinotropic polypeptide and glucagon-like peptide-1 modulate beta-cell chromatin structure. J Biol Chem. 2009;284:12896–904.
Kinzler KW, Vogelstein B. Lessons from hereditary colorectal cancer. Cell. 1996;87:159–70.
Klose RJ, Bird AP. Genomic DNA methylation: the mark and its mediators. Trends Biochem Sci. 2006;31:89–97.
Kota J, Chivukula RR, O'Donnell KA, et al. Therapeutic microRNA delivery suppresses tumorigenesis in a murine liver cancer model. Cell. 2009;137:1005–17.
Ley TJ, Ding L, Walter MJ, et al. DNMT3A mutations in acute myeloid leukemia. N Engl J Med. 2010;363:2424–33.
Lister R, Pelizzola M, Dowen RH, et al. Human DNA methylomes at base resolution show widespread epigenomic differences. Nature. 2009;462:315–22.
Lu Q, Kaplan M, Ray D, et al. Demethylation of ITGAL (CD11a) regulatory sequences in systemic lupus erythematosus. Arthritis Rheum. 2002;46:1282–91.
Lu Q, Wu A, Ray D, et al. DNA methylation and chromatin structure regulate T cell perforin gene expression. J Immunol. 2003;170:5124–32.

Lu Q, Wu A, Richardson BC. Demethylation of the same promoter sequence increases CD70 expression in lupus T cells and T cells treated with lupus-inducing drugs. J Immunol. 2005;174:6212–19.

Mau T, Yung R. Potential of epigenetic therapies in non-cancerous conditions. Front Genet. 2014;5:438.

Mayer W, Niveleau A, Walter J, et al. Demethylation of the zygotic paternal genome. Nature. 2000;403:501–2.

Milagro FI, Campion J, Garcia-Diaz DF, et al. High fat diet-induced obesity modifies the methylation pattern of leptin promoter in rats. J Physiol Biochem. 2009;65:1–9.

Nakano K, Whitaker JW, Boyle DL, et al. DNA methylome signature in rheumatoid arthritis. Ann Rheum Dis. 2013;72:110–17.

Nam EJ, Yoon H, Kim SW, et al. MicroRNA expression profiles in serous ovarian carcinoma. Clin Cancer Res. 2008;14:2690–5.

Newman SA, Muller GB. Epigenetic mechanisms of character origination. J Exp Zool. 2000;288:304–17.

Nile CJ, Read RC, Akil M, et al. Methylation status of a single CpG site in the IL6 promoter is related to IL6 messenger RNA levels and rheumatoid arthritis. Arthritis Rheum. 2008;58:2686–93.

Noer A, Boquest AC, Collas P. Dynamics of adipogenic promoter DNA methylation during clonal culture of human adipose stem cells to senescence. BMC Cell Biol. 2007;8:18.

Ozdag H, Teschendorff AE, Ahmed AA, et al. Differential expression of selected histone modifier genes in human solid cancers. BMC Genomics. 2006;7:90.

Park JK, Kogure T, Nuovo GJ, et al. miR-221 silencing blocks hepatocellular carcinoma and promotes survival. Cancer Res. 2011;71:7608–16.

Perera F, Tang WY, Herbstman J, et al. Relation of DNA methylation of 5′-CpG island of ACSL3 to transplacental exposure to airborne polycyclic aromatic hydrocarbons and childhood asthma. PLoS One. 2009;4:e4488.

Picascia A, Grimaldi V, Pignalosa O, et al. Epigenetic control of autoimmune diseases: from bench to bedside. Clin Immunol. 2015;157:1–15.

Pogribny IP, Beland FA. DNA hypomethylation in the origin and pathogenesis of human diseases. Cell Mol Life Sci. 2009;66:2249–61.

Pokrywka M, Kiec-Wilk B, Polus A, et al. DNA methylation in obesity. Postepy Hig Med Dosw (Online). 2014;68:1383–91.

Ptak C, Petronis A. Epigenetics and complex disease: from etiology to new therapeutics. Annu Rev Pharmacol Toxicol. 2008;48:257–76.

Qin H, Zhu X, Liang J, et al. MicroRNA-29b contributes to DNA hypomethylation of CD4+ T cells in systemic lupus erythematosus by indirectly targeting DNA methyltransferase 1. J Dermatol Sci. 2013;69:61–7.

Ray-Gallet D, Almouzni G. Nucleosome dynamics and histone variants. Essays Biochem. 2010;48:75–87.

Renz H, von Mutius E, Brandtzaeg P, et al. Gene-environment interactions in chronic inflammatory disease. Nat Immunol. 2011;12:273–7.

Riddihough G, Zahn LM. Epigenetics. What is epigenetics? Introduction. Science. 2010;330:611.

Ronn T, Ling C. DNA methylation as a diagnostic and therapeutic target in the battle against Type 2 diabetes. Epigenomics. 2015;7:451–60.

Sanchez R, Zhou MM. The role of human bromodomains in chromatin biology and gene transcription. Curr Opin Drug Discov Devel. 2009;12:659–65.

Sandoval J, Esteller M. Cancer epigenomics: beyond genomics. Curr Opin Genet Dev. 2012;22:50–5.

Santos F, Hendrich B, Reik W, et al. Dynamic reprogramming of DNA methylation in the early mouse embryo. Dev Biol. 2002;241:172–82.

Seligson DB, Horvath S, Shi T, et al. Global histone modification patterns predict risk of prostate cancer recurrence. Nature. 2005;435:1262–6.

Stepanow S, Reichwald K, Huse K, et al. Allele-specific, age-dependent and BMI-associated DNA methylation of human MCHR1. PLoS One. 2011;6:e17711.

Tabatabaei-Malazy O, Larijani B, Abdollahi M. Targeting metabolic disorders by natural products. J Diabetes Metab Disord. 2015;14:57.
Talaat RM, Mohamed SF, Bassyouni IH, et al. Th1/Th2/Th17/Treg cytokine imbalance in systemic lupus erythematosus (SLE) patients: correlation with disease activity. Cytokine. 2015;72:146–53.
Valdespino V, Valdespino PM. Potential of epigenetic therapies in the management of solid tumors. Cancer Manag Res. 2015;7:241–51.
Wang KC, Chang HY. Molecular mechanisms of long noncoding RNAs. Mol Cell. 2011;43:904–14.
Wijermans P, Lubbert M, Verhoef G, et al. Low-dose 5-aza-2′-deoxycytidine, a DNA hypomethylating agent, for the treatment of high-risk myelodysplastic syndrome: a multicenter phase II study in elderly patients. J Clin Oncol. 2000;18:956–62.
Wolffe AP, Matzke MA. Epigenetics: regulation through repression. Science. 1999;286:481–6.
Xie H, Wang M, Bonaldo Mde F, et al. Epigenomic analysis of Alu repeats in human ependymomas. Proc Natl Acad Sci U S A. 2010;107:6952–7.
Yang BT, Dayeh TA, Kirkpatrick CL, et al. Insulin promoter DNA methylation correlates negatively with insulin gene expression and positively with HbA(1c) levels in human pancreatic islets. Diabetologia. 2011;54:360–7.
Zhang Z, Zhang R. Epigenetics in autoimmune diseases: pathogenesis and prospects for therapy. Autoimmun Rev. 2015;14:854–63.
Zhang Y, Zhao M, Sawalha AH, et al. Impaired DNA methylation and its mechanisms in CD4(+) T cells of systemic lupus erythematosus. J Autoimmun. 2013;41:92–9.
Zhao M, Wang Z, Yung S, et al. Epigenetic dynamics in immunity and autoimmunity. Int J Biochem Cell Biol. 2015;67:65–74.
Zhernakova A, Withoff S, Wijmenga C. Clinical implications of shared genetics and pathogenesis in autoimmune diseases. Nat Rev Endocrinol. 2013;9:646–59.

Nian Dong Department of Pulmonary Medicine, The First affiliated Hospital, Wenzhou Medical University, Wenzhou, China.

Lin Shi, Zhongshan Hospital, Shanghai Institute of Clinical Bioinformatics; Fudan University Center for Clinical Bioinformatics; Biomedical Research Center of Fudan University Zhongshan Hospital, China.

Chengshui Chen, Department of Pulmonary Medicine, The First affiliated Hospital, Wenzhou Medical University, Wenzhou, China.

Wenhuan Ma, MD, PhD, Associate Professor, Vice President of Zhabei District Hospital of Traditional Chinese Medicine, Deputy Director of Shanghai Research Center of Traditional Chinese Medicine for Community Health Service, Master Supervisor of Shanghai University of Chinese Medicine, member of Professional Committee of Respiratory Disease in Chinese Association of the Integration of Traditional and Western Medicine, Shanghai Association of the Integration of Traditional and Western Medicine. Clinical specialty: Combined Therapy of Chinese Medicine with Western Medicine for Cardiovascular and Respiratory Diseases.

Xiangdong Wang, Department of Pulmonary Medicine, The First affiliated Hospital, Wenzhou Medical University, Wenzhou, China.
Zhongshan Hospital, Shanghai Institute of Clinical Bioinformatics; Fudan University Center for Clinical Bioinformatics; Biomedical Research Center of Fudan University Zhongshan Hospital, China.

Chapter 6
Proteomic Profiling: Data Mining and Analyses

Lan Zhang, Wei Zhu, Yong Zeng, Jigang Zhang, and Hong-Wen Deng

Abstract Proteomics, the large scale study of proteins, provides a complementary approach to genomics in exploring biological phenomena. With modern development of Mass Spectrometry-based technologies, proteomics has evolved into a powerful analytical platform for life science researchers, and has advanced our understanding of the complex and dynamic nature of proteins. In the clinical field, proteomics studies have been widely applied in identifying biomarkers, monitoring disease status, and assessing treatment effect. In this chapter, an overview of current proteomics profiling is introduced from four perspectives: collecting protein samples with appropriate experimental approaches, characterizing protein features with advanced mass spectrometry-based technologies, annotating protein information with publicly available databases, and interpreting protein functions with bioinformatics analyses. We also give an example of how proteomics research workflow is applied in breast cancer studies.

Keywords Proteomics • Mass spectrometry technologies • Database and standard • Bioinformatics analysis

---

L. Zhang • J. Zhang
Center for Bioinformatics and Genomics, Department of Biostatistics and Bioinformatics, Tulane University, New Orleans, LA 70112, USA

W. Zhu
Center for Bioinformatics and Genomics, Department of Biostatistics and Bioinformatics, Tulane University, New Orleans, LA 70112, USA

College of Life Sciences, Hunan Normal University, Changsha, Hunan 410081, People's Republic of China

Y. Zeng
Center for Bioinformatics and Genomics, Department of Biostatistics and Bioinformatics, Tulane University, New Orleans, LA 70112, USA

College of Life Sciences and Bioengineering, Beijing Jiaotong University, Beijing 100044, China

H.-W. Deng (✉)
Center for Bioinformatics and Genomics, Department of Biostatistics and Bioinformatics, Tulane University School of Public Health and Tropical Medicine, New Orleans, LA, USA
e-mail: hdeng2@tulane.edu

X. Wang et al. (eds.), *Application of Clinical Bioinformatics*,
Translational Bioinformatics 11, DOI 10.1007/978-94-017-7543-4_6

Abbreviations

2-DE	Two-dimensional gel electrophoresis
AP-MS	Affinity purification coupled with mass spectrometry
BLAST	Basic local alignment search tool
CDS	Coding sequences
DE	Differentially expressed
ES	Enrichment score
ESI	Electrospray ionization
FTICR	Fourier-transform ion cyclotron resonance
GC-MS	Gas chromatography-mass spectrometry
GO	Gene ontology
GPMDB	Global proteome machine database
GSEA	Gene set enrichment analysis
HGNC	Hugo gene nomenclature committee
HUPO-PSI	Hupo proteomics standards initiative
KNN	K-nearest neighbors
LC-MS	Liquid chromatography-mass spectrometry
LOWESS	Locally weighted scatterplot smoothing
LSA	Least-squares adaptive
MALDI	Matrix-assisted laser desorption/ionization
MALDI-TOF	Matrix assisted laser desorption ionization time-of-flight
MIAPE	Minimum information about a proteomics experiment
MOPED	Model organism protein expression database
MOWSE	Molecular weight search
MS	Mass spectrometry
MSE	Mass spectrometry with elevated energy
MudPIT	Multidimensional protein identification technology
PLGEM	Power law global error model
PLGS	Proteinlynx global server
PPIN	Protein-protein interaction network
PPIs	Protein–protein interactions
PSEA	Protein set enrichment analysis
PTMs	Post translational modifications
RPLC	Reversed-phase liquid chromatography
SAM	Significance analysis of microarray
SDS-PAGE	Sodium dodecyl sulfate polyacrylamide gel electrophoresis
SILAC	Stable isotope labeling by amino acids in cell culture
SpI	Spectral index
TOF	Time-of-flight
XML	Extensible markup language

6.1 Introduction

In the past, the genome has been the principal focus for explaining the molecular basis of cell functions. Since the survival of cells usually depends on a multitude of metabolic and regulatory pathways, studying biological process at the genome level is far from enough. Due to complex biological modifications such as transcriptional splicing, post-transcriptional splicing, translational modifications and translational regulation, the expression or function of proteins cannot be precisely predicted from analysis of nucleic acids. Compared with mRNA and DNA level data analysis, proteomic analysis is more efficient in identifying phenotype-related genomic features, since largely proteins determine traits.

Over the past two decades, technologies in proteome research have improved dramatically. Two classic methods, two-dimensional gel electrophoresis (2-DE) and antibody microarray have played significant roles in proteomics studies. However, these two approaches cannot meet the requirements of large-scale protein identification and accurate quantification. Remarkably, the invention and rapid development of mass spectrometry (MS) technologies has provided a powerful platform for proteomics profiling and made it possible to identify and quantify whole proteins in biological systems. Taking advantage of the incredible advances in instrument performance, methods efficiency and computational power, MS-based technologies play an irreplaceable role in proteomics researches. MS is composed of an ion source (converts analyte molecules into gas-phase ions), a mass analyzer (separates ionized analytes according to mass-to-charge (m/z) ratio), and a detector (records the quantity of ions at each m/z value) (Han et al. 2008). After the characteristic m/z ratio of a molecular species is determined, the compounds of a given sample can be identified by matching obtained information with standard databases (Feist and Hummon 2015).

Two fundamental strategies are currently employed for identification and characterization of proteins by MS approaches. In the Top-Down strategy, intact protein ions or large protein fragments are subjected to gas-phase fragmentation ready for MS analysis, so that more complete protein backbone and credible post-translational modifications (PTMs) could be preserved (Pesavento et al. 2006; Du et al. 2006). This strategy simplifies the sample preparation process and provides higher sequence coverage over peptide-level studies. In Bottom-Up strategies, proteins are proteolytically digested into peptides prior to mass analysis, and thus peptide-level information could offer a basis for comprehensive protein identification. With its exhaustive analysis of samples, the Bottom-Up approach is more suitable for high-complexity samples for large-scale analysis, due to its superior front-end separations and higher sensitivity. Since the Bottom-Up method is more widely used in the proteomics field, we will focus mainly on describing the workflow and subsequent data analysis of Bottom-Up strategies in this chapter.

With the rapid growth of proteomics, there has been a notable increase in publications of clinical proteomics field. Clinical proteomics represents the

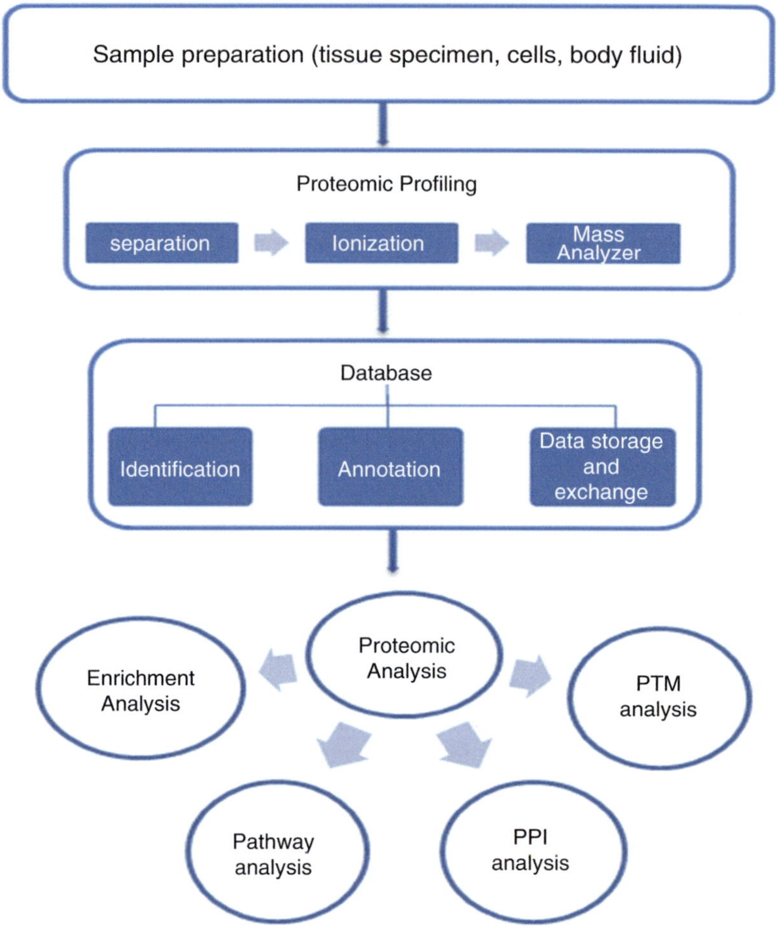

Fig. 6.1 General workflow of proteomics study

comprehensive study of proteins present in medical specimens, such as body fluids and tissues, in both qualitative and quantitative manners. By exploring the differences between specimens from healthy and diseased individuals, important disease biomarkers may be discovered, and such biomarkers could be used to monitor treatment effect as well. Generally, proteomics studies contain several experimental procedures including specimen preparation, peptide/protein profiling, and data analysis. The general workflow of most proteome research is shown in Fig. 6.1. Although the field of proteomics has advanced tremendously, there are still some technical limitations such as low coverage rate. Despite these challenges, proteomics studies have generated invaluable information for understanding cellular functions and facilitated the identification of biomarkers critical for the detection, prognosis, diagnosis, and treatment of diseases.

6.2 Sample Preparation

Regardless of the diversity of MS-based analysis approaches, high quality sample preparation is critical for a successful proteomics experiment. Generally, samples for proteomics studies are gained from the organism *in vivo,* cultured tissue or cell line *in vitro*, after which whole proteins can be isolated from tissue/cell lysis. Necessary purification and enrichment will be applied to collect proteins before proteolytic digestion and subsequent profiling by liquid chromatography coupled with mass spectrometry (LC-MS). Highly efficient lysis and digestions of samples, as well as the removal of contaminants while keeping sample loss minimal, are always the final goals of strategic optimization. The general workflow of sample preparation procedure is shown in Fig. 6.2.

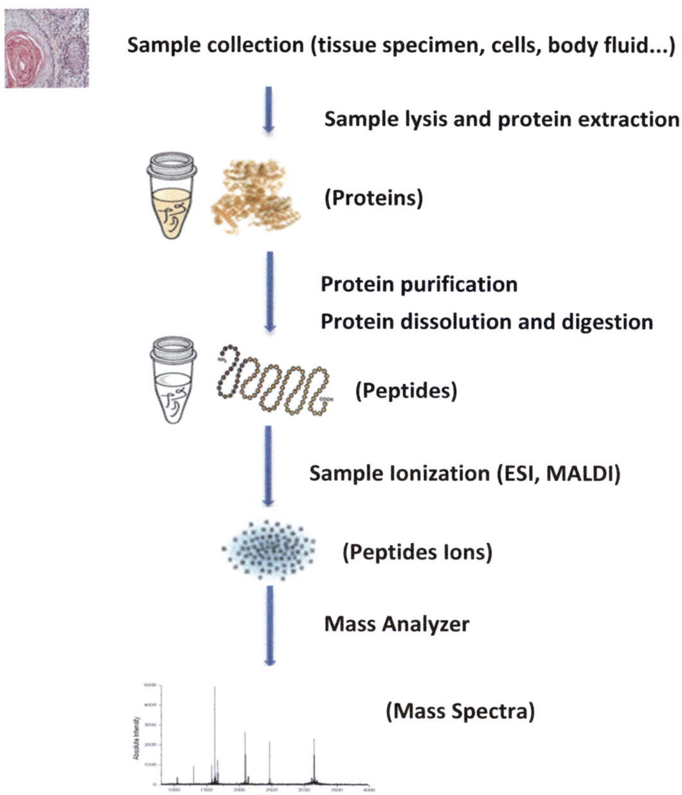

Fig. 6.2 General workflow of sample preparation prior to mass spectrometry analysis

6.2.1 Sample Collection

Important factors such as appropriate selection of material types and optimal establishment of experimental conditions must be considered prior to preparing samples. Since the quality of the samples will influence the overall pattern of biomarker discovery, well-defined clinical specimens are very important for successful clinical proteomics profiling.

Generally, there are three major types of clinical specimens: body fluid (e.g., serum/plasma, cerebrospinal fluid, urine, saliva, tears, lymph, etc.), tissues (e.g., liver, muscle, brain, etc.), and cells (e.g., blood cells) (Silberring and Ciborowski 2010). Each of these sources has its own advantages and disadvantages for biomarker discovery in clinical proteomics analysis. For instance, body fluids are convenient and easy to use, and proteins secreted in body fluids may reflect a wide range of pathophysiological conditions during disease development. Nevertheless, complex pretreatment processes are needed to remove the primary existing highly abundant proteins, since they may mask the proteins of interest. Tissues are widely used for clinical proteomics as well, and lesion samples are commonly used as initial screening materials to find the direct causes of a disease. However, the heterogeneous cell types and cell stages in tissue biopsies may lead to difficulty in identifying biomarkers, especially for cancer studies. Generally, tumor tissues are heterogeneous, which contain various cell types such as stromal cells, immune cells and extracellular matrix proteins (Lu et al. 2012) and it may blind the true protein alterations in cancer cells. Thus, methods effectively coupling laser-capture microdissection (LCM) with MS have been developed. Since LCM could isolates specific subpopulations of tissue cells under direct microscopic visualization and increases the homogeneity of histologically enriched cell populations, the linkage of LCM and MS provides a robust profiling of target sample from solid tumors (Wisniewski et al. 2011).

6.2.2 Samples Lysis and Protein Extraction

Chemical lysis, chemical extraction and mechanical disruption are the most common strategies performed to acquire proteins from clinical specimens (Raynie 2010; Canas et al. 2007). Choosing a lysis buffer depends greatly on detergent attributes, as detergents with higher critical micelle concentration and lower micelle molecular weight could effectively solubilize proteins and be readily removed (Feist and Hummon 2015). Mechanical disruption such as gentle rocking, cell scraping and sonication could be arranged in the order of violence degree from easy to hard. Due to the diversity in organismal samples, the appropriate choice of lysis buffer and mechanical disruption varies in accordance with different physical and chemical features of target proteins.

6.2.3 *Protein Purification*

High yield of contaminants is prevalent in a variety of clinical samples and can lead to research deviation. The release of abundant cellular compositions from lysate, such as lipids and nucleotides, might disturb chromatographic separation and spectral signals for protein identification (Feist and Hummon 2015). For the in-gel digestion method, proteins are solubilized with detergents prior to separation by sodium dodecyl sulfate polyacrylamide gel electrophoresis (SDS-PAGE) (Wisniewski et al. 2009). Detergents are commonly incompatible with reverse-phase separations, and may even harm mass spectrometry instruments and columns irreversibly. In addition, some detergents also interfere with enzymatic digestion (Zhou et al. 2012). Thus, removing contaminants and detergents is a necessary step of protein purification following protein extraction. Classic protein precipitation techniques including acetone precipitation (Buxton et al. 1979), trichloroacetic acid precipitation (Arnold and Ulbrich-Hofmann 1999), chloroform-methanol mixture (Wessel and Flugge 1984) and ethyl acetate (Yeung and Stanley 2010), have been proven effective in protein purification. Specifically, the chloroform-methanol method is more efficient for membrane proteins, while acetone precipitation sequesters mostly water-soluble proteins (Ferro et al. 2000).

6.2.4 *Protein Dissolution and Digestion*

After pollution elimination, purified protein precipitate is collected in pellet form by centrifuging and then dried to some extent. The compact structure of the pellet impairs sufficient exposure of proteins to the dissolution buffer, so fulfilling suspension of a pellet in dissolution buffer becomes essential to obtain a sufficient amount of protein. The most widespread strategies include shaking, vortexing and sonication (Burgess 2009; Isaacson et al. 2006).

Protein digestion with trypsin is the most crucial step in sample-preparation, followed by LC-MS to identify amino acid sequences and investigate PTMs (Hustoft et al. 2011). Well-defined specificity of trypsin protease guarantees precise digestion of peptides, and thus yields reliable detection of molecular weights of peptides in the following LC-MS strategies (Matthiesen and Mutenda 2007). There are two mature approaches to prepare peptide cleavage for MS-based proteome analysis: in-gel digestion and in-solution digestion. In-gel solution fractionates proteins by employing SDS-PAGE then digesting them in gel (Shevchenko et al. 1996). In-gel digestion economizes analytical time in MS analysis while achieving exhaustive and specific protein information (Granvogl et al. 2007), and excluding contaminants from the sample prior to MS analysis (Weiner et al. 1972). In-solution digestion extracts proteins with strong chaotropic reagents, and the precipitation and digestion of proteins are performed under denaturing conditions in the liquid phase (Wisniewski et al. 2009). In this process, protein samples receive highly-automated handling but may suffer from incomplete dissolution.

6.2.5 *Sample Separation*

MS-based proteomics studies depend on optimized separation technologies to separate complex biological samples into several simple components. Sufficient separation is necessary both for high-efficiency identification of peptide fragments and sensitive detection of low-abundance proteins. Two of the most widely used separation approaches in proteomics areas are 2-DE and liquid chromatography (LC).

6.2.5.1 Gel-Based

2-DE has been employed in proteomics as a mature separation technology for decades, and it is still used to study low-complexity proteomes and intact proteins with PTMs (Zhang et al. 2013). Based on the isoelectric point and molecular weight, 2-DE is able to separate thousands of proteins from a complex mixture sample on a single gel (Zhang et al. 2013). 2-DE is a relatively simple separation procedure, and the gel provides a good carrier matrix for the safe storage of proteins (Shevchenko et al. 2006). After the 2-DE separation, several different protein visualization methods such as Colloidal Coomassie blue, silver staining and fluorescence are available for protein detection (Rabilloud and Lelong 2011). The density of stained spots reflects the amount of proteins, and the comparison of staining intensity between spots provides a measure of relative quantification.

6.2.5.2 Liquid Chromatography-Based

LC isolates different components of samples on the basis of their individual affinity features with the stationary and mobile phases. With appropriate experimental design, different peptides can be sufficiently separated according to particular features of retention time in a specific column. In proteomics studies, the most common filling materials of LC columns include ion exchange, reverse phase, affinity and hybrid materials (Yates et al. 2009). Specifically, reverse phase liquid chromatography (RPLC) plays an essential role in LC-MS based proteomics. RPLC chromatographic column is packed with nonpolar material as stationary phase, thus the hydrophobic molecules in the mobile phase tend to bind with the column, resulting in preferred elution and outflow of hydrophilic molecules (Shen and Smith 2002). Since its mobile phase tandem with electrospray ionization (ESI) possesses high resolution, efficiency, reproducibility, and compatibility, analytical RPLC is widely used as the single phase or as the last dimension of multidimensional separation such as multidimensional protein identification technology (MudPIT) (Yates et al. 2009) before mass analysis (Opiteck and Jorgenson 1997).

6.2.5.3 Gas Chromatography-Based

In addition to LC-MS, gas chromatography-mass spectrometry (GC-MS) has been used in proteomics research as well. Subjected to a carrier gas (e.g. helium); the molecules are separated based on their volatility and bond characteristics in the gas chromatography stage. After the sample is converted into gas phase ions that are sorted based on their mass to charge ratios, the representative electrical signals can be translated into a meaningful mass spectrum. Generally, analytes for GC-MS include gases and naturally volatile substances, as well as compounds which require derivatization (Schauer et al. 2005). Since GC-MS is effective for the analysis of steroids, diglycerides, mono-, di- and trisaccharides, it is more popularly applied in metabolic studies rather than proteomics research (Rohloff 2015).

6.2.5.4 Protein Microarray Technologies

As one complementary technique to MS-based proteomics analysis, protein microarray has been broadly involved in protein profiling (Puig-Costa et al. 2011). Based on different immobilized molecules (antigens or antibodies), protein microarrays are classified into two categories: forward-phase microarrays and reverse-phase microarrays (Chandra et al. 2011). In forward-phase microarrays, a library of antibodies is immobilized onto the chip surface and then probed with a mixture of proteins. Antibody microarrays, which are effective at detecting variations of protein expression with a relatively large dynamic range, are the most common forward-phase microarrays (Berrade et al. 2011). Forward-phase microarrays arrays have been widely implemented in the detection of antibody-antigen interactions, biomarker detection (Shafer et al. 2007), immunological studies (Robinson et al. 2002) and PTMs studies (Chen et al. 2007). Conversely, in reverse-phase microarrays arrays the protein lysates are extracted from target cells, tissues or serum samples onto a slide, and the arrays are then incubated to antibodies against the interest proteins. In clinical researches, reverse phase microarrays have been successfully used for analyses of PTMs (LaBaer and Ramachandran 2005), signal transduction in living cells (Chan et al. 2004), and for studying cell signaling pathways in cancer (Sheehan et al. 2005; Grubb et al. 2003).

6.3 Mass Spectrometry Technologies

6.3.1 Ionizing Technologies

Proteins and peptides are polar, nonvolatile, and thermally unstable molecules, so MS-based analytical strategies demand an ionization technique to gasify analyte without extensive degradation. Herein, we will briefly introduce MALDI (Matrix-

Assisted Laser Desorption/Ionization) (Karas and Hillenkamp 1988) and Electrospray Ionization (ESI) (Fenn et al. 1989) which are two widely used ionization methods in proteomics studies.

In the MALDI method, analyte molecules are dispersed with a large amount of matrix forming co-crystals, which will be rapidly sublimated by the release of heat and energy provided by laser source. MALDI-generated ions are predominantly singly charged, which makes MALDI applicable to top-down analysis of high-molecular-weight proteins. Due to its simplicity, excellent mass accuracy, high resolution and sensitivity, Matrix Assisted Laser Desorption Ionization Time-of-Flight (MALDI-TOF) is broadly used to identify proteins by what is referred to as peptide-mass mapping (Medzihradszky and Chalkley 2015) or peptide-mass fingerprinting (Thiede et al. 2005). MALDI-MS technology is commonly used in conjunction with 2-DE for prior protein purification and fractionation, but is less appropriate for peptide chromatography due to difficult sample handling.

ESI is a technique which produces charged gas-phase ions from peptide solution by imposing high voltage and electrospray (Yates et al. 2009). Coupled with capillary electrophoresis or liquid chromatography for molecular fractionation prior to mass spectrometric analysis, ESI-MS can be an effective technique capable of analyzing a wide range of molecules in a complex biological sample (Matthiesen and Mutenda 2007). The analysis can either be processed online (directly convey the liquid eluting from the LC system to an electrospray) or offline (collect fractions for later analysis). With very little fragmentation, the solution-phase information can be better reflected in gas-phase. These advantages make ESI an ideal choice of ion source in LC-MS proteome profiling.

6.3.2 Mass Analyzer

The mass analyzer is an integral part of MS technology because it can store and separate ions based on the mass-to-charge (m/z) ratios from mixture samples. For proteomics research, four broad types of mass analyzers are commonly recognized as alternatives: Quadrupole, Ion Trap, Time-of-Flight (TOF) and Fourier-Transform Ion Cyclotron Resonance (FTICR) mass analyzers. Each mass analyzer features a unique design and performance, including differences in mass range, resolution, sensitivity, ion transmission and dynamic range. TOF mass spectrometers determine m/z ratio by measuring the mass-dependent time of ions with different masses from the ion source to the detector (Chernushevich et al. 2001). Ion trap mass spectrometers can store an ion of interest and eject all other ions simultaneously, then fragment the precursor ion to produce sequence information (Matthiesen and Mutenda 2007). So far, TOF and Ion-Trap instruments have dominated proteomics studies and aided in a number of meaningful discoveries. These analyzers can also work together in tandem to achieve reinforced performance and satisfy specific needs by making use of the advantages of each approach (Table 6.1).

Table 6.1 Tandem mass spectrometers for proteomics

Instrument	Q-q-Q	Q-q-LIT	Q-q-TOF	TOF-TOF	LTQ-Orbitrap
Ion source	ESI	ESI	ESI; MALDI	MALDI	ESI; MALDI
Tandem capability	MS^2	MS^n*	MS^2	MS^2	MS^n*
Features	Low resolution;	Limited resolution;	High resolution;	High resolution;	High resolution;
	Medium mass accuracy;	Low mass accuracy;	High mass accuracy;	High mass accuracy;	High mass accuracy;
	High sensitivity;	High sensitivity;	High sensitivity;	High sensitivity;	High sensitivity;
	Wide dynamic range;	Limited dynamic range	Wide dynamic range;	Wide dynamic range;	Wide dynamic range;
	Moderate scan rate;	Low cost;	Moderate scan rate;	Fast scan rate;	Moderate scan rate;
		Fast scan rate;			High cost;
Major applications	Quantification in SRM mode;	Quantification in SRM mode;	Protein identification;	Protein identification by PMF or PFF;	Top-down proteomics;
	PTM detection;	PTM detection;	PTM identification;		PTM characterization;
	Targeted proteomics;		Intact protein analysis;		Protein identification and quantification;

Note:
Q-q-Q Triple Quadrupole mass spectrometer, *Q-q-LIT* Hybrid Quadrupole-Linear ion trap mass spectrometer, *Q-q-TOF* Hybrid Quadrupole-Time-of-flight mass spectrometer, *TOF-TOF* Tandem Time-of-flight mass spectrometer, *LTQ-Orbitrap* Hybrid Linear ion trap-Orbitrap mass spectrometer, *SRM* Selected reaction monitoring, *PMF* Peptide mass fingerprinting, *PFF* Peptide-fragment fingerprinting, *n** Number of subsequent mass assessments ($2 < n < 13$)

6.4 Quantitative Mass Spectrometry in Proteomics

The goal of quantitative proteomics is to measure dynamic changes of protein and PTMs abundances under altered conditions. There are two mainstream quantitative methods in MS-based proteomics: relative and absolute quantitative proteomics. Relative quantitation estimates the expression ratios of detecting proteins against an internal reference protein, while absolute quantitation describes the expression amount or concentration of each protein in a given sample (Elliott et al. 2009). Generated from LC-MS experiments, differential analysis can be carried out using both labeling and label-free techniques.

6.4.1 Labeling-Based Technique

Stable isotope-labeling is a straightforward and powerful quantitative method for MS-based proteomics studies. Different protein samples are labeled distinctively then combined into a mixture. Subsequently, the pooled mixture is taken through the sample preparation step and analyzed by following LC-MS methods (Zhu et al. 2010).

Stable-isotope tags have been introduced to proteins either via metabolic marking using heavy amino acids (Conrads et al. 2002), via chemical reactions using isotope-coded affinity tags or similar reagents (Yao et al. 2001), or enzymatically via transfer of ^{18}O from water to peptides (Yao et al. 2001). The metabolic labeling of stable isotopes imports isotopic-marked material into cell media during protein synthesis to overcome processing errors. The SILAC (stable isotope labeling of amino acids in cell culture) has emerged as a popular alternative in which only specific amino acids like arginine and lysine are labeled (Ong et al. 2002). Comparatively, these post-biosynthetic labeling strategies, including chemical labeling and enzymatic labeling, work for any sample at either the protein or the peptide level. The relative quantification of peptides is measured by comparing peptide pairs marked as heavy or light, then protein levels are estimated from statistical evaluation of the peptide ratios (Yates et al. 2009). When a known concentration labeled synthetic peptide is added to the sample, the stable isotope-labeling technology can identify the absolute measurement of protein or peptide abundance (Gerber et al. 2003).

6.4.2 Label-Free Technique

The label-free technique is an alternative methodology for quantitative proteomics study in complex biological samples. In general, label-free quantitative approaches possess two different quantification methods: chromatographic peak areas and spectral counting. There is a linear correlation between chromatographic peak area and abundance of measured peptides, so peptides can be relatively quantified by comparing chromatographic peak areas at specific retention times between LC-MS runs (Chelius and Bondarenko 2002; Tang et al. 2004). On the contrary, since the number of fragment-ion spectra for peptides mirrors the amount of the corresponding protein proportionally, the spectral counting method can estimate a relative quantification of protein through counting and comparing them between different samples (Liu et al. 2004). Compared with isotope labeling methods, label-free quantitative proteomics techniques provide rapid and economic measurement of protein expression levels. The label-free approach has a better dynamic range and proteome coverage for peptide identification; however, quantification accuracy and reproducibility are inferior to those produced by labeling-based strategies (Li et al. 2012).

6.5 Protein Identification from Mass Spectrometry Data

During the past few decades, profit from the application of mass spectrometry analysis to protein samples, the field of proteomics is developing rapidly. Large volumes of experimental data was generated base on MS proteomic platform (Riffle and Eng 2009). To systematically identify whole proteins from the raw data, multiple search engines in conjunction with several well-established protein sequence databases are applied. In the following sections, we will briefly introduce some common databases and frequently used search engines for protein identification.

6.5.1 Common Protein Sequence Databases

6.5.1.1 UniProt

UniProt (http://www.uniprot.org/) is a comprehensive, high-quality and freely accessible database of protein sequence and functional information which combines the Swiss-Prot, TrEMBL and PIR-PSD databases into a single resource (Apweiler et al. 2004). It contains four core sub-databases such as: Protein knowledgebase (UniProtKB), Sequence clusters (UniRef), Sequence archive (UniParc) and Proteomes. It has excellent tools for dataset retrieval such as BLAST (Basic Local Alignment Search Tool), Align and Retrieve/ID mapping. Many entries with large amounts of biological and sequencing information are derived from research literature or genome sequencing projects. This database serves as a basic and curated sequence resource for protein prediction and for planning new experiments (Hinz and UniProt 2010). Small datasets can be directly downloaded from the UniProt web site. However, for downloading complete datasets, UniProt FTP site (ftp://ftp.uniprot.org/) is recommended.

6.5.1.2 Swiss-Prot

Swiss-Prot is a leading and comprehensive universal protein sequence database which was integrated into UniProt and can be accessed via the official website of UniProt (http://www.uniprot.org/). Currently, it contains more than 540,000 manually reviewed and annotated entries from numerous species. It is a non-redundant database, which means that all reports for a given protein are merged into a single entry. It is also highly integrated with other databases (Apweiler et al. 2004; Gasteiger et al. 2001).

6.5.1.3 TrEMBL

As an essential part of UniProt database, TrEMBL (http://www.uniprot.org/) currently contains more than 50 million automatically annotated and not reviewed entries. It is a computer-annotated protein sequence database complementing the Swiss-Prot Protein Knowledgebase. TrEMBL consists of computer annotated translation of the coding sequences (CDs) incorporated in public databases such as EMBL/GenBank/DDBJ Nucleotide Sequence Databases and also protein sequences extracted from the literature or submitted to UniProtKB/Swiss-Prot (Schneider et al. 2004). TrEMBL strictly follows the Swiss-Prot format and conventions (Apweiler et al. 2004).

6.5.1.4 RefSeq

NCBI's Reference Sequence database (http://www.ncbi.nlm.nih.gov/refseq/) contains large amounts of sequencing information for DNA, RNA and protein from plasmids, organelles, viruses, archaea, bacteria, and eukaryotes. The aim of the project is to provide a non-redundant collection of references for nucleotide and protein sequences. Each reference sequence is constructed wholly from sequence data submitted to the International Nucleotide Sequence Database Collaboration. As a fundamental database with genetic and functional information, RefSeq provides reference standards for multiple purposes, such as genome annotation or reporting locations of sequence variation in medical records. The RefSeq database can be easily retrieved in several different ways including Nucleotide, Protein, and Map Viewer. Sequence information included in RefSeq database can be searched via BLAST and downloaded from the RefSeq FTP site.

6.5.2 *Software for Protein Identification*

6.5.2.1 Mascot

Mascot is a database search engine, which can identify proteins by matching mass spectrometry raw data to known peptide sequence databases (Perkins et al. 1999; Koenig et al. 2008). Mascot is widely used in proteomics research. Customers can easily get access to a series of Mascot software and temporarily use them on the Matrix Science website (http://www.matrixscience.com/). However, for large scale data analysis and routine work, researchers must purchase the software and obtain a license to run in-house. Mascot is based on the Molecular Weight Search (MOWSE) algorithm (Pappin et al. 1993) as well as probability-based scoring (Fenyo 2000).

6.5.2.2 SEQUEST

SEQUEST is a tandem MS data analysis program for protein identification. SEQUEST identifies proteins by matching experimental tandem mass spectra to known protein/peptide sequences database. A cross-correlation function is calculated between the measured fragment mass spectrum and the protein sequences in the database, and it is used to score the proteins in the database (Eng et al. 1994). SEQUEST supports the use of information from several fragment mass spectra in the database search and shows good sensitivity and flexibility in handling data generated by different types of mass spectrometers (Griffin et al. 1995; Sadygov et al. 2004). Modern proteome research based on tandem mass spectrometer will produce a large scale of tandem mass spectra data. Identifying such a data collection requires automation and stability, and SEQUEST is the first software to fill that need.

6.5.2.3 PLGS

PLGS (ProteinLynx Global SERVER) is developed by Waters Corporation. As a fully integrated Mass-Informatics platform for quantitative and qualitative proteomics research, it plays a central role in data analysis in the Waters proteomics system. PLGS always comes together with instruments from Waters, and it has the selectivity and specificity required for MS^E data analysis. Based on open system architecture, PLGS has automatic workflow for high throughput data processing. It can minimize the false positive rate by setting strict statistical filters. The friendly operation interface of PLGS allows users to set individual parameters including database importing, protein modification selecting, and threshold setting. The project and database management tools included in PLGS provide functions for results visualization and reporting.

6.5.2.4 PEAKS

PEAKS is commonly used for peptide identification through *de novo* peptide sequencing, thus extracting amino acid sequence information without the use of databases. Based on a progressive model and algorithm, PEAKS computes the best peptide sequences whose fragment ions can best match the peaks in the MS/MS spectrum. PEAKS provides a complete sequence for each peptide, confidence scores on individual amino acid assignments, and simple reporting for high-throughput analysis. As a well-known *de novo* sequencing software, PEAKS has the ability to compare results of multiple search engines (Ma et al. 2003).

6.6 Proteomics Data Analysis

MS-based proteomics is widely used and has become an invaluable tool for profiling system-wide protein identification, Protein-protein interactions (PPIs) and Post-translational modifications (PTMs). Proteomics studies have been applied for various purposes in clinical field, for example, biomarker discovery in diseases and study of drug responses. With the rising amount of published proteomics research, the need for in-depth analysis of large-scale proteomics datasets is becoming more and more urgent. Therefore, many computational techniques, databases and tools have been developed to interpret proteomics data.

6.6.1 Singular Interpretation of Protein List

Irrespective of the employed method, the main result of the large amount of published proteomics studies is a list of identified proteins, all of which need to be interpreted in different ways for diverse research topics. Usually, the first step of functional analysis is to map the protein name to a unique identifier. Unlike gene names which have been widely standardized by HGNC (HUGO Gene Nomenclature Committee) (Povey et al. 2001), protein names can differ between databases and may even change in different versions of the same database (Malik et al. 2010). Given that many analysis tools only accept specific protein identifiers as input, it is of utmost importance to standardize protein names before subsequent analysis. Platforms such as Uniprot (Bairoch et al. 2009) and Ensembl (Hubbard et al. 2009) are useful in normalizing protein names. In addition, PICR (Cote et al. 2007) and CRONOS (Waegele et al. 2009) are instrumental in connecting protein names to corresponding gene identifiers. Furthermore, it is also essential to check the non-redundancy of protein lists, as repetitive entries may lead to biased result in some analysis tools.

Reviewing extensive amounts of literature to interpret a given protein list is still common, and several comprehensive databases like Uniprot (Bairoch et al. 2009) can be easily accessed to explore the associated function for each protein entry. Moreover, several available repositories can be used to search for previously carried-out proteomics experiments to compare the proteomics results to other relevant experiments. Databases like PRIDE (Jones et al. 2006), PepSeeker (McLaughlin et al. 2006), PeptideAtlas (Deutsch et al. 2008), GPM (Craig et al. 2004) have been proven useful in this regard. Although such an interpretation provides us with an easy and intuitive way to learn more about protein function, it has its own limitations. Due to the intricate hypotheses for different experiment design and proteomics datasets, merely comparing experimental findings to manually collected information could introduce bias. In addition, some potentially essential hidden relationships between the members of a protein list may be overlooked.

6.6.2 Comparative Proteomics Analysis

By identifying proteins that differ in their abundance between two or more experimental states, comparative proteomics has become a powerful tool to explore protein expression profiling response to diverse biological systems, such as different cell developmental stages or disease conditions. However, the reliability and utility of comparative proteomics analysis is highly dependent on accurate and rigorous measurement of quantitative changes. Since proteomics data is usually subject to a number of challenging analytical issues like experimental noise and low coverage rate, cares should be taken when dealing with such data (Listgarten and Emili 2005).

6.6.2.1 Data Normalization

Typically, in order to get a more accurate estimate of the underlying biological effects being measured, normalization is performed to remove systematic biases from the data before subsequent statistical inference. The normalization process may help to adjust the variability due to sample preparation or equipment conditions. Several normalization methods have been used in proteomics studies. The simplest one is the global normalization method, in which the raw protein expression level is normalized against that of a well-conserved protein (Karpievitch et al. 2012). In addition, owing to the similarity of data features between gene expression and protein data, numerous normalization approaches such as locally weighted scatterplot smoothing (LOWESS) regression and quantile normalization, have been widely borrowed from gene expression studies (Karpievitch et al. 2012).

6.6.2.2 Missing Values

Missing issues are common in MS expression data, and considerable proportions of missing observations have been observed in proteomics studies (Karpievitch et al. 2012). There are several causes for missing events, such as low concentrations (hard to be detected) of the peptide, or absence of the peptide in the sample. Typically one cannot clearly determine the missing mechanism, which further complicates the problem. There are two major categories: missing at random (MAR) and missing not at random (MNAR). For the MAR situation, a simple imputation process is commonly used that replace missing values by a constant or a randomly selected value (e.g., impute with mean, median, etc.). For more complex situations, several other methods are applied, like KNN (K-nearest neighbors) (Troyanskaya et al. 2001), LSA (Least-squares adaptive) (Bo et al. 2004), which estimate missing values based on the expression profiles of other proteins with similar intensity within the dataset. Because of the complicated missing mechanisms in proteomics, no individual method is guaranteed to be the best solution.

6.6.2.3 Differential Expression Analysis

Conventional proteomics studies often include comparison of protein expression profiles under two or more different conditions (e.g., normal versus disease). Statistical approaches such as ANOVA (analysis of variance) or regression are commonly applied in comparative proteomics (Wang et al. 2012) to determine which subsets of proteins are differentially expressed (DE) with a pre-defined statistical threshold. Typically, a test statistic, which reflects how much a feature discriminates between classes, will be generated by a specific test, and the P values for DE protein adjusted for multiple testing be calculated based on the null distribution (theoretical or permutation-based) (Pendarvis et al. 2009). A variety of methods originally designed to compare microarray datasets have been used in proteomics data as well. For instance, the SAM (Significance Analysis of Microarray) method and the PLGEM (Power Law Global Error Model) are now widely applied in comparative proteomics analysis (Bin Goh and Wong 2014; Roxas and Li 2008).

6.6.3 Enrichment Analysis

Various functional databases contain experimentally proven or otherwise inferred connections between the genes or proteins and their specific functions. Generally, these functions belong to a certain controlled vocabulary, which means they have a clearly described meaning defined and recognized by domain specialists. The GO (Ashburner et al. 2000) is a project for consistent descriptions of gene and gene product attributes across species. Each GO contains an identifier and a term belongs to one of the three GO categories: "biological process," "molecular function" or "cellular component." It is worth mentioning that GO is organized in a hierarchical way, which can be represented as a tree structure. Normally, an initial step for functional interpretation of the acquired protein list is to link the protein identifier with its associated GO. For instance, one can use the AmiGO (Carbon et al. 2009), which is a user-friendly platform, to search the GO associated with a particular gene/protein.

6.6.3.1 Annotation Term Enrichment Analysis

Generally, the subsequent step after GO annotation is GO enrichment analysis, which determines whether a specific GO term is enriched in a particular group of biological processes, functions or cellular compartments. It involves comparing the frequency of individual functional annotations within a reference list, and the enrichment score can be statistically tested (e.g., hyper-geometric, binomial or Chi-square tests, etc.). The result will be a p value, which can be used as the cutoff to measure significance of over- or under-representation of a GO term. Since the number of functional terms for enrichment test is usually large, an adjusted multiple p-value (e.g. Benjamini-Hochberg correction (Benjamini and Hochberg 1995)) is often provided.

A wide range of software is available for GO enrichment analysis, and an extensive list can be found at (http://geneontology.org/). DAVID is a popular meta-tool (Dennis et al. 2003) that aids with GO enrichment analysis. It applies a modified Fisher exact p-value to determine whether a GO term is enriched in a given proteomics dataset with reference dataset as background (Malik et al. 2010). It enables the researcher to gather information about over- or under-representation, and to understand the biological meaning behind large lists of proteins. In proteomics studies, the GO-term enrichment analysis has been applied in numerous contexts. The advantage of annotation term enrichment compared to the singular analysis of individual protein lists is that it can summarize the functional properties on a global scale instead of individual protein entries.

Careful attention should be paid to certain issues while performing GO enrichment analysis. The first concern is the choice of reference dataset, which is either predefined by the tool (e.g., the human proteome), or can be selected manually (e.g., all identified proteins). The second concern is the choice of cutoff to determine which proteins can be retained in the list. In addition, the overrepresentation analysis is usually inclined to suffer from limited discriminative power.

6.6.3.2 Set Enrichment Analysis

The classical overrepresentation analyses strictly rely on a pre-defined quantitative threshold selecting proteins to be included, and ignore the expression level alterations when calculating the functional enrichment score. Thus, set-based enrichment analyses were developed to evaluate the significance of predefined protein lists. As a powerful method to determine functional significance of gene groups, gene set enrichment analysis (GSEA) was originally developed in the gene expression data (Subramanian et al. 2005). Without modification, this method has been applied in the proteomics field (Clutterbuck et al. 2011), and a modified approach called PSEA (Protein set enrichment analysis) was developed to study differential protein expression based on SpI (Spectral index) from label-free quantitative proteomics in breast cancer research (Cha et al. 2010).

The basis of set-based enrichment analysis is to determine whether the pre-defined gene subsets (e.g., based on a common functional annotation) are distributed in a ranked list (e.g., generated by quantitative feature of expression data) randomly. The pre-defined annotation terms are commonly gained from libraries such as GO (Ashburner et al. 2000) or MSigDB (Liberzon et al. 2011). Then, the ranked protein list is generated based on the quantitative differences between case and control group and the enrichment score (ES) is calculated based on specified algorithms. Usually, empirical distribution of ES is determined by permutations, and it is used to test the statistical significance of observed ES.

The set-based enrichment analysis has many advantages compared with traditional overrepresentation analysis. There is no requirement of an arbitrary threshold to distinguish significantly differential proteins, and it provides more statistical power. In addition, set enrichment analysis could detect coordinated changes of gene products, which may reveal some noteworthy proteins and protein groups important for disease status.

6.6.3.3 Pathway Analysis

Biological pathways usually describe a series of chemical reactions in the cell that lead to a specific biological outcome. Proteins involved in these chemical interactions and those that have a regulatory effect play important roles in pathway analysis. Rather than merely taking the gene-centric view of GO-based analyses, pathway analyses provide us with more insight into biological mechanisms (Schmidt et al. 2014).

Several comprehensive databases such as KEGG (Kanehisa et al. 2014), Reactome (Croft et al. 2011) and Ingenuity Pathways Knowledge Base (Ficenec et al. 2003) contain a large amount of pathway information. These instrumental tools are typically derived from intracellular reactions like metabolic signaling pathways. In addition to the above comprehensive resources, numerous highly specific databases have been developed as well. For instance, the PANTHER (Thomas et al. 2003), GenMAPP (Salomonis et al. 2007) and PID (Schaefer et al. 2009) mainly focus on signal transduction processes. Lately, several databases were created which include pathways active in disease. Netpath

(Kandasamy et al. 2010), for example, can help researchers to extract cancer relevant pathways from a complex dataset.

A comprehensive overview of hundreds biological pathway-related resources and molecular interaction-related resources can be found on the Pathguide website (http://pathguide.org) (Bader et al. 2006). This powerful resource could help researchers to select the optimal database for their studied biological systems, and it can be recommended as a starting point for proteomics pathway analysis. Similar to the GO term annotation, the identification of pathways under diverse biological conditions is highly dependent on the algorithm used, such as topology-driven approaches and multivariate approach. For instance, the "PathNet" (Dutta et al. 2012) algorithm incorporates both differential expression information of genes and connectivity information from canonical pathways to investigate relations among diverse pathways. As various biological components simultaneously participate in multiple biological processes, it is difficult to make a clear distinction between individual pathways, and even popular curated pathway databases show limited overlap. With more advanced experimental techniques and bioinformatics tools, this could be improved over time.

6.6.4 Protein-Protein Interaction Analysis

Protein-protein interactions (PPIs) represent non-random physical contact between two or more proteins, and thus form smaller or larger complexes in a space- and time-dependent manner. Since physical contact between proteins could trigger conformational changes or PTMs that modulate the activity of those proteins, PPIs can help us gain further insight into biological processes. PPIs are not static or permanent, but experience continuous reassembly and turnover. PPIs are usually regulated based on many factors, such as specific cell-type, developmental stage of the cell, cell-cycle phase, external stimulus or signal and the presence of other proteins. Therefore, the network of PPIs can provide powerful information to reveal and explain important functional modules within proteins. For instance, if proteins are neighbors in the PPI network and found to be co-regulated in the differential expression list, it may suggest that they work together to play important roles in the affected biological processes.

PPIs are essential to development and homeostasis of biological mechanisms, and many human diseases can be traced to aberrant PPIs. Thus, the inhibition of these aberrant associations is of great clinical significance. Because of the diverse nature of PPIs, the successful design of therapeutics requires detailed knowledge of each system at a molecular level. Several recent studies have identified and characterized specific interactions from various disease systems, and many of the key PPIs are known to participate in disease-associated signaling pathways (Ryan and Matthews 2005).

6.6.4.1 Techniques to Perform PPIs

Plentiful experimental techniques have been developed to measure PPIs, among which the Y2H (yeast two-hybrid) screening (Fields and Song 1989) and AP-MS (affinity purification coupled with mass spectrometry) (Rigaut et al. 1999) are most popular approaches. To measure whether two proteins physically interact with each other, modified yeast strains are used in Y2H system to express a "bait-protein" (fused to a DNA-binding domain) and a "prey-protein" (fused to a transcription activation domain), which, if they interact, trigger the expression of a reporter gene. In an AP-MS experiment, the protein of interest is attached to a larger protein fragment (the "tag"). As the tagged protein can be purified easily from the cell extract, proteins binding to the tagged protein are co-purified and could be subsequently identified by MS. Large-scale AP-MS experiments have been applied to study yeast and human PPIs. Although the high-throughput experimental methods present many advantages over traditional approaches, they still have some limitations.

6.6.4.2 Commonly Used Databases for PPIs Analyses

Large databases documenting protein interactions are publicly available for several organisms, and HPRD (Prasad et al. 2009), IntAct (Kerrien et al. 2007), MINT (Persico et al. 2005), MIPS (Mewes et al. 2004), DIP (Xenarios et al. 2002) and BioGRID (Breitkreutz et al. 2008) are some of the most commonly used resources (shown in Table 6.2). Among these, HPRD (http://www.hprd.org/) can depict and incorporate interaction networks, domain structure, post-translational modifications, and associated disease for human proteins. IntAct (http://www.ebi.ac.uk/intact/) is an analysis tool for molecular interaction data, which gathers useful information from previously published results or user-submitted data. MINT (http://mint.bio.uniroma2.it/mint/Welcome.do) is an interactive database based on experimentally verified PPIs, usually extracting data from technical literature. MIPS (http://mips.helmholtz-muenchen.de/proj/ppi/) contains manually curated high-quality PPI data collected from the scientific literature by expert curators. DIP (http://dip.doe-mbi.ucla.edu/dip/Main.cgi) interprets signaling or regulatory pathways, and can also detect protein interactions at the cellular level. BioGRID (http://thebiogrid.org/) is a curated biological database of protein-protein interactions, genetic interactions, chemical interactions, and post-translational modifications. In addition, several meta-bases such as APID (Prieto and Rivas 2006), I2D (Brown and Jurisica 2007) and STRING (Jensen et al. 2009) are widely used in PPI analyses (shown in Table 6.3). For instance, STRING (http://string-db.org/), a powerful meta-database including data from many curated databases, serves as a very popular tool for interaction network analysis of proteomic data, incorporating both interactions and pathway information to form an easy-to-use

Table 6.2 List of commonly used PPI database

Database	Url	PPIs	Major source
HPRD	http://www.hprd.org/	41,327	Manually extracted from the literature
IntAct	http://www.ebi.ac.uk/intact/	531,946	Literature curation or direct user submissions
MINT	http://mint.bio.uniroma2.it/mint/Welcome.do	241,458	Mainly focus on experimentally verified PPIs
MIPS	http://mips.helmholtz-muenchen.de/proj/ppi/	–	High-quality data collected from the scientific literature
DIP	http://dip.doe-mbi.ucla.edu/dip/Main.cgi	79,646	Experimentally determined interactions
BioGrid	http://www.thebiogrid.org/	–	Comprehensive literature curation

"–" indicates no accurate updated statistics for protein-protein interactions in the resource

Table 6.3 List of commonly used meta-PPIs database and PPI visualizing tools

Source	Url	Major source
Meta-PPIs database		
STRING	http://string-db.org/	Experimental and predicted PPIs
I2D	http://ophid.utoronto.ca/ophidv2.204/	Integration of known, experimental and predicted PPIs
APID	http://bioinfow.dep.usal.es/apid/index.htm	Computational and known experimentally validated PPIs
PPI visualizing source		
Cytoscape	http://www.cytoscape.org/	Visualizing complex networks, integrating networks with attribute data
Osprey	http://biodata.mshri.on.ca/osprey/servlet/Index	Visualization of complex interaction networks
Visant	http://visant.bu.edu/	Visual analyses of metabolic networks

web interface. Moreover, tools like Cytoscape (Shannon et al. 2003), Osprey (Breitkreutz et al. 2003) and VisANT (Hu et al. 2004) are popular open-source programs for visualizing protein-protein interaction networks (PPINs) (shown in Table 6.3).

Since the covered experimental data sets and criteria for PPIs vary widely, cares should be taken in selecting the most suitable database for proteomic analysis. In addition, one must be very careful to select appropriate parameters and types of interaction data, especially when incorporating the predicted interactions. Checking GO term coherence is a good way to assess the reliability of an edge in a PPIN, that is, whether the two proteins on both ends of the edge could be annotated to an informative GO term in common (Chua and Wong 2008). Another way to assess the reliability of an edge in a PPIN is based on the hypothesis that proteins are more likely to share common neighbors in the PPIN if they interact.

6.6.4.3 PPIs Analysis Methods

Most PPIs analysis methods include identifying protein complexes by mining modular or dense sub-networks from PPI networks. In disease-related research, four sequential steps are usually conducted to generate a biological hypothesis on target cells (e.g. cancer cells) through PPIN construction and analysis (Srihari et al. 2015). The starting point is to define the seed proteins, which should be the molecules of major interest and will be the skeleton of the PPIN. Typical choices are differentially expressed proteins observed in a given experiment, or proteins known to be involved in the disease process. The second step is to determine interactions between proteins. The interacting partners of proteins are commonly identified from curated databases as described above, and the reliability of these interactions needs to be assessed. The third step is the PPIN construction and visualization process. The PPIN will be constructed based on high-confidence evidence, and several algorithms allow the creation of a visual representation of the built network. The final step is to use bioinformatics tools to extract meaningful biological information from the network. Although the strategy seems straightforward, these methods are restricted by limitations in existing PPI datasets, particularly the lack of sufficient interactions between proteins and the presence of a large number of false-positive interactions. Therefore, increasing interaction coverage by integrating PPI datasets from multiple studies and reducing noise by assessing the reliabilities of interactions are crucial for accurate PPIN analysis.

6.6.5 Post-translational Modifications Analysis

Generally, PTMs represent covalent events that change the properties of a protein with proteolytic cleavage or adding a modifying group to amino acids. Variations at PTMs level exponentially escalate the complexity of the proteome relative to both the transcriptome and genome. The most common PTMs include phosphorylation, glycosylation, ubiquitination, nitrosylation, methylation, acetylation, and lipidation (Mann and Jensen 2003).

Far from being mere "decorations," PTMs play a key role in functional proteomics. Not only in isolation but also in coordination, PTMs can influence numerous properties of proteins including enzymatic activity, protein interactions and subcellular location. Therefore, PTMs enable signaling and regulatory mechanisms that modulate a given protein's cellular function. For instance, kinase cascades are turned on and off by the reversible addition and removal of phosphate groups in the cell signaling process, and the ubiquitination process marks cyclins for destruction at defined time points which are important for the cell cycle (Mann and Jensen 2003). Table 6.4 summarizes the major features of some important PTMs.

PTMs are vital cellular control mechanisms, and they play significant roles in various disease conditions, especially cancer. There are cases in which mutations of

Table 6.4 Major features of important PTMs

PTM type	Major features
Phosphorylation	Reversible, activation/inactivation of enzyme activity, modulation of molecular interactions, signaling, etc.
Glycosylation	Structural and functional roles in membrane and secreted proteins, cell-cell recognition/signaling, etc.
Ubiquitination	Destruction signal, cell localization, promote/prevent protein interactions, etc.
Nitrosylation	Regulating signal transduction, regulate enzyme activities, etc.
Acetylation	Protein stability, protection of N terminus, regulation of protein–DNA interactions, etc.
Methylation	Regulation of gene expression, etc.
Lipidation	Protein function, protein subcellular localization, etc.

the post-translational target sites were found to be involved in disease directly. One example is a loss of N-linked glycosylation in the prion protein, a variant which causes numerous clinical symptoms such as early-onset dementia, cerebral atrophy, and hypometabolism (Grasbon-Frodl et al. 2004). In addition to these examples, systematic studies implicating PTMs in disease are now facilitated by the rapid growth of databases knowledge.

6.6.5.1 Techniques to Perform PTMs

Thousands PTMs sites have been discovered and the functional importance of several types of modifications have been revealed with novel enrichment strategies (Minguez et al. 2012), ranging from small chemical modifications to the addition of complete proteins. Generally, the modification analysis is achieved by sequence comparison between obtained experimental data to a known amino acid. Isolation of the precisely processed proteins is usually the first step for direct analysis of PTMs studies. Once a protein has been isolated, a variety of techniques can be used to determine the modified amino acids. From Edman degradation to the MS method, the sensitivity for detection of PTMs has increased with the advances in techniques (Mann and Jensen 2003). Since MS measures fragmentation pattern of peptides and molecular weight, it is a good general method for identifying modifications that changing molecular weight. In MS-based PTMs analysis, it is important to generate sufficient peptide fragmentation information for peptide identification and site localization. In some cases, MS can measure the precise molecular weight of the intact protein, especially if the protein is not too heterogeneous and its mass is not too large. Once the masses of the non-modified and modified amino acid residues add up to the measured intact molecular weight, the protein is completely characterized.

6.6.5.2 PTMs Analysis Methods

The network of proteins that regulate each other through PTMs has become a valuable factor in the interpretation of experimental protein lists, as PTMs are unique to the protein level and constitute a major regulation process. For instance, PTMs could provide information on the correlation between a modifying enzyme and its substrate, which may be impossible to identify through other analysis levels (Zhao and Jensen 2009). In addition, the crosstalk between PTMs regulators makes such biological processes more complex.

Although our current knowledge of PTMs and their various roles is still incomplete, methods to systematically identify modifications have resulted in several PTMs modification databases. Historically, the most studied PTMs is phosphorylation, which can be used as an example of approaches to the general prediction of PTMs. When dealing with large lists of experimentally detected phosphorylation sites, it is generally of interest to determine which of the sites are novel. Databases like Uniprot (Bairoch et al. 2009), dbPTM (Lee et al. 2006), Phospho.ELM (Dinkel et al. 2011), PHOSIDA (Gnad et al. 2011), and PhosphoSitePlus (Hornbeck et al. 2004) have been developed as useful resources to study phosphorylation across organisms. Tools for the study of phosphorylation sites have largely fallen into two general approaches: enzyme-specific and enzyme-independent (Schwartz et al. 2009). In the enzyme-specific method, analysis tools are commonly based on the principle that each kinase has its own unique sequence specificity. Incorporating kinase-substrate data available from variable resources such as literature, databases, and combinatorial peptide library screens, these tools have identified enzyme-specific signatures that can be used to predict other substrates of a particular kinase (Blom et al. 2004). In the enzyme-independent method, analysis tools are commonly based on MS data, which contains only phosphorylation sites without regard to the responsible enzyme. With specific algorithms like neural networks (Ingrell et al. 2007) or support vector machines (Gnad et al. 2007), the enzyme-independent tools do not need to model the properties of substrate recognition.

Typical PTMs analyses incorporate various analyses such as GO enrichment, pathway analysis and PPI analysis as mentioned above. For instance, the NetWorKIN software (Linding et al. 2007) integrates PPI information from STRING software on linear kinase motifs to provide kinase–substrate relationships in large-scale data (Olsen and Mann 2013). In addition, promising efforts have been made to develop resources to study PTMs, such as PTMcode (http://ptmcode.embl.de) (Minguez et al. 2013), which aims to collect known and predicted PTMs associations to provide a framework for experimental or computational analysis on various scales. Although robust MS-based proteomic workflows for large-scale PTMs identification and quantification have been developed, a complete inventory of sites has not been constructed for PTMs. In addition, PTMs coverage still needs to be improved, especially in clinical proteomics. The major challenge of the field lies in the systematic pipelines for follow-up analysis and functional interpretation of PTMs data. However, with the growing availability of the technology, increasingly in-depth interpretation of PTMs can be achieved, and PTMs information could give more precise therapies targeted at the molecular nature of a given disease.

6.7 Public Proteome Repositories and Standards

As proteins are the direct executors of biological function in organisms, proteomics study plays a significant role in the post-genome era. From 2D-gel to LC-MS, the rapid development of new technologies has made it possible to investigate whole proteins in proteomics study, and modern proteomics research is based on LC-MS generated large scale data. To facilitate the dissemination of these data, centralized data repositories have been developed that make the data and results accessible to proteomics researchers and biologists alike (Riffle and Eng 2009). In order to facilitate data comparison, exchange and verification in or between public repositories, the common standards for data representation in proteomics must be established. In the fields of mass spectrometry and protein-protein interaction study, much progress has been made in developing common standards for data storage, sharing and exchange (Orchard et al. 2003). The most well-known international consortium of proteomics research associations, the Human Proteome Organization (HUPO), was launched in 2001. The organization promotes the development and awareness of proteomics research and facilitates scientific collaborations between members and initiative. Its goal is to gain a comprehensive understanding of the human proteome (Huber 2003).

6.7.1 Public Proteomics Repositories

6.7.1.1 PRIDE

The PRIDE (http://www.ebi.ac.uk/pride/) is a centralized, standards compliant, public data repository, which includes information on protein and peptide identifications, post-translational modifications, and supporting spectral evidence. Protein and peptide identifications in this database have been described in previous scientific literature. Data generated from different species, tissues and subcellular locations (perhaps under specific disease conditions) can be uploaded, downloaded or viewed via a single, centralized web interface (Jones et al. 2006). PRIDE supports the submission of data from different platforms. However, data prepared to be uploaded to PRIDE database should obey strict proteomics data standards (Riffle and Eng 2009). Many tools allow researchers to achieve standards compliance for data generated by many different platforms. By the end of 2014, PRIDE had accumulated data for 41,835 proteins, 269,806 unique peptides, and about 101 million spectra (Perez-Riverol et al. 2015). Currently, datasets from a total of 51,922 assays and 3233 projects were centralized in PRIDE. It is one of the most popular proteomic data repositories and has played an important role in Human Proteome Project (HPP) (Chen et al. 2015).

6.7.1.2 Global Proteome Machine Database

The Global Proteome Machine Database (GPMDB) is a resource for collecting diverse tandem proteomics data and open source software, and it also includes peptide and protein identifications that are important for further MS computational research (Craig et al. 2004). The database allows worldwide research scientists to use its proteomics data and tools for the purpose of proteome research. Raw data submitted by researchers or downloaded from other databases will be reprocessed. XML (Extensible Markup Language) files including protein or peptide identification information will be uploaded and stored in GPMDB. By the end of 2014, GPMDB database spanned a total of 136,373 proteins, 1,786,698 peptides, and 1020 million spectra (Perez-Riverol et al. 2015; Chen et al. 2015). The GPMDB plays an important role in proteome research and can be accessed at (http://gpmdb.thegpm.org/index.html).

6.7.1.3 PeptideAtlas

The PeptideAtlas (http://www.peptideatlas.org/) is one of the largest and most well-curated protein expression data resources. It serves as a compendium of peptides observed with tandem mass spectrometry methods from multiple species. It also contains a growing set of software tools and underlying platforms for proteomics data analysis and visualization. It stores various formats of output files and metadata from MS-based experiments (Deutsch 2010). PeptideAtlas supports raw data submissions from users, which will be reprocessed through a uniform analysis and validation pipeline. The results are loaded into a database, and the information derived from the raw data is returned to the community for identification and statistical analysis purposes. Users can search the PeptideAtlas web interface by protein accession, peptide sequence, gene name, keyword, or phrase (Perez-Riverol et al. 2015). Search results are displayed on a page with associated summary statistics. PeptideAtlas can help plan targeted proteomics experiments, improve genome annotation, and support data mining projects (Chen et al. 2015; Deutsch 2010).

6.7.1.4 MOPED

MOPED (Model Organism Protein Expression Database) is a proteomics repository that integrates protein expression information from human specimens and several other model organisms (Kolker et al. 2012). It provides protein-level expression data, meta-analysis capabilities, and quantitative data from standard analyses based on mass spectrometry. It also provides new estimates of protein abundance and concentration, and statistical summaries from experiments. The web interface contains six main panels: "protein absolute expression", "protein relative expression", "gene relative expression", "pathways", "experiments", and "visualization" for different data handling purposes. Additionally, a suite of tools for data searching and visualization are available. With rapid development in recent years, MOPED has grown into a repository containing more than 17,000 proteins, 250,000 unique peptides, and approximately 15 million spectra (Kolker et al. 2012; Perez-Riverol et al. 2015; Chen et al. 2015). As a significant public proteomics database, MOPED provides abundant information on complex biological processes and thus benefits fundamental biological or medical investigation. The MOPED database can be accessed at (http://moped.proteinspire.org).

6.7.1.5 Human Proteinpedia

Human Proteinpedia (http://www.humanproteinpedia.org/) is a public resource for proteomics data storing, integrating and exchanging (Kandasamy et al. 2009). The distributed annotation system of Human Proteinpedia allows the researchers to contribute and maintain protein annotations (Kandasamy et al. 2009). Human Proteinpedia integrates diverse features of the human proteome including post-translational modifications, subcellular localization, protein-protein interactions, and expression of proteins in multiple human tissues and cell lines (Perez-Riverol et al. 2015; Kandasamy et al. 2009).

6.7.1.6 Tranche

Tranche (http://www.proteomexchange.org/databases/tranche) is a data repository for storing, sharing information for proteomics researchers. As a widely-used database, Tranche hosts several kinds of data. Indeed, it plays a crucial role in proteomics field. It allows researchers to use and disseminate both data and software. A client tool is required to upload and download datasets. Tranche provides interfaces for PRIDE, Human Proteinpedia, and PeptideAtlas to store and disseminate large MS-based data files (Smith et al. 2011).

In addition to the resources mentioned above, there are some other important databases. Table 6.5 lists the most popular public proteomics databases and their corresponding websites.

Table 6.5 List of frequently used proteomics database

Database	Website
Proteomics IDEntifications database (PRIDE)	http://www.ebi.ac.uk/pride/archive/
Human Protein Reference Database (HPRD)	http://www.hprd.org/
PeptideAtlas	http://www.peptideatlas.org/
Human Proteinpedia	http://www.humanproteinpedia.org/
Tranche	http://www.proteomexchange.org/databases/tranche
Global Proteome Machine Database (GPMDB)	http://gpmdb.thegpm.org/
Model Organism Protein Expression Database (MOPED)	https://www.proteinspire.org/MOPED/mopedviews/proteinExpressionDatabase.jsf
Protein Abundance Across Organisms (PaxDb)	http://pax-db.org/#!home
Integrated Proteome Resources (iProX)	http://www.iprox.org/index

6.7.2 Public Proteomics Standards

6.7.2.1 MS Raw Data Unification

Different models of instruments were applied for MS data acquisition. An important consideration is how to get the data provided by multiple MS software into mzML format. The mzXML is a XML-based common file format for proteomics data, and it provides a standard container for MS and MS/MS data. There are many free and commercial software packages that support mzML format. However, in order to export the acquired MS data as mzML format, some applications must be used in conjunction with additional translators and transformation utilities. ProteoWizard and OpenMS are two popular programs for MS raw data handling.

6.7.2.2 Qualitative and Quantitative Proteomics

The purpose of qualitative proteomics studies is to identify peptides or proteins, which can normally be identified by analyzing raw MS data. As mentioned in the previous section, there are many tools used for peptide and protein identification. Mascot, SEQUEST or PEAKS are frequently-used programs for peptide and protein identification. Furthermore, based on rapidly developing experimental technology and software packages, quantitative proteomics study can be achieved based on variety of quantitative techniques including multiple labeling or label-free approaches. For instance, MaxQuant and PLGS are well equipped for handling quantitative proteomics data based on label or label-free methods, respectively.

6.7.2.3 Integration of Public Standards and Local or Third Party Tools

Many researchers have their own bioinformatics capabilities and homemade tools for specific data analysis in proteomics study. We will briefly introduce how public standards can be integrated into the tools produced by proteome bioinformatics individuals or groups. Many popular commercial or free software packages have their own native file formats for data storage or reporting. Since 2002, when the HUPO-PSI (HUPO Proteomics Standards Initiative) was founded, the uniform public standards have become more carefully established for data reporting and exchange. More and more existing software tools for proteomics are following HUPO-PSI standards. In order to improve the efficiency of proteomics data use and exchange, in-house software development in many proteomics labs should adhere to the HUPO-PSI standards as well (Medina-Aunon et al. 2013).

6.7.2.4 Reporting, Uploading and Exchanging Data

MIAPE (Minimum Information about a Proteomics Experiment) guidelines were published by HUPO-PSI in 2007. According to MIAPE, formalized information should be reported when publishing a dataset. Two websites provide assistance to guide users to create a MIAPE compliant report: MIAPEGelDB (Robin et al. 2008) and Proteo-Red MIAPE web repository (Martinez-Bartolome et al. 2010). Proteo-Red MIAPE web toolkit (Medina-Aunon et al. 2011) is a website capable of linking the latest versions of the HUPO-PSI XML schemas to the Proteo-Red MIAPE web repository in an automated, accessible, and comprehensive way. It covers multiple data formats such as mzML, mzIdentML, and PRIDE XML (Medina-Aunon et al. 2013). When uploading data to either a public repository or local database, the most important thing is that the experiment has been reported using data standards so it can be shared and verified. General information on MIAPE modules and the corresponding data exchange formats can be found in Table 6.6.

Table 6.6 MIAPE modules and data exchange formats

Techniques	Guidelines	Format
MS	MS (MIAPE-MS)	mzData, mzML, traML
	Identifications (MIAPE-MSI)	mzIdentML
	Quantitation (MIAPE-Quant)	mzQuantML
Molecular interactions	Interactions (MIMIx)	PSI-MI XML,
Protein separation	Gel electrophoresis (MIAPE-GE)	PSI-MI XML
	Gel informatics (MIAPE-GI)	None
Sample processing	Column chromatography (MIAPE-CC)	spML
	CE (MIAPE-CE)	

6.8 Proteomics Applied in Breast Cancer Study

The application of high-throughput gene expression technology contributes greatly to the identification of breast cancer-related genes and subtypes. Over the past decade, numerous global proteomic analyses have been conducted to promote the molecular understanding of breast tumor progression and provide important prognostic information for breast cancer. For instance, Barry L. Karger's group performed a comprehensive proteomic study on human breast cancer epithelial cells across 18 samples (Cha et al. 2010). The clinical material for this study included 9 breast tissue samples from healthy women (without breast cancer history), and 9 breast cancer tissues obtained from either mastectomy or surgical excision. All tissues were of high quality and were prepared strictly according to robust experimental protocols. Highly enriched populations of epithelial cells were produced by laser capture micro-dissection and used in subsequent proteomic analysis. Through common sample preparation process (cell lysis, protein separation and in-gel digestion), the protein digests were analyzed by LC-MS/MS strategy. Generated from MS/MS scans, the raw files were searched against Swiss-Prot annotation database. Of the 18 samples, 12,970 unique peptides and 2588 proteins were identified. The highly differentially expressed proteins were considered as potential candidate biomarkers. Both annotation term enrichment analysis and set-based enrichment analysis were performed. With the GOMiner tool, the over-represented functional categories for differentially expressed proteins were determined. With protein set enrichment analysis (PSEA), 35 MSigDB derived protein sets were found to be significant. The combination of the above two approaches allowed researchers to find many confirmatory biological findings important to malignant breast epithelial cells, and also revealed extensive insight into the molecular participants involved in tumorigenesis signaling cascades.

Studies incorporating PPI information were widely applied in breast cancer research as well. For instance, Ideker's et al. applied a protein network-based approach to analyze the expression profile of two cohorts of breast cancer patients, and proved that protein subnetwork markers could improve prediction of cancer outcome (Chuang et al. 2007). Intensive PPIN information was first gathered from the database, yeast two-hybrid experiments as well as the literature. Then, in order to integrate the gene expression and network data sets, they overlaid the expression value of each gene on its corresponding proteins in the network. Based on a mature scoring and searching algorithm, the subnetworks with discriminative activities were identified. With the PPIN based strategy, they successfully found that markers identified through the subnetwork are more reproducible, and more accurate in classifying metastatic versus non-metastatic breast tumors.

References

Apweiler R, Bairoch A, Wu CH. Protein sequence databases. Curr Opin Chem Biol. 2004;8 (1):76–80. doi:10.1016/j.cbpa.2003.12.004.

Arnold U, Ulbrich-Hofmann R. Quantitative protein precipitation from guanidine hydrochloride-containing solutions by sodium deoxycholate/trichloroacetic acid. Anal Biochem. 1999;271 (2):197–9. doi:10.1006/abio.1999.4149, S0003-2697(99)94149-0 [pii].

Ashburner M, Ball CA, Blake JA, Botstein D, Butler H, Cherry JM, Davis AP, Dolinski K, Dwight SS, Eppig JT, Harris MA, Hill DP, Issel-Tarver L, Kasarskis A, Lewis S, Matese JC, Richardson JE, Ringwald M, Rubin GM, Sherlock G, Gene Ontology C. Gene ontology: tool for the unification of biology. Nat Genet. 2000;25(1):25–9.

Bader GD, Cary MP, Sander C. Pathguide: a pathway resource list. Nucleic Acids Res. 2006;34: D504–6. doi:10.1093/nar/gkj126.

Bairoch A, Consortium U, Bougueleret L, Altairac S, Amendolia V, Auchincloss A, Argoud-Puy G, Axelsen K, Baratin D, Blatter M-C, Boeckmann B, Bolleman J, Bollondi L, Boutet E, Quintaje SB, Breuza L, Bridge A, deCastro E, Ciapina L, Coral D, Coudert E, Cusin I, Delbard G, Dornevil D, Roggli PD, Duvaud S, Estreicher A, Famiglietti L, Feuermann M, Gehant S, Farriol-Mathis N, Ferro S, Gasteiger E, Gateau A, Gerritsen V, Gos A, Gruaz-Gumowski N, Hinz U, Hulo C, Hulo N, James J, Jimenez S, Jungo F, Junker V, Kappler T, Keller G, Lachaize C, Lane-Guermonprez L, Langendijk-Genevaux P, Lara V, Lemercier P, Le Saux V, Lieberherr D, Lima TO, Mangold V, Martin X, Masson P, Michoud K, Moinat M, Morgat A, Mottaz A, Paesano S, Pedruzzi I, Phan I, Pilbout S, Pillet V, Poux S, Pozzato M, Redaschi N, Reynaud S, Rivoire C, Roechert B, Schneider M, Sigrist C, Sonesson K, Staehli S, Stutz A, Sundaram S, Tognolli M, Verbregue L, Veuthey A-L, Yip L, Zuletta L, Apweiler R, Alam-Faruque Y, Antunes R, Barrell D, Binns D, Bower L, Browne P, Chan WM, Dimmer E, Eberhardt R, Fedotov A, Foulger R, Garavelli J, Golin R, Horne A, Huntley R, Jacobsen J, Kleen M, Kersey P, Laiho K, Leinonen R, Legge D, Lin Q, Magrane M, Martin MJ, O'Donovan C, Orchard S, O'Rourke J, Patient S, Pruess M, Sitnov A, Stanley E, Corbett M, di Martino G, Donnelly M, Luo J, van Rensburg P, Wu C, Arighi C, Arminski L, Barker W, Chen Y, Hu Z-Z, Hua H-K, Huang H, Mazumder R, McGarvey P, Natale DA, Nikolskaya A, Petrova N, Suzek BE, Vasudevan S, Vinayaka CR, Yeh LS, Zhang J. The Universal Protein resource (UniProt) 2009. Nucleic Acids Res. 2009;37:D169–74. doi:10.1093/nar/gkn664.

Benjamini Y, Hochberg Y. Controlling the false discovery rate – a practical and powerful approach to multiple testing. J R Stat Soc Ser B Methodol. 1995;57(1):289–300.

Berrade L, Garcia AE, Camarero JA. Protein microarrays: novel developments and applications. Pharm Res. 2011;28(7):1480–99. doi:10.1007/s11095-010-0325-1.

Bin Goh WW, Wong L. Computational proteomics: designing a comprehensive analytical strategy. Drug Discov Today. 2014;19(3):266–74. doi:10.1016/j.drudis.2013.07.008.

Blom N, Sicheritz-Ponten T, Gupta R, Gammeltoft S, Brunak S. Prediction of post-translational glycosylation and phosphorylation of proteins from the amino acid sequence. Proteomics. 2004;4(6):1633–49. doi:10.1002/pmic.200300771.

Bo TH, Dysvik J, Jonassen I. LSimpute: accurate estimation of missing values in microarray data with least squares methods. Nucleic Acids Res. 2004;32(3):e34. doi:10.1093/nar/gnh026.

Breitkreutz BJ, Stark C, Tyers M. Osprey: a network visualization system. Genome Biol. 2003;4 (3):1–4. doi:10.1186/gb-2003-4-3-r22.

Breitkreutz B-J, Stark C, Reguly T, Boucher L, Breitkreutz A, Livstone M, Oughtred R, Lackner DH, Bahler J, Wood V, Dolinski K, Tyers M. The BioGRID interaction database: 2008 update. Nucleic Acids Res. 2008;36:D637–40. doi:10.1093/nar/gkm1001.

Brown KR, Jurisica I. Unequal evolutionary conservation of human protein interactions in interologous networks. Genome Biol. 2007;8(5):R95. doi:10.1186/gb-2007-8-5-r95.

Burgess RR. Protein precipitation techniques. Methods Enzymol. 2009;463:331–42. doi:10.1016/S0076-6879(09)63020-2, S0076-6879(09)63020-2 [pii].

Buxton TB, Crockett JK, Moore 3rd WL, Moore Jr WL, Rissing JP. Protein precipitation by acetone for the analysis of polyethylene glycol in intestinal perfusion fluid. Gastroenterology. 1979;76(4):820–4. doi:S001650857900072X [pii].

Canas B, Pineiro C, Calvo E, Lopez-Ferrer D, Gallardo JM. Trends in sample preparation for classical and second generation proteomics. J Chromatogr A. 2007;1153(1–2):235–58. doi: S0021-9673(07)00091-X [pii], 10.1016/j.chroma.2007.01.045.

Carbon S, Ireland A, Mungall CJ, Shu S, Marshall B, Lewis S, Ami GOH, Web Presence Working G. AmiGO: online access to ontology and annotation data. Bioinformatics. 2009;25(2):288–9. doi:10.1093/bioinformatics/btn615.

Cha S, Imielinski MB, Rejtar T, Richardson EA, Thakur D, Sgroi DC, Karger BL. In situ proteomic analysis of human breast cancer epithelial cells using laser capture microdissection: annotation by protein set enrichment analysis and gene ontology. Mol Cell Proteomics. 2010;9 (11):2529–44. doi:10.1074/mcp.M110.000398.

Chan SM, Ermann J, Su L, Fathman CG, Utz PJ. Protein microarrays for multiplex analysis of signal transduction pathways. Nat Med. 2004;10(12):1390–6. doi:nm1139 [pii], 10.1038/nm1139.

Chandra H, Reddy PJ, Srivastava S. Protein microarrays and novel detection platforms. Expert Rev Proteomics. 2011;8(1):61–79. doi:10.1586/epr.10.99.

Chelius D, Bondarenko PV. Quantitative profiling of proteins in complex mixtures using liquid chromatography and mass spectrometry. J Proteome Res. 2002;1(4):317–23.

Chen S, Zheng T, Shortreed MR, Alexander C, Smith LM. Analysis of cell surface carbohydrate expression patterns in normal and tumorigenic human breast cell lines using lectin arrays. Anal Chem. 2007;79(15):5698–702. doi:10.1021/ac070423k.

Chen T, Zhao J, Ma J, Zhu Y. Web resources for mass spectrometry-based proteomics. Genomics Proteomics Bioinformatics. 2015;13(1):36–9. doi:10.1016/j.gpb.2015.01.004.

Chernushevich IV, Loboda AV, Thomson BA. An introduction to quadrupole-time-of-flight mass spectrometry. J Mass Spectrom. 2001;36(8):849–65. doi:10.1002/jms.207 [pii], 10.1002/jms.207.

Chua HN, Wong L. Increasing the reliability of protein interactomes. Drug Discov Today. 2008;13 (15–16):652–8. doi:10.1016/j.drudis.2008.05.004.

Chuang H-Y, Lee E, Liu Y-T, Lee D, Ideker T. Network-based classification of breast cancer metastasis. Mol Syst Biol. 2007;3:140. doi:10.1038/msb4100180.

Clutterbuck AL, Smith JR, Allaway D, Harris P, Liddell S, Mobasheri A. High throughput proteomic analysis of the secretome in an explant model of articular cartilage inflammation. J Proteomics. 2011;74(5):704–15. doi:10.1016/j.jprot.2011.02.017.

Conrads TP, Issaq HJ, Veenstra TD. New tools for quantitative phosphoproteome analysis. Biochem Biophys Res Commun. 2002;290(3):885–90. doi:10.1006/bbrc.2001.6275, S0006291X01962758 [pii].

Cote RG, Jones P, Martens L, Kerrien S, Reisinger F, Lin Q, Leinonen R, Apweiler R, Hermjakob H. The Protein Identifier Cross-Referencing (PICR) service: reconciling protein identifiers across multiple source databases. BMC Bioinf. 2007;8:401. doi:10.1186/1471-2105-8-401.

Craig R, Cortens JP, Beavis RC. Open source system for analyzing, validating, and storing protein identification data. J Proteome Res. 2004;3(6):1234–42. doi:10.1021/pr049882h.

Croft D, O'Kelly G, Wu G, Haw R, Gillespie M, Matthews L, Caudy M, Garapati P, Gopinath G, Jassal B, Jupe S, Kalatskaya I, Mahajan S, May B, Ndegwa N, Schmidt E, Shamovsky V, Yung C, Birney E, Hermjakob H, D'Eustachio P, Stein L. Reactome: a database of reactions, pathways and biological processes. Nucleic Acids Res. 2011;39:D691–7. doi:10.1093/nar/gkq1018.

Dennis G, Sherman BT, Hosack DA, Yang J, Gao W, Lane HC, Lempicki RA. DAVID: database for annotation, visualization, and integrated discovery. Genome Biol. 2003;4(9):1–11. doi:10.1186/gb-2003-4-9-r60.

Deutsch EW. The PeptideAtlas project. Methods Mol Biol. 2010;604:285–96. doi:10.1007/978-1-60761-444-9_19.

Deutsch EW, Lam H, Aebersold R. PeptideAtlas: a resource for target selection for emerging targeted proteomics workflows. Embo Rep. 2008;9(5):429–34. doi:10.1038/embor.2008.56.

Dinkel H, Chica C, Via A, Gould CM, Jensen LJ, Gibson TJ, Diella F. Phospho.ELM: a database of phosphorylation sites-update 2011. Nucleic Acids Res. 2011;39:D261–7. doi:10.1093/nar/gkq1104.

Du Y, Parks BA, Sohn S, Kwast KE, Kelleher NL. Top-down approaches for measuring expression ratios of intact yeast proteins using Fourier transform mass spectrometry. Anal Chem. 2006;78(3):686–94. doi:10.1021/ac050993p.

Dutta B, Wallqvist A, Reifman J. PathNet: a tool for pathway analysis using topological information. Source Code Biol Med. 2012;7(1):10. doi:10.1186/1751-0473-7-10.

Elliott MH, Smith DS, Parker CE, Borchers C. Current trends in quantitative proteomics. J Mass Spectrom. 2009;44(12):1637–60. doi:10.1002/jms.1692.

Eng JK, McCormack AL, Yates JR. An approach to correlate tandem mass spectral data of peptides with amino acid sequences in a protein database. J Am Soc Mass Spectrom. 1994;5(11):976–89. doi:10.1016/1044-0305(94)80016-2.

Feist P, Hummon AB. Proteomic challenges: sample preparation techniques for microgram-quantity protein analysis from biological samples. Int J Mol Sci. 2015;16(2):3537–63. doi:10.3390/ijms16023537, ijms16023537 [pii].

Fenn JB, Mann M, Meng CK, Wong SF, Whitehouse CM. Electrospray ionization for mass spectrometry of large biomolecules. Science. 1989;246(4926):64–71.

Fenyo D. Identifying the proteome: software tools. Curr Opin Biotechnol. 2000;11(4):391–5. doi:10.1016/S0958-1669(00)00115-4.

Ferro M, Seigneurin-Berny D, Rolland N, Chapel A, Salvi D, Garin J, Joyard J. Organic solvent extraction as a versatile procedure to identify hydrophobic chloroplast membrane proteins. Electrophoresis. 2000;21(16):3517–26. doi:10.1002/1522-2683(20001001)21:16<3517::AID-ELPS3517>3.0.CO;2-H [pii], 10.1002/1522-2683(20001001)21:16<3517::AID-ELPS3517>3.0.CO;2-H.

Ficenec D, Osborne M, Pradines J, Richards D, Felciano R, Cho RJ, Chen RO, Liefeld T, Owen J, Ruttenberg A, Reich C, Horvath J, Clark T. Computational knowledge integration in biopharmaceutical research. Brief Bioinform. 2003;4(3):260–78.

Fields S, Song OK. A novel genetic system to detect protein protein interactions. Nature. 1989;340(6230):245–6. doi:10.1038/340245a0.

Gasteiger E, Jung E, Bairoch A. SWISS-PROT: connecting biomolecular knowledge via a protein database. Curr Issues Mol Biol. 2001;3(3):47–55.

Gerber SA, Rush J, Stemman O, Kirschner MW, Gygi SP. Absolute quantification of proteins and phosphoproteins from cell lysates by tandem MS. Proc Natl Acad Sci U S A. 2003;100(12):6940–5. doi:10.1073/pnas.0832254100, 0832254100 [pii].

Gnad F, Ren S, Cox J, Olsen JV, Macek B, Oroshi M, Mann M. PHOSIDA (phosphorylation site database): management, structural and evolutionary investigation, and prediction of phosphosites. Genome Biol. 2007;8(11):1–13. doi:10.1186/gb-2007-8-11-r250.

Gnad F, Gunawardena J, Mann M. PHOSIDA 2011: the posttranslational modification database. Nucleic Acids Res. 2011;39:D253–60. doi:10.1093/nar/gkq1159.

Granvogl B, Ploscher M, Eichacker LA. Sample preparation by in-gel digestion for mass spectrometry-based proteomics. Anal Bioanal Chem. 2007;389(4):991–1002. doi:10.1007/s00216-007-1451-4.

Grasbon-Frodl E, Lorenz H, Mann U, Nitsch RM, Windl O, Kretzschmar HA. Loss of glycosylation associated with the T183A mutation in human prion disease. Acta Neuropathol. 2004;108(6):476–84. doi:10.1007/s00401-004-0913-4.

Griffin PR, MacCoss MJ, Eng JK, Blevins RA, Aaronson JS, Yates 3rd JR. Direct database searching with MALDI-PSD spectra of peptides. Rapid Commun Mass Spectrom RCM. 1995;9(15):1546–51. doi:10.1002/rcm.1290091515.

Grubb RL, Calvert VS, Wulkuhle JD, Paweletz CP, Linehan WM, Phillips JL, Chuaqui R, Valasco A, Gillespie J, Emmert-Buck M, Liotta LA, Petricoin EF. Signal pathway profiling of prostate cancer using reverse phase protein arrays. Proteomics. 2003;3(11):2142–6. doi:10.1002/pmic.200300598.

Han X, Aslanian A, Yates 3rd JR. Mass spectrometry for proteomics. Curr Opin Chem Biol. 2008;12(5):483–90. doi:10.1016/j.cbpa.2008.07.024, S1367-5931(08)00117-8 [pii].

Hinz U, UniProt C. From protein sequences to 3D-structures and beyond: the example of the UniProt knowledgebase. Cell Mol Life Sci CMLS. 2010;67(7):1049–64. doi:10.1007/s00018-009-0229-6.

Hornbeck PV, Chabra I, Kornhauser JM, Skrzypek E, Zhang B. Phosphosite: a bioinformatics resource dedicated to physiological protein phosphorylation. Proteomics. 2004;4(6):1551–61. doi:10.1002/pmic.200300772.

Hu ZJ, Mellor J, Wu J, DeLisi C. VisANT: an online visualization and analysis tool for biological interaction data. BMC Bioinf. 2004;5:1–8. doi:10.1186/1471-2105-5-17.

Hubbard TJP, Aken BL, Ayling S, Ballester B, Beal K, Bragin E, Brent S, Chen Y, Clapham P, Clarke L, Coates G, Fairley S, Fitzgerald S, Fernandez-Banet J, Gordon L, Graf S, Haider S, Hammond M, Holland R, Howe K, Jenkinson A, Johnson N, Kahari A, Keefe D, Keenan S, Kinsella R, Kokocinski F, Kulesha E, Lawson D, Longden I, Megy K, Meidl P, Overduin B, Parker A, Pritchard B, Rios D, Schuster M, Slater G, Smedley D, Spooner W, Spudich G, Trevanion S, Vilella A, Vogel J, White S, Wilder S, Zadissa A, Birney E, Cunningham F, Curwen V, Durbin R, Fernandez-Suarez XM, Herrero J, Kasprzyk A, Proctor G, Smith J, Searle S, Flicek P. Ensembl 2009. Nucleic Acids Res. 2009;37:D690–7. doi:10.1093/nar/gkn828.

Huber LA. Is proteomics heading in the wrong direction? Nat Rev Mol Cell Biol. 2003;4(1):74–80. doi:10.1038/nrm1007.

Hustoft HK, Reubsaet L, Greibrokk T, Lundanes E, Malerod H. Critical assessment of accelerating trypsination methods. J Pharm Biomed Anal. 2011;56(5):1069–78. doi:10.1016/j.jpba.2011.08.013, S0731-7085(11)00451-1 [pii].

Ingrell CR, Miller ML, Jensen ON, Blom N. NetPhosYeast: prediction of protein phosphorylation sites in yeast. Bioinformatics. 2007;23(7):895–7. doi:10.1093/informatics/btm020.

Isaacson T, Damasceno CM, Saravanan RS, He Y, Catala C, Saladie M, Rose JK. Sample extraction techniques for enhanced proteomic analysis of plant tissues. Nat Protoc. 2006;1(2):769–74. doi:nprot.2006.102 [pii], 10.1038/nprot.2006.102.

Jensen LJ, Kuhn M, Stark M, Chaffron S, Creevey C, Muller J, Doerks T, Julien P, Roth A, Simonovic M, Bork P, von Mering C. STRING 8-a global view on proteins and their functional interactions in 630 organisms. Nucleic Acids Res. 2009;37:D412–16. doi:10.1093/nar/gkn760.

Jones P, Cote RG, Martens L, Quinn AF, Taylor CF, Derache W, Hermjakob H, Apweiler R. PRIDE: a public repository of protein and peptide identifications for the proteomics community. Nucleic Acids Res. 2006;34(Database issue):D659–63. doi:10.1093/nar/gkj138.

Kandasamy K, Keerthikumar S, Goel R, Mathivanan S, Patankar N, Shafreen B, Renuse S, Pawar H, Ramachandra YL, Acharya PK, Ranganathan P, Chaerkady R, Keshava Prasad TS, Pandey A. Human Proteinpedia: a unified discovery resource for proteomics research. Nucleic Acids Res. 2009;37(Database issue):D773–81. doi:10.1093/nar/gkn701.

Kandasamy K, Mohan SS, Raju R, Keerthikumar S, Kumar GSS, Venugopal AK, Telikicherla D, Navarro JD, Mathivanan S, Pecquet C, Gollapudi SK, Tattikota SG, Mohan S, Padhukasahasram H, Subbannayya Y, Goel R, Jacob HKC, Zhong J, Sekhar R, Nanjappa V, Balakrishnan L, Subbaiah R, Ramachandra YL, Rahiman BA, Prasad TSK, Lin J-X, Houtman JCD, Desiderio S, Renauld J-C, Constantinescu SN, Ohara O, Hirano T, Kubo M, Singh S, Khatri P, Draghici S, Bader GD, Sander C, Leonard WJ, Pandey A. NetPath: a public resource of curated signal transduction pathways. Genome Biol. 2010;11(1):1–9. doi:10.1186/gb-2010-11-1-r3.

Kanehisa M, Goto S, Sato Y, Kawashima M, Furumichi M, Tanabe M. Data, information, knowledge and principle: back to metabolism in KEGG. Nucleic Acids Res. 2014;42(D1):D199–205. doi:10.1093/nar/gkt1076.

Karas M, Hillenkamp F. Laser desorption ionization of proteins with molecular masses exceeding 10,000 daltons. Anal Chem. 1988;60(20):2299–301.

Karpievitch YV, Dabney AR, Smith RD. Normalization and missing value imputation for label-free LC-MS analysis. BMC Bioinf. 2012;13 Suppl 16:S5. doi:10.1186/1471-2105-13-s16-s5.

Kerrien S, Alam-Faruque Y, Aranda B, Bancarz I, Bridge A, Derow C, Dimmer E, Feuermann M, Friedrichsen A, Huntley R, Kohler C, Khadake J, Leroy C, Liban A, Lieftink C, Montecchi-Palazzi L, Orchard S, Risse J, Robbe K, Roechert B, Thorneycroft D, Zhang Y, Apweiler R, Hermjakob H. IntAct – open source resource for molecular interaction data. Nucleic Acids Res. 2007;35:D561–5. doi:10.1093/nar/gkl958.

Koenig T, Menze BH, Kirchner M, Monigatti F, Parker KC, Patterson T, Steen JJ, Hamprecht FA, Steen H. Robust prediction of the MASCOT score for an improved quality assessment in mass spectrometric proteomics. J Proteome Res. 2008;7(9):3708–17. doi:10.1021/pr700859x.

Kolker E, Higdon R, Haynes W, Welch D, Broomall W, Lancet D, Stanberry L, Kolker N. MOPED: model organism protein expression database. Nucleic Acids Res. 2012;40(Database issue):D1093–9. doi:10.1093/nar/gkr1177, gkr1177 [pii].

LaBaer J, Ramachandran N. Protein microarrays as tools for functional proteomics. Curr Opin Chem Biol. 2005;9(1):14–9. doi:S1367-5931(04)00165-6 [pii], 10.1016/j.cbpa.2004.12.006.

Lee T-Y, Huang H-D, Hung J-H, Huang H-Y, Yang Y-S, Wang T-H. dbPTM: an information repository of protein post-translational modification. Nucleic Acids Res. 2006;34:D622–7. doi:10.1093/nar/gkj083.

Li Z, Adams RM, Chourey K, Hurst GB, Hettich RL, Pan C. Systematic comparison of label-free, metabolic labeling, and isobaric chemical labeling for quantitative proteomics on LTQ Orbitrap Velos. J Proteome Res. 2012;11(3):1582–90. doi:10.1021/pr200748h.

Liberzon A, Subramanian A, Pinchback R, Thorvaldsdottir H, Tamayo P, Mesirov JP. Molecular signatures database (MSigDB) 3.0. Bioinformatics. 2011;27(12):1739–40. doi:10.1093/bioinformatics/btr260.

Linding R, Jensen LJ, Ostheimer GJ, van Vugt MATM, Jorgensen C, Miron IM, Diella F, Colwill K, Taylor L, Elder K, Metalnikov P, Nguyen V, Pasculescu A, Jin J, Park JG, Samson LD, Woodgett JR, Russell RB, Bork P, Yaffe MB, Pawson T. Systematic discovery of in vivo phosphorylation networks. Cell. 2007;129(7):1415–26. doi:10.1016/j.cell.2007.05.052.

Listgarten J, Emili A. Statistical and computational methods for comparative proteomic profiling using liquid chromatography-tandem mass spectrometry. Mol Cell Proteomics. 2005;4(4):419–34. doi:10.1074/mcp.R500005-MCP200.

Liu H, Sadygov RG, Yates 3rd JR. A model for random sampling and estimation of relative protein abundance in shotgun proteomics. Anal Chem. 2004;76(14):4193–201. doi:10.1021/ac0498563.

Lu P, Weaver VM, Werb Z. The extracellular matrix: a dynamic niche in cancer progression. J Cell Biol. 2012;196(4):395–406. doi:10.1083/jcb.201102147.

Ma B, Zhang K, Hendrie C, Liang C, Li M, Doherty-Kirby A, Lajoie G. PEAKS: powerful software for peptide de novo sequencing by tandem mass spectrometry. Rapid Commun Mass Spectrom RCM. 2003;17(20):2337–42. doi:10.1002/rcm.1196.

Malik R, Dulla K, Nigg EA, Korner R. From proteome lists to biological impact--tools and strategies for the analysis of large MS data sets. Proteomics. 2010;10(6):1270–83. doi:10.1002/pmic.200900365.

Mann M, Jensen ON. Proteomic analysis of post-translational modifications. Nat Biotechnol. 2003;21(3):255–61. doi:10.1038/nbt0303-255.

Martinez-Bartolome S, Medina-Aunon JA, Jones AR, Albar JP. Semi-automatic tool to describe, store and compare proteomics experiments based on MIAPE compliant reports. Proteomics. 2010;10(6):1256–60. doi:10.1002/pmic.200900367.

Matthiesen R, Mutenda KE. Introduction to proteomics. Methods Mol Biol. 2007;367:1–35. doi:1-59745-275-0:1 [pii], 10.1385/1-59745-275-0:1.

McLaughlin T, Siepen JA, Selley J, Lynch JA, Lau KW, Yin H, Gaskell SJ, Hubbard SJ. PepSeeker: a database of proteome peptide identifications for investigating fragmentation patterns. Nucleic Acids Res. 2006;34:D649–54. doi:10.1093/nar/gkj066.

Medina-Aunon JA, Martinez-Bartolome S, Lopez-Garcia MA, Salazar E, Navajas R, Jones AR, Paradela A, Albar JP. The ProteoRed MIAPE web toolkit: a user-friendly framework to connect and share proteomics standards. Mol Cell Proteomics. 2011;10(10):M111 008334. doi:10.1074/mcp.M111.008334.

Medina-Aunon JA, Krishna R, Ghali F, Albar JP, Jones AJ. A guide for integration of proteomic data standards into laboratory workflows. Proteomics. 2013;13(3–4):480–92. doi:10.1002/pmic.201200268.

Medzihradszky KF, Chalkley RJ. Lessons in de novo peptide sequencing by tandem mass spectrometry. Mass Spectrom Rev. 2015;34(1):43–63.

Mewes HW, Amid C, Arnold R, Frishman D, Guldener U, Mannhaupt G, Munsterkotter M, Pagel P, Strack N, Stumpflen V, Warfsmann J, Ruepp A. MIPS: analysis and annotation of proteins from whole genomes. Nucleic Acids Res. 2004;32:D41–4. doi:10.1093/nar/gkh092.

Minguez P, Parca L, Diella F, Mende DR, Kumar R, Helmer-Citterich M, Gavin A-C, van Noort V, Bork P. Deciphering a global network of functionally associated post-translational modifications. Mol Syst Biol. 2012;8:599. doi:10.1038/msb.2012.31.

Minguez P, Letunic I, Parca L, Bork P. PTMcode: a database of known and predicted functional associations between post-translational modifications in proteins. Nucleic Acids Res. 2013;41(D1):D306–11. doi:10.1093/nar/gks1230.

Olsen JV, Mann M. Status of large-scale analysis of post-translational modifications by mass spectrometry. Mol Cell Proteomics. 2013;12(12):3444–52.

Ong SE, Blagoev B, Kratchmarova I, Kristensen DB, Steen H, Pandey A, Mann M. Stable isotope labeling by amino acids in cell culture, SILAC, as a simple and accurate approach to expression proteomics. Mol Cell Proteomics. 2002;1(5):376–86.

Opiteck GJ, Jorgenson JW. Two-dimensional SEC/RPLC coupled to mass spectrometry for the analysis of peptides. Anal Chem. 1997;69(13):2283–91.

Orchard S, Hermjakob H, Apweiler R. The proteomics standards initiative. Proteomics. 2003;3(7):1374–6. doi:10.1002/pmic.200300496.

Pappin DJC, Hojrup P, Bleasby AJ. Rapid identification of proteins by peptide-mass finger printing (Vol 3, Pg 327, 1993). Curr Biol. 1993;3(7):487–487.

Pendarvis K, Kumar R, Burgess SC, Nanduri B. An automated proteomic data analysis workflow for mass spectrometry. BMC Bioinf. 2009;10. doi:10.1186/1471-2105-10-s11-s17.

Perez-Riverol Y, Alpi E, Wang R, Hermjakob H, Vizcaino JA. Making proteomics data accessible and reusable: current state of proteomics databases and repositories. Proteomics. 2015;15(5–6):930–49. doi:10.1002/pmic.201400302.

Perkins DN, Pappin DJ, Creasy DM, Cottrell JS. Probability-based protein identification by searching sequence databases using mass spectrometry data. Electrophoresis. 1999;20(18):3551–3567. doi:10.1002/(SICI)1522-2683(19991201)20:18<3551::AID-ELPS3551>3.0.CO;2–2 [pii], 10.1002/(SICI)1522-2683(19991201)20:18<3551::AID-ELPS3551>3.0.CO;2–2.

Persico M, Ceol A, Gavrila C, Hoffmann R, Florio A, Cesareni G. HomoMINT: an inferred human network based on orthology mapping of protein interactions discovered in model organisms. BMC Bioinf. 2005;6:1–12. doi:10.1186/1471-2105-6-s4-s21.

Pesavento JJ, Mizzen CA, Kelleher NL. Quantitative analysis of modified proteins and their positional isomers by tandem mass spectrometry: human histone H4. Anal Chem. 2006;78(13):4271–80. doi:10.1021/ac0600050.

Povey S, Lovering R, Bruford E, Wright M, Lush M, Wain H. The HUGO Gene Nomenclature Committee (HGNC). Hum Genet. 2001;109(6):678–80. doi:10.1007/s00439-001-0615-0.

Prasad TSK, Goel R, Kandasamy K, Keerthikumar S, Kumar S, Mathivanan S, Telikicherla D, Raju R, Shafreen B, Venugopal A, Balakrishnan L, Marimuthu A, Banerjee S, Somanathan DS, Sebastian A, Rani S, Ray S, Kishore CJH, Kanth S, Ahmed M, Kashyap MK, Mohmood R,

Ramachandra YL, Krishna V, Rahiman BA, Mohan S, Ranganathan P, Ramabadran S, Chaerkady R, Pandey A. Human protein reference database-2009 update. Nucleic Acids Res. 2009;37:D767–72. doi:10.1093/nar/gkn892.

Prieto C, Rivas JDL. APID: agile protein interaction DataAnalyzer. Nucleic Acids Res. 2006;34: W298–302. doi:10.1093/nar/gkl128.

Puig-Costa M, Oliveras-Ferraros C, Flaquer S, Llopis-Puigmarti F, Pujol-Amado E, Martin-Castillo B, Vazquez-Martin A, Cufi S, Ortiz R, Roig J, Codina-Cazador A, Menendez JA. Antibody microarray-based technology to rapidly define matrix metalloproteinase (MMP) signatures in patients undergoing resection for primary gastric carcinoma. J Surg Oncol. 2011;104(1):106–9. doi:10.1002/jso.21887.

Rabilloud T, Lelong C. Two-dimensional gel electrophoresis in proteomics: a tutorial. J Proteomics. 2011;74(10):1829–41. doi:10.1016/j.jprot.2011.05.040, S1874-3919(11)00254-5 [pii].

Raynie DE. Modern extraction techniques. Anal Chem. 2010;82(12):4911–16. doi:10.1021/ac101223c.

Riffle M, Eng JK. Proteomics data repositories. Proteomics. 2009;9(20):4653–63. doi:10.1002/pmic.200900216.

Rigaut G, Shevchenko A, Rutz B, Wilm M, Mann M, Seraphin B. A generic protein purification method for protein complex characterization and proteome exploration. Nat Biotechnol. 1999;17(10):1030–2. doi:10.1038/13732.

Robin X, Hoogland C, Appel RD, Lisacek F. MIAPEGelDB, a web-based submission tool and public repository for MIAPE gel electrophoresis documents. J Proteomics. 2008;71(2):249–51. doi:10.1016/j.jprot.2008.06.005.

Robinson WH, DiGennaro C, Hueber W, Haab BB, Kamachi M, Dean EJ, Fournel S, Fong D, Genovese MC, de Vegvar HE, Skriner K, Hirschberg DL, Morris RI, Muller S, Pruijn GJ, van Venrooij WJ, Smolen JS, Brown PO, Steinman L, Utz PJ. Autoantigen microarrays for multiplex characterization of autoantibody responses. Nat Med. 2002;8(3):295–301. doi:10.1038/nm0302-295, nm0302-295 [pii].

Rohloff J. Analysis of phenolic and cyclic compounds in plants using derivatization techniques in combination with GC-MS-based metabolite profiling. Molecules. 2015;20(2):3431–62. doi:10.3390/molecules20023431.

Roxas BAP, Li Q. Significance analysis of microarray for relative quantitation of LC/MS data in proteomics. BMC Bioinf. 2008;9:1–17. doi:10.1186/1471-2105-9-187.

Ryan DP, Matthews JM. Protein-protein interactions in human disease. Curr Opin Struct Biol. 2005;15(4):441–6. doi:10.1016/j.sbo.2005.06.001.

Sadygov RG, Cociorva D, Yates 3rd JR. Large-scale database searching using tandem mass spectra: looking up the answer in the back of the book. Nat Methods. 2004;1(3):195–202. doi:10.1038/nmeth725.

Salomonis N, Hanspers K, Zambon AC, Vranizan K, Lawlor SC, Dahlquist KD, Doniger SW, Stuart J, Conklin BR, Pico AR. GenMAPP 2: new features and resources for pathway analysis. BMC Bioinf. 2007;8:1–12. doi:10.1186/1471-2105-8-217.

Schaefer CF, Anthony K, Krupa S, Buchoff J, Day M, Hannay T, Buetow KH. PID: the pathway interaction database. Nucleic Acids Res. 2009;37:D674–9. doi:10.1093/nar/gkn653.

Schauer N, Steinhauser D, Strelkov S, Schomburg D, Allison G, Moritz T, Lundgren K, Roessner-Tunali U, Forbes MG, Willmitzer L, Fernie AR, Kopka J. GC-MS libraries for the rapid identification of metabolites in complex biological samples. FEBS Lett. 2005;579(6):1332–7. doi:10.1016/j.febslet.2005.01.029.

Schmidt A, Forne I, Imhof A. Bioinformatic analysis of proteomics data. BMC Syst Biol. 2014;8 Suppl 2:S3. doi:10.1186/1752-0509-8-s2-s3.

Schneider M, Tognolli M, Bairoch A. The Swiss-Prot protein knowledgebase and ExPASy: providing the plant community with high quality proteomic data and tools. Plant Physiol Biochem PPB/Societe francaise de physiologie vegetale. 2004;42(12):1013–21. doi:10.1016/j.plaphy.2004.10.009.

Schwartz D, Chou MF, Church GM. Predicting protein post-translational modifications using meta-analysis of proteome scale data sets. Mol Cell Proteomics. 2009;8(2):365–79. doi:10.1074/mcp.M800332-MCP200.

Shafer MW, Mangold L, Partin AW, Haab BB. Antibody array profiling reveals serum TSP-1 as a marker to distinguish benign from malignant prostatic disease. Prostate. 2007;67(3):255–67. doi:10.1002/pros.20514.

Shannon P, Markiel A, Ozier O, Baliga NS, Wang JT, Ramage D, Amin N, Schwikowski B, Ideker T. Cytoscape: a software environment for integrated models of biomolecular interaction networks. Genome Res. 2003;13(11):2498–504. doi:10.1101/gr.1239303.

Sheehan KM, Calvert VS, Kay EW, Lu Y, Fishman D, Espina V, Aquino J, Speer R, Araujo R, Mills GB, Liotta LA, Petricoin 3rd EF, Wulfkuhle JD. Use of reverse phase protein microarrays and reference standard development for molecular network analysis of metastatic ovarian carcinoma. Mol Cell Proteomics. 2005;4(4):346–55. doi:T500003-MCP200 [pii], 10.1074/mcp.T500003-MCP200.

Shen Y, Smith RD. Proteomics based on high-efficiency capillary separations. Electrophoresis. 2002;23(18):3106–24. doi:10.1002/1522-2683(200209)23:18<3106::AID-ELPS3106>3.0.CO;2-Y.

Shevchenko A, Wilm M, Vorm O, Mann M. Mass spectrometric sequencing of proteins silver-stained polyacrylamide gels. Anal Chem. 1996;68(5):850–8.

Shevchenko A, Tomas H, Havlis J, Olsen JV, Mann M. In-gel digestion for mass spectrometric characterization of proteins and proteomes. Nat Protoc. 2006;1(6):2856–60. doi: nprot.2006.468 [pii], 10.1038/nprot.2006.468.

Silberring J, Ciborowski P. Biomarker discovery and clinical proteomics. Trac Trends Anal Chem. 2010;29(2):128–40. doi:10.1016/j.trac.2009.11.007.

Smith BE, Hill JA, Gjukich MA, Andrews PC. Tranche distributed repository and ProteomeCommons.org. Methods Mol Biol. 2011;696:123–45. doi:10.1007/978-1-60761-987-1_8.

Srihari S, Yong CH, Patil A, Wong L. Methods for protein complex prediction and their contributions towards understanding the organisation, function and dynamics of complexes. FEBS Lett. 2015;589(19 Pt A):2590–602. doi:10.1016/j.febslet.2015.04.026.

Subramanian A, Tamayo P, Mootha VK, Mukherjee S, Ebert BL, Gillette MA, Paulovich A, Pomeroy SL, Golub TR, Lander ES, Mesirov JP. Gene set enrichment analysis: a knowledge-based approach for interpreting genome-wide expression profiles. Proc Natl Acad Sci U S A. 2005;102(43):15545–50. doi:10.1073/pnas.0506580102.

Tang K, Page JS, Smith RD. Charge competition and the linear dynamic range of detection in electrospray ionization mass spectrometry. J Am Soc Mass Spectrom. 2004;15(10):1416–23. doi:S1044030504003150 [pii], 10.1016/j.jasms.2004.04.034.

Thiede B, Hohenwarter W, Krah A, Mattow J, Schmid M, Schmidt F, Jungblut PR. Peptide mass fingerprinting. Methods. 2005;35(3):237–47. doi:S1046-2023(04)00205-1 [pii], 10.1016/j.ymeth.2004.08.015.

Thomas PD, Campbell MJ, Kejariwal A, Mi HY, Karlak B, Daverman R, Diemer K, Muruganujan A, Narechania A. PANTHER: a library of protein families and subfamilies indexed by function. Genome Res. 2003;13(9):2129–41. doi:10.1101/gr.772403.

Troyanskaya O, Cantor M, Sherlock G, Brown P, Hastie T, Tibshirani R, Botstein D, Altman RB. Missing value estimation methods for DNA microarrays. Bioinformatics. 2001;17(6):520–5. doi:10.1093/bioinformatics/17.6.520.

Waegele B, Dunger-Kaltenbach I, Fobo G, Montrone C, Mewes HW, Ruepp A. CRONOS: the cross-reference navigation server. Bioinformatics. 2009;25(1):141–3. doi:10.1093/bioinformatics/btn590.

Wang X, Anderson GA, Smith RD, Dabney AR. A hybrid approach to protein differential expression in mass spectrometry-based proteomics. Bioinformatics. 2012;28(12):1586–91. doi:10.1093/bioinformatics/bts193.

Weiner AM, Platt T, Weber K. Amino-terminal sequence analysis of proteins purified on a nanomole scale by gel electrophoresis. J Biol Chem. 1972;247(10):3242–51.

Wessel D, Flugge UI. A method for the quantitative recovery of protein in dilute solution in the presence of detergents and lipids. Anal Biochem. 1984;138(1):141–3. doi:0003-2697(84) 90782-6 [pii].

Wisniewski JR, Zougman A, Nagaraj N, Mann M. Universal sample preparation method for proteome analysis. Nat Methods. 2009;6(5):359–62. doi:10.1038/nmeth.1322, nmeth.1322 [pii].

Wisniewski JR, Ostasiewicz P, Mann M. High recovery FASP applied to the proteomic analysis of microdissected formalin fixed paraffin embedded cancer tissues retrieves known colon cancer markers. J Proteome Res. 2011;10(7):3040–9. doi:10.1021/pr200019m.

Xenarios I, Salwinski L, Duan XQJ, Higney P, Kim SM, Eisenberg D. DIP, the Database of Interacting Proteins: a research tool for studying cellular networks of protein interactions. Nucleic Acids Res. 2002;30(1):303–5. doi:10.1093/nar/30.1.303.

Yao X, Freas A, Ramirez J, Demirev PA, Fenselau C. Proteolytic 18O labeling for comparative proteomics: model studies with two serotypes of adenovirus. Anal Chem. 2001;73 (13):2836–42.

Yates JR, Ruse CI, Nakorchevsky A. Proteomics by mass spectrometry: approaches, advances, and applications. Annu Rev Biomed Eng. 2009;11:49–79. doi:10.1146/annurev-bioeng-061008-124934.

Yeung YG, Stanley ER. Rapid detergent removal from peptide samples with ethyl acetate for mass spectrometry analysis. Curr Protoc Protein Sci. Chapter 16: unit 16. 2010;12. doi:10.1002/0471140864.ps1612s59.

Zhang Y, Fonslow BR, Shan B, Baek MC, Yates 3rd JR. Protein analysis by shotgun/bottom-up proteomics. Chem Rev. 2013;113(4):2343–94. doi:10.1021/cr3003533.

Zhao Y, Jensen ON. Modification-specific proteomics: strategies for characterization of post-translational modifications using enrichment techniques. Proteomics. 2009;9(20):4632–41. doi:10.1002/pmic.200900398.

Zhou JY, Dann GP, Shi T, Wang L, Gao X, Su D, Nicora CD, Shukla AK, Moore RJ, Liu T, Camp 2nd DG, Smith RD, Qian WJ. Simple sodium dodecyl sulfate-assisted sample preparation method for LC-MS-based proteomics applications. Anal Chem. 2012;84(6):2862–7. doi:10.1021/ac203394r.

Zhu W, Smith JW, Huang CM. Mass spectrometry-based label-free quantitative proteomics. J Biomed Biotechnol. 2010;2010:840518. doi:10.1155/2010/840518.

Chapter 7
Targeted Metabolomics: The Next Generation of Clinical Chemistry!

Klaus M. Weinberger and Marc Breit

Abstract Targeted metabolomics, i.e. the quantitation of predefined sets of endogenous metabolites selected for their relevance in metabolism, has emerged as a new and particularly informative discipline in functional genomics although its roots in diagnosing inborn disorders of metabolism in neonates go back much further than those of genomics or proteomics. Because of its unique capabilities in depicting actual physiological and pathophysiological conditions instead of just predispositions or risk factors, it seems ideally suited for complementing the currently established diagnostic platform technologies (enzyme assays, ion-selective electrodes, immunoassays, and molecular diagnostics) in a synergistic fashion. Of course, both technical and content-related prerequisites have to be met before a new technology can make any inroads in clinical practice and, so, this chapter discusses the development of metabolomics since the early twentieth century, the renaissance of clinical biochemistry in areas like neonatal screening and oncology, the most promising new indications, in which diagnostically relevant metabolic biomarker signatures have been identified and – partly – also validated and, eventually, selected risks and opportunities that have to be kept in mind when trying to promote this area of research and development. Bottom line: there is substantial reason to believe that targeted metabolomics can be the new platform technology in clinical chemistry if the community succeeds in taking advantage of the obvious strengths of this discipline and in avoiding some of the pitfalls that have hindered clinical acceptance for other varieties of functional genomics.

K.M. Weinberger (✉)
Research Group for Clinical Bioinformatics, Institute of Electrical and Biomedical Engineering (IEBE), University for Health Sciences, Medical Informatics and Technology (UMIT), 6060 Hall in Tirol, Austria

sAnalytiCo Ltd, Forsyth House, Cromac Square, Belfast BT2 8LA, UK

Weinberger & Weinberger Life Sciences Consulting, Weidach 82, 6414 Mieming, Austria
e-mail: klaus.weinberger@sanalytico.com

M. Breit
Research Group for Clinical Bioinformatics, Institute of Electrical and Biomedical Engineering (IEBE), University for Health Sciences, Medical Informatics and Technology (UMIT), 6060 Hall in Tirol, Austria

Institute of Health Care Engineering with European Notified Body of Medical Devices, Graz University of Technology, Stremayrgasse 16, 8010 Graz, Austria

X. Wang et al. (eds.), *Application of Clinical Bioinformatics*,
Translational Bioinformatics 11, DOI 10.1007/978-94-017-7543-4_7

Keywords Clinical chemistry • Targeted metabolomics • Multiparametric biomarkers • Chemometrics • Biochemical interpretation

Abbreviations

AAA	aromatic amino acids
AD	Alzheimer's disease
AMD	age-related macular degeneration
AUC	area under the curve
BCAA	branched-chain amino acids
CKD	chronic kidney disease
CNS	central nervous system
COPD	chronic obstructive pulmonary disease
CV	coefficient of variation
DBS	dried blood spots
DN	diabetic nephropathy
DNA	deoxyribonucleic acid
DoE	design of experiments
EBV	Epstein-Barr virus
ELISA	enzyme-linked immunosorbent assays
EMA	European Medicines Agency
FDA	Food and Drug Administration
FRET	fluorescence resonance energy transfer
GO	Gene Ontology
GWAS	genome-wide association studies
HCV	hepatitis C virus
HDL	high-density lipoprotein
HTA	health technology assessment
IP	intellectual property
IVD	*in vitro* diagnostics
KEGG	Kyoto Encyclopedia of Genes and Genomes
LDL	low-density lipoprotein
LIMS	laboratory information management system
MSEA	metabolite set enrichment analyses
M2-PK	M2 isoform of pyruvate kinase
NBS	newborn screening
NGS	next-generation sequencing
NMR	nuclear magnetic resonance
PCA	principle components analysis
PCR	polymerase chain reaction
PDE	phosphodiesterase
PLS-DA	partial least squares discriminant analysis
PPV	positive predictive value
P4	predictive, preventive, personalized, and participatory

RIA	radio-immunoassay
RNA	ribonucleic acid
ROC	receiver operating characteristics
RSD	relative standard deviation
R4S	Region 4 Stork
SAM	standard addition method
SID	stable isotope dilution
SMRT	single molecule real-time
SVM	support vector machine
TCM	traditional chinese medicine
TDM	therapeutic drug monitoring
T2D	type II diabetes
ZDF	Zucker diabetic fatty

7.1 Origins of Clinical Chemistry

The roots of human efforts to detect and understand disease can be traced back almost as far as there is a preserved written record of ancient civilizations. The famous Egyptian Edwin Smith Papyrus (Cunha 1949; Jex 1951) dating from c. 1600 BC, a copy of an older text often attributed to the politician-priest, architect and later god (!) Imhotep (c. 2600 BC) although this source seems highly questionable (Blomstedt 2014), is probably the oldest treatise that tries to disentangle medical science from paranormal beliefs or 'magic'. While mainly describing surgical procedures for traumatic injuries (a 'military surgeon's manual'), the text delineates a clear process of examination (including inspection, palpation, and olfaction) leading to diagnosis, and prognosis (Stiefel et al. 2006).

A similarly structured, albeit far more superstitious, approach to observe signs and symptoms of a disease and derive diagnoses and prognoses is presented by the Babylonian scholar Esagil-kin-apli of Borsippa (c. 1000 BC) in his 'Diagnostic Handbook', Sakikkū (Fales 2010). In Greece, philosophers like Hippocrates (c. 460–370 BC), Pythagoras (c. 570–495 BC), and others recognized inherited traits as important factors for certain illnesses and introduced terms like acute vs. chronic diseases or endemic vs. epidemic appearance, which are still used in the same sense (Begbie 1872; Kempf 1904; Roberts 1990). Eventually, antiquity's medical progress seems to have culminated in the works of Galen of Pergamon (129 – c. 216 AD), both a philosopher and physician whose most important contributions lay in the field of anatomy and in the extension of Hippocrates' theory of humors, which he linked to human temperaments. Today, he is best remembered for his detailed diagnostic accounts on the so-called Antonine Plague, most likely a smallpox pandemic that struck the Roman Empire in 165–168 AD (Littman and Littman 1973; Mattern 2011).

While most of the Greek and Roman achievements – including the field of medicine – did not outlast the turmoil of the Völkerwanderung and the religiously dominated, 'dark' Middle Ages in Europe, Persian scientists preserved and significantly extended the body of Greco-Roman medical knowledge. Most notably, the physician, polymath, and extremely prolific author Ibn Sīnā (latinized: Avicenna, c. 980–1037 AD), one of the masterminds of the so-called Golden Islamic Age, developed the concept of a syndrome in his influential encyclopedia Al-Qanun fi al-Tibb ('Canon of Medicine') and made important methodological contributions like the quantitative assessment of experiments and studies (Qayumi 1998). He also recognized the infectious etiology of many diseases (including sexual transmission) and laid the foundations for pharmacology as a medical discipline by defining a catalogue of criteria for testing new drugs. Even the importance of physical exercise for maintaining health is discussed in surprisingly concise and 'modern' terms (Choopani and Emtiazy 2015). It is primarily due to his works that medical knowledge made it back to Europe as the 'Canon' was used as the main textbook at many European universities from the fourteenth to the sixteenth century, sometimes even until the early eighteenth century, e.g. in Padova, and served as the basis for further research and progress during the Renaissance and the Age of Enlightenment (Abdel-Halim 2014).

Fairly independent from this stream of developments, the traditional Chinese Medicine (TCM) also formed a highly structured and systematic approach to diagnostics, which is based on the pillars of interrogation (wèn), inspection (wàng), palpation (qiè), auscultation and olfaction (wén). Of course, this set – maybe except for olfaction – is the core of the standard anamnesis and examination practiced in Western medicine as well (Maciocia 1989).

In contrast to this protracted evolution, the role of analytical chemistry in medicine is fairly young. Of course, the molecular basis of an illness is also analyzed when a physician smells volatile fermentation products of bacteria infesting a wound, the presence of acetone in the breath of a ketotic person, or tastes the sweetness of glucose secreted in the urine of a diabetic, which has been done since antiquity. However, the basic concept of clinical chemistry, which is considered self-evident today, i.e. that the concentration of a certain biochemical in a body fluid can be indicative of a particular disease or pathophysiological state, has only been phrased in a concise and scientifically credible manner (and also reduced to practice) in the early twentieth century.

This paradigm shift was pioneered by Sir Archibald Garrod (1857–1936), a British physician who already contributed significantly to the one-gene-one-enzyme-hypothesis (Beadle and Tatum 1941; Yanofsky 2005; Nobel Prize for George Beadle and Edward Tatum in 1958) when working on the chemical pathology of inborn errors of metabolism, e.g. of alkaptonuria (Garrod 1899, 1902) and cystinuria (Garrod and Hurtley 1906). His findings, which were based on optimized and standardized analytical procedures (Garrod and Hurtley 1905), paved the way to a fruitful link of genetics, analytical chemistry and pathophysiology (Garrod 1911) and made him the true 'godfather of clinical chemistry'. Garrod's work also directly led to today's population-based screening programs for inherited metabolic disorders (see Sect. 7.3.2). Although it took the best part of the twentieth century to develop and implement the mass spectrometric platforms that are now used for newborn screening (Chace et al. 2002, 2003; Liebl et al. 2003; Röschinger et al. 2003), Garrod's series of four Croonian Lectures delivered to the Royal College of Physicians of London in June of 1908 (Garrod 1908) are still considered the starting shot for this prototypic kind of multiparametric metabolic diagnostics.

7.2 Technological Revolutions in the Twentieth Century

The next logical step was to extend the analytical portfolio to also cover biopolymers but substantial progress towards a diagnostic utilization was only made for two of the three main classes of polymeric biomolecules, namely proteins and nucleic acids. Despite their undoubted biological importance, oligo- and polysaccharides – with a few exceptions, e.g. in microbiology or certain lysosomal storage disorders – still haven't found their way into routine diagnostics.

7.2.1 *Proteins*

Proteins with catalytic activity – first termed 'ferments', later 'enzymes' as suggested by Wilhelm Kühne (1877) – had been known since Anselme Payen's discovery of amylase (then called diastase) in malt (Payen and Persoz 1833). Their use in enzyme assays to specifically detect and quantify metabolites based on various physico-chemical readouts (spectrophotometric, fluorometric, chemiluminescent, etc.) is straightforward, and the quantitation of glucose using hexokinase or glucose oxidase is still the single most frequently applied diagnostic test. Of course, the same setting (in reverse) can also be used to determine an enzymatic activity in a biological sample, and once enzyme levels in peripheral blood were recognized as indicators for organ-specific tissue damage, e.g. the transaminases for liver damage in viral hepatitis (De Ritis et al. 1955, 1956, 1957), these assays became a core element of the diagnostic repertoire.

Of course, proteins with other functions (structural, regulatory, transport, immune, etc.) are not amenable to this detection strategy. However, with the advent of sequencing techniques for peptides, either chemical (Sanger 1945, 1949; Edman 1950; Ryle et al. 1955; Edman and Begg 1967; first Nobel Prize for Frederick Sanger in 1958), or using mass spectrometry (Biemann et al. 1966; Falick et al. 1993; Biemann 2014), and the determination of the three-dimensional structure of proteins, first for myoglobin (Kendrew and Perutz 1957; Kendrew et al. 1960; Nobel Prize for John Kendrew and Max Perutz in 1962), the interest in protein function and their potential role in disease became overwhelming and, so, additional techniques for their characterization and quantitation were direly needed.

The first step towards this end was polyacrylamide gel electrophoresis (Raymond and Weintraub 1959), which – in its refined form (Laemmli 1970) – became one of the standard procedures in biomedical research, particularly when combined with a blotting technique for the antibody-based identification of individual bands, the so-called Western Blot (Towbin et al. 1979). Still, the routine diagnostic detection of proteins is rarely based on gel electrophoresis but rather applies immunoassays without a previous separation step, e.g. radio-immunoassays (RIA; Yalow and Berson 1960) or enzyme-linked immunosorbent assays (ELISA; Engvall and Perlmann 1971; Van Weemen and Schuurs 1971). The performance of these immunoassays directly depends on the availability and characteristics (specificity, affinity/avidity, stability, and other properties) of suitable antibodies, and it was only with the invention of cell culture strategies to generate monoclonal antibodies (Köhler and Milstein 1975; Nobel Prize for Georges Köhler and César Milstein in 1984) that standardized immunoassays against virtually any protein could be developed and – equally important for the routine application – automated for high throughput on robotic laboratory systems. These immunoassays, typically designed as a so-called sandwich ELISA with two different antibodies for capturing and detecting the target antigen, and the enzyme assays described above still constitute the core of the analytical portfolio in today's clinical chemistry and medical microbiology/virology.

7.2.2 Nucleic Acids

Deoxyribonucleic acid (DNA) was already identified in 1869 by Friedrich Miescher (Miescher 1869; Miescher-Rüsch 1871; reviewed by Dahm 2005), and its chemical building blocks, the nucleotides adenine, thymine, guanine, and cytosine, were also known early on (Levene and Jacobs 1912; Levene 1919) but the molecule was generally considered too simple to be the carrier of genetic information. However, in the mid twentieth century, a period of rapid progress in microbiology, biochemistry and molecular biology led to the identification of DNA as the 'transforming principle' and, thus, the hereditary material (Griffith 1928; Avery et al. 1944; Hershey and Chase 1952; Nobel Prize for Alfred Hershey in 1969). Briefly afterwards, based on X-ray crystallography data on the one hand (Franklin and Gosling 1953; Wilkins et al. 1953) and the analysis of the relative amounts of the four nucleotides in the DNA of various species on the other (the so-called 'Chargaff rules'; Vischer and Chargaff 1947; Vischer et al. 1949; Chargaff et al. 1950), Jim Watson and Francis Crick succeeded in elucidating the famous double-helix structure (Watson and Crick 1953a, b; Nobel Prize for James Watson, Francis Crick, and Maurice Wilkins in 1962), which directly led to the mechanism of semi-conservative replication and the central dogma of molecular biology (Crick 1956, 1970). Still in the same decade, the triplett nature of the genetic code was suggested based on theoretical considerations (Gamow and Ycas 1955; Gamow et al. 1956) and, eventually, the code itself was experimentally solved (Crick et al. 1961; Nirenberg and Matthaei 1961; Lengyel et al. 1961; Matthaei et al. 1962; Leder and Nirenberg 1964; Khorana 1965; Nobel Prize for Marshall Nirenberg, Har Gobind Khorana, and Robert Holley in 1968).

To harness the information content, analytical techniques for determining DNA sequences were required, and the two major approaches to this end were chemical sequencing (Maxam and Gilbert 1977) and the now 'classic' Sanger sequencing using dideoxynucleotides as chain terminators (Sanger and Coulson 1975; Nobel Prize for Frederick Sanger (his second), Paul Berg, and Walter Gilbert in 1980). Sequence-specific detection of DNA fragments separated by gel electrophoresis, the so-called Southern Blot (Southern 1975), and its simplified variant without chromatographic separation, the dot blot, were then the first assay types to usher in the era of molecular diagnostics, i.e. the detection and characterization of nucleic acids for diagnostic purposes, soon after followed by the analogous method for ribonucleic acid (RNA) fragments, the Northern Blot (Alwine et al. 1977). However, while allowing for the highly specific identification of nucleic acids (primarily used for detecting viral infections), both blotting techniques have relevant limits in terms of sensitivity and their quantitative properties; nevertheless, semi-quantitative estimates of DNA and RNA amounts have been successfully used in clinical routine, e.g. in viral load testing.

Still, these shortcomings may explain why the next technical innovations in this area had such a tremendous (and rapid) impact. The invention of the polymerase chain reaction (PCR) by Kary Mullis (Saiki et al. 1985; Mullis et al. 1986; Mullis and Faloona 1987; Nobel Prize for Kary Mullis in 1993), the first and still only method that achieves target amplification instead of signal amplification, immediately overcame the sensitivity issues (being able to detect a single molecule) with dramatic clinical consequences, e.g. for improved safety of blood and organ donations (Fishman and Rubin 1998; Bihl et al. 2007), to name just one of the most important applications. Based on this principle and the introduction of fluorescent dyes coupled to the terminators, Sanger sequencing became much more efficient ('cycle sequencing') and already set the stage for what would eventually become the Human Genome Project (Strauss et al. 1986; Hood et al. 1987; Kaiser et al. 1989).

Equally important was the improvement of DNA quantitation by introducing a kinetic analysis of the PCR ('real time PCR' or 'qPCR'), first with conventional intercalating dyes (Higuchi et al. 1993) and then with specialized bi-labeled probes each carrying a 'reporter' and a 'quencher' dye suitable for fluorescence resonance energy transfer (FRET), the so-called 'TaqMan' probes (Heid et al. 1996). As with PCR itself, it only took a few years until the first diagnostic assays were implemented, in this case for viral load testing, e.g. for Epstein-Barr virus (EBV; Kimura et al. 1999) or hepatitis B virus (HBV; Weinberger et al. 2000), and the method was soon also combined with a reverse transcriptase step to quantify RNA viruses like hepatitis C virus (HCV; Martell et al. 1999) or relevant viral transcripts, e.g. of Epstein-Barr virus (Weinberger et al. 2004).

By now, qualitative and quantitate assays of molecular diagnostics represent the second major pillar of state-of-the-art laboratory medicine, and – given the incredible rate, at which new genetic information is generated by next-generation sequencing (NGS) platforms such as 454 pyrosequencing (Margulies et al. 2005; Wheeler et al. 2008; Green et al. 2010; Zheng et al. 2010), Solexa/Illumina reversible termination sequencing (Furey et al. 1998; Osborne et al. 2000; Bentley et al. 2008), SOLiD sequencing by ligation (Valouev et al. 2008; McKernan et al. 2009), Pacific Bio single molecule real-time (SMRT) sequencing (Eid et al. 2009; Chin et al. 2011; Rasko et al. 2011), or ion torrent semiconductor sequencing (Rothberg et al. 2011; Mellmann et al. 2011; Vogel et al. 2012) and mined by bioinformatics tools (Crowgey et al. 2015; De Brevern et al. 2015) – its importance and market volume will certainly continue to grow in the foreseeable future.

7.3 Renaissance of Clinical Biochemistry

Some of the most commonly determined – and publicly recognized – parameters in clinical chemistry are metabolites (or, in the jargon: 'substrates' – the use of enzyme assays to measure metabolites clearly left its marks here); as mentioned above, glucose is the most frequently detected analyte at all (Newman and Turner 2005), creatinine is still the basis of the assessment of renal function despite severe analytical and diagnostic shortcomings (Breit and Weinberger 2016), and many patients actually know and discuss their cholesterol level – even high-density lipoprotein (HDL) and low-density lipoprotein (LDL) cholesterol – as a cardiovascular risk factor (Gotto 1997). In contrast, for the last few decades, the focus of research and development for new diagnostic markers has been on proteins and – probably even more so – on nucleic acids. This bias created a situation in the pharmaceutical industry and also at regulatory authorities, in which the term 'biomarker', either in safety assessment or in patient stratification, was considered almost synonymous with genomic and proteomic biomarkers, but was hardly associated with metabolites (Robinson et al. 2008). However, there are two fields of biomedical research that have recently demonstrated the potential of 'classical' biochemistry for drug development and also for diagnostic innovation:

7.3.1 Warburg Hypothesis

The first one is the rediscovery, experimental confirmation and, finally, reinterpretation of the Warburg hypothesis. Otto Warburg (1883–1970) had recognized that, even in the presence of sufficient levels of oxygen, tumor cells have a peculiar kind of energy metabolism characterized by high glucose consumption, low to absent levels of respiration, and a substantial lactate production, the so-called anaerobic glycolysis (Warburg et al. 1927). These findings led him to postulate that a mitochondrial defect was a causative factor in tumorigenesis (Warburg 1956a, b). Despite his preeminent scientific reputation – he had received the Nobel Prize in 1931 for identifying the 'respiration ferment' (Warburg 1925, 1928) and is generally considered one of the greatest biochemists of the twentieth century – this groundbreaking contribution to oncology almost completely sank into oblivion for a couple of decades until Erich Eigenbrodt's and Sybille Mazurek's systematic work on the M2 isoform of pyruvate kinase (M2-PK) renewed the interest in this topic and elucidated basic regulatory mechanisms of energy metabolism in tumors (Eigenbrodt et al. 1977, 1983, 1997, 1998; Mazurek et al. 1998, 2000; Mazurek and Eigenbrodt 2003). Based on their findings, a specialized database for tumor metabolism was created (www.metabolic-database.com) and a diagnostic test for M2-PK was developed by ScheBo Biotech (Gießen, Germany), which is commercially available, e.g. as a screening test for colorectal cancer (Hardt et al. 2004).

Still, recognition of this remarkable body of work was somewhat reluctant, and it took another wave of high-ranking publications from Lew Cantley's team at Harvard, Matt Vanderheiden's at MIT, Ralph Deberardinis' at UTSW, and Craig Thompson's at Sloan Kettering to really reposition the Warburg hypothesis in the focus of attention (Christofk et al. 2008a, b; Dang et al. 2009; Dang 2010; Mullen et al. 2011; Jiang and Deberardinis 2012; Son et al. 2013). Today, many major pharmaceutical companies have research groups specializing in oncometabolomics (again) or, at least, entered alliances to boost progress in this area (Sborov et al. 2015). As a result, new drug targets in energy metabolism have been validated (Sotgia et al. 2013) and, based on the insight that these metabolic alterations may be common to many different types of cancer, drug candidates with potentially very broad anti-tumor efficacy have been suggested (Flaveny et al. 2015). Even clinical development programs specifically targeting cancer metabolism, particularly glycolysis and glutaminolysis, are already underway, e.g. by Agios Pharmaceuticals (Cambridge, MA).

7.3.2 *Newborn Screening*

The second remarkable success story of clinical biochemistry in the last two decades – and this time with a clear diagnostic focus – was the extension and global dissemination of neonatal screening programs for inborn errors of metabolism. While this topic brings us full circle with regard to the remarks on Archibald Garrod's work in Sect. 7.1, it also illustrates how a combination of technical innovations (both preanalytical and analytical), biochemical insights, and new approaches to data management and mining were required to render this ambitious project successful.

The most influential preanalytical improvement was definitely the introduction of dried blood spots (DBS) as an alternative sample type in the Scottish phenylketonuria screening program by Bob Guthrie (Guthrie and Susi 1963) who demonstrated that capillary blood blotted and dried on filter paper was an extremely stable source of biomolecules, could be sent to central laboratories by regular mail as it did not require any special handling (e.g. cooling), and was compatible with various analytical methods for proteins (Orfanos et al. 1977), enzymatic activities (Orfanos et al. 1978; Orfanos et al. 1980a), small peptides (Orfanos et al. 1980b), and amino acids (Guthrie 1969; Vollmer et al. 1990). Because of its striking benefits, particularly the simple sample logistics for large-scale studies, e.g. in epidemiology (Parker and Cubitt 1999), the use of DBS was soon extended to a variety of molecular diagnostics assays (Schwartz et al. 1990; Raskin et al. 1992; Cassol et al. 1992).

The actual breakthroughs, however, that overcame the analytical bottleneck of monoparametric assays for single diseases and allowed for an extension of neonatal screening platforms to cover a wide range of genetic disorders at a single blow were: (a) the availability of robust and sufficiently selective and sensitive tandem (more specifically: triple quadrupole) mass spectrometers, and (b) the use of stable isotope dilution (SID) to achieve quantitative results for amino acids and acylcarnitines on these instruments (Millington et al. 1992; Chace et al. 1993, 2002, 2003; Van Hove et al. 1993; Röschinger et al. 2000, 2003 Rolinski et al. 2000; Chace 2001; Fingerhut et al. 2001). Yet, based on multiparametric assays for these two classes of metabolites and simple, hypothesis-driven but – at that time – highly innovative chemometrics to derive diagnostic signatures for each condition, the portfolio was not only expanded to 30 or more congenital disorders but the diagnostic performance was significantly improved (Chace et al. 1998). Of course, given the low to very low prevalence of the individual diseases – ranging from roughly one in 5000 live births for medium-chain acyl-CoA dehydrogenase deficiency in Northern Germany (Sander et al. 2001) down to one in several hundred thousand live births for many others, the greatest challenge is to design tests with outstanding specificity to warrant at least acceptable positive predictive values (PPV) since the PPV strongly depends on the prevalence of the condition:

$$PPV = \frac{sensitivity \times prevalence}{sensitivity \times prevalence + (1 - specificity) \times (1 - prevalence)} \quad (7.1)$$

This is illustrated by plotting the PPV versus the prevalence at a given (high) sensitivity (Fig. 7.1) and may explain why various attempts have been made to improve the diagnostic criteria, e.g. by machine learning techniques (Baumgartner et al. 2004) or the Region 4 Stork (R4S) interpretive tools, which are part of a global initiative to collect newborn screening (NBS) data (McHugh et al. 2011; Marquardt et al. 2012; Hall et al. 2014). To just elaborate on one example, in the latter study conducted on more than 170,000 subjects, the R4S tools plus second-tier tests were able to reduce the rate of false-positives from 0.26 to 0.02 %, which corresponded to the overall PPV rising from 10 % to more than 50 % (Hall et al. 2014), a potentially decisive difference in health technology assessment (HTA) studies.

By now, population-based screening programs are implemented in most industrialized and many developing countries, albeit with different diagnostic portfolios, and are prototypic examples for the way, in which metabolomics could further evolve – if not revolutionize – clinical chemistry.

7.4 Promising New Indications

7.4.1 Motivations for Clinical Mass Spectrometry

Besides neonatal screening, which is typically conducted in a highly centralized fashion with few centers servicing large geographic areas, mass spectrometry has been (and could further be) introduced in clinical laboratories essentially for two reasons:

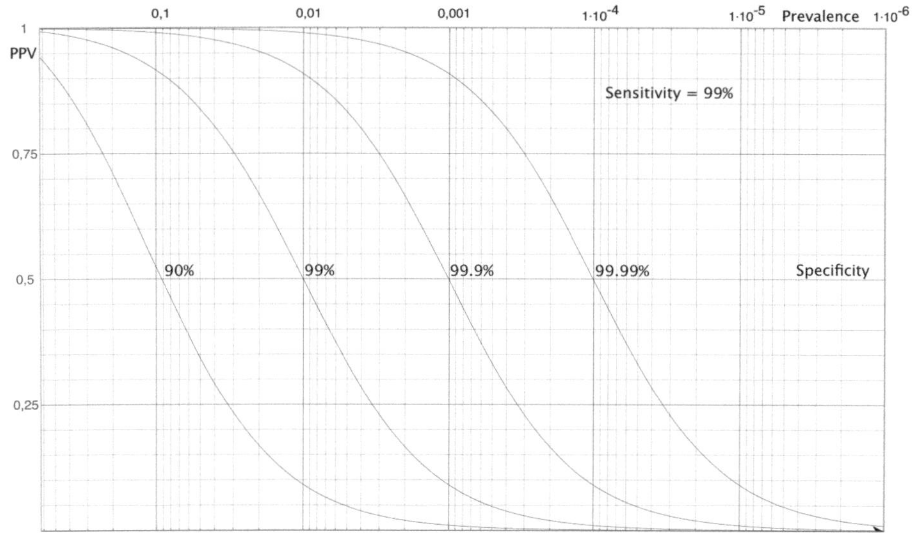

Fig. 7.1 Positive predictive values for rare diseases. The relationship between the positive predictive value (PPV) and the prevalence of a diagnosed disease (see Eq. 7.1) is demonstrated by plotting selected graphs for specificities ranging from 90 to 99.99 % at a given (reasonably high) sensitivity of 99 %. Note that even at the highest specificity level, the PPV for a disease with a prevalence of 1 in 100,000 is below 10 %

First and quite obvious, to replace analytically flawed assays, often immunoassays for small molecules, which have inherent issues (limited specificity, steric hindrance preventing the set-up of a sandwich ELISA, etc.) if antibodies are available at all; most endogenous metabolites such as amino acids, organic acids, simple carbohydrates, and many classes of lipids are ubiquitous in animals and, thus, no natural antibodies can be raised against them. The most relevant and well-established examples for this category are therapeutic drug monitoring (TDM) for immunosuppressants (Koal et al. 2004; Ceglarek et al. 2004, 2006; Seger et al. 2009), antiretroviral drugs (Koal et al. 2005, 2006), antibiotics (Koal et al. 2006; König et al. 2013; Zander et al. 2015), or antidepressants (Berm et al. 2015), and assays for vitamin D and its metabolites (Vogeser 2010; Van den Ouweland et al. 2013). In addition, the last few years have seen the launch of multiparametric tandem mass spectrometric assays for other clinically relevant metabolites such as steroid hormones (Stero***IDQ***™ from Biocrates, Innsbruck, Austria; MassChrom® Steroids from Chromsystems, Gräfelfing, Germany), bile acids (Bile Acids Kit from Biocrates), or Catecholamines (from ChromSystems), and the Stero***IDQ***™ kit has already demonstrated that such an assay can be developed not just as a set of reagents but as a fully integrated solution and can meet the quality and regulatory requirements for a European registration as a medical device for *in vitro* diagnostics (IVD; the CE/IVD mark according to the directive 98/79/EC).

Second, and this is certainly more promising and more challenging at the same time, mass spectrometry could facilitate the introduction of novel diagnostic content, e.g. of biomarker signatures that have already been identified on this platform and would not have to go through a technology transfer process with its inherent risks of poor analytical performance or of a failure to combine the necessary parameters in one assay with potentially grave consequences for the intellectual property (IP) situation. Any addition to the diagnostic repertoire, however, requires outstanding scientific evidence, preferably substantiated by translational research and validated by replication in independent clinical cohorts (Breit and Weinberger 2016). This is even more important since confidence in and acceptance for omics-derived biomarkers has suffered dramatically from the early (and extremely well-published) attempts to identify proteomic signatures for various conditions, particularly in oncology (Liotta and Petricoin 2000; Petricoin et al. 2002a, b, c), none of which could ever be validated, let alone be translated into a diagnostic assay. Unfortunately, the field of metabolomics has seen similar examples, again in oncology, with the protracted controversy over sarcosine as a prostate cancer marker (Sreekumar et al. 2009; Jentzmik et al. 2010; Struys et al. 2010; and many more). Thus, the critical question is whether there are metabolic biomarkers mature enough for clinical applications and, if yes, for which indications.

7.4.2 Diabetes

The simple, actually trivial, answer to this question is that metabolomics should be a particularly helpful tool for characterizing metabolic disorders; it goes without saying that a condition like type II diabetes (T2D), in which large, metabolically active organs (liver, skeletal muscle, etc.) are ‘misbehaving’, has a more profound and much more easily detectable impact on the metabolome than, say, a minuscule early-stage tumor whose – doubtlessly altered – metabolism (see Sect. 7.3.1) would have to be spotted against the background of an overwhelming majority of metabolically normal healthy tissue.

As a matter of fact, this theoretical assumption has been confirmed by a long series of metabolomics studies on diabetic animal models and clinical cohorts. It would lead much too far to discuss all the individual findings but studies in various rodent models, e.g. db/db mice (Hummel et al. 1966) or Zucker diabetic fatty (ZDF) rats, have shown repeatedly that – among many other metabolic alterations – the aromatic and the branched-chain amino acids (AAA and BCAA, respectively) are early and sensitive markers for diabetes (Altmaier et al. 2008; Weinberger 2008; Giesbertz et al. 2015). These parameters proved to be significantly more sensitive than conventional (monoparametric) markers used at that time, and helped in prioritizing a phosphodiesterase (PDE) 4 inhibitor as a new anti-diabetic drug candidate, which is now in clinical development with very promising results (Altmaier et al. 2008; Vollert et al. 2012; Wouters et al. 2012).

The same findings were obtained in several human cohorts (Weinberger 2008; Gieger et al. 2008; Suhre et al. 2010; Lu et al. 2013a, b; Duranton et al. 2014), and it turned out that these alterations were not just indicative of incident diabetes but also very early events in prediabetes (Kulkarni 2012; Dzien & Weinberger, unpublished data), which is, of course, the much more pressing diagnostic need. Eventually, a large-scale study on the Framingham Offspring cohort demonstrated beyond any reasonable doubt that elevated levels of AAA and BCAA were even predictive of future diabetes as much as 12 years before onset of the disease (Wang et al. 2011). Thus, T2D risk assessment and monitoring the progress of the metabolic syndrome towards incident diabetes is clearly one of the most advanced and promising indications for a diagnostic metabolomics assay as it fulfills all the relevant criteria (huge unmet diagnostic need, repeatedly validated biomarkers, mechanistic understanding from translational research, and excellent 'kitability' as demonstrated by products that are already on the market, e.g. the Absolute***IDQ***® P180 or the MetaDis***IDQ***® kits from Biocrates).

7.4.3 Chronic Kidney Disease

The second indication, for which there is a similarly compelling set of data, is closely associated with the diabetes pandemic, namely chronic kidney disease (CKD). Again, it would go beyond the scope of this chapter to discuss any details here but a characteristic set of metabolic changes, e.g. in dimethylarginine metabolism, the urea cycle, tryptophan catabolism, and oxidative stress have been identified in relevant animal models and repeatedly confirmed in different human cohorts, either in large population-based or in dedicated clinical studies (Boudonck et al. 2009a, b; Lundin and Weinberger 2010; Lundin et al. 2011; Duranton et al. 2014; Pena et al. 2015; Breit and Weinberger 2016). Beyond these individual findings, machine learning techniques were used to create classifiers based on metabolite panels measured in plasma or urine, and both classifiers were not just diagnostic, i.e. associated with the renal function at the time of sampling, but also predictive of the renal function at a follow-up examination 2 years later (Nkuipou-Kenfack et al. 2014).

7.4.4 Neurology

The greatest interest of the clinical community (but also of patients' organizations and the public) can certainly be observed in the fields of neurology and oncology but both have their specific challenges. In neurology, the most actively pursued indication is Alzheimer's disease (AD), for which the currently available diagnostic armamentarium is still particularly poor (Lewczuk et al. 2015; Laske et al. 2015). Since 2008, various studies attempted to find metabolic markers for AD (Barba et al. 2008) but the last few years really saw an explosion of publications on precisely this topic (69 hits in PubMed from 2011 to August 2015), some of them by the very pioneers of metabolomics (Kaddurah-Daouk et al. 2011, 2013 Orešic et al. 2011; Han et al. 2011). However, important questions regarding the use of serum or plasma metabolomics for conditions affecting the central nervous system (CNS) remain unanswered or, at least, undiscussed in most of these papers, e.g. how permeable is the blood-brain barrier for all the metabolites analyzed, i.e. how much of the changes found in peripheral blood is actually due to CNS metabolism, and what is the extent of disease-related changes of this permeability? Of course, it would mean a truly Sisyphean labor to address these issues in a systematic fashion, and a diagnostic signature can also be perfectly valid and clinically valuable if it is only based on statistical evidence and the underlying mechanisms remain unclear. Yet, the questions raised above (and many others) may have significant repercussions for the kind of controls one would have to use in biomarker studies to ensure appropriate diagnostic specificity (assuming that the AD community is not just hunting for a systemic inflammation marker).

7.4.5 Oncology

As mentioned above, in oncology, the general expectation from new omics technologies was somewhat soured by the tough lessons learned from proteomics but then reinvigorated by the renaissance of the Warburg hypothesis (see above). In any case, the (perceived) diagnostic need in this area is greater than in any other medical discipline and, so, a plethora of studies have been conducted to find markers or signatures for various types of cancer (far too many to be discussed in any detail in this chapter; selected recent reviews by Bezabeh et al. 2014; Halama 2014; Lloyd et al. 2015; Patel and Ahmed 2015).

Still, one has to keep in mind that there is no such thing as the 'generic biomarker for a certain type of cancer' identified by simplistic study designs like 'cancer vs. healthy control'; both clinical oncologists and the pharmaceutical industry rather focus on predictive companion diagnostic markers, which could identify patients who are likely to benefit from a particular drug or treatment regimen (Aichler et al. 2014; Michels et al. 2014), whereas the supreme goal of early diagnosis (allowing for curative surgery or even radiotherapy) seems extremely ambitious given the sensitivity considerations discussed above (very small tumor, lots of healthy surrounding tissue). This situation would only change if the tumor did indeed synthesize specific compounds that are not found in healthy tissues instead of just more or less of a ubiquitous metabolite, and the first highly promising example for this scenario is the identification of 2-hydroxyglutarate as an alternative product of the mutated isocitrate dehydrogenase often found in tumors, which makes it a prototypic oncometabolite (Dang et al. 2009; Gross et al. 2010; Ward et al. 2010).

7.4.6 *Others*

Of course, the literature is also full of other reports on biomarker discovery by metabolomics; to list just a few very recent examples, their targeted indications range from different types of lung cancer (Wikoff et al. 2015; Fahrmann et al. 2015) to primary biliary cirrhosis and autoimmune hepatitis (Lian et al. 2015), and from fetal chromosomal aberrations (Pinto et al. 2015) to Sjögren's syndrome (Kageyama et al. 2015), not to mention mycobacterial infections (Mirsaeidi et al. 2015), venous thrombosis (Deguchi et al. 2015), breast cancer (Mishra and Ambs 2015), Snyder-Robinson syndrome (Abela et al. 2016), inflammatory bowel disease (Sands 2015), Takayasu arteritis (Guleria et al. 2015), or preeclampsia (Koster et al. 2015).

In summary, many areas of clinical research try to mimic the success stories of metabolic biomarker discovery and validation in diabetology and nephrology, and these efforts get facilitated by the broader dissemination of standardized assays, kit products, or – at least – validated protocols including preanalytical recommendations. Yet, none of these areas have reached a similar level of maturity and credibility as the two applications discussed above in greater detail, so it seems likely that prediabetes and chronic kidney disease/diabetic nephropathy (DN) may be the first two (of presumably many) instances of new metabolomics-derived content to appear on the requisition slips of clinical core labs.

7.5 Technological and Bioinformatics Challenges

It is quite evident that all of the trends and innovations described and predicted above cannot work without the appropriate design of experiments (DoE), a solid framework for data management and workflow control, e.g. in a laboratory information management system (LIMS), data analysis incorporating quality control steps and sophisticated statistics and, eventually, biochemical interpretation of the results, either in the discovery and validation of new markers or in the clinical assessment of routine measurements. The bioinformatics tools addressing these needs specifically for targeted metabolomics are described in detail (and from a more technical angle) in Chap. 8 of this book while the following section will try and highlight a couple of challenges and pitfalls.

7.5.1 *Clinical Translation*

The first of these remarks may sound trivial but the authors feel that it is of the utmost importance to stress it once again: innovations in bioanalytics and bioinformatics cannot overcome the need for diligent scientific work, or – in other words – omics platforms and data mining tools will not replace careful study design and detail-oriented experimental work but rather make them more efficient and more informative.

This is particularly true for large-scale, population-based studies aiming at the identification of new diagnostic biomarker signatures. Certainly, the statistical significance levels of many of the published results are impressive but so is the complexity of the study populations, the bioanalytical procedures, and – thus – the generated data sets. In the end, even p-values of 10^{-100} or 10^{-200} are only measures of a likelihood, could be caused by unrecognized flaws in the study design or patients' documentation, systematic biases of the lab methods, or other intrinsic structures of the data, and will not substitute an independent replication of the results. In genomics studies, particularly in genome-wide association studies (GWAS), this has long been recognized, and no reasonable journal would accept GWAS data for publication unless the results were confirmed in a second, independent cohort.

In metabolomics, the same standard has been set by the first examples of GWAS on metabolic traits (Gieger et al. 2008; Illig et al. 2010), a combination that has been outstandingly fruitful over the last few years and identified both significant and meaningful links of genetics and metabolism (Suhre et al. 2011; Nicholson et al. 2011; Ried et al. 2013, 2014; Draisma et al. 2015). More recently, this approach was also extended to epigenetics (Petersen et al. 2014; Pfeiffer et al. 2015). In the wake of these studies, metabolic biomarker discovery and validation was conducted in similar cohorts (and often by the same investigators) yielding highly credible and convincing results, on which diagnostic product development could well be founded (Wang-Sattler et al. 2012; Floegel et al. 2013; Goek et al. 2013; Würtz et al. 2015). Yet, and this critical remark is unfortunately necessary: the vast majority of metabolomics studies does not yet meet these criteria regarding the study design, and – one a more basic level – regarding the wording of the publications; it goes without saying that not every single study can be conducted in cohorts of thousands of well-documented subjects (quite often, this is even a *contradictio in eo ipso*) but then it would at least be important to call the results 'biomarker candidates' instead of 'biomarkers' and refrain from far-reaching claims about the clinical utility of such markers but rather recommend independent validation, even if the statistical results are highly significant.

7.5.2 Standardization of Quantitative Analytics

The assessment of an assay in clinical chemistry typically happens at two different but closely interrelated levels (neglecting the commercial aspects for now): one addressing the analytical characteristics and the other scrutinizing the actual diagnostic performance. The latter aspect has already been briefly discussed in Sect. 7.3 with regards to rare diseases, and the accepted parameter for doing so is the area under the curve (AUC) in a receiver operating characteristics (ROC) analysis, which still allows to choose the individual cut-off value in a way that either sensitivity or specificity are optimized depending on the clinical priorities (or a balanced compromise is found). Of course, the typically digital diagnostic result directly depends on the quality of the underlying analytics. In the case of quantitative assays measuring continuous variables, the key performance parameters are precision, describing the reproducibility, and accuracy, describing the correct determination of the true value. While precision can easily be determined for each laboratory, instrument, or operator by standard procedures, essentially repeated analyses of the same sample and calculation of the coefficient of variation (CV) or relative standard deviation (RSD), e.g. following the 'Guidance for Industry Bioanalytical method validation' issued by the American Food and Drug Administration (FDA) or the 'Guideline on bioanalytical method validation' issued by the European Medicines Agency (EMA), the assessment of accuracy in metabolomics is a far greater challenge. If no orthogonal gold standard method is available, the most straightforward approach would be to spike known concentrations in an authentic matrix and compare the results with the expected values. However, since the analytes of interest are all ubiquitous, endogenous metabolites, one would have to create such a matrix artificially, either bottom-up by composing a synthetic matrix from chemically defined ingredients, or top-down by depletion of blood fluids on activated carbon, but the resulting liquids are usually no fully appropriate substitute in terms of mimicking the preanalytical and analytical complexity of actual serum or plasma samples. As an alternative, one has to resort to the standard addition method (SAM) to determine the true value but that also has its limitations, e.g. it would require that any matrix effects must not vary with the analyte-to-matrix ratio, which is actually difficult to claim.

Yet, assuming these technicalities have been satisfactorily solved, both parameters still have a different meaning in the development and life cycle of a diagnostic test. In the daily routine of general practitioners, the vast majority of patients repeatedly see the same physician who always sends his samples to the same lab. In this setting, analytical precision is of paramount importance in order to follow the course of a patient's condition, e.g. to determine disease progression or monitor therapeutic efficacy. However, in clinical research and the actual product development and registration of a diagnostic test, multicenter studies and comparison/integration of data across multiple sources is indispensable. Here, accuracy is a *conditio sine qua non*, which can only be efficiently controlled by interlaboratory comparison of proficiency in round robin tests or, at the very least, by broadly available standard materials (e.g. as part of kit products), against which each lab can check their own calibrations.

Unfortunately, these considerations mean that the community is facing a kind of chicken-and-egg problem: standards, kits, and – even more so – round robin tests typically only become available once a test is fully developed and established in a reasonable number of labs while accuracy and standardization would have been most direly needed for the clinical research identifying and validating the new marker.

7.5.3 Biochemical Annotation and Interpretation

A third aspect to be discussed here is the role of biochemical background knowledge. As already briefly mentioned in Sect. 7.4, diagnostic markers and tests can be perfectly valid if they are only based on statistical significance and lack a mechanistic explanation, theoretically even if they were unidentified peaks in a non-targeted profiling approach (with the discussed impact on study design, selection of appropriate controls, diagnostic specificity, etc.). Yet, and this is particularly true for metabolomics, the utilization of the detailed understanding of many biochemical pathways opens up additional possibilities for hypothesis-driven (or -supported) data mining strategies that can improve the statistical power of studies and/or reduce the number of false-positive findings.

These options have recently been reviewed in detail (Enot et al. 2011; Breit et al. 2015) and are also covered in greater depth in Chap. 8 of this book. However, two brief comments should be made at this point: Over the last century, biochemistry has elucidated the majority of relevant enzymatic and non-enzymatic steps in human metabolism (the map is certainly not 100 % complete but offers a very good coverage), both in terms of reaction mechanisms (substrates and products, enzymes and cofactors involved, localization, often even X-ray crystallographic structures) and of quantitative characteristics (kinetics and energetics of the reactions, homeostatic equilibria, etc.). Based on this understanding, one can use the rate-limiting step of a pathway as a surrogate read-out if the analytical coverage is limited or combine metabolites belonging to the same pathway in metabolite set enrichment analyses (MSEA) – a far more informed and more highly resolved kind of ontology than the Gene Ontology (GO) typically used in genomics or proteomics studies. From a strictly statistical angle, the latter approach still has its shortcomings because it relies on rather 'traditional' definitions of biochemical pathways, e.g. in the Kyoto Encyclopedia of Genes and Genomes (KEGG). Here, metabolic pathways are defined as entities of greatly different magnitude and complexity (often based on the history of their discovery) and, thus, there is a bias towards overestimating the importance of complex pathways while even pathways consisting of just of few reactions can have major implications for the overall metabolic situation (compare, for instance, glycerophospholipid metabolism to the urea cycle). Moreover, using this classification for statistics does not take into account that different pathways have varying degrees of redundancy; the more alternative routes metabolism can take, the less dramatic are the repercussions of a single event (up- or downregulation of enzyme expression, gain or loss of function mutations, etc.) for the entire network. Thus, enrichment analyses do have a huge potential in this context but cannot replace expert assessment of the findings.

To this end, mapping quantitative metabolomics data onto pathways offers one more avenue of scrutinizing biostatistical results, which – as discussed earlier in this section – are only probabilistic in nature and, thus, prone to a residual risk of false positives. As soon as a certain enzyme metabolizes several substrates (things are a little more complex in the reverse situation), it is always a sensible plausibility check to see whether the trends for these substrates are the same. To name just the simplest example: of course, there could be various reasons why the three BCAA valine, leucine, and isoleucine show a different behavior in a certain study but, as they are essential, i.e. not synthesized in the body, and the first step in their catabolism is catalyzed by the same enzyme, in most cases they will show very similar trends. So, deviations from this rule should trigger a thorough investigation of potential analytical issues (and the same holds true for many aspects of lipid metabolism and other pathways).

Also, this thought loops directly back to the enrichment analyses: groups of metabolites that aren't necessarily part of a classical pathway but share biochemical properties or reactions make for extremely informative sets in MSEA, e.g. saturated, monounsaturated, or polyunsaturated fatty acids; short-, middle-, or long-chain acylcarnitines; branched-chain, aromatic, glucogenic, or ketogenic amino acids, and various ratios thereof. For kit-based targeted metabolomics, such an approach – albeit on a limited scale – has already been implemented in commercially available software (the Met***IDQ***™ suite accompanying the MetaDis***IDQ***® kit from Biocrates; Gruber et al. 2012; Then et al. 2013).

7.5.4 Important Caveats in Supervised Statistics

Another important pitfall in data analysis has become markedly more relevant in recent years. In the early phase of metabolomics – or 'metabonomics' in the terminology of many nuclear magnetic resonance (NMR) enthusiasts (Nicholson et al. 1999), the biomedical aspects of many studies may not have been perfectly designed but, at least, the bioanalytics and the chemometrics were usually conducted by real experts; after all, this was a new technology, and methodological progress was considered as important as biological content. Since metabolomics has found a wider distribution (either through kits or the services offered by core facilities), it starts to be viewed as just another experimental tool that can be used by investigators who are more interested in the content than the methods. While, in principle, this is a thoroughly positive development, it leaves many researchers alone with the problem of how to manage and analyze their metabolomics data and, in this situation, they may be tempted to use multivariate statistics and other bioinformatics tools, which are now freely available online (see Chap. 8) but which they do not fully understand. In this context, some of the most critical steps are outlier detection and handling, scaling and normalization, imputation of missing values, and the far too liberal use of supervised data mining that let the user generate impressive scores plots of discriminant analyses or ROC curves with outstanding AUC from rather poor and noisy data sets.

To illustrate this point more concisely, the authors analyzed an extremely noisy urinary metabolomics data set (poor sample quality, many missing values, many analytes around or below their limit of detection) from ten db/db mice and ten heterozygous healthy controls (db/-) (Altmaier et al. 2008) on the MetaboAnalyst platform (Xia et al. 2009, 2012, 2015) using standard settings (log normalization, Pareto scaling) and could easily generate a very nice principle components analysis (PCA) scores plot and equally convincing class probability prediction plots based on partial least squares discriminant analysis (PLS-DA), a Random Forests approach, or a linear support vector machine (SVM), each suggesting excellent separation of the two cohorts. In addition, the ROC analyses also yielded close-to-optimal curves with AUC values greater than 0.99 for combinations of three to 80 metabolites in the actual classifiers (Fig. 7.2). Now, although this data set had

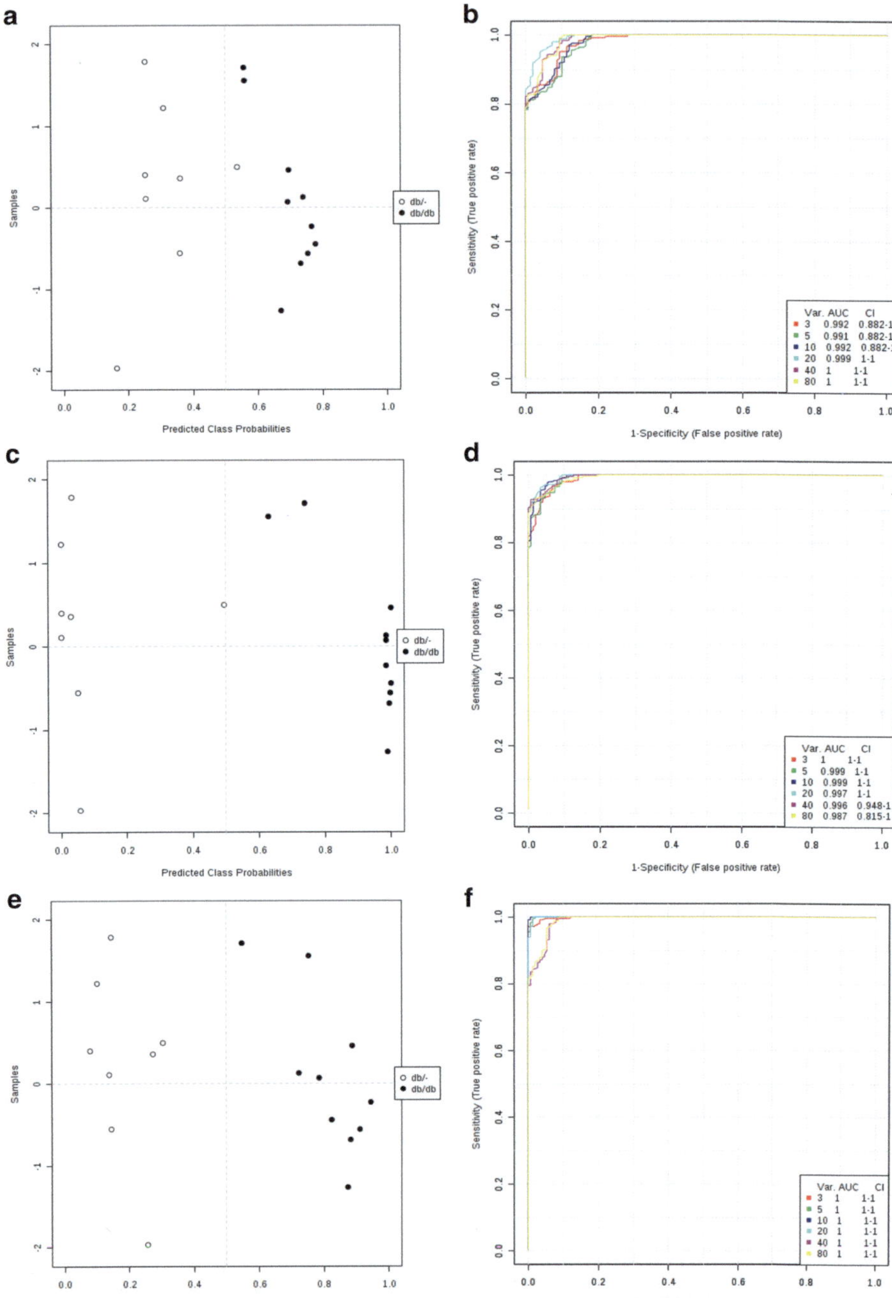

Fig. 7.2 Supervised multivariate statistics, part 1. Three different classification algorithms, partial least squares discriminant analysis (PLS-DA, panels **a** and **b**), random forests (panels **c** and **d**), and linear support vector machine (SVM, panels **e** and **f**), all available on the MetaboAnalyst website (Xia et al. 2015) were applied to an extremely noisy metabolomics data set with many missing values and many concentrations at or below the limit of detection (for details, see Sect. 7.5.4).

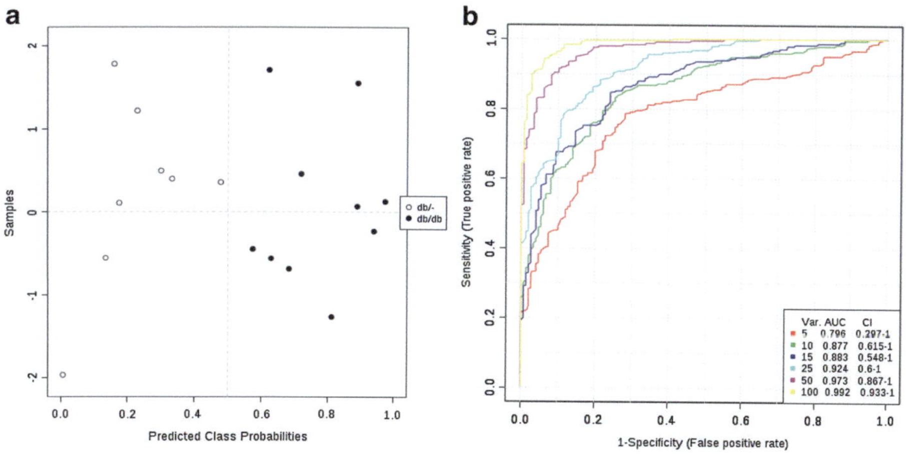

Fig. 7.3 Supervised multivariate statistics, part 2. The linear SVM approach used for panels **e** and **f** in Fig. 7.2 was applied to a data set of the same dimensions (>200 parameters, 10 samples per group, see Sect. 7.5.4) but consisting of random numbers. Again, the class probability prediction plot (panel **a**) suggests a perfect separation of both cohorts, and the ROC analysis yields excellent AUC values ranging from 0.796 to 0.992 for classifiers consisting of 5–100 parameters (panel **b**) (See Sect. 7.5.4)

been studied extensively by experts in statistics, biochemistry, and diabetology who did not find much meaningful content (in contrast to the extremely informative serum and plasma data from the same mice), there was still the possibility of some hidden gems that everybody had missed so far. Therefore, the same analyses were conducted for a set of random numbers (again >200 parameters, 10 samples per cohort): the ROC curves still had an AUC between 0.796 with five variables and 0.992 with 100 variables in the classifiers (Fig. 7.3), and the platform calculated an average accuracy of 0.981 after 100 cross-validations!

Thus, it should be acknowledged as crucial to give users of such online services sound introductions to the available methods including their strengths and weaknesses, and warn them about the risks of overestimating the 'diagnostic performance' of a metabolic signature, actually of any high-dimensional data set; plus, as a general rule, supervised multivariate statistics should only be applied in combination with a thorough biochemical plausibility check of the results, e.g. by pathway mapping or enrichment analyses (see above). However, it is probably too optimistic to expect such a simple recommendation to have any noticeable effect. Therefore, these aspects must be checked in the peer-review process for

Fig. 7.2 (continued) Each of the tools generated highly discriminating class probability predictions (panels **a**, **c**, and **e**) and extremely optimistic receiver operating characteristics (ROC) curves (panels **b**, **d**, and **f**) with areas under the curves (AUC) of very close to 1. Note that, in several cases, the numbers given for the AUC do obviously not match the graphs, particularly when the AUC equals exactly 1

publications to enforce a more responsible manner of presenting data from metabolomics studies.

7.5.5 Biological Specificity

The final remark in this section goes back to the issue of diagnostic specificity and the selection of appropriate controls. Everybody who analyzes metabolomics data sets on a regular basis and in various indications will sooner or later realize that there is an intermediary level of complexity between the individual metabolite concentrations and the – typically multifactorial – clinical condition (of course, for the monogenic disorders detected in neonatal screening, this relationship is much more direct). In most cases, several metabolites show alterations that belong to one pathway or basic pathomechanism, and several of these findings form the biochemical signature of a complex disease. This concept of 'intermediary phenotypes' was first described by Karsten Suhre in the context of GWAS on metabolic traits (Illig et al. 2010) and holds true for many clinically relevant indications, e.g. diabetes (Altmaier et al. 2008; Weinberger 2008; Gieger et al. 2008) or chronic kidney disease (Lundin and Weinberger 2010; Lundin et al. 2011; Duranton et al. 2014; reviewed by Breit and Weinberger 2016).

To stick to the latter example, CKD presents itself (amongst others) with significant alterations in dimethylarginine metabolism, the urea cycle, tryptophan catabolism, or oxidative stress levels (see also Sect. 7.4), each of which has in turn a typical signature of concentration changes of individual metabolites. To elaborate on only two of these: tryptophan metabolism in CKD is characterized by extremely low levels of tryptophan and increased turn-over to kynurenine and serotonin, i.e. elevated ratios of these two products to tryptophan (Lundin et al. 2011; Goek et al. 2012, 2013; Breit and Weinberger 2016); oxidative stress on the other hand is reflected by an elevated ratio of methionine-sulfoxide to methionine and, subsequently, by an impaired enzymatic activity of phenylalanine hydroxylase due to low levels of its oxidation-sensitive cofactor tetrahydrobiopterin, i.e. a decreased ratio of tyrosine to phenylalanine (Weinberger 2008; Sonntag et al. 2008; Breit and Weinberger 2016) (Fig. 7.4).

Yet, some of these fundamental mechanisms of pathobiochemistry such as membrane damage, mitochondrial leakage, dysregulated autophagy, oxidative stress, endothelial dysfunction, inflammation, and others play a central role in many different diseases. By far most prominently, oxidative stress has been observed in neurodegenerative conditions ranging from Alzheimer's (Coppedè and Migliore 2015; Rosales-Corral et al. 2015) to Parkinson's disease (Blesa et al. 2015) and from age-related macular degeneration (AMD; Blasiak et al. 2014) to multiple sclerosis (Haider 2015) but it is, at the same time, a hallmark of chronic obstructive pulmonary disease (COPD; Domej et al. 2014), non-alcoholic steatohepatitis (Lim et al. 2015) and atherosclerosis (Husain et al. 2015), diabetes (Keane et al. 2015) and CKD (see above), not to mention its

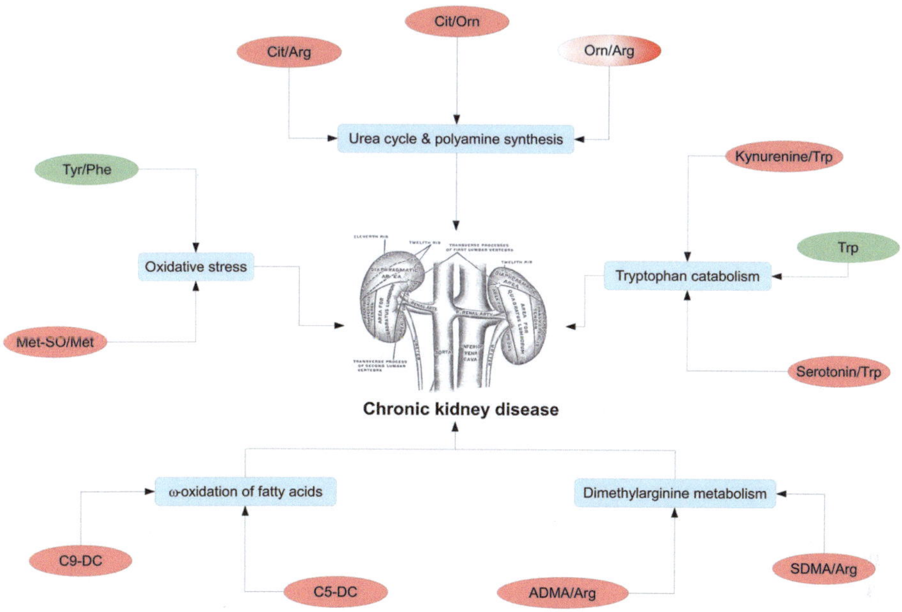

Fig. 7.4 Systems diagnostics. Simplified overview of the main metabolic alterations found in chronic kidney disease (CKD; reviewed by Breit and Weinberger 2016); for many of the changes, which have been validated so far, there is an intermediary level of complexity linking the metabolic phenotype to the clinical diagnosis: several individual metabolites represent a particular pathway or basic biochemical pathomechanism, e.g. oxidative stress, and several of these entities characterize the actual disease (See Sect. 7.5.5)

Janus-faced role in many types of cancer (Ramesh et al. 2014; Manda et al. 2015). Only slightly exaggerated, one gets the impression that, today, it would warrant a high-ranking publication if somebody found a disease, in which oxidative stress was actually not involved.

In return, this clearly implicates that many metabolic changes are not likely to be specific for a particular disease but that a diagnostic marker panel would have to consist of several such signatures to yield the desired specificity. More concisely, this also means that the appropriate controls for studies on any of the diseases listed in the last paragraph are not just healthy individuals but patients with other conditions, in which (a subset of) the same molecular mechanisms may play a role. Such a reductionist view on pathobiochemistry aiming at some kind of 'systems diagnostics' would, thus, have far-reaching consequences for the study design in clinical research, but also for regulatory strategies, the daily diagnostic practice and, eventually, the reimbursement of diagnostic assays.

7.6 Conclusions

All of the above-mentioned aspects warrant a couple of technical and content-related conclusions.

First of all, there is plenty of evidence that metabolomics has ushered in a new era of clinical biochemistry; metabolic analyses had actually preceded immunoassays and molecular diagnostics by decades but may now make a comeback with highly informative, multiparametric biomarker panels for some of the most pressing diagnostic needs. These panels may actually even trigger a rather radical paradigm shift towards diagnosing sets of basic pathomechanisms ('systems diagnostics') instead of the traditional diagnostic entities.

From a technical perspective, the basic requirements for this development seem to be met (sufficiently sensitive and selective tandem mass spectrometers are available, stable isotope dilution allows for highly accurate quantitation, etc.) but there is still a long way to go to reach the same level of robustness, ease-of-use, and automation as the established platforms, and this holds both for the hardware (instruments) and the software (consumables, bioinformatics, etc.). However, manufacturers of mass spectrometers and of metabolomics kits have recognized the gap and are making significant progress towards standardized integrated solutions that could serve as prototypes of new diagnostic tools.

In this context, robustness does not only refer to technical stability but also to the role of various confounders. Metabolomics' greatest advantage, namely being so close to the functional end-point of biochemical processes instead of just depicting predispositions like most genetic analyses, can also represent a major challenge: metabolic markers may be rather sensitive to circadian rhythm, diet, medication, physical activity or, for that matter, preanalytical sample handling. While such influences cannot be entirely avoided, the last few years saw some highly systematic work towards understanding these factors in a quantitive fashion (Yu et al. 2012; Floegel et al. 2013, 2014; Jaremek et al. 2013; Mathew et al. 2014; Breier et al. 2014; Anton et al. 2015) paving the way for chemometric tools to compensate for the additional variability.

Thus, there is definitely reason to be optimistic: the content identified by metabolomics studies is often extremely compelling, the first few biomarker panels for important indications have been successfully validated, and the development of more robust instruments, suitable laboratory automation platforms, standardized consumables, and appropriate software certainly moves in a promising direction although a seamless integration of these elements may still take some time. Even regulatory hurdles have already been overcome, as demonstrated by the CE/IVD-labeled Stero***IDQ***™ kit, and – commercially just as important – the first patents for concise, diagnostically relevant metabolite signatures have been granted (e.g., Lundin and Weinberger 2010) after many futile attempts – even by global companies – to patent all possible subsets of absurdly long Markush lists.

Still, this summary would not be complete without a word of warning. The failure of early attempts to identify cancer biomarkers in proteomics, which were prematurely heralded as clinical breakthroughs, has afflicted lasting damage to the entire field, and this lesson must be taken seriously if metabolomics is to succeed as the new platform in clinical chemistry. Of course, enthusiasm and a certain pioneering spirit are necessary to promote a new technology but nothing undermines scientific credibility more than making far-reaching medical claims based on questionable study designs and inappropriate data analytics that cannot be substantiated in validation studies. So, it is up to the metabolomics community to act responsibly and realize the huge potential that this technology holds – if they do, the next generation of patients will benefit from a radically new way of laboratory diagnostics, which should be a cornerstone of Leroy Hood's vision of predictive, preventive, personalized, and participatory (P4) medicine.

References

Abdel-Halim RES. The role of Ibn Sina (Avicenna)'s medical poem in the transmission of medical knowledge to medieval Europe. Urol Ann. 2014;6(1):1.

Abela L, Simmons L, Steindl K, Schmitt B, Mastrangelo M, Joset P, et al. N(8)-acetylspermidine as a potential plasma biomarker for Snyder-Robinson syndrome identified by clinical metabolomics. J Inherit Metab Dis. 2016;39(1):131–7. doi:10.1007/s10545-015-9876-y. Epub 2015 Jul 15.

Aichler M, Luber B, Lordick F, Walch A. Proteomic and metabolic prediction of response to therapy in gastric cancer. World J Gastroenterol. 2014;20(38):13648–57. doi:10.3748/wjg.v20.i38.13648. Review.

Altmaier E, Ramsay SL, Graber A, Mewes HW, Weinberger KM, Suhre K. Bioinformatics analysis of targeted metabolomics--uncovering old and new tales of diabetic mice under medication. Endocrinology. 2008;149(7):3478–89. doi:10.1210/en.2007-1747. Epub 2008 Mar 27.

Alwine JC, Kemp DJ, Stark GR. Method for detection of specific RNAs in agarose gels by transfer to diazobenzyloxymethyl-paper and hybridization with DNA probes. Proc Natl Acad Sci U S A. 1977;74(12):5350–4.

Anton G, Wilson R, Yu ZH, Prehn C, Zukunft S, Adamski J, et al. Pre-analytical sample quality: metabolite ratios as an intrinsic marker for prolonged room temperature exposure of serum samples. PLoS One. 2015;10(3):e0121495. doi:10.1371/journal.pone.0121495. eCollection2015.

Avery OT, MacLeod CM, McCarty M. Studies on the chemical nature of the substance inducing transformation of pneumococcal types induction of transformation by a desoxyribonucleic acid fraction isolated from pneumococcus type III. J Exp Med. 1944;79(2):137–58.

Barba I, Fernandez-Montesinos R, Garcia-Dorado D, Pozo D. Alzheimer's disease beyond the genomic era: nuclear magnetic resonance (NMR) spectroscopy-based metabolomics. J Cell Mol Med. 2008;12(5A):1477–85. doi:10.1111/j.1582-4934.2008.00385.x. Epub 2008 June 28. Review.

Baumgartner C, Böhm C, Baumgartner D, Marini G, Weinberger K, Olgemöller B, Liebl B, Roscher AA. Supervised machine learning techniques for the classification of metabolic disorders in newborns. Bioinformatics. 2004;20(17):2985–96. Epub 2004 June 4.

Beadle GW, Tatum EL. Genetic control of biochemical reactions in Neurospora. Proc Natl Acad Sci U S A. 1941;27(11):499–506.

Begbie JW. Hippocrates: his life and writings. Br Med J. 1872;2(626):709–11.

Bentley DR, Balasubramanian S, Swerdlow HP, Smith GP, Milton J, Brown CG, et al. Accurate whole human genome sequencing using reversible terminator chemistry. Nature. 2008;456(7218):53–9. doi:10.1038/nature07517.

Berm EJ, Hak E, Postma M, Boshuisen M, Breuning L, Brouwers JR, et al. Effects and cost-effectiveness of pharmacogenetic screening for CYP2D6 among older adults starting therapy with nortriptyline or venlafaxine: study protocol for a pragmatic randomized controlled trial (CYSCEtrial). Trials. 2015;16:37. doi:10.1186/s13063-015-0561-0.

Bezabeh T, Ijare OB, Nikulin AE, Somorjai RL, Smith IC. MRS-based metabolomics in cancer research. Magn Reson Insights. 2014;7:1–14. doi:10.4137/MRI.S13755.eCollection2014. Review.

Biemann K. Laying the groundwork for proteomics: mass spectrometry from 1958 to 1988. J Proteomics. 2014;107:62–70. doi:10.1016/j.jprot.2014.01.008. Epub 2014 Jan 18. Review.

Biemann K, Cone C, Webster BR, Arsenault GP. Determination of the amino acid sequence in oligopeptides by computer interpretation of their high-resolution mass spectra. J Am Chem Soc. 1966;88(23):5598–606.

Bihl F, Castelli D, Marincola F, Dodd RY, Brander C. Transfusion-transmitted infections. J Transl Med. 2007;5:25.

Blasiak J, Petrovski G, Veréb Z, Facskó A, Kaarniranta K. Oxidative stress, hypoxia, and autophagy in the neovascular processes of age-related macular degeneration. Biomed Res Int. 2014;2014:768026. doi:10.1155/2014/768026. Epub 2014 Feb 23.

Blesa J, Trigo-Damas I, Quiroga-Varela A, Jackson-Lewis VR. Oxidative stress and Parkinson's disease. Front Neuroanat. 2015;9:91. doi:10.3389/fnana.2015.00091.eCollection2015. Review.

Blomstedt P. Imhotep and the discovery of cerebrospinal fluid. Anat Res Int. 2014;2014:256105. doi:10.1155/2014/256105. Epub 2014 Mar 13.

Boudonck KJ, Mitchell MW, Német L, Keresztes L, Nyska A, Shinar D, Rosenstock M. Discovery of metabolomics biomarkers for early detection of nephrotoxicity. Toxicol Pathol. 2009a;37 (3):280–92. doi:10.1177/0192623309332992.

Boudonck KJ, Rose DJ, Karoly ED, Lee DP, Lawton KA, Lapinskas PJ. Metabolomics for early detection of drug-induced kidney injury: review of the current status. Bioanalysis. 2009b;1 (9):1645–63. doi:10.4155/bio.09.142. Review.

Breier M, Wahl S, Prehn C, Fugmann M, Ferrari U, Weise M, et al. Targeted metabolomics identifies reliable and stable metabolites in human serum and plasma samples. PLoS One. 2014;9(2):e89728. doi:10.1371/journal.pone.0089728.eCollection2014.

Breit M, Weinberger KM. Metabolic biomarkers for chronic kidney disease. Arch Biochem Biophys. 2016;589:62–80. doi:10.1016/j.abb.2015.07.018. Epub 2015 Jul 31.

Breit M, Baumgartner C, Weinberger KM. Data handling and analysis in metabolomics. In: Khanmohammadi M, editor. Current applications of chemometrics. New York: Nova; 2015. p. 181–203.

Cassol S, Salas T, Gill MJ, Montpetit M, Rudnik J, Sy CT, O'Shaughnessy MV. Stability of dried blood spot specimens for detection of human immunodeficiency virus DNA by polymerase chain reaction. J Clin Microbiol. 1992;30(12):3039–42.

Ceglarek U, Lembcke J, Fiedler GM, Werner M, Witzigmann H, Hauss JP, Thiery J. Rapid simultaneous quantification of immunosuppressants in transplant patients by turbulent flow chromatography combined with tandem mass spectrometry. Clin Chim Acta. 2004;346 (2):181–90.

Ceglarek U, Casetta B, Lembcke J, Baumann S, Fiedler GM, Thiery J. Inclusion of MPA and in a rapid multi-drug LC-tandem mass spectrometric method for simultaneous determination of immunosuppressants. Clin Chim Acta. 2006;373(1–2):168–71. Epub 2006 May 17.

Chace DH. Mass spectrometry in the clinical laboratory. Chem Rev. 2001;101(2):445–77. Review.

Chace DH, Millington DS, Terada N, Kahler SG, Roe CR, Hofman LF. Rapid diagnosis of phenylketonuria by quantitative analysis for phenylalanine and tyrosine in neonatal blood spots by tandem mass spectrometry. Clin Chem. 1993;39(1):66–71.

Chace DH, Sherwin JE, Hillman SL, Lorey F, Cunningham GC. Use of phenylalanine-to-tyrosine ratio determined by tandem mass spectrometry to improve newborn screening for phenylketonuria of early discharge specimens collected in the first 24 hours. Clin Chem. 1998;44 (12):2405–9.

Chace DH, Kalas TA, Naylor EW. The application of tandem mass spectrometry to neonatal screening for inherited disorders of intermediary metabolism. Annu Rev Genomics Hum Genet. 2002;3:17–45. Epub 2002 Apr 15. Review.

Chace DH, Kalas TA, Naylor EW. Use of tandem mass spectrometry for multianalyte screening of dried blood specimens from newborns. Clin Chem. 2003;49(11):1797–817.

Chargaff E, Magasanik B, Vischer E, Green C, Doniger R, Elson D. Nucleotide composition of pentose nucleic acids from yeast and mammalian tissues. J Biol Chem. 1950;186(1):51–67.

Chin CS, Sorenson J, Harris JB, Robins WP, Charles RC, Jean-Charles RR, Bullard J, Webster DR, Kasarskis A, Peluso P, Paxinos EE, Yamaichi Y, Calderwood SB, Mekalanos JJ, Schadt EE, Waldor MK. The origin of the Haitian cholera outbreak strain. N Engl J Med. 2011;364 (1):33–42. doi:10.1056/NEJMoa1012928. Epub 2010 Dec 9.

Choopani R, Emtiazy M. The concept of lifestyle factors, based on the teaching of avicenna (ibn sina). Int J Prev Med. 2015;6:30. doi:10.4103/2008-7802.154772.eCollection2015.

Christofk HR, Vander Heiden MG, Wu N, Asara JM, Cantley LC. Pyruvate kinase M2 is a phosphotyrosine-binding protein. Nature. 2008a;452(7184):181–6. doi:10.1038/nature06667.

Christofk HR, Vander Heiden MG, Harris MH, Ramanathan A, Gerszten RE, Wei R, et al. The M2 splice isoform of pyruvate kinase is important for cancer metabolism and tumour growth. Nature. 2008b;452(7184):230–3. doi:10.1038/nature06734.

Coppedè F, Migliore L. DNA damage in neurodegenerative diseases. Mutat Res. 2015;776:84–97. doi:10.1016/j.mrfmmm.2014.11.010. Epub 2014 Dec 9. Review.

Crick F. Central dogma of molecular biology. Nature. 1970;227(5258):561–3.

Crick F. Ideas on protein synthesis. Francis Harry Compton Crick Papers. Wellcome Library for the History and Understanding of Medicine. 1956 [cited 2015 Sep 1]. Available from: http://profiles.nlm.nih.gov/ps/access/SCBBFT.pdf

Crick FH, Barnett L, Brenner S, Watts-Tobin RJ. General nature of the genetic code for proteins. Nature. 1961;192:1227–32.

Crowgey EL, Kolb A, Wu CH. Development of bioinformatics pipeline for analyzing clinical pediatric NGS data. AMIA Jt Summits Transl Sci Proc. 2015;2015:207–11. eCollection 2015.

Cunha F. The Edwin Smith surgical papyrus. Am J Surg. 1949;78(2):277.

Dahm R. Friedrich Miescher and the discovery of DNA. Dev Biol. 2005;278(2):274–88. Review.

Dang CV. Glutaminolysis: supplying carbon or nitrogen or both for cancer cells? Cell Cycle. 2010;9(19):3884–6. Epub 2010 Oct 9.

Dang L, White DW, Gross S, Bennett BD, Bittinger MA, Driggers EM, et al. Cancer-associated IDH1 mutations produce 2-hydroxyglutarate. Nature. 2009;462(7274):739–44. doi:10.1038/nature08617.

de Brevern AG, Meyniel JP, Fairhead C, Neuvéglise C, Malpertuy A. Trends in IT innovation to build a next generation bioinformatics solution to manage and analyse biological big data produced by NGS Technologies. Biomed Res Int. 2015;2015:904541. doi:10.1155/2015/904541. Epub 2015 June 1.

De Ritis F, Coltorti M, Giusti G. Transaminase activity of the blood in viral hepatitis. Boll Soc Ital Biol Sper. 1955;31(5):394–6. Italian.

De Ritis F, Coltorti M, Giusti G. Serum and liver transaminase activities in experimental virus hepatitis in mice. Science. 1956;124(3210):32.

De Ritis F, Coltorti M, Giusti G. An enzymic test for the diagnosis of viral hepatitis; the transaminase serum activities. Clin Chim Acta. 1957;2(1):70–4.

Deguchi H, Banerjee Y, Trauger S, Siuzdak G, Kalisiak E, Fernández JA, et al. Acylcarnitines are anticoagulants that inhibit factor Xa and are reduced in venous thrombosis, based on metabolomics data. Blood. 2015;126(13):1595–600. doi:10.1182/blood-2015-03-636761. Epub 2015 Jul 14.

Domej W, Oettl K, Renner W. Oxidative stress and free radicals in COPD–implications and relevance for treatment. Int J Chron Obstruct Pulmon Dis. 2014;9:1207–24. doi:10.2147/COPD.S51226.eCollection2014.

Draisma HH, Pool R, Kobl M, Jansen R, Petersen AK, Vaarhorst AA, et al. Genome-wide association study identifies novel genetic variants contributing to variation in blood metabolite levels. Nat Commun. 2015;6:7208. doi:10.1038/ncomms8208.

Duranton F, Lundin U, Gayrard N, Mischak H, Aparicio M, Mourad G, et al. Plasma and urinary amino acid metabolomic profiling in patients with different levels of kidney function. Clin J Am Soc Nephrol. 2014;9(1):37–45.

Edman P. Method for determination of the amino acid sequence in peptides. Acta Chem Scand. 1950;4(7):283–93.

Edman P, Begg G. A protein sequenator. Eur J Biochem. 1967;1(1):80–91.

Eid J, Fehr A, Gray J, Luong K, Lyle J, Otto G, et al. Real-time DNA sequencing from single polymerase molecules. Science. 2009;323(5910):133–8.

Eigenbrodt E, Mostafa MA, Schoner W. Inactivation of pyruvate kinase type M2 from chicken liver by phosphorylation, catalyzed by a cAMP-independent protein kinase. Hoppe Seylers Z Physiol Chem. 1977;358(8):1047–55.

Eigenbrodt E, Leib S, Kramer W, Friis RR, Schoner W. Structural and kinetic differences between the M2 type pyruvate kinases from lung and various tumors. Biomed Biochim Acta. 1983;42 (11–12):S278–82.

Eigenbrodt E, Basenau D, Holthusen S, Mazurek S, Fischer G. Quantification of tumor type M2 pyruvate kinase (Tu M2-PK) in human carcinomas. Anticancer Res. 1997;17(4B):3153–6.

Eigenbrodt E, Kallinowski F, Ott M, Mazurek S, Vaupel P. Pyruvate kinase and the interaction of amino acid and carbohydrate metabolism in solid tumors. Anticancer Res. 1998;18 (5A):3267–74.

Engvall E, Perlmann P. Enzyme-linked immunosorbent assay (ELISA) quantitative assay of immunoglobulin G. Immunochemistry. 1971;8(9):871–4.

Enot DP, Haas B, Weinberger KM. Bioinformatics for mass spectrometry-based metabolomics. Methods Mol Biol. 2011;719:351–75.

Fahrmann JF, Kim K, DeFelice BC, Taylor SL, Gandara DR, Yoneda KY, et al. Investigation of metabolomic blood biomarkers for detection of adenocarcinoma lung cancer. Cancer Epidemiol Biomarkers Prev. 2015;24(11):1716–23. doi:10.1158/1055-9965.EPI-15-0427. Epub 2015 Aug 17.

Fales FM. Chapter 2: mesopotamia. Handb Clin Neurol. 2010;95:15–27. doi:10.1016/S0072-9752 (08)02102-7.

Falick AM, Hines WM, Medzihradszky KF, Baldwin MA, Gibson BW. Low-mass ions produced from peptides by high-energy collision-induced dissociation in tandem mass spectrometry. J Am Soc Mass Spectrom. 1993;4(11):882–93. doi:10.1016/1044-0305(93)87006-X.

Fingerhut R, Röschinger W, Muntau AC, Dame T, Kreischer J, Arnecke R, Superti-Furga A, Troxler H, Liebl B, Olgemöller B, Roscher AA. Hepatic carnitine palmitoyltransferase I deficiency: acylcarnitine profiles in blood spots are highly specific. Clin Chem. 2001;47 (10):1763–8.

Fishman JA, Rubin RH. Infection in organ-transplant recipients. N Engl J Med. 1998;338 (24):1741–51. Review.

Flaveny CA, Griffett K, El-Gendy Bel D, Kazantzis M, Sengupta M, Amelio AL. Broad anti-tumor activity of a small molecule that selectively targets the Warburg effect and lipogenesis. Cancer Cell. 2015;28(1):42–56. doi:10.1016/j.ccell.2015.05.007. Epub 2015 June 25.

Floegel A, Stefan N, Yu Z, Mühlenbruch K, Drogan D, Joost HG, et al. Identification of serum metabolites associated with risk of type 2 diabetes using a targeted metabolomic approach. Diabetes. 2013;62(2):639–48.

Floegel A, Wientzek A, Bachlechner U, Jacobs S, Drogan D, Prehn C, et al. Linking diet, physical activity, cardiorespiratory fitness and obesity to serum metabolite networks: findings from a population-based study. Int J Obes (Lond). 2014;38(11):1388–96. doi:10.1038/ijo.2014.39. Epub 2014 Mar 10.

Franklin RE, Gosling RG. Evidence for 2-chain helix in crystalline structure of sodium deoxyribonucleate. Nature. 1953;172(4369):156–7.

Furey WS, Joyce CM, Osborne MA, Klenerman D, Peliska JA, Balasubramanian S. Use of fluorescence resonance energy transfer to investigate the conformation of DNA substrates bound to the Klenow fragment. Biochemistry. 1998;37(9):2979–90.

Gamow G, Ycas M. Statistical correlation of protein and ribonucleic acid composition. Proc Natl Acad Sci U S A. 1955;41(12):1011–19.

Gamow G, Rich A, Ycas M. The problem of information transfer from the nucleic acids to proteins. Adv Biol Med Phys. 1956;4:23–68.

Garrod AE. Alkaptonuria: a simple method for the extraction of homogentisinic acid from the urine. J Physiol. 1899;23(6):512–14.

Garrod A. The incidence of alkaptonuria: a study in chemical individuality. Lancet. 1902;160 (4137):1616–20.

Garrod A. The croonian lectures on inborn errors of metabolism. Lancet. 1908;172(4427):1–7.

Garrod AE. Where chemistry and medicine meet. Br Med J. 1911;1(2633):1413.

Garrod AE, Hurtley WH. On the estimation of homogentisic acid in urine by the method of wolkow and Baumann. J Physiol. 1905;33(3):206–10.

Garrod AE, Hurtley WH. Concerning cystinuria. J Physiol. 1906;34(3):217–23.

Gieger C, Geistlinger L, Altmaier E, Hrabé de Angelis M, Kronenberg F, Meitinger T, et al. Genetics meets metabolomics: a genome-wide association study of metabolite profiles in human serum. PLoS Genet. 2008;4(11):e1000282.

Giesbertz P, Padberg I, Rein D, Ecker J, Höfle AS, Spanier B, Daniel H. Metabolite profiling in plasma and tissues of ob/ob and db/db mice identifies novel markers of obesity and type 2 diabetes. Diabetologia. 2015;58(9):2133–43. doi:10.1007/s00125-015-3656-y. Epub 2015 June 10.

Goek ON, Döring A, Gieger C, Heier M, Koenig W, Prehn C, et al. Serum metabolite concentrations and decreased GFR in the general population. Am J Kidney Dis. 2012;60(2):197–206.

Goek ON, Prehn C, Sekula P, Römisch-Margl W, Döring A, Gieger C, et al. Metabolites associate with kidney function decline and incident chronic kidney disease in the general population. Nephrol Dial Transplant. 2013;28(8):2131–8.

Gotto Jr AM. Cholesterol management in theory and practice. Circulation. 1997;96(12):4424–30. Review.

Green RE, Krause J, Briggs AW, Maricic T, Stenzel U, Kircher M, et al. A draft sequence of the Neandertal genome. Science. 2010;328(5979):710–22.

Griffith F. The significance of pneumococcal types. J Hyg (Lond). 1928;27(2):113–59.

Gross S, Cairns RA, Minden MD, Driggers EM, Bittinger MA, Jang HG, et al. Cancer-associated metabolite 2-hydroxyglutarate accumulates in acute myelogenous leukemia with isocitrate dehydrogenase 1 and 2 mutations. J Exp Med. 2010;207(2):339–44. doi:10.1084/jem.20092506. Epub 2010 Feb 8.

Gruber AC, Lundin U, Dzien A, Weinberger KM. From hyperphagic rodents to diabetic complications: targeted metabolomics in preclinical and clinical diabetology. J Comput Sci Syst Biol. 2012;5:1. http://dx.doi.org/10.4172/0974-7230.S1.04.

Guleria A, Misra DP, Rawat A, Dubey D, Khetrapal CL, Bacon P, et al. NMR-based serum metabolomics discriminates Takayasu Arteritis from healthy individuals: a proof-of-principle study. J Proteome Res. 2015;14(8):3372–81. doi:10.1021/acs.jproteome.5b00422. Epub 2015 June 29.

Guthrie R. Screening for phenylketonuria. Triangle. 1969;9(3):104–9.

Guthrie R, Susi A. A simple phenylalanine method for detecting phenylketonuria in large populations of newborn infants. Pediatrics. 1963;32:338–43.

Haider L. Inflammation, iron, energy failure, and oxidative stress in the pathogenesis of multiple sclerosis. Oxid Med Cell Longev. 2015;2015:725370. doi:10.1155/2015/725370. Epub 2015 May 27.

Halama A. Metabolomics in cell culture–a strategy to study crucial metabolic pathways in cancer development and the response to treatment. Arch Biochem Biophys. 2014;564:100–9. doi:10.1016/j.abb.2014.09.002. Epub 2014 Sep 10. Review.

Hall PL, Marquardt G, McHugh DM, Currier RJ, Tang H, Stoway SD, Rinaldo P. Postanalytical tools improve performance of newborn screening by tandem mass spectrometry. Genet Med. 2014;16(12):889–95. doi:10.1038/gim.2014.62. Epub 2014 May 29.

Han X, Rozen S, Boyle SH, Hellegers C, Cheng H, Burke JR, et al. Metabolomics in early Alzheimer's disease: identification of altered plasma sphingolipidome using shotgun lipidomics. PLoS One. 2011;6(7):e21643. doi:10.1371/journal.pone.0021643. Epub 2011 July 11.

Hardt PD, Mazurek S, Toepler M, Schlierbach P, Bretzel RG, Eigenbrodt E, Kloer HU. Faecal tumour M2 pyruvate kinase: a new, sensitive screening tool for colorectal cancer. Br J Cancer. 2004;91(5):980–4.

Heid CA, Stevens J, Livak KJ, Williams PM. Real time quantitative PCR. Genome Res. 1996;6(10):986–94.

Hershey AD, Chase M. Independent functions of viral protein and nucleic acid in growth of bacteriophage. J Gen Physiol. 1952;36(1):39–56.

Higuchi R, Fockler C, Dollinger G, Watson R. Kinetic PCR analysis: real-time monitoring of DNA amplification reactions. Biotechnology (N Y). 1993;11(9):1026–30.

Hood LE, Hunkapiller MW, Smith LM. Automated DNA sequencing and analysis of the human genome. Genomics. 1987;1(3):201–12. Review.

Hummel KP, Dickie MM, Coleman DL. Diabetes, a new mutation in the mouse. Science. 1966;153(3740):1127–8.

Husain K, Hernandez W, Ansari RA, Ferder L. Inflammation, oxidative stress and renin angiotensin system in atherosclerosis. World J Biol Chem. 2015;6(3):209–17. doi:10.4331/wjbc.v6.i3.209. Review.

Illig T, Gieger C, Zhai G, Römisch-Margl W, Wang-Sattler R, Prehn C, et al. A genome-wide perspective of genetic variation in human metabolism. Nat Genet. 2010;42(2):137–41.

Jaremek M, Yu Z, Mangino M, Mittelstrass K, Prehn C, Singmann P, et al. Alcohol-induced metabolomic differences in humans. Transl Psychiatry. 2013;3:e276.

Jentzmik F, Stephan C, Miller K, Schrader M, Erbersdobler A, Kristiansen G, et al. Sarcosine in urine after digital rectal examination fails as a marker in prostate cancer detection and identification of aggressive tumours. Eur Urol. 2010;58(1):12–8; discussion 20–1. doi:10.1016/j.eururo.2010.01.035. Epub 2010 Jan 26.

Jex HS. The Edwin Smith Surgical Papyrus: first milestone in the march of medicine. Merck Rep. 1951;60(2):20–2.

Jiang L, Deberardinis RJ. Cancer metabolism: when more is less. Nature. 2012;489(7417):511–12. doi:10.1038/489511a.

Kaddurah-Daouk R, Rozen S, Matson W, Han X, Hulette CM, Burke JR, et al. Metabolomic changes in autopsy-confirmed Alzheimer's disease. Alzheimers Dement. 2011;7(3):309–17. doi:10.1016/j.jalz.2010.06.001. Epub 2010 Nov 13.

Kaddurah-Daouk R, Zhu H, Sharma S, Bogdanov M, Rozen SG, Matson W, et al. Alterations in metabolic pathways and networks in Alzheimer's disease. Transl Psychiatry. 2013;3:e244. doi:10.1038/tp.2013.18.

Kageyama G, Saegusa J, Irino Y, Tanaka S, Tsuda K, Takahashi S, et al. Metabolomics analysis of saliva from patients with primary Sjögren's syndrome. Clin Exp Immunol. 2015;182(2):149–53. doi:10.1111/cei.12683. Epub 2015 Sep 15.

Kaiser RJ, MacKellar SL, Vinayak RS, Sanders JZ, Saavedra RA, Hood LE. Specific-primer-directed DNA sequencing using automated fluorescence detection. Nucleic Acids Res. 1989;17(15):6087–102.

Keane KN, Cruzat VF, Carlessi R, de Bittencourt Jr PI, Newsholme P. Molecular events linking oxidative stress and inflammation to insulin resistance and ß-cell dysfunction. Oxid Med Cell Longev. 2015;2015:181643. doi:10.1155/2015/181643. Epub 2015 July 14. Review.

Kempf EJ. From Hippocrates to Galen. Med Library Hist J. 1904;2(4):282–307.

Kendrew JC, Perutz MF. X-ray studies of compounds of biological interest. Annu Rev Biochem. 1957;26:327–72.

Kendrew JC, Dickerson RE, Strandberg BE, Hart RG, Davies DR, Phillips DC, Shore VC. Structure of myoglobin: a three-dimensional Fourier synthesis at 2 A. resolution. Nature. 1960;185(4711):422–7.

Khorana HG. Polynucleotide synthesis and the genetic code. Fed Proc. 1965;24(6):1473–87. Review.

Kimura H, Morita M, Yabuta Y, Kuzushima K, Kato K, Kojima S, Matsuyama T, Morishima T. Quantitative analysis of Epstein-Barr virus load by using a real-time PCR assay. J Clin Microbiol. 1999;37(1):132–6.

Koal T, Deters M, Casetta B, Kaever V. Simultaneous determination of four immunosuppressants by means of high speed and robust on-line solid phase extraction-high performance liquid chromatography-tandem mass spectrometry. J Chromatogr B Analyt Technol Biomed Life Sci. 2004;805(2):215–22.

Koal T, Burhenne H, Römling R, Svoboda M, Resch K, Kaever V. Quantification of antiretroviral drugs in dried blood spot samples by means of liquid chromatography/tandem mass spectrometry. Rapid Commun Mass Spectrom. 2005;19(21):2995–3001.

Koal T, Deters M, Resch K, Kaever V. Quantification of the carbapenem antibiotic ertapenem in human plasma by a validated liquid chromatography-mass spectrometry method. Clin Chim Acta. 2006;364(1–2):239–45. Epub 2005 Aug 10.

Köhler G, Milstein C. Continuous cultures of fused cells secreting antibody of predefined specificity. Nature. 1975;256(5517):495–7.

König K, Kobold U, Fink G, Leinenbach A, Dülffer T, Thiele R, et al. Quantification of vancomycin in human serum by LC-MS/MS. Clin Chem Lab Med. 2013;51(9):1761–9. doi:10.1515/cclm-2013-0142.

Koster MP, Vreeken RJ, Harms AC, Dane AD, Kuc S, Schielen PC, et al. First-trimester serum acylcarnitine levels to predict preeclampsia: a metabolomics approach. Dis Markers. 2015;2015:857108. doi:10.1155/2015/857108. Epub 2015 June 4.

Kühne W. [Über das Sekret des Pankreas (1876), Heidelberg Nat]. Med. Verhandl. 1877;1:233–35.

Kulkarni RN. Identifying biomarkers of subclinical diabetes. Diabetes. 2012;61(8):1925–6. doi:10.2337/db12-0599.

Laemmli UK. Cleavage of structural proteins during the assembly of the head of bacteriophage T4. Nature. 1970;227(5259):680–5.

Laske C, Sohrabi HR, Frost SM, López-de-Ipiña K, Garrard P, Buscema M, et al. Innovative diagnostic tools for early detection of Alzheimer's disease. Alzheimers Dement. 2015;11 (5):561–78. doi:10.1016/j.jalz.2014.06.004. Epub 2014 Nov 15. Review.

Leder P, Nirenberg MW. RNA codewords and protein synthesis, 3. On the nucleotide sequence of a cysteine and a leucine codeword. Proc Natl Acad Sci U S A. 1964;52:1521–9.

Lengyel P, Speyer JF, Ochoa S. Synthetic polynucleotides and the amino acid code. Proc Natl Acad Sci U S A. 1961;47:1936–42.

Levene PA. The structure of yeast nucleic acid. Stud Rockefeller Inst Med Res. 1919;30:221.

Levene PA, Jacobs WA. On the structure of thymus nucleic acid. J Biol Chem. 1912;12 (3):411–20.

Lewczuk P, Mroczko B, Fagan A, Kornhuber J. Biomarkers of Alzheimer's disease and mild cognitive impairment: a current perspective. Adv Med Sci. 2015;60(1):76–82. doi:10.1016/j.advms.2014.11.002. Epub 2014 Dec 9. Review.

Lian JS, Liu W, Hao SR, Chen DY, Wang YY, Yang JL, Jia HY, Huang JR. A serum metabolomic analysis for diagnosis and biomarker discovery of primary biliary cirrhosis and autoimmune hepatitis. Hepatobiliary Pancreat Dis Int. 2015;14(4):413–21.

Liebl DJ, Morris CJ, Henkemeyer M, Parada LF. mRNA expression of ephrins and Eph receptor tyrosine kinases in the neonatal and adult mouse central nervous system. J Neurosci Res. 2003;71(1):7–22.

Lim S, Oh TJ, Koh KK. Mechanistic link between nonalcoholic fatty liver disease and cardiometabolic disorders. Int J Cardiol. 2015;201:408–14. doi:10.1016/j.ijcard.2015.08.107. [Epub ahead of print] Review.

Liotta L, Petricoin E. Molecular profiling of human cancer. Nat Rev Genet. 2000;1(1):48–56. Review.

Littman RJ, Littman ML. Galen and the Antonine plague. Am J Philol. 1973;94:243–55.

Lloyd SM, Arnold J, Sreekumar A. Metabolomic profiling of hormone-dependent cancers: a bird's eye view. Trends Endocrinol Metab. 2015;26(9):477–85. doi:10.1016/j.tem.2015.07.001. Epub 2015 Aug 1. Review.

Lu J, Xie G, Jia W, Jia W. Metabolomics in human type 2 diabetes research. Front Med. 2013a;7 (1):4–13. doi:10.1007/s11684-013-0248-4. Epub 2013 Feb 2. Review.

Lu J, Xie G, Jia W, Jia W. Insulin resistance and the metabolism of branched-chain amino acids. Front Med. 2013b;7(1):53–9. doi:10.1007/s11684-013-0255-5. Epub 2013 Feb 6. Review.

Lundin U, Weinberger K (Inventors). Biocrates life sciences AG (Assignee). New biomarkers for assessing kidney diseases. International patent WO/2010/139341. Published 2010 Dec 09.

Lundin U, Modre-Osprian R, Weinberger KM. Targeted metabolomics for clinical biomarker discovery in multifactorial diseases. In: Ikehara K, editor. Advances in the study of genetic disorders. Croatia: InTech; 2011. p. 81–98.

Maciocia G. The foundations of Chinese medicine. London: Churchill Livingstone; 1989. p. 221.

Manda G, Isvoranu G, Comanescu MV, Manea A, Debelec Butuner B, Korkmaz KS. The redox biology network in cancer pathophysiology and therapeutics. Redox Biol. 2015;5:347–57. doi:10.1016/j.redox.2015.06.014 [Epub ahead of print].

Margulies M, Egholm M, Altman WE, Attiya S, Bader JS, Bemben LA, et al. Genome sequencing in microfabricated high-density picolitre reactors. Nature. 2005;437(7057):376–80.

Marquardt G, Currier R, McHugh DM, Gavrilov D, Magera MJ, Matern D, et al. Enhanced interpretation of newborn screening results without analyte cutoff values. Genet Med. 2012;14(7):648–55. doi:10.1038/gim.2012.2. Epub 2012 Feb 16.

Martell M, Gómez J, Esteban JI, Sauleda S, Quer J, Cabot B, Esteban R, Guardia J. High-throughput real-time reverse transcription-PCR quantitation of hepatitis C virus RNA. J Clin Microbiol. 1999;37(2):327–32.

Mathew S, Krug S, Skurk T, Halama A, Stank A, Artati A, et al. Metabolomics of Ramadan fasting: an opportunity for the controlled study of physiological responses to food intake. J Transl Med. 2014;12:161. doi:10.1186/1479-5876-12-161.

Mattern S. Galen and his patients. Lancet. 2011;378(9790):478–9.

Matthaei JH, Jones OW, Martin RG, Nirenberg MW. Characteristics and composition of RNA coding units. Proc Natl Acad Sci U S A. 1962;48:666–77.

Maxam AM, Gilbert W. A new method for sequencing DNA. Proc Natl Acad Sci. 1977;74 (2):560–4.

Mazurek S, Eigenbrodt E. The tumor metabolome. Anticancer Res. 2003;23(2A):1149–54.

Mazurek S, Grimm H, Wilker S, Leib S, Eigenbrodt E. Metabolic characteristics of different malignant cancer cell lines. Anticancer Res. 1998;18(5A):3275–82.

Mazurek S, Grimm H, Oehmke M, Weisse G, Teigelkamp S, Eigenbrodt E. Tumor M2-PK and glutaminolytic enzymes in the metabolic shift of tumor cells. Anticancer Res. 2000;20 (6D):5151–4.

McHugh D, Cameron CA, Abdenur JE, Abdulrahman M, Adair O, Al Nuaimi SA, et al. Clinical validation of cutoff target ranges in newborn screening of metabolic disorders by tandem mass spectrometry: a worldwide collaborative project. Genet Med. 2011;13(3):230–54. doi:10.1097/GIM.0b013e31820d5e67.

McKernan KJ, Peckham HE, Costa GL, McLaughlin SF, Fu Y, Tsung EF, et al. Sequence and structural variation in a human genome uncovered by short-read, massively parallel ligation sequencing using two-base encoding. Genome Res. 2009;19(9):1527–41. doi:10.1101/gr.091868.109. Epub 2009 June 22.

Mellmann A, Harmsen D, Cummings CA, Zentz EB, Leopold SR, Rico A, et al. Prospective genomic characterization of the German enterohemorrhagic Escherichia coli O104:H4 outbreak by rapid next generation sequencing technology. PLoS One. 2011;6(7):e22751. doi:10.1371/journal.pone.0022751. Epub 2011 July 20.

Michels J, Obrist F, Castedo M, Vitale I, Kroemer G. PARP and other prospective targets for poisoning cancer cell metabolism. Biochem Pharmacol. 2014;92(1):164–71. doi:10.1016/j.bcp.2014.08.026. Epub 2014 Sep 6. Review.

Miescher F. [Aus dem wissenschaftlichen Briefwechsel von F. Miescher]. Letter I. To Wilhelm His. Tübingen. 1869 Feb 26. In: His W, et al., editors. [Die Histochemischen und Physiologischen Arbeiten von Friedrich Miescher]. Leipzig: F. C. W. Vogel; 1897;1. p. 33–8.

Miescher-Rüsch F. Ueber die chemische Zusammensetzung der Eiterzellen. Med Chem Unters. 1871;4:441–60.

Millington DS, Terada N, Chace DH, Chen YT, Ding JH, Kodo N, Roe CR. The role of tandem mass spectrometry in the diagnosis of fatty acid oxidation disorders. Prog Clin Biol Res. 1992;375:339–54.

Mirsaeidi M, Banoei MM, Winston BW, Schraufnagel DE. Metabolomics: applications and promise in Mycobacterial disease. Ann Am Thorac Soc. 2015;12(9):1278–87. doi:10.1513/AnnalsATS.201505-279PS.

Mishra P, Ambs S. Metabolic signatures of human breast cancer. Mol Cell Oncol. 2015;2(3). pii: e992217.

Mullen AR, Wheaton WW, Jin ES, Chen PH, Sullivan LB, Cheng T, et al. Reductive carboxylation supports growth in tumour cells with defective mitochondria. Nature. 2011;481(7381):385–8. doi:10.1038/nature10642.

Mullis KB, Faloona FA. Specific synthesis of DNA in vitro via a polymerase-catalyzed chain reaction. Methods Enzymol. 1987;155:335–50.

Mullis K, Faloona F, Scharf S, Saiki R, Horn G, Erlich H. Specific enzymatic amplification of DNA in vitro: the polymerase chain reaction. Cold Spring Harb Symp Quant Biol. 1986;51(Pt 1):263–73.

Newman JD, Turner AP. Home blood glucose biosensors: a commercial perspective. Biosens Bioelectron. 2005;20(12):2435–53. Epub 2005 Jan 18. Review.

Nicholson JK, Lindon JC, Holmes E. 'Metabonomics': understanding the metabolic responses of living systems to pathophysiological stimuli via multivariate statistical analysis of biological NMR spectroscopic data. Xenobiotica. 1999;29(11):1181–9. Review.

Nicholson G, Rantalainen M, Li JV, Maher AD, Malmodin D, Ahmadi KR, et al. A genome-wide metabolic QTL analysis in Europeans implicates two loci shaped by recent positive selection. PLoS Genet. 2011;7(9):e1002270. doi:10.1371/journal.pgen.1002270. Epub 2011 Sep 8.

Nirenberg MW, Matthaei JH. The dependence of cell-free protein synthesis in E. coli upon naturally occurring or synthetic polyribonucleotides. Proc Natl Acad Sci U S A. 1961;47:1588–602.

Nkuipou-Kenfack E, Duranton F, Gayrard N, Argilés À, Lundin U, Weinberger KM, et al. Assessment of metabolomic and proteomic biomarkers in detection and prognosis of progression of renal function in chronic kidney disease. PLoS One. 2014;9(5):e96955.

Orešic M, Hyötyläinen T, Herukka SK, Sysi-Aho M, Mattila I, Seppänan-Laakso T, et al. Metabolome in progression to Alzheimer's disease. Transl Psychiatry. 2011;1:e57. doi:10.1038/tp.2011.55.

Orfanos AP, Murphey WH, Guthrie R. A simple fluorometric assay of protoporphyrin in erythrocytes (EPP) as a screening test for lead poisoning. J Lab Clin Med. 1977;89(3):659–65.

Orfanos AP, Naylor EW, Guthrie R. Micromethod for estimating adenosine deaminase activity in dried blood spots on filter paper. Clin Chem. 1978;24(4):591–4.

Orfanos AP, Naylor EW, Guthrie R. Fluorometric micromethod for determination of arginase activity in dried blood spots on filter paper. Clin Chem. 1980a;26(8):1198–200.

Orfanos AP, Naylor EW, Guthrie R. Ultramicromethod for estimation of total glutathione in dried blood spots on filter paper. Anal Biochem. 1980b;104(1):70–4.

Osborne MA, Furey WS, Klenerman D, Balasubramanian S. Single-molecule analysis of DNA immobilized on microspheres. Anal Chem. 2000;72(15):3678–81.

Parker SP, Cubitt WD. The use of the dried blood spot sample in epidemiological studies. J Clin Pathol. 1999;52(9):633–9. Review.

Patel S, Ahmed S. Emerging field of metabolomics: big promise for cancer biomarker identification and drug discovery. J Pharm Biomed Anal. 2015;107:63–74. doi:10.1016/j.jpba.2014.12.020. Epub 2014 Dec 22. Review.

Payen A, Persoz JF. Memoir on diastase, the principal products of its reactions, and their applications to the industrial arts. In Annales de Chimie et de Physique. 1833;53:73–92.

Pena MJ, de Zeeuw D, Mischak H, Jankowski J, Oberbauer R, Woloszczuk W, et al. Prognostic clinical and molecular biomarkers of renal disease in type 2 diabetes. Nephrol Dial Transplant. 2015;30 Suppl 4:iv86–95. doi:10.1093/ndt/gfv252. Review.

Petersen AK, Zeilinger S, Kastenmüller G, Römisch-Margl W, Brugger M, Peters A, et al. Epigenetics meets metabolomics: an epigenome-wide association study with blood serum metabolic traits. Hum Mol Genet. 2014;23(2):534–45. doi:10.1093/hmg/ddt430. Epub 2013 Sep 6.

Petricoin EF, Ardekani AM, Hitt BA, Levine PJ, Fusaro VA, Steinberg SM, et al. Use of proteomic patterns in serum to identify ovarian cancer. Lancet. 2002a;359(9306):572–7.

Petricoin EF, Zoon KC, Kohn EC, Barrett JC, Liotta LA. Clinical proteomics: translating benchside promise into bedside reality. Nat Rev Drug Discov. 2002b;1(9):683–95. Review.

Petricoin 3rd EF, Ornstein DK, Paweletz CP, Ardekani A, Hackett PS, Hitt BA, et al. Serum proteomic patterns for detection of prostate cancer. J Natl Cancer Inst. 2002c;94(20):1576–8.

Pfeiffer L, Wahl S, Pilling LC, Reischl E, Sandling JK, Kunze S, et al. DNA methylation of lipid-related genes affects blood lipid levels. Circ Cardiovasc Genet. 2015;8(2):334–42. doi:10.1161/CIRCGENETICS.114.000804. Epub 2015 Jan 12.

Pinto J, Almeida LM, Martins AS, Duarte D, Domingues MR, Barros AS, et al. Impact of fetal chromosomal disorders on maternal blood metabolome: toward new biomarkers? Am J Obstet Gynecol. 2015;213(6):841.e1–841.e15. doi:10.1016/j.ajog.2015.07.032. Epub 2015 Jul 26.

Qayumi AK. Avicenna: a bright star from the east. J Invest Surg. 1998;11(4):243–4. Review.

Ramesh A, Varghese SS, Doraiswamy J, Malaiappan S. Role of sulfiredoxin in systemic diseases influenced by oxidative stress. Redox Biol. 2014;2C:1023–8. doi:10.1016/j.redox.2014.09.002. [Epub ahead of print] Review.

Raskin S, Phillips 3rd JA, Kaplan G, McClure M, Vnencak-Jones C. Cystic fibrosis genotyping by direct PCR analysis of Guthrie blood spots. PCR Methods Appl. 1992;2(2):154–6.

Rasko DA, Webster DR, Sahl JW, Bashir A, Boisen N, Scheutz N, et al. Origins of the E. coli strain causing an outbreak of hemolytic-uremic syndrome in Germany. N Engl J Med. 2011;365(8):709–17. doi:10.1056/NEJMoa1106920. Epub 2011 July 27.

Raymond S, Weintraub L. Acrylamide gel as a supporting medium for zone electrophoresis. Science. 1959;130(3377):711.

Ried JS, Baurecht H, Stückler F, Krumsiek J, Gieger C, Heinrich J, et al. Integrative genetic and metabolite profiling analysis suggests altered phosphatidylcholine metabolism in asthma. Allergy. 2013;68(5):629–36. doi:10.1111/all.12110. Epub 2013 Mar 1.

Ried JS, Shin SY, Krumsiek J, Illig T, Theis FJ, Spector TD, et al. Novel genetic associations with serum level metabolites identified by phenotype set enrichment analyses. Hum Mol Genet. 2014;23(21):5847–57. doi:10.1093/hmg/ddu301. Epub 2014 June 13.

Roberts CS. The case of Richard Cabot. In: Walker HK, Hall WD, Hurst JW, editors. Clinical methods: the history, physical, and laboratory examinations. 3rd ed. Boston: Butterworths; 1990.

Robinson S, Pool R, Giffin R. Forum on drug discovery, development, and translation. Emerging safety science: workshop summary. Washington, DC: National Academies Press (US); 2008.

Rolinski B, Arnecke R, Dame T, Kreischer J, Olgemöller B, Wolf E, et al. The biochemical metabolite screen in the Munich ENU Mouse Mutagenesis Project: determination of amino acids and acylcarnitines by tandem mass spectrometry. Mamm Genome. 2000;11(7):547–51.

Rosales-Corral S, Tan DX, Manchester L, Reiter RJ. Diabetes and Alzheimer disease, two overlapping pathologies with the same background: oxidative stress. Oxid Med Cell Longev. 2015;2015:985845. doi:10.1155/2015/985845. Epub 2015 Feb 26. Review.

Röschinger W, Muntau AC, Duran M, Dorland L, IJlst L, Wanders RJ, Roscher AA. Carnitine-acylcarnitine translocase deficiency: metabolic consequences of an impaired mitochondrial carnitine cycle. Clin Chim Acta. 2000;298(1–2):55–68.

Röschinger W, Olgemöller B, Fingerhut R, Liebl B, Roscher AA. Advances in analytical mass spectrometry to improve screening for inherited metabolic diseases. Eur J Pediatr. 2003;162 Suppl 1:S67–76. Epub 2003 Nov 14. Review.

Rothberg JM, Hinz W, Rearick TM, Schultz J, Mileski W, Davey M, et al. An integrated semiconductor device enabling non-optical genome sequencing. Nature. 2011;475 (7356):348–52. doi:10.1038/nature10242.

Ryle AP, Sanger F, Smith LF, Kitai R. The disulphide bonds of insulin. Biochem J. 1955;60 (4):541–56.

Saiki RK, Scharf S, Faloona F, Mullis KB, Horn GT, Erlich HA, Arnheim N. Enzymatic amplification of beta-globin genomic sequences and restriction site analysis for diagnosis of sickle cell anemia. Science. 1985;230(4732):1350–4.

Sander S, Janzen N, Janetzky B, Scholl S, Steuerwald U, Schäfer J, Sander J. Neonatal screening for medium chain acyl-CoA deficiency: high incidence in Lower Saxony (northern Germany). Eur J Pediatr. 2001;160(5):318–19.

Sands BE. Biomarkers of inflammation in inflammatory Bowel disease. Gastroenterology. 2015;149(5):1275–1285.e2. doi:10.1053/j.gastro.2015.07.003. Epub 2015 Jul 9.

Sanger F. The free amino groups of insulin. Biochem J. 1945;39(5):507–15.

Sanger F. The terminal peptides of insulin. Biochem J. 1949;45(5):563–74.

Sanger F, Coulson AR. A rapid method for determining sequences in DNA by primed synthesis with DNA polymerase. J Mol Biol. 1975;94(3):441–8.

Sborov DW, Haverkos BM, Harris PJ. Investigational cancer drugs targeting cell metabolism in clinical development. Expert Opin Investig Drugs. 2015;24(1):79–94. Epub 2014 Sep 16.

Schwartz EI, Khalchitsky SE, Eisensmith RC, Woo SL. Polymerase chain reaction amplification from dried blood spots on Guthrie cards. Lancet. 1990;336(8715):639–40.

Seger C, Tentschert K, Stöggl W, Griesmacher A, Ramsay SL. A rapid HPLC-MS/MS method for the simultaneous quantification of cyclosporine A, tacrolimus, sirolimus and everolimus in human blood samples. Nat Protoc. 2009;4(4):526–34. doi:10.1038/nprot.2009.25.

Son J, Lyssiotis CA, Ying H, Wang X, Hua S, Ligorio M, et al. Glutamine supports pancreatic cancer growth through a KRAS-regulated metabolic pathway. Nature. 2013;496(7443):101–5.

Sonntag D, Koal T, Ramsay SL, Dammeier S, Weinberger KM, Unterwurzacher I (Inventors). Biocrates life sciences AG (Assignee). Inflammation and oxidative stress level assay. International patent WO/2008/145384. Published 2008 Dec 4.

Sotgia F, Martinez-Outschoorn UE, Lisanti MP. Cancer metabolism: new validated targets for drug discovery. Oncotarget. 2013;4(8):1309–16.

Southern EM. Detection of specific sequences among DNA fragments separated by gel electrophoresis. J Mol Biol. 1975;98(3):503–17.

Sreekumar A, Poisson LM, Rajendiran TM, Khan AP, Cao Q, Yu J, et al. Metabolomic profiles delineate potential role for sarcosine in prostate cancer progression. Nature. 2009;457 (7231):910–14. doi:10.1038/nature07762.

Stiefel M, Shaner A, Schaefer SD. The Edwin Smith Papyrus: the birth of analytical thinking in medicine and otolaryngology. Laryngoscope. 2006;116(2):182–8.

Strauss EC, Kobori JA, Siu G, Hood LE. Specific-primer-directed DNA sequencing. Anal Biochem. 1986;154(1):353–60.

Struys EA, Heijboer AC, van Moorselaar J, Jakobs C, Blankenstein MA. Serum sarcosine is not a marker for prostate cancer. Ann Clin Biochem. 2010;47(Pt 3):282. doi:10.1258/acb.2010.009270. Epub 2010 Mar 16.

Suhre K, Meisinger C, Döring A, Altmaier E, Belcredi P, Gieger C, et al. Metabolic footprint of diabetes: a multiplatform metabolomics study in an epidemiological setting. PLoS One. 2010;5 (11):e13953.

Suhre K, Shin SY, Petersen AK, Mohney RP, Meredith D, Wägele B, et al. Human metabolic individuality in biomedical and pharmaceutical research. Nature. 2011;477(7362):54–60.

Then C, Wahl S, Kirchhofer A, Grallert H, Krug S, Kastenmüller G, et al. Plasma metabolomics reveal alterations of sphingo- and glycerophospholipid levels in non-diabetic carriers of the transcription factor 7-like 2 polymorphism rs7903146. PLoS One. 2013;8(10):e78430. doi:10. 1371/journal.pone.0078430.eCollection2013.

Towbin H, Staehelin T, Gordon J. Electrophoretic transfer of proteins from polyacrylamide gels to nitrocellulose sheets: procedure and some applications. Proc Natl Acad Sci U S A. 1979;76 (9):4350–4.

Valouev A, Ichikawa J, Tonthat T, Stuart J, Ranade S, Peckham H, et al. A high-resolution, nucleosome position map of C. elegans reveals a lack of universal sequence-dictated positioning. Genome Res. 2008;18(7):1051–63. doi:10.1101/gr.076463.108. Epub 2008 May 13.

van den Ouweland JM, Vogeser M, Bächer S. Vitamin D and metabolites measurement by tandem mass spectrometry. Rev Endocr Metab Disord. 2013;14(2):159–84. doi:10.1007/s11154-013-9241-0.

Van Hove JL, Zhang W, Kahler SG, Roe CR, Chen YT, Terada N, Chace DH, Iafolla AK, Ding JH, Millington DS. Medium-chain acyl-CoA dehydrogenase (MCAD) deficiency: diagnosis by acylcarnitine analysis in blood. Am J Hum Genet. 1993;52(5):958–66.

Van Weemen BK, Schuurs AH. Immunoassay using antigen-enzyme conjugates. FEBS Lett. 1971;15(3):232–6.

Vischer E, Chargaff E. The separation and characterization of purines in minute amounts of nucleic acid hydrolysates. J Biol Chem. 1947;168(2):781.

Vischer E, Zamenhof S, Chargaff E. Microbial nucleic acids; the desoxypentose nucleic acids of avian tubercle bacilli and yeast. J Biol Chem. 1949;177(1):429–38.

Vogel U, Szczepanowski R, Claus H, Jünemann S, Prior K, Harmsen D. Ion torrent personal genome machine sequencing for genomic typing of Neisseria meningitidis for rapid determination of multiple layers of typing information. J Clin Microbiol. 2012;50(6):1889–94. doi:10. 1128/JCM.00038-12. Epub 2012 Mar 29.

Vogeser M. Quantification of circulating 25-hydroxyvitamin D by liquid chromatography-tandem mass spectrometry. J Steroid Biochem Mol Biol. 2010;121(3–5):565–73. doi:10.1016/j.jsbmb. 2010.02.025. Epub 2010 Mar 4.

Vollert S, Kaessner N, Heuser A, Hanauer G, Dieckmann A, Knaack D, et al. The glucose-lowering effects of the PDE4 inhibitors roflumilast and roflumilast-N-oxide in db/db mice. Diabetologia. 2012;55(10):2779–88. doi:10.1007/s00125-012-2632-z. Epub 2012 July 13.

Vollmer DW, Jinks DC, Guthrie R. Isocratic reverse-phase liquid chromatography assay for amino acid metabolic disorders using eluates of dried blood spots. Anal Biochem. 1990;189 (1):115–21.

Wang TJ, Larson MG, Vasan RS, Cheng S, Rhee EP, McCabe E, et al. Metabolite profiles and the risk of developing diabetes. Nat Med. 2011;17(4):448–53. doi:10.1038/nm.2307. Epub 2011 Mar 20.

Wang-Sattler R, Yu Z, Herder C, Messias AC, Floegel A, He Y, et al. Novel biomarkers for pre-diabetes identified by metabolomics. Mol Syst Biol. 2012;8:615.

Warburg O. Iron, the oxygen-carrier of respiration-ferment. Science. 1925;61(1588):575–82.

Warburg O. The chemical constitution of respiration ferment. Science. 1928;68(1767):437–43.

Warburg O. On the origin of cancer cells. Science. 1956a;123(3191):309–14.

Warburg O. On respiratory impairment in cancer cells. Science. 1956b;124(3215):269–70.

Warburg O, Wind F, Negelein E. The metabolism of tumors in the body. J Gen Physiol. 1927;8 (6):519–30.

Ward PS, Patel J, Wise DR, Abdel-Wahab O, Bennett BD, Coller HA, et al. The common feature of leukemia-associated IDH1 and IDH2 mutations is a neomorphic enzyme activity converting alpha-ketoglutarate to 2-hydroxyglutarate. Cancer Cell. 2010;17(3):225–34. doi:10.1016/j.ccr.2010.01.020. Epub 2010 Feb 18.

Watson JD, Crick FH. The structure of DNA. Cold Spring Harb Symp Quant Biol. 1953a;18:123–31.

Watson JD, Crick FH. Molecular structure of nucleic acids. Nature. 1953b;171(4356):737–8.

Weinberger KM. Metabolomics in diagnosing metabolic diseases. Ther Umsch. 2008;65(9):487–91. doi:10.1024/0040-5930.65.9.487.Review.German.

Weinberger KM, Wiedenmann E, Böhm S, Jilg W. Sensitive and accurate quantitation of hepatitis B virus DNA using a kinetic fluorescence detection system (TaqMan PCR). J Virol Methods. 2000;85(1–2):75–82.

Weinberger B, Plentz A, Weinberger KM, Hahn J, Holler E, Jilg W. Quantitation of Epstein-Barr virus mRNA using reverse transcription and real-time PCR. J Med Virol. 2004;74(4):612–18.

Wheeler DA, Srinivasan M, Egholm M, Shen Y, Chen L, McGuire A, et al. The complete genome of an individual by massively parallel DNA sequencing. Nature. 2008;452(7189):872–6.

Wikoff WR, Hanash S, DeFelice B, Miyamoto S, Barnett M, Zhao Y, et al. Diacetylspermine is a novel prediagnostic serum biomarker for non-small-cell lung cancer and has additive performance with pro-surfactant protein B. J Clin Oncol. 2015;33(33):3880–6. doi:10.1200/JCO.2015.61.7779. Epub 2015 Aug 17.

Wilkins MHF, Stokes AR, Wilson HR. Molecular structure of nucleic acids: molecular structure of deoxypentose nucleic acids. Nature. 1953;171(4356):738–40.

Wouters EF, Bredenbröker D, Teichmann P, Brose M, Rabe KF, Fabbri LM, Göke B. Effect of the phosphodiesterase 4 inhibitor roflumilast on glucose metabolism in patients with treatment-naive, newly diagnosed type 2 diabetes mellitus. J Clin Endocrinol Metab. 2012;97(9):E1720–5. doi:10.1210/jc.2011-2886. Epub 2012 June 20.

Würtz P, Havulinna AS, Soininen P, Tynkkynen T, Prieto-Merino D, Tillin T, et al. Metabolite profiling and cardiovascular event risk: a prospective study of 3 population-based cohorts. Circulation. 2015;131(9):774–85. doi:10.1161/CIRCULATIONAHA.114.013116. Epub 2015 Jan 8.

Xia J, Psychogios N, Young N, Wishart DS. MetaboAnalyst: a web server for metabolomic data analysis and interpretation. Nucleic Acids Res. 2009;37(Web Server issue):W652–60. doi:10.1093/nar/gkp356. Epub 2009 May 8.

Xia J, Mandal R, Sinelnikov IV, Broadhurst D, Wishart DS. MetaboAnalyst 2.0–a comprehensive server for metabolomic data analysis. Nucleic Acids Res. 2012;40(Web Server issue):W127–33.

Xia J, Sinelnikov IV, Han B, Wishart DS. MetaboAnalyst 3.0-making metabolomics more meaningful. Nucleic Acids Res. 2015;43(W1):W251–7. doi:10.1093/nar/gkv380. Epub 2015 Apr 20.

Yalow RS, Berson SA. Immunoassay of endogenous plasma insulin in man. J Clin Invest. 1960;39:1157–75.

Yanofsky C. The favorable features of tryptophan synthase for proving Beadle and Tatum's one gene-one enzyme hypothesis. Genetics. 2005;169(2):511–16.

Yu Z, Zhai G, Singmann P, He Y, Xu T, Prehn C, et al. Human serum metabolic profiles are age dependent. Aging Cell. 2012;11(6):960–7. doi:10.1111/j.1474-9726.2012.00865.x. Epub 2012 Aug 27.

Zander J, Maier B, Suhr A, Zoller M, Frey L, Teupser D, Vogeser M. Quantification of piperacillin, tazobactam, cefepime, meropenem, ciprofloxacin and linezolid in serum using an isotope dilution UHPLC-MS/MS method with semi-automated sample preparation. Clin Chem Lab Med. 2015;53(5):781–91. doi:10.1515/cclm-2014-0746.

Zheng Z, Advani A, Melefors O, Glavas S, Nordström H, Ye W, Engstrand L, Andersson AF. Titration-free massively parallel pyrosequencing using trace amounts of starting material. Nucleic Acids Res. 2010;38(13):e137. doi:10.1093/nar/gkq332. Epub 2010 Apr 30.

Chapter 8
Clinical Bioinformatics for Biomarker Discovery in Targeted Metabolomics

Marc Breit, Christian Baumgartner, Michael Netzer, and Klaus M. Weinberger

Abstract In this chapter, methods of clinical bioinformatics in targeted metabolomics are discussed, with an emphasis on the discovery of metabolic biomarkers. The reader is introduced to general aspects such as initiatives in metabolomics standardization, regulatory guidelines and software validation, and is presented an overview of the bioinformatics workflow in metabolomics. Engineering-based concepts of clinical bioinformatics in supporting the storage and automated analysis of samples, the integration of data in public repositories, and in the management of data using metabolomics application software are discussed. Chemometrics algorithms for data processing are summarized, modalities of biostatistics and data analysis presented, as well as data mining and machine learning approaches, aiming at the discovery of biomarkers in targeted metabolomics. Methods of data interpretation in the context of annotated

M. Breit (✉)
Research Group for Clinical Bioinformatics, Institute of Electrical and Biomedical Engineering (IEBE), University for Health Sciences, Medical Informatics and Technology (UMIT), 6060 Hall in Tirol, Austria

Institute of Health Care Engineering with European Notified Body of Medical Devices, Graz University of Technology, Stremayrgasse 16, A-8010 Graz, Austria
e-mail: m.breit@umit.at

C. Baumgartner, Ph.D.
Institute of Health Care Engineering with European Notified Body of Medical Devices, Graz University of Technology, Stremayrgasse 16, A-8010 Graz, Austria
e-mail: Christian.Baumgartner@TUGraz.at

M. Netzer
Research Group for Clinical Bioinformatics, Institute of Electrical and Biomedical Engineering (IEBE), University for Health Sciences, Medical Informatics and Technology (UMIT), 6060 Hall in Tirol, Austria

K.M. Weinberger
Research Group for Clinical Bioinformatics, Institute of Electrical and Biomedical Engineering (IEBE), University for Health Sciences, Medical Informatics and Technology (UMIT), 6060 Hall in Tirol, Austria

sAnalytiCo Ltd., Forsyth House, Cromac Square, Belfast BT2 8LA, UK

Weinberger & Weinberger Life Sciences Consulting, Weidach 82, 6414 Mieming, Austria

X. Wang et al. (eds.), *Application of Clinical Bioinformatics*,
Translational Bioinformatics 11, DOI 10.1007/978-94-017-7543-4_8

biochemical pathways are suggested, theoretical concepts of metabolic modeling and engineering are introduced, and the *in-silico* modeling and simulation of molecular processes is briefly touched. Finally, a short outlook on future perspectives in the application of clinical bioinformatics in targeted metabolomics is given, e.g. on the development of integrated mass spectrometry solutions, ready for routine clinical usage in laboratory medicine, or on the application of concepts of artificial intelligence in laboratory automation – liquid handling robots, autonomously performing experiments and generating hypotheses.

Keywords Clinical bioinformatics • Metabolic biomarker discovery • Metabolic modeling • Metabolomics application software • Targeted metabolomics

Abbreviations

ACToR	Aggregated Computational Toxicology Resource
ANN	artificial neural network
ANOVA	analysis-of-variance
ArMet	Architecture for a Metabolomics Experiment
ATP	adenosine triphosphate
BBMRI	Biobanking and Biomolecular Resources Research Infrastructure
BGI	Beijing Genomics Institute
ChEBI	Chemical Entities of Biological Interest
COSMOS	COordination Of Standards In MetabOlomicS
CO_2	carbon dioxide
CV	coefficient of variation
C2	acetylcarnitine
C5	valerylcarnitine
DBSCAN	density-based spatial clustering of applications with noise
DRCC	Data Repository and Coordination Centre
ELIXIR	European life-sciences Infrastructure for biological Information
EMA	European Medicines Agency
EPA	Environmental Protection Agency
FDA	Food and Drug Administration
GAMP	Good Automated Manufacturing Practice
GWAS	genome-wide association studies
HMDB	Human Metabolome Database
H_2O	water
ICH	International Conference on Harmonization
ISA	Investigation Study Assay
ISO	International Organization of Standardization
ISPE	International Society for Pharmaceutical Engineering
JDAMP	Joint Committee on Atomic and Molecular Physical Data
KDD	knowledge discovery in databases
KEGG	Kyoto Encyclopedia of Genes and Genomes

LIMS	laboratory information management systems
LLOQ	lower limit of quantitation
MBRole	Metabolite Biological Role
MeMo	Metabolic Modeling
MFC	maximum fold change
MIBBI	Minimum Information for Biological and Biomedical investigations
MIAMET	Minimum Information About METabolomics experiments
MS	mass spectrometry
MSEA	Metabolite Set Enrichment Analysis
MSI	Metabolomics Standards Initiative
NADH	nicotinamide adenine dinucleotide
netCDF	Network Common Data Format
NH_3	ammonia
ODE	ordinary differential equation
OMIM	Online Mendelian Inheritance in Man
ORA	overrepresentation analysis
PCA	principal component analysis
PLS	partial least squares
PRIMe	Platform for RIKEN Metabolomics
QC	quality control
RCMRC	Regional Comprehensive Metabolomics Research Cores
RF	random forest
SFR	stacked feature ranking
SMPDB	Small Molecule Pathway Database
SOP	Standard Operating Procedure
SSP	single sample profiling
SVM	support vector machines
T3DB	Toxin Target/Target Database
XML	eXtensible Markup Language.

8.1 Introduction

Despite the fact that this book as a whole is dedicated to the topic of clinical bioinformatics, and the term is already suggested and widely discussed in the existing literature (Chang 2005; Trent 2008; Wang and Liotta 2011; Bellazzi et al. 2012), the authors of this chapter briefly deal with a definition and a delimitation of related fields, to provide the reader with a picture that is as complete as possible. According to the existing literature, clinical bioinformatics can be defined as 'clinical application of bioinformatics-associated sciences and technologies to understand molecular mechanisms and potential therapies for human diseases' (Wang and Liotta 2011; Bellazzi et al. 2012). Furthermore, clinical bioinformatics is closely interrelated with other fields of computational life sciences, such as bioinformatics (Hogeweg 2011), biomedical informatics (Bernstam et al. 2010), computational biology (Bourne et al. 2015; Fogg and Kovats 2015;

Nussinov 2015; Nussinov et al. 2015), health informatics (Mettler and Raptis 2012), systems biology (Kitano 2002), and translational bioinformatics (Butte 2008) – often leading to overlaps in definitions.

In this chapter, focus is put on the computational, algorithmic, and technological aspects of clinical bioinformatics in targeted metabolomics research as well as on commercial applications, with a slight emphasis on the search for novel biomarker candidates in metabolism. To obtain a holistic picture of current developments in the field of targeted metabolomics – together with a discussion on the history and perspectives of clinical chemistry, the motivation for clinical mass spectrometry (MS), and a summary on promising new indications (i.e., diabetes, chronic kidney disease, neurology, and oncology) – it should be read together with the previous chapter of this book, '*Targeted metabolomics: the next generation of clinical chemistry!?*' (see Chap. 7).

In the context of (targeted) metabolomics research, the application of clinical bioinformatics can be understood as the development and interplay of a multitude of software applications, computational methods, algorithms, or modalities, easing research and product development, such as bioinformatics software applications, chemometrics algorithms, biostatistical modalities, methods of artificial intelligence or mathematical modeling. Subsequently, the reader will be gradually introduced to (a) general aspects, as metabolomics standardization initiatives and regulatory requirements for software validation, (b) the handling and automated preparation of samples, metabolomics databases and software applications for data management, (c) algorithms and modalities for data processing, analysis and mining, and (d) computational methods for the interpretation, modeling and simulation of data.

8.2 Standardization, Guidelines, and Workflows

8.2.1 Metabolomics Standardization and Initiatives

Acknowledging the ever increasing amount of data and the multitude of research projects in metabolomics during the past decade, a variety of initiatives for the standardization in metabolomics has been suggested. These initiatives may be categorized into the three different levels of (a) data and file formats, (b) technically, bioinformatics-oriented projects and architectures, and (c) high-level, strategic and coordinative initiatives, yet, the borders are sometimes blurred, not always allowing for a unique assignment.

On a low, primarily data-oriented level, the usage of different data and file formats in metabolomics is suggested, partly already being suggested and established in other fields of research. Those data formats are for example the Joint Committee on Atomic and Molecular Physical Data (JCAMP)-DX (http://www.jcamp-dx.org/) and the Network Common Data Format (netCDF) (http://

www.unidata.ucar.edu/software/netcdf/), both historically used in MS data storage, or the mzXML (Pedrioli et al. 2004; Lin et al. 2005) and mzData formats (http://www.proteomecommons.org), taken over from proteomics research.

Primarily technically and bioinformatics-oriented architectures and projects are e.g. the Architecture for a Metabolomics Experiment (ArMet) (Jenkins et al. 2004), Metabolic Modeling (MeMo) (Spasić et al. 2006, the Investigation, Study, Assay (ISA)-TAB tab delimited format (Sansone et al. 2008, 2012), Galaxy (Goecks et al. 2010), MetaboLights (Steinbeck et al. 2012; Haug et al. 2014) or the METLIN metabolite database (Tautenhahn et al. 2012).

Suggestions for higher level, strategic initiatives, are e.g. the Minimum Information About METabolomics experiments (MIAMET) approach (Bino et al. 2004), the Metabolomics Standards Initiative (MSI) (MSI Board Members 2007), the Minimum Information for Biological and Biomedical investigations (MIBBI) consortium (Taylor et al. 2008), or the COordination Of Standards In MetabOlomicS (COSMOS) initiative (Salek et al. 2013, 2015).

In addition to the consideration of independent proposals, the currently evolving collaboration between international initiatives (also of related fields) is an observation worth mentioning (Salek et al. 2015). Metabolomics standardization initiatives are developing around the globe and, at least to date, have substantially arrived on a political and governmental scale. In Europe, collaboration is impelled, e.g. between the European life-sciences Infrastructure for biological Information (ELIXIR) initiative (http://www.elixir-europe.org/), the Biobanking and Biomolecular Resources Research Infrastructure (BBMRI) (http://www.bbmri.eu/), and the BioMedbridges consortium (http://www.biomedbridges.eu/). Collaboration is further intensified with North American initiatives, e.g. the North American hub for metabolomics related research, including six Regional Comprehensive Metabolomics Research Cores (RCMRC), and the Data Repository and Coordination Centre (DRCC) (http://www.nih.gov/news/health/sep2012/od-19.htm), or the Canadian Human Metabolome Database (HMDB) (Wishart et al. 2013). Important examples in the Asian region include the Platform for RIKEN Metabolomics (PRIMe) in Japan (Sakurai et al. 2013) and the Chinese Beijing Genomics Institute (BGI) (http://www.genomics.cn/).

The list of projects and initiatives presented here certainly mirrors only a part of the complete range, but the reader will hopefully obtain a feeling for the dynamics in this field, regionally as well as on a global, interlinking scale. When trying to abstract the common goal behind the different initiatives, this leads to the objective of establishing a gold standard, which finds broad acceptance and even more importantly – usage – in the community.

8.2.2 Regulatory Guidelines and Software Validation

When developing new methods or products in metabolomics research, different regulatory guidances need to be taken into account, usually issued either by standardization institutions, e.g. the International Organization of Standardization (ISO), or regulatory authorities such as the Food and Drug Administration (FDA) or the European Medicines Agency (EMA).

As a groundwork for establishing a company quality management system, the guidance of the International Organization of Standardization (ISO), the 'ISO 9001:2008 Quality management systems – Requirements' (ISO, 2008) should be considered (the new version, ISO 9001:2015 is only available as a final draft so far). For the definition of standard operating procedures (SOPs) – which are an essential tool for quality assurance (QA) in chemical analytics – the 'Guidance for Preparing Standard Operating Procedures (SOPs)' (Environmental Protection Agency; EPA 2007) is recommended as guidance of choice. The validation of bioanalytical methods needs to be performed in accordance with the 'Guidance for Industry – Bioanalytical Method Validation' (FDA 2001). As an international ethical and scientific quality standard for performing research on human subjects, the 'Guidance for Industry – E6 Good Clinical Practice: Consolidated Guidance' (International Conference on Harmonization; ICH 1996) should be considered.

With special regard to the engineering of bioinformatics software applications and the implementation of computational methodologies in metabolomics research, two different guidelines are recommended by regulatory authorities as a minimum requirement. This is required as groundwork for the software development process and the validation of software (in the case of software being intended as part of a higher class medical product, additional efforts for validation would be required). As a general guidance for all bioinformatics- and software-based developments, the 'General Principles of Software Validation; Final Guidance for Industry and FDA Staff' (FDA 2002) must be followed. With a stronger focus on the engineering of applications in laboratory automation, 'The Good Automated Manufacturing Practice (GAMP) – Guide for Validation of Automated Systems in Pharmaceutical Manufacture' (International Society for Pharmaceutical Engineering; ISPE 2008) should be considered as a recommended guidance of choice.

Albeit an unwritten law, it is strongly recommended that regulatory bodies be involved in any development process as early as possible, increasing the probability of a later successful validation or qualification of new product developments (Baumgartner, personal communication).

8.2.3 Bioinformatics Workflow in Targeted Metabolomics

Metabolomics process and workflow concepts are suggested in different manners in the existing literature, e.g. for metabolomics experiments in cancer studies (Beger 2013), for the discovery of metabolic biomarkers (Baumgartner and Graber 2008; Baumgartner et al. 2011) or with regard to the complete pipeline for the development of new diagnostics or biomarkers (Phillips et al. 2006). When trying to put the different proposed workflow concepts together – with a focus on clinical studies and the discovery of metabolic biomarkers – and by trying to assemble the essential steps mentioned, the following could be defined:

(a) Experimental study design, including the definition of the research hypothesis and the obtainment of ethical approval
(b) Clinical study execution, including sample collection and quality-controlled storage
(c) The execution of bioanalytical methods, including sample preparation and sample analysis
(d) The integration and management of data by means of bioinformatics applications
(e) The processing and analysis of data, primarily by means of chemometrics algorithms
(f) The interpretation of findings with regard to biochemical plausibility and the generation of new hypotheses
(g) The validation of findings through independent clinical studies
(h) The qualification of new diagnostics or biomarkers for routine usage in laboratory medicine

Clinical bioinformatics and software-based support are considered as being important throughout the complete workflow and all different phases, yet, the focus in this chapter is put on three different central areas, namely (a) data integration and management, (b) data processing and analysis, and (c) data interpretation and metabolic modeling, which will now be presented and discussed in greated detail in the subsequent sections.

With regard to the application in metabolomics research, clinical bioinformatics plays an essential role in both major conceptual schools of thought, untargeted and targeted metabolomics (Patti et al. 2012). Even if, according to the objective of this chapter, focus is put on the bioinformatics workflow supporting targeted metabolomics (Weinberger and Graber 2005; Weinberger et al. 2005; Weinberger 2008), for methods of clinical bioinformatics, it is not always possible to define borders clearly, and some of the presented methods serve in both areas.

8.3 Data Acquisition, Integration, and Management

8.3.1 Sample Handling and Laboratory Automation

In the acquisition of data in metabolomics experiments, clinical bioinformatics is used to store and link clinical patient information (of course, with an appropriate level of anonymization) with information on single biological samples. Essentially, in the traceability of samples throughout the workflow, the ability to identify samples and measurement results at any point in time, and to link it back to clinical patient information, is indispensable – which can be solved by clinical bioinformatics through the usage of unique barcode identifiers. In the past years, more and more comprehensive biobanking initiatives have popped up, aiming to provide comprehensive solutions for the handling of samples and integration with clinical information and molecular measurement data (Yuille et al. 2008; Harris et al. 2012; Navis et al. 2014; van Ommen et al. 2015).

Regarding the preparation of samples, the engineering of software-based methods has become a key factor in the automation of sample preparation, usually being realized in the context of dedicated metabolomics solutions of liquid handling robotics. As prominent manufacturers of liquid handling robotics systems, Tecan (Männedorf, Switzerland) and Hamilton Robotics (Bonaduz, Switzerland) may be mentioned. A comprehensive review on the developments in the automation of LC-MS based methods, used in proteomics as well as metabolomics, is given in the literature (Vogeser and Kirchhoff 2011). With a special focus on the automated preparation of a ready-to-use targeted metabolomics kit (Absolute***IDQ***® p180 kit, Biocrates Life Sciences AG, Innsbruck, Austria), an application method utilizing a comprehensive liquid robotics system (Hamilton Robotics, Bonaduz, Switzerland) was engineered (Breit et al. 2011). The developed software method handles the different steps of automation, including (a) loading and transport steps of sample vials, plates, and reagents, (b) the handling of liquids, i.e., samples and reagents, based on optimized liquid classes, and (c) the logging and tracking of errors and status information during automated pipetting and preparation of samples.

8.3.2 Metabolomics Databases and Public Repositories

As in other fields of applied computer science, in metabolomics research, databases (as well as data warehouses) play an essential role in the storage and integration of data – of course, with different kinds of structures, complexities and performance requirements, depending on the intended usage. In addition to the multitude of proprietary solutions, a definitely notable and quite impressive selection of public repositories exists, in which knowledge related to metabolomics is nowadays collected and annotated. Subsequently, a short list of the different categories of public repositories is provided, partly as it has been suggested in the literature (Wishart 2012), together with selected examples.

As a comprehensive source of knowledge, general purpose metabolomics databases are suggested such as the Human Metabolome Database (Wishart et al. 2009). For the annotation of information on the interconnection of metabolites, metabolic pathway databases were presented, such as the Kyoto Encyclopedia of Genes and Genomes (KEGG) (Kanehisa and Goto 2000), the BioCyc Collection of databases (Karp et al. 2005), or Reactome (Joshi-Tope et al. 2005). Focusing on the collection of metabolic molecules, compound databases are provided, e.g. Chemical Entities of Biological Interest (ChEBI) (Degtyarenko et al. 2008) or ChemSpider (Pence and Williams 2010). Spectral databases such as the Golm Metabolome Database (Kopka et al. 2005) or Metlin (Smith et al. 2005) support the user in the identification of compounds by comparison to reference spectra. Aiming at the storage of information on pharmaceutical products and their molecular targets, drug databases have been established, e.g. DrugBank (Wishart et al. 2006) or the Small Molecule Pathway Database (SMPDB) (Frolkis et al. 2010). Collecting data on toxic compounds, dedicated databases were presented in the literature, for example the Aggregated Computational Toxicology Resource (ACToR) (Judson et al. 2008) or the Toxin Target/Target Database (T3DB) (Lim et al. 2010). With a special focus on monogenic diseases and their pathophysiology, databases such as Online Mendelian Inheritance in Man (OMIM) (Hamosh et al. 2000) are integrating genetic and medical knowledge accordingly.

8.3.3 Metabolomics Application Software

As a next central area in the application of clinical bioinformatics in metabolomics, the development of software solutions for the management of data in metabolomics projects is considered. The spectrum of research-based, publicly available metabolomics application software is even broader than that of metabolomics databases, and it would be virtually impossible to provide an exhaustive picture of this area; thus, the reader may be referred to the appropriate literature reviewing these solutions (e.g. Sugimoto et al. 2012), to get a more comprehensive overview. At this point, commercially available software supporting MS-based metabolomics experiments and the topic of metabolomics laboratory information management systems (LIMS) are shortly discussed.

The development of metabolomics LIMS was introduced in the literature as a special challenge (Wishart 2007) – integrating information from different sources and ideally enabling the management of data throughout the complete workflow in metabolomics. Software managing the metabolomics workflow was introduced, both developed through research projects and commercial solutions. Software applications originating from the research community, in this case metabolomics LIMS systems, are for example SetupX (Scholz and Fiehn 2007) or Sesame (Markley et al. 2007).

As commercial solutions, software provided by manufacturers of MS systems may be noted, being partly comprehensive solutions and partly a collection of different software applications. Examples of this kind of software are: (a) Analyst® and Multiquant™ by AB Sciex (Framingham, MA), (b) MassHunter, Mass Profiler Professional (MPP) and Pathway Architect by Agilent (Santa Clara, CA, USA), (c) AMIX and ProfileAnalysis by Bruker (Billerica, MA, USA), (d) TurboMass by PerkinElmer (Waltham, MA), (e) Profiling solution by Shimadzu (Kyoto, Japan), (f) Xcalibur™ by Thermo Fisher Scientific (Waltham, MA), or (g) MassLynx™ by Waters (Milford, MA).

In addition to these applications, commercial providers of metabolomics solutions who build their business models around products, contract research or biomarker discovery services, are offering software applications, aiming at supporting the tasks of the complete workflow. Metabolon (Durham, NC, USA), which offers services focusing on untargeted metabolomics for example, provides the following set of software applications: (a) mLIMS™, for proprietary sample and data management, (b) Metabolyzer™, for metabolite identification and peak integration, (c) IonTracker™, quality control (QC) and identification of novel metabolites, (d) Cross-set integrator™, QC of peak integration (quantitation), and (e) VPhil™, QC tool for chemical spectral analyst data curation.

Biocrates (Innsbruck, Austria), pioneer in the commercialization of targeted metabolomics solutions offering ready-to-use reagents kits and providing contract research services, has also developed its own set of bioinformatics software, namely (a) Met***IDQ***™, a targeted metabolomics application software, (b) StatPack, for full statistical analysis of results, (c) and RatioExplorer, for comprehensive metabolomics data interpretation. As an example of a dedicated targeted metabolomics software, it is worth noting that Met***IDQ***™ supports a major part of the bioinformatics enabled workflow in targeted metabolomics, ranging from the integration of clinical patient data and sample information to the storage of measurement data and the biochemical interpretation of results in the context of biochemical pathways. The development of Met***IDQ***™ was started roughly a decade ago (Breit et al. 2006), at this time primarily consisting of the LIMS part – supporting project and sample management, the administration of standardized operating procedures and general settings, as MS configuration parameters, analyzed sample types or metabolite classes.

8.4 Data Processing, Analysis, and Mining

8.4.1 Data Processing and Chemometrics Methods

The data processing methods and algorithms presented in this section should probably be more correctly defined as chemometrics or cheminformatics than as bioinformatics methods (Enot et al. 2011; Breit et al. 2015a); still, in the software-based implementation of those modalities, they emerge as an essential building block of the clinical bioinformatics mosaic.

Metabolomics data, obtained through MS experiments, need to undergo a processing, due to their bandwidth in biological and physical data characteristics, i.e. the variety in their abundances, ionization behavior, polarity, and solubility. Furthermore, MS data show a broad variety in their mass domain (mass-to-charge ratio, m/z), time domain (retention time), and ion intensities. Also, the diversity of analytical MS platforms with different chromatographic performances, ionization efficiencies, mass analyzer resolutions, or ion detector sensitivities leads to the necessity of data processing.

The crude MS signals are subject to noise, resulting from chemical noise, sample preparation errors or electronic noise. Different approaches were introduced to process those signals. Methods for baseline correction were suggested, using e.g. polynomial models (Bylund 2001). Additionally, noise reduction methods such as smoothing filters (Zhu et al. 2003; Jonsson et al. 2005) or wavelet transformations (Zhao et al. 2006; Karpievitch et al. 2007) are presented. Yet, those methods should be used carefully, since they are known to introduce a potential bias (Listgarten and Emili 2005; Fredriksson et al. 2009).

For the detection and extraction of features from the given raw signal or peaks expected to represent chemical compounds, methods of data compression are suggested (Katajamaa and Oresic 2007). Fundamental approaches in this area are mass binning (Wang et al. 2003; Beckmann et al. 2008) or peak picking. In peak picking, different techniques were introduced, such as pattern classification (Tibshirani et al. 2004), extraction of derivatives (Vivó-Truyols et al. 2005; Fredriksson et al. 2009), refinement of heuristics (Morris et al. 2005), peak shape model approximations (Tan et al. 2006), or wavelet applications (Zhao et al. 2006; Karpievitch et al. 2007). Especially in large scale experiments, data may be subject to systemic drifts, which can be corrected: on the mass domain, through calibration of mass binning, and on the time domain, using non-parametric alignment methods. A correction of occurring isotopes can be achieved through isotope convolution (Eibl et al. 2008).

Due to existing inter-sample variance – i.e. differences in sample concentrations or homogeneity, degradation over time, or analytical drifts – a normalization of data prior to further statistical treatment should be considered (Goodacre et al. 2004; Enot et al. 2008). Inter-sample normalization can be achieved through the estimation of a scaling factor, e.g. by sub-selection of peaks/signals (Wang et al. 2003; Warrack et al. 2009), mapping of intensity distributions (Dieterle et al. 2006; Torgrip et al. 2008), introduction of class information (Enot et al. 2008), or a summation of measurements (Torgrip et al. 2008). In addition, the normalization according to biological properties is suggested, e.g. in cell culture or tissue samples (e.g. by cell count or dry weight) or urinary substances (by creatinine or urine volume) (Warrack et al. 2009). Yet, this approach is questioned in the literature (Enot et al. 2011; Breit et al. 2015a), and its usage should be decided according to the underlying clinical question.

8.4.2 Biostatistics and Data Analysis

In the preparation of metabolic data for biostatistical analysis, two basic issues need to be considered: data transformations and the handling of missing values. Data transformation becomes necessary, due to non-normal distributions of concentrations, small intra-parameter dynamic ranges, and large inter-parameter dynamic ranges. As a solution, e.g. log transformations are introduced in the literature (Purohit et al. 2004; Listgarten and Emili 2005; Lu and King 2009). With respect to the handling of missing values, different basic approaches are suggested: (a) to discard features (Bijlsma et al. 2006; Enot et al. 2008), (b) to impute missing values using pre-specified values – which is known to have serious limitations (Jain et al. 2008), or (c) to use advanced multivariate methods – which is recommended if data point estimation is expected to provide additional gain (Stacklies et al. 2007).

The variance of data in MS based metabolomics experiments is basically influenced by a biological variance of metabolites and an analytical variance. Biological variance can yield a broad bandwidth, with low coefficients of variation (CVs), e.g. in case of strict homeostatic control, but eventually reaching high CVs, e.g. in urine (Crews et al. 2009; Parsons et al. 2009). Analytical variance ranges up to 15 % for quantitative parameters, but can be optimized to minimum levels of 2–5 % for selected analytes. For the validation of bioanalytical methods, limits are demanded, in the dynamic range, of CVs of less than 15 %, and for the lower limit of quantitation (LLOQ), of CVs of less than 20 % (FDA 2001).

Features in metabolomics data yield correlations resulting from the interdependence of abundances, specific chemical characteristics or analytical platforms (Draper et al. 2009). Aiming at an examination of correlation due to biological mechanisms, different network-based approaches have been suggested in the literature, linking metabolite correlations to reaction networks (Camacho et al. 2005; Mendes et al. 2005; Steuer 2006), or specifying structural properties of biochemical networks using topological network descriptors (Müller et al. 2011).

General biostatistical methods used for the analysis of metabolic data, such as principal component analysis (PCA), cluster analysis, partial least squares (PLS) analysis, random forest (RF) models, or conventional statistical tests (e.g. Student's *t*-test), are widely discussed in the literature, to which the reader may be referred for a more detailed introduction (e.g. Sugimoto et al. 2012).

With regard to the dynamic analysis of longitudinal time-course metabolic concentration data, a variety of approaches is suggested in the literature. The essential methods suggested for dynamic data analysis can be classified into methods based on fundamental models, on predefined basic functions, on dimension reduction, on multivariate time series, on analysis-of-variance (ANOVA), or on imposing smoothness (Smilde et al. 2010). For the analysis of periodic and oscillating data, Fourier analysis, wavelet transformations, or principal component analysis (PCA) with wavelets have been introduced (Bakshi 1998; Smilde et al. 2010). For more details on further specialized methods, the reader may again be referred to the according literature (Mishina et al. 1993; Jansen et al. 2004; Smilde et al. 2005; Berk et al. 2011; Jansen et al. 2012; Stanberry et al. 2013; Breit et al. 2015b).

8.4.3 Data Mining and Machine Learning

In addition to the more traditional biostatistical concepts, methods of the closely interrelated fields of artificial intelligence, machine learning and data mining respectively knowledge discovery in databases (KDD) are increasingly used in the analysis of metabolic data.

The subject of artificial intelligence was first suggested and discussed on a broad basis in the literature around half a century ago (Turing 1950; McCarthy et al. 1955; Solomonoff 1957, 1964; Samuel 1959). Closely related to it, the field of machine learning has developed continuously, and is categorized into three different basic approaches: supervised, unsupervised and reinforcement learning (Alpaydin 2009; Domingos 2012; Sebag 2014; Ren et al. 2015). Popular supervised machine learning methods include, for instance, support vector machines (SVMs; Byvatov and Schneider 2003) and artificial neural networks (ANNs; Bicciato 2004). Partition-based algorithms such as K-Means (Dudik et al. 2015), density-based methods such as density-based spatial clustering of applications with noise (DBSCAN; Dudik et al. 2015), and hierarchical clustering methods (Gambin and Slonimski 2005) are important representatives of unsupervised methods. Also, the process of knowledge discovery in databases, and within it the step of data mining, is widely discussed in the literature (Fayyad et al. 1996). For all of those areas, please refer to the appropriate literature for more detailed information.

With respect to the practical application of those methods in metabolomics experiments, currently the discovery of metabolic biomarkers is probably the most dynamic and promising topic for which methods are developed (Lehmann and Romano 2005; Enot et al. 2006; Baumgartner et al. 2010; Breit et al. 2015b). In the metabolic biomarker discovery process, an important data mining step is the task of feature selection, aiming to obtain a set of highly discriminating features: wrappers (the combination of a search strategy and a learning algorithm), embedded methods (the selection of features is built into the learning algorithm), and filters (the calculation of scores to select a set of discriminating features) are three important categories of feature selection methods (Saeys et al. 2007). For instance, a rule-based filter approach (associative voting) to find biomarker candidates in prostate cancer data was proposed in the literature (Osl et al. 2008). In addition to these categories, ensemble-based feature selection strategies rely on the combination of different feature selection methods (Saeys et al. 2007). A feature selection modality termed stacked feature ranking (SFR) that relies on the combination of different feature rankings using a two-level architecture with a suggestion and a decision layer, was introduced (Netzer et al. 2009). Using this approach, the authors were able to identify highly discriminating volatile organic marker candidates in exhaled breath of patients with liver disease. Recently, new network-based methods have also been described. A new network type called 'ratio network' to study kinetic changes of putative biomarkers from time series cohort studies using targeted MS/MS profiling, was introduced (see Fig. 8.1; Netzer et al. 2011). Furthermore, a ranking of features was proposed in the literature, based on different topological descriptors using correlation and ratio networks, and by evaluating the discriminatory ability using classification methods (Netzer et al. 2012). Here, the authors were able to identify biomarker candidates for obesity using quantitative targeted MS/MS. For further methods – in the application of methods of bioinformatics, data mining and machine learning in metabolic biomarker discovery – please refer to a comprehensive overview, provided in a recent review (Baumgartner et al. 2011).

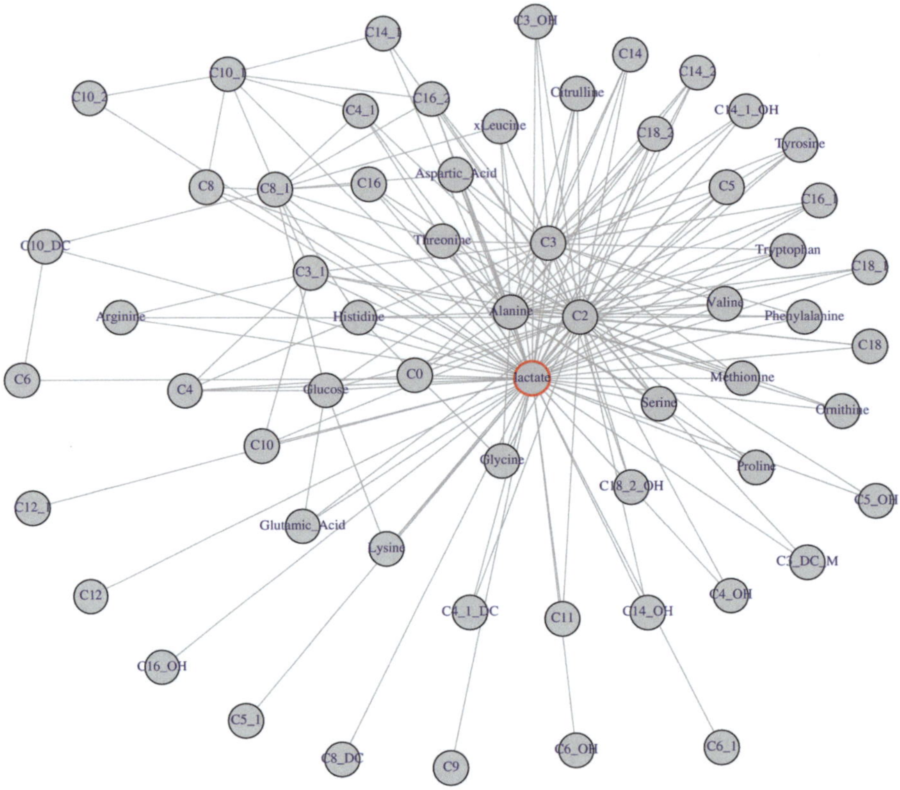

Fig. 8.1 **Ratio network**. Ratio network, inferred to analyze metabolic processes in the organism during physical activity (Redrawn after Netzer et al. 2011). Note that the figure was generated using the R (R Development Core Team 2008) packages 'igraph' (http://igraph.org/) and 'BiomarkeR' (Netzer et al. 2011)

8.5 Data Interpretation, Modeling, and Simulation

8.5.1 Pathway Visualization and Network-Based Interpretation

From the perspective of the clinical bioinformatics workflow in targeted metabolomics, subsequent to the analysis of data by the means of univariate or multivariate statistical approaches, as well as other advanced computational methods, findings – usually a set of selected metabolites – need to undergo further checks for plausibility, analytically, but especially biochemically. Concepts of biochemical data interpretation, usually in the context of annotated metabolic pathways, are widely discussed in the literature (e.g. Weckwerth and Morgenthal 2005; Xia et al. 2009; Enot et al. 2011; Breit et al. 2015a), and for the completeness of the workflow perspective in the chapter, just a short repeated summary of some selected basic approaches is provided.

Based on annotated biochemical knowledge, nowadays provided through metabolic databases and public repositories (see Sect. 8.3.2), an intuitive, often used approach is the shell-wise exploration of metabolites and their molecular, enzymatic reactions. The biochemical verification of findings in metabolomics experiments, in the context of analyzing single enzymatic reactions, was first successfully presented in monogenic disorders in neonates (Röschinger et al. 2003), but later also in multifactorial diseases (Suhre et al. 2011; Suhre and Gieger 2012). The introduction of ratios (of product and substrate concentrations), using hypothesis-driven knowledge, suggested in genome-wide association studies (GWAS) (Gieger et al. 2008; Illig et al. 2010), could provide proof of an improvement of statistical significance of findings by many orders of magnitude.

Originating from transcriptomics experiments (Mootha et al. 2003; Subramanian et al. 2005; Chagoyen and Pazos 2013), the concept of enrichment analysis was adapted in metabolomics research; in mapping statistically significant metabolites on metabolic pathways, with a subsequent statistics-based pathway ranking. Basic approaches introduced in the literature include overrepresentation analysis (ORA) and set enrichment analysis (Huang da et al. 2009; Enot et al. 2011; Chagoyen and Pazos 2013), as well as the approach of single sample profiling (SSP) (Xia et al. 2012). Software tools enabling this type of interpretation are Metabolite Set Enrichment Analysis (MSEA) (Xia and Wishart 2010), Metabolite Biological Role (MBRole) (Chagoyen and Pazos 2011) or MetaboAnalyst (Xia et al. 2012). Crucial in this type of biochemical interpretation are a selection of species-specific pathways (to exclude potential false positives), as well as a consideration of the biological importance of the underlying pathway in a functional assessment (Breit et al. 2015a).

Considering the annotated knowledge on biochemical pathways as one integrated source of knowledge – across the somehow artificial borders of closed metabolic pathways – the concept of route finding is applied. Route finding offers the possibility to identify different kinds of routes between two selected metabolites: the shortest route, routes up to a maximum length, node-disjoint paths (routes not sharing a certain metabolite), or edge-disjoint paths (routes not sharing a certain enzyme). Noteable examples of software providing metabolic route finding are MetaRoute (Blum and Kohlbacher 2008), the MetPath module of the Met***IDQ***™ suite (Enot et al. 2011) or Metabolic Route Search and Design (MRSD) (Xia et al. 2011). For a reduction of complexity and false positive results, common cofactors and small inorganic molecules need to be excluded, e.g. adenosine triphosphate (ATP), nicotinamide adenine dinucleotide (NADH), carbon dioxide (CO_2), water (H_2O) or ammonia (NH_3).

Furthermore, in recent years, more innovative network inference and visualization tools have been suggested (Emmert-Streib 2013; Emmert-Streib et al. 2014; Tripathi et al. 2014). Common objective behind all the different approaches of biochemical interpretation, generally described in the literature, is their utilization for the generation of new hypotheses; and furthermore, to check findings for biological and medical plausibility, since findings potentially could be skewed through redundancies in metabolism, through drug compounds disturbing signals, or through analytical or statistical artifacts.

8.5.2 Metabolic Modeling and Engineering

Theoretical biochemical foundations, for the bioinformatics-based modeling of metabolic interactions – in this case probably better categorized under the term computational metabolomics, than under clinical bioinformatics – were laid approximately 150 years ago, by introducing the 'law of mass action' (Guldberg and Waage 1864, 1867, 1879). This concept, together with its popular adaptation – the Michaelis-Menten model of enzyme-catalyzed single-substrate reactions (Michaelis and Menten 1913) – and the equilibrium constant (Devlin 2006; Nelson and Cox 2008) – describing the homeostatic behavior of the metabolic system – are substantial building blocks for the modeling of metabolic interactions (Voit et al. 2015).

Fundamental types of metabolic modeling approaches discussed and reviewed in the literature include, for example, qualitative models, models of flux balance analysis, or kinetic models using ordinary differential equations (ODEs) (Rios-Estepa and Lange 2007; Steuer and Junker 2009). Furthermore, intermediate approaches are suggested, e.g. by approximating local mechanisms through parametric linear representations (Steuer and Junker 2009). The practical application of metabolic modeling is established as a crucial part in metabolic engineering (the optimization of cellular processes), e.g. in the production of pharmaceuticals or nourishments (Stephanopoulos 1999; Nielsen 2001).

An alternative approach to modeling of kinetic regulatory mechanisms in metabolism was recently introduced in the literature; in this case, based on an empiric deduction of *de facto* kinetic response patterns – which were obtained through the analysis of quantitated time-course concentration data, measured under *in-vivo* conditions (Breit et al. 2015b). In a setting of a cycle ergometry performance test in which workload was incrementally increased (every 3 min by 25 Watts, up to the maximum individual performance), data of 47 test persons were analyzed, both male and female, either being professional alpine skiers or amateur endurance athletes. Utilizing a mass spectrometry-based targeted metabolomics approach, quantitative concentration data of 110 metabolites were measured (from the classes, acylcarnitines, amino acids, and sugars), of which the 30 most analyzed reliable and robust metabolites were considered for data analysis. Dynamic metabolic biomarker candidates could be selected, based on maximum fold changes (MFCs) in preprocessed, longitudinal concentrations, combined with an examination of p-values of statistical hypothesis testing. For each of the 30 analyzed

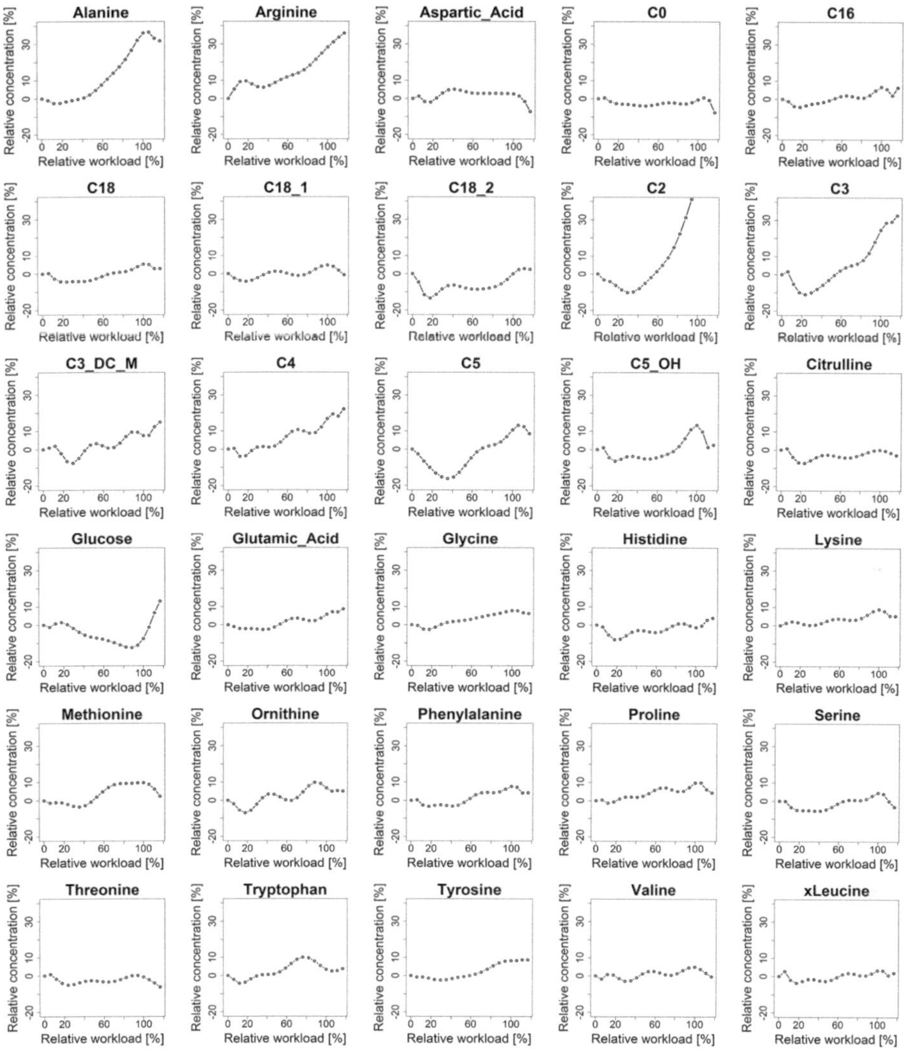

Fig. 8.2 Metabolite kinetic signatures. Metabolite kinetic patterns, in response to external perturbations, caused by a cycle ergometry performance test (Redrawn after Breit et al. 2015b)

metabolites, a specific kinetic signature could be characterized – based on a mathematical modeling approach using polynomial fitting, with a degree of nine chosen (Fig. 8.2). Based on a hierarchical cluster analysis, common basic kinetic response patterns could be specified, i.e. sustained, early, late and delayed response patterns. Sustained response patterns, with primarily indifferent, constant concentration over time, were yielded by a majority of metabolites. An early response pattern, with a major change in concentration at the beginning of exercise, was shown in the time-course data of valerylcarnitine (C5), showing a moderate effect size (MFC = 1.38, $p < 0.001$). A late response pattern, was, amongst others, shown by acetylcarnitine (C2), with the highest value in dynamic change in concentration (MFC = 1.97, $p < 0.001$). A delayed response, similar to a L-curve/hockey-stick function, with a steep increase in concentration after the end of exercise, was yielded by glucose, showing a moderate effect size (MFC = 1.32, $p < 0.001$; see Fig. 8.2).

8.5.3 *In-Silico Biology*

Closely attached to the modeling of metabolic mechanisms are concepts termed as *in-silico* biology (Palsson 2000), systems biology (Wolkenhauer 2001), or computational systems biology (Kitano 2002), depending on the definition chosen. These concepts aim to integrate modeling and simulation of interactions on all molecular levels, e.g. genomics, transcriptomics, proteomics, and metabolomics, and range from single cell models up to the modeling of a complete organism.

The basic idea lies in the utilization of fundamental concepts of systems theory (Wiener 1948; Bertalanffy 1968) and an adaptation of those to the application in biology (Mesarovic 1968). This idea found a revival at the beginning of the twentieth century (Voit 2000; Kitano 2001), aiming at the computationally supported understanding of biological systems as a whole, i.e. regarding their structure, dynamics, control methods, and design methods (Kitano 2002).

The *in-silico* modeling and simulation of mechanisms on different molecular levels is widely discussed in the literature, e.g. by performing a sensitivity analysis on a theoretical model of the TNFα-mediated NF-κB signaling cascade (Breit 2004), by simulating the mitochondrial fatty acid β-oxidation (Modre et al. 2009), or in the context of the e-cell project (Takahashi et al. 2003; Smolen et al. 2004; Nishino et al. 2013).

From the perspective of clinical bioinformatics, here, challenges lie in the intelligent integration of the different 'omics' data together with clinical information, as well as the development and implementation of computational methods, enabling and supporting this systems-based approach to modeling and simulation of molecular processes – and in consequence to deduct implications on metabolism.

8.6 Discussion

8.6.1 *Conclusions*

As discussed in the previous chapter of this book (see Chap. 7), clinical biochemistry is currently experiencing a renaissance, with dedicated driving forces behind it, i.e. a revival in the examination of the Warburg hypothesis, the establishment of newborn screening programs on a broad basis, and technological improvements in mass spectrometry – making it ready for the application in routine clinical analysis. In parallel, methods of clinical bioinformatics – resulting from the disruptive innovations in computer sciences in the past decade(s) – are enabling levels of workflow automation, data integration, and technological usability, which, 20 years ago, molecular researchers would have dreamed of.

In this chapter, emphasis has been put on the impact of clinical bioinformatics in targeted metabolomics, supporting the search for new metabolic markers and the development of new diagnostics, hopefully soon finding application in the very promising fields of diabetes and chronic kidney disease, as well as potentially in neurology or cancer (see Chap. 7). Apart from the primarily clinical focus of this chapter, most of the discussed bioinformatics and computational concepts and methodologies are also valid for other major application areas of targeted (and untargeted) metabolomics research, e.g. pharmaceutical R&D and drug discovery, animal health and veterinary research, bioprocessing and cell cultures, nutrition and consumer goods, or plant metabolism.

In summary – even if this assessment will potentially not be shared by all (clinical) bioinformaticians – bioinformatics software and methods in targeted metabolomics (as well as in other areas of molecular research) might primarily be considered as a service, easing the work of other involved persons such as physicians, chemical analysts or biochemists. In addition to the 'playground' of technological innovation, one should always bear in mind the clinical benefit of patients, i.e. improvements in the diagnosis, prognosis and theranosis of diseases (DeNardo and DeNardo 2012).

8.6.2 Future Perspectives

Finally, some thoughts on future perspectives of clinical bioinformatics in targeted metabolomics, and what potential further innovations in this field might be.

As also suggested in the previous chapter of this book (see Chap. 7), a major challenge – in the transition of basic findings in targeted metabolomics research into clinical applications – lies in the standardization and automation of analytical assays, together with integrated software solutions, validated for usage in clinical applications. To overcome this hurdle, different companies are working on integrated solutions, combining hardware, software, and reagents into one integrated system, so called 'black-box' clinical analyzers, ready for the routine usage in clinical laboratories. These developments are to a certain extent driven by large companies, namely the suppliers of mass spectrometer systems (see Sect. 8.3.3), often in collaboration with reagent suppliers; and with bioinformatics solutions as an essential building block, enabling the processing of samples and tracking of clinical information throughout the complete targeted metabolomics workflow, from clinical questions to the analysis of samples and a quantified concentration value of a single metabolite.

With regard to, say, slightly more 'futuristic' perspectives, the miniaturization of sample analysis is still an intense topic of research, with a major focus on the development of microfluidics devices (Kraly et al. 2009) or lab-on-a-chip solutions (Trietsch et al. 2011). With reference to the integration of data (from different omics sources, together with clinical information), the step-wise breakthrough of cloud computing – now also arriving at the end-user (or patient) – will probably ease the access and strengthen the usage of public data repositories. Advancements in the developments of new algorithms and methodologies will further accelerate the *in-silico* modeling and simulation of metabolic processes. Probably the most exciting field in clinical bioinformatics, at least from the computational perspective, is the application and development of methods of machine learning and artificial intelligence in molecular research. Five years ago, in genomics analysis, a robotics system has been presented, able to autonomously perform experiments and deduce new hypotheses from generated findings: '*An integrated Laboratory Robotic System for Autonomous Discovery of Gene Function*' (Sparkes et al. 2010). To conclude with a visionary application in targeted metabolomics – which would be a robotics system, autonomously performing metabolomics experiments and deducting new hypotheses in the examination of fundamental metabolic processes, as well as for the clinically oriented discovery of novel metabolic biomarkers.

References

Alpaydin E. Introduction to machine learning. 2nd ed. Cambridge: The MIT Press; 2009.

Bakshi BR. Multiscale pca with application to multivariate statistical process monitoring. AIChE J. 1998;44:1596–610.

Baumgartner C, Graber A. Data mining and knowledge discovery in metabolomics. In: Masseglia F, Poncelet P, Teisseire M, editors. Successes and new directions in data mining. London: Information Science Reference; 2008. p. 141–66.

Baumgartner C, Lewis GD, Netzer M, Pfeifer B, Gerszten RE. A new data mining approach for profiling and categorizing kinetic patterns of metabolic biomarkers after myocardial injury. Bioinformatics. 2010;26(14):1745–51.

Baumgartner C, Osl M, Netzer M, Baumgartner D. Bioinformatic-driven search for metabolic biomarkers in disease. J Clin Bioinforma. 2011;1(1):2. doi:10.1186/2043-9113-1-2.

Beckmann M, Parker D, Enot DP, Duval E, Draper J. High-throughput, nontargeted metabolite fingerprinting using nominal mass flow injection electrospray mass spectrometry. Nat Protoc. 2008;3(3):486–504.

Beger RD. A review of applications of metabolomics in cancer. Metabolites. 2013;3(3):552–74. doi:10.3390/metabo3030552.

Bellazzi R, Masseroli M, Murphy S, Shabo A, Romano P. Clinical bioinformatics: challenges and opportunities. BMC Bioinformatics. 2012;13(Suppl 14):S1. doi:10.1186/1471-2105-13-S14-S1. Epub 2012 Sep 7.

Berk M, Ebbels T, Montana G. A statistical framework for biomarker discovery in metabolomic time course data. Bioinformatics. 2011;27(14):1979–85.

Bernstam EV, Smith JW, Johnson TR. What is biomedical informatics? J Biomed Inform. 2010;43 (1):104–10. doi:10.1016/j.jbi.2009.08.006. Epub 2009 Aug 13. Review.

Bertalanffy L. General system theory: foundations, development, applications. New York: George Braziller; 1968.

Bicciato S. Artificial neural network technologies to identify biomarkers for therapeutic intervention. Curr Opin Mol Ther. 2004;6(6):616–23.

Bijlsma S, Bobeldijk I, Verheij ER, Ramaker R, Kochhar S, Macdonald IA, van Ommen B, Smilde AK. Large-scale human metabolomics studies: a strategy for data (pre-) processing and validation. Anal Chem. 2006;78(2):567–74.

Bino RJ, Hall RD, Fiehn O, Kopka J, Saito K, Draper J, et al. Potential of metabolomics as a functional genomics tool. Trends Plant Sci. 2004;9(9):418–25.

Blum T, Kohlbacher O. MetaRoute: fast search for relevant metabolic routes for interactive network navigation and visualization. Bioinformatics. 2008;24(18):2108–9.

Board Members MSI, Sansone SA, Fan T, Goodacre R, Griffin JL, Hardy NW, et al. The metabolomics standards initiative. Nat Biotechnol. 2007;25(8):846–8.

Bourne PE, Brenner SE, Eisen MB. Ten years of PLoS computational biology: a decade of appreciation and innovation. PLoS Comput Biol. 2015;11(6), e1004317. doi:10.1371/journal.pcbi.1004317.

Breit M. Sensitivity analysis of biological pathways. Master's thesis. Hall in Tirol: University for Health Sciences Medical Informatics and Technology (UMIT), 2004.

Breit M, Graber A, Tilg B. Development of an integrated bioinformatics platform for the identification of metabolic markers. Presented at: BMT annual meeting 2006; 2006 Sept 6–9; Zurich, Switzerland.

Breit M, Bichteler F, Urban M, Bellus TH, Winter A, Weinberger KM. Standardized preparation of a mass spectrometry-based research kit for targeted metabolomics on a liquid handling robot. Poster session presented at: Advances in Separation Technology (AST2011), European Lab Automation (ELA2011); 2011 June 30–July 1; Hamburg, Germany.

Breit M, Baumgartner C, Weinberger KM. Data handling and analysis in metabolomics. In: Khanmohammadi M, editor. Current applications of chemometrics. New York: Nova Science Publishers; 2015a. p. 181–203.

Breit M, Netzer M, Weinberger KM, Baumgartner C. Modeling and classification of kinetic patterns of dynamic metabolic biomarkers in physical activity. PLoS Comput Biol. 2015b;11 (8):e1004454. doi:10.1371/journal.pcbi.1004454. eCollection 2015.

Butte AJ. Translational bioinformatics: coming of age. J Am Med Inform Assoc. 2008;15 (6):709–14. doi:10.1197/jamia.M2824. Epub 2008 Aug 28.

Bylund D. Chemometric tools for enhanced performance in liquid chromatography-mass spectrometry. Comprehensive Summaries of Uppsala Dissertations from the Faculty of Science and Technology. Uppsala: Uppsala University; 2001.

Byvatov E, Schneider G. Support vector machine applications in bioinformatics. Appl Bioinformatics. 2003;2(2):67–77.

Camacho D, de la Fuente A, Mendes P. The origin of correlations in metabolomics data. Metabolomics. 2005;1(1):53–63.

Chagoyen M, Pazos F. MBRole: enrichment analysis of metabolomic data. Bioinformatics. 2011;27(5):730–1.

Chagoyen M, Pazos F. Tools for the functional interpretation of metabolomic experiments. Brief Bioinform. 2013;14(6):737–44.

Chang PL. Clinical bioinformatics. Chang Gung Med J. 2005;28(4):201–11. Review.

Crews B, Wikoff WR, Patti GJ, Woo HK, Kalisiak E, Heideker J, Siuzdak G. Variability analysis of human plasma and cerebral spinal fluid reveals statistical significance of changes in mass spectrometry-based metabolomics data. Anal Chem. 2009;81(20):8538–44.

da Huang W, Sherman BT, Lempicki RA. Bioinformatics enrichment tools: paths toward the comprehensive functional analysis of large gene lists. Nucleic Acids Res. 2009;37(1):1–13. doi:10.1093/nar/gkn923. Epub 2008 Nov 25.

Degtyarenko K, de Matos P, Ennis M, Hastings J, Zbinden M, McNaught A, et al. ChEBI: a database and ontology for chemical entities of biological interest. Nucleic Acids Res. 2008;36 (Database issue):D344–50. Epub 2007 Oct 11.

DeNardo GL, DeNardo SJ. Concepts, consequences, and implications of theranosis. Semin Nucl Med. 2012;42(3):147–50. doi:10.1053/j.semnuclmed.2011.12.003.

Devlin TM, editor. Textbook of biochemistry with clinical correlations. 6th ed. Hoboken: Wiley-Liss; 2006.

Dieterle F, Ross A, Schlotterbeck G, Senn H. Probabilistic quotient normalization as robust method to account for dilution of complex biological mixtures. Application in 1H NMR metabonomics. Anal Chem. 2006;78:4281–90.

Domingos P. A few useful things to know about machine learning. Commun ACM. 2012;55 (10):78–87.

Draper J, Enot DP, Parker D, Beckmann M, Snowdon S, Lin W, Zubair H. Metabolite signal identification in accurate mass metabolomics data with MZedDB, an interactive m/z annotation tool utilising predicted ionisation behaviour 'rules'. BMC Bioinformatics. 2009;10:227.

Dudik JM, Kurosu A, Coyle JL, Sejdic E. A comparative analysis of DBSCAN, K-means, and quadratic variation algorithms for automatic identification of swallows from swallowing accelerometry signals. Comput Biol Med. 2015;59:10–8. doi:10.1016/j.compbiomed.2015.01.007. Epub 2015 Jan 17.

Eibl G, Bernardo K, Koal T, Ramsay SL, Weinberger KM, Graber A. Isotope correction of mass spectrometry profiles. Rapid Commun Mass Spectrom. 2008;22(14):2248–52.

Emmert-Streib F. Structural properties and complexity of a new network class: Collatz step graphs. PLoS One. 2013;8(2), e56461.

Emmert-Streib F, Zhang SD, Hamilton P. Dry computational approaches for wet medical problems. J Transl Med. 2014;12:26.

Enot DP, Beckmann M, Overy D, Draper J. Predicting interpretability of metabolome models based on behavior, putative identity, and biological relevance of explanatory signals. Proc Natl Acad Sci U S A. 2006;103(40):14865–70.

Enot DP, Lin W, Beckmann M, Parker D, Overy DP, Draper J. Preprocessing, classification modeling and feature selection using flow injection electrospray mass spectrometry metabolite fingerprint data. Nat Protoc. 2008;3(3):446–70.

Enot DP, Haas B, Weinberger KM. Bioinformatics for mass spectrometry-based metabolomics. Methods Mol Biol. 2011;719:351–75. doi:10.1007/978-1-61779-027-0_16.

EPA – Environmental Protection Agency. Guidance for preparing standard operating procedures (SOPs) (G-6). Washington, DC: Office of Environmental Information; 2007.

Fayyad U, Piatetsky-Shapiro G, Smyth P. The KDD process for extracting useful knowledge from volumes of data. Commun ACM. 1996;39(11):27–34.

FDA – Food and Drug Administration. Guidance for industry – bioanalytical method validation. Rockville: Center for Drug Evaluation and Research (CDER); 2001.

FDA – Food and Drug Administration. General principles of software validation; final guidance for industry and FDA staff. Rockville: Center for Biologics Evaluation and Research (CBER); 2002.

Fogg CN, Kovats DE. Computational biology: moving into the future one click at a time. PLoS Comput Biol. 2015;11(6):e1004323. doi:10.1371/journal.pcbi.1004323. eCollection 2015 Jun.

Fredriksson MJ, Petersson P, Axelsson BO, Bylund D. An automatic peak finding method for LC-MS data using Gaussian second derivative filtering. J Sep Sci. 2009;32(22):3906–18.

Frolkis A, Knox C, Lim E, Jewison T, Law V, Hau DD, Liu P, Gautam B, Ly S, Guo AC, Xia J, Liang Y, Shrivastava S, Wishart DS. SMPDB: the small molecule pathway database. Nucleic Acids Res. 2010;38(Database issue):D480–7. doi:10.1093/nar/gkp1002. Epub 2009 Nov 30.

Gambin A, Slonimski PP. Hierarchical clustering based upon contextual alignment of proteins: a different way to approach phylogeny. C R Biol. 2005;328(1):11–22.

Gieger C, Geistlinger L, Altmaier E, Hrabé de Angelis M, Kronenberg F, Meitinger T. Genetics meets metabolomics: a genome-wide association study of metabolite profiles in human serum. PLoS Genet. 2008;4(11), e1000282.

Goecks J, Nekrutenko A, Taylor J. Galaxy Team. Galaxy: a comprehensive approach for supporting accessible, reproducible, and transparent computational research in the life sciences. Genome Biol. 2010;11(8):R86. doi:10.1186/gb-2010-11-8-r86. Epub 2010 Aug 25.

Goodacre R, Vaidyanathan S, Dunn WB, Harrigan G, Kell DB. Metabolomics by numbers: acquiring and understanding global metabolite data. Trends Biotechnol. 2004;22:245–52.

Guldberg CM, Waage P. Studier i affiniteten (Translation: Studies on affinities.) Forhandlinger i Videnskabs-Selskabet i Christiania; 1864.

Guldberg CM, Waage P. Études sur les affinites chimiques (Translation: Studies on chemical affinities.) Christiania: Brøgger & Christie; 1867.

Guldberg CM, Waage P. Über die chemische Affinität (Translation: On chemical affinity.) Erdmann's Journal für practische Cehmie. 1879;127:69–114.

Hamosh A, Scott AF, Amberger J, Valle D, McKusick VA. Online Mendelian Inheritance in Man (OMIM). Hum Mutat. 2000;15(1):57–61.

Harris JR, Burton P, Knoppers BM, Lindpaintner K, Bledsoe M, Brookes AJ, et al. Toward a roadmap in global biobanking for health. Eur J Hum Genet. 2012;20(11):1105–11.

Haug K, Salek RM, Conesa P, Hastings J, de Matos P, Rijnbeek M, et al. MetaboLights--an open-access general-purpose repository for metabolomics studies and associated meta-data. Nucleic Acids Res. 2014;41(Database issue):D781–6. doi:10.1093/nar/gks1004. Epub 2012 Oct 29.

Hogeweg P. The roots of bioinformatics in theoretical biology. PLoS Comput Biol. 2011;7(3), e1002021. doi:10.1371/journal.pcbi.1002021. Epub 2011 Mar 31.

ICH – International Conference on Harmonization. Guidance for industry – E6 good clinical practice: consolidated guidance. Rockville: Center for Drug Evaluation and Research (CDER); 1996.

Illig T, Gieger C, Zhai G, Römisch-Margl W, Wang-Sattler R, Prehn C, et al. A genome-wide perspective of genetic variation in human metabolism. Nat Genet. 2010;42:137–41.

ISO – International Organization for Standardization. ISO 9001:2008 quality management systems – requirements. Geneva: ISO Headquarters; 2008.

ISPE – International Society for Pharmaceutical Engineering. The good automated manufacturing practice (GAMP) – guide for validation of automated systems in pharmaceutical manufacture. Tampa: ISPE Headquarters; 2008.

Jain RB, Caudill SP, Wang RY, Monsell E. Evaluation of maximum likelihood procedures to estimate left censored observations. Anal Chem. 2008;80(4):1124–32.

Jansen JJ, Hoefsloot HC, Boelens HF, Van Der Greef J, Smilde AK. Analysis of longitudinal metabolomics data. Bioinformatics. 2004;20(15):2438–46.

Jansen JJ, Szymanska E, Hoefsloot HC, Jacobs DM, Strassburg K, Smilde AK. Between metabolite relationships: an essential aspect of metabolic change. Metabolomics. 2012;8(3):422–32.

Jenkins H, Hardy N, Beckmann M, Draper J, Smith AR, Taylor J, et al. A proposed framework for the description of plant metabolomics experiments and their results. Nat Biotechnol. 2004;22 (12):1601–6.

Jonsson P, Johansson AI, Gullberg J, Trygg J, Grung B. High-throughput data analysis for detecting and identifying differences between samples in GC/MS-based metabolomic analyses. Anal Chem. 2005;77(17):5635–42.

Joshi-Tope G, Gillespie M, Vastrik I, D'Eustachio P, Schmidt E, de Bono B, et al. Reactome: a knowledgebase of biological pathways. Nucleic Acids Res. 2005;33(Database issue):D428–32.

Judson R, Richard A, Dix D, Houck K, Elloumi F, Martin M, et al. ACToR—aggregated computational toxicology resource. Toxicol Appl Pharmacol. 2008;233(1):7–13.

Kanehisa M, Goto S. KEGG: kyoto encyclopedia of genes and genomes. Nucleic Acids Res. 2000;28(1):27–30.

Karp PD, Ouzounis CA, Moore-Kochlacs C, Goldovsky L, Kaipa P, Ahrén D, Tsoka S, Darzentas N, Kunin V, López-Bigas N. Expansion of the BioCyc collection of pathway/genome databases to 160 genomes. Nucleic Acids Res. 2005;33((19)):6083–9. Print 2005.

Karpievitch YV, Hill EG, Smolka AJ, Morris JS, Coombes KR, Baggerly KA, Almeida JS. PrepMS: TOF MS data graphical preprocessing tool. Bioinformatics. 2007;23(2):264–5.

Katajamaa M, Oresic M. Data processing for mass spectrometry-based metabolomics. J Chromatogr A. 2007;1158(1-2):318–28.

Kitano H. Systems biology: toward system-level understanding of biological systems. In: Kitano H, editor. Foundations of systems biology. Cambridge, MA: MIT Press; 2001. p. 1–29.

Kitano H. Computational systems biology. Nature. 2002;420(6912):206–10. Review.

Kopka J, Schauer N, Krueger S, Birkemeyer C, Usadel B, Bergmüller E, Dörmann P, Weckwerth W, Gibon Y, Stitt M, Willmitzer L, Fernie AR, Steinhauser D. GMD@CSB.DB: the golm metabolome database. Bioinformatics. 2005;21(8):1635–8. Epub 2004 Dec 21.

Kraly JR, Holcomb RE, Guan Q, Henry CS. Review: microfluidic applications in metabolomics and metabolic profiling. Anal Chim Acta. 2009;653(1):23–35. doi:10.1016/j.aca.2009.08.037. Epub 2009 Sep 1. Review.

Lehmann EL, Romano JP. Testing statistical hypotheses. 3rd ed. New York: Springer; 2005.

Lim E, Pon A, Djoumbou Y, Knox C, Shrivastava S, Guo AC, Neveu V, Wishart DS. T3DB: a comprehensively annotated database of common toxins and their targets. Nucleic Acids Res. 2010;38(Database issue):D781–6. doi:10.1093/nar/gkp934. Epub 2009 Nov 6.

Lin SM, Zhu L, Winter AQ, Sasinowski M, Kibbe WA. What is mzXML good for? Expert Rev Proteomics. 2005;2(6):839–45.

Listgarten J, Emili A. Statistical and computational methods for comparative proteomic profiling using liquid chromatography-tandem mass spectrometry. Mol Cell Proteomics. 2005;4 (4):419–34.

Lu C, King RD. An investigation into the population abundance distribution of mRNAs, proteins, and metabolites in biological systems. Bioinformatics. 2009;25(16):2020–7.

Markley JL, Anderson ME, Cui Q, Eghbalnia HR, Lewis IA, Hegeman AD, et al. New bioinformatics resources for metabolomics. Pac Symp Biocomput. 2007;157–68.

McCarthy J, Minsky ML, Rochester N, Shannon CE. A proposal for the Dartmouth summer research project on artificial intelligence. 1955 Aug 31 [cited 2015 Apr 07] Available from: http://jmc.stanford.edu/articles/dartmouth/dartmouth.pdf

Mendes P, Camacho D, de la Fuente A. Modelling and simulation for metabolomics data analysis. Biochem Soc Trans. 2005;33(Pt 6):1427–9.

Mesarovic MD. System theory and biology – view of a theoretician. In: Mesarovic MD, editor. Systems theory and biology. New York: Springer; 1968. p. 59–87.

Mettler T, Raptis DA. What constitutes the field of health information systems? Fostering a systematic framework and research agenda. Health Informatics J. 2012;18(2):147–56. doi:10.1177/1460458212452496.

Michaelis L, Menten ML. Die Kinetik der Invertinwirkung. (Translation: The kinetics of invertase activity.). Biochem Z. 1913;49:333–69.

Mishina EV, Straubinger RM, Pyszczynski NA, Jusko WJ. Enhancement of tissue delivery and receptor occupancy of methylprednisolone in rats by a liposomal formulation. Pharm Res. 1993;10(10):1402–10.

Modre-Osprian R, Osprian I, Tilg B, Schreier G, Weinberger KM, Graber A. Dynamic simulations on the mitochondrial fatty acid beta-oxidation network. BMC Syst Biol. 2009;3:2. doi:10.1186/1752-0509-3-2.

Mootha VK, Lindgren CM, Eriksson KF, Subramanian A, Sihag S, Lehar J, et al. PGC-1alpha-responsive genes involved in oxidative phosphorylation are coordinately downregulated in human diabetes. Nat Genet. 2003;34(3):267–73.

Morris JS, Coombes KR, Koomen J, Baggerly KA, Kobayashi R. Feature extraction and quantification for mass spectrometry in biomedical applications using the mean spectrum. Bioinformatics. 2005;21(9):1764–75.

Müller LAJ, Kugler KG, Netzer M, Graber A, Dehmer M. A network-based approach to classify the three domains of life. Biol Direct. 2011;6:53.

Navis GJ, Blankestijn PJ, Deegens J, De Fijter JW, Homan van der Heide JJ, Rabelink T, et al. The biobank of nephrological diseases in the Netherlands cohort: the string of pearls initiative collaboration on chronic kidney disease in the university medical centers in the Netherlands. Nephrol Dial Transplant. 2014;29(6):1145–50.

Nelson DL, Cox MM. Lehninger principles of biochemistry. 5th ed. New York: W. H. Freeman and Company; 2008.

Netzer M, Millonig G, Osl M, Pfeifer B, Praun S, Villinger J, Vogel W, Baumgartner C. A new ensemble-based algorithm for identifying breath gas marker candidates in liver disease using ion molecule reaction mass spectrometry. Bioinformatics. 2009;25(7):941–7.

Netzer M, Weinberger KM, Handler M, Seger M, Fang X, Kugler KG, Graber A, Baumgartner C. Profiling the human response to physical exercise: a computational strategy for the identification and kinetic analysis of metabolic biomarkers. J Clin Bioinforma. 2011;1(1):34. doi:10.1186/2043-9113-1-34.

Netzer M, Kugler KG, Müller LA, Weinberger KM, Graber A, Baumgartner C, Dehmer M. A network-based feature selection approach to identify metabolic signatures in disease. J Theor Biol. 2012;310:216–22. doi:10.1016/j.jtbi.2012.06.003. Epub 2012 Jul 4.

Nielsen J. Metabolic engineering. Appl Microbiol Biotechnol. 2001;55(3):263–83. Review.

Nishino T, Yachie-Kinoshita A, Hirayama A, Soga T, Suematsu M, Tomita M. Dynamic simulation and metabolome analysis of long-term erythrocyte storage in adenine-guanosine solution. PLoS One. 2013;8(8), e71060.

Nussinov R. Advancements and challenges in computational biology. PLoS Comput Biol. 2015;11(1):e1004053. doi:10.1371/journal.pcbi.1004053. eCollection 2015 Jan.

Nussinov R, Bonhoeffer S, Papin JA, Sporns O. From "what is?" to "what Isn't?" computational biology. PLoS Comput Biol. 2015;11(7):e1004318. doi:10.1371/journal.pcbi.1004318. eCollection 2015 Jul.

Osl M, Dreiseitl S, Pfeifer B, Weinberger K, Klocker H, Bartsch G, Schäfer G, Tilg B, Graber A, Baumgartner C. A new rule-based algorithm for identifying metabolic markers in prostate cancer using tandem mass spectrometry. Bioinformatics. 2008;24(24):2908–14.

Palsson B. The challenges of in silico biology. Nat Biotechnol. 2000;18(11):1147–50.

Parsons HM, Ekman DR, Collette TW, Viant MR. Spectral relative standard deviation: a practical benchmark in metabolomics. Analyst. 2009;134(3):478–85.

Patti GJ, Yanes O, Siuzdak G. Innovation: Metabolomics: the apogee of the omics trilogy. Nat Rev Mol Cell Biol. 2012;13(4):263–9. doi:10.1038/nrm3314. Review.

Pedrioli PG, Eng JK, Hubley R, Vogelzang M, Deutsch EW, Raught B, et al. A common open representation of mass spectrometry data and its application to proteomics research. Nat Biotechnol. 2004;22(11):1459–66.

Pence HE, Williams A. ChemSpider: an online chemical information resource. J Chem Educ. 2010;87(11):1123–4.

Phillips KA, Van Bebber S, Issa AM. Diagnostics and biomarker development: priming the pipeline. Nat Rev Drug Discov. 2006;5(6):463–9.

Purohit PV, Rocke DM, Viant MR, Woodruff DL. Discrimination models using variance-stabilizing transformation of metabolomic NMR data. OMICS. 2004;8(2):118–30.

R Development Core Team. R: A language and environment for statistical computing. Vienna: R Foundation for Statistical Computing; 2008.

Ren S, Hinzman AA, Kang EL, Szczesniak RD, Lu LJ. Computational and statistical analysis of metabolomics data. Metabolomics. 2015;11(6):1492–513.

Rios-Estepa R, Lange BM. Experimental and mathematical approaches to modeling plant metabolic networks. Phytochemistry. 2007;68(16):2351–74.

Röschinger W, Olgemöller B, Fingerhut R, Liebl B, Roscher AA. Advances in analytical mass spectrometry to improve screening for inherited metabolic diseases. Eur J Pediatr. 2003;162 Suppl 1:S67–76.

Saeys Y, Inza I, Larrañaga P. A review of feature selection techniques in bioinformatics. Bioinformatics. 2007;23(19):2507–17. Epub 2007 Aug 24.

Sakurai T, Yamada Y, Sawada Y, Matsuda F, Akiyama K, Shinozaki K, Hirai MY, Saito K. PRIMe Update: innovative content for plant metabolomics and integration of gene expression and metabolite accumulation. Plant Cell Physiol. 2013;54(2), e5. doi:10.1093/pcp/pcs184. Epub 2013 Jan 3.

Salek RM, Haug K, Steinbeck C. Dissemination of metabolomics results: role of MetaboLights and COSMOS. Gigascience. 2013;2(1):8. doi:10.1186/2047-217X-2-8.

Salek RM, Neumann S, Schober D, Hummel J, Billiau K, Kopka J, et al. COordination of Standards in MetabOlomicS (COSMOS): facilitating integrated metabolomics data access. Metabolomics. 2015;11(6):1587–97. Epub 2015 May 26.

Samuel AL. Some studies in machine learning using the game of checkers. IBM J Res Dev. 1959;3 (3):210–29.

Sansone SA, Rocca-Serra P, Brandizi M, Brazma A, Field D, Fostel J, et al. The first RSBI (ISA-TAB) workshop: "can a simple format work for complex studies?". OMICS. 2008;12 (2):143–9. doi:10.1089/omi.2008.0019.

Sansone SA, Rocca-Serra P, Field D, Maguire E, Taylor C, Hofmann O, et al. Toward interoperable bioscience data. Nat Genet. 2012;44(2):121–6. doi:10.1038/ng.1054.

Scholz M, Fiehn O. SetupX—a public study design database for metabolomic projects. Pac Symp Biocomput. 2007;169–80.

Sebag M. A tour of machine learning: An AI perspective. AI Commun. 2014;27(1):11–23.

Smilde AK, Jansen JJ, Hoefsloot HC, Lamers RJA, Van Der Greef J, Timmerman ME. ANOVA-simultaneous component analysis (ASCA): a new tool for analyzing designed metabolomics data. Bioinformatics. 2005;21(13):3043–8.

Smilde AK, Westerhuis JA, Hoefsloot HCJ, Bijlsma S, Rubingh CM, Vis DJ, et al. Dynamic metabolomic data analysis: a tutorial review. Metabolomics. 2010;6(1):3–17.

Smith CA, O'Maille G, Want EJ, Qin C, Trauger SA, Brandon TR, Custodio DE, Abagyan R, Siuzdak G. METLIN: a metabolite mass spectral database. Ther Drug Monit. 2005;27 (6):747–51.

Smolen P, Hardin PE, Lo BS, Baxter DA, Byrne JH. Simulation of Drosophila circadian oscillations, mutations, and light responses by a model with VRI, PDP-1, and CLK. Biophys J. 2004;86(5):2786–802.

Solomonoff RJ. An inductive inference machine. IRE Convention Record, Section on Information Theory. 1957;2:56–62.

Solomonoff RJ. A formal theory of inductive inference. Part I. Information and control. 1964;7 (1):1–22.

Sparkes A, King RD, Aubrey W, Benway M, Byrne E, Clare A, et al. An integrated laboratory robotic system for autonomous discovery of gene function. J Assoc Lab Autom. 2010;15 (1):33–40.

Spasić I, Dunn WB, Velarde G, Tseng A, Jenkins H, Hardy N, et al. MeMo: a hybrid SQL/XML approach to metabolomic data management for functional genomics. BMC Bioinformatics. 2006;7:281.

Stacklies W, Redestig H, Scholz M, Walther D, Selbig J. pcaMethods—a bioconductor package providing PCA methods for incomplete data. Bioinformatics. 2007;23(9):1164–7.

Stanberry L, Mias GI, Haynes W, Higdon R, Snyder M, Kolker E. Integrative analysis of longitudinal metabolomics data from a personal multi-omics profile. Metabolites. 2013;3 (3):741–60.

Steinbeck C, Conesa P, Haug K, Mahendraker T, Williams M, Maguire E, et al. MetaboLights: towards a new COSMOS of metabolomics data management. Metabolomics. 2012;8 (5):757–60. Epub 2012 Sep 25.

Stephanopoulos G. Metabolic fluxes and metabolic engineering. Metab Eng. 1999;1(1):1–11. Review.

Steuer R. Review: on the analysis and interpretation of correlations in metabolomic data. Brief Bioinform. 2006;7(2):151–8.

Steuer R, Junker BH. Computational models of metabolism: stability and regulation in metabolic networks. In: Rice SA, editor. Advances in chemical physics, vol. 142. Hoboken: Wiley; 2009.

Subramanian A, Tamayo P, Mootha VK, Mukherjee S, Ebert BL, Gillette MA, et al. Gene set enrichment analysis: a knowledge-based approach for interpreting genome-wide expression profiles. Proc Natl Acad Sci U S A. 2005;102(43):15545–50. Epub 2005 Sep 30.

Sugimoto M, Kawakami M, Robert M, Soga T, Tomita M. Bioinformatics tools for mass spectroscopy-based metabolomic data processing and analysis. Curr Bioinform. 2012;7 (1):96–108.

Suhre K, Gieger C. Genetic variation in metabolic phenotypes: study designs and applications. Nat Rev Genet. 2012;13(11):759–69. doi:10.1038/nrg3314. Epub 2012 Oct 3. Review.

Suhre K, Shin SY, Petersen AK, Mohney RP, Meredith D, Wägele B, et al. Human metabolic individuality in biomedical and pharmaceutical research. Nature. 2011;477(7362):54–60.

Takahashi K, Ishikawa N, Sadamoto Y, Sasamoto H, Ohta S, Shiozawa A, Miyoshi F, Naito Y, Nakayama Y, Tomita M. E-Cell 2: multi-platform E-Cell simulation system. Bioinformatics. 2003;19(13):1727–9.

Tan CS, Ploner A, Quandt A, Lehtiö J, Pawitan Y. Finding regions of significance in SELDI measurements for identifying protein biomarkers. Bioinformatics. 2006;22(12):1515–23.

Tautenhahn R, Cho K, Uritboonthai W, Zhu Z, Patti GJ, Siuzdak G. An accelerated workflow for untargeted metabolomics using the METLIN database. Nat Biotechnol. 2012;30(9):826–8. doi:10.1038/nbt.2348.

Taylor CF, Field D, Sansone SA, Aerts J, Apweiler R, Ashburner M, et al. Promoting coherent minimum reporting guidelines for biological and biomedical investigations: the MIBBI project. Nat Biotechnol. 2008;26(8):889–96. doi:10.1038/nbt.1411.

Tibshirani R, Hastie T, Narasimhan B, Soltys S, Shi G, Koong A, Le QT. Sample classification from protein mass spectrometry, by peak probability contrasts. Bioinformatics. 2004;20 (17):3034–44.

Torgrip RJO, Aberg KM, Alm E, Schuppe-Koistinen I, Lindberg J. A note on normalization of biofluid 1D 1H-NMR data. Metabolomics. 2008;4:114–21.

Trent RJ, editor. Clinical bioinformatics. Totowa: Humana; 2008.

Trietsch SJ, Hankemeier T, Van der Linden HJ. Lab-on-a-chip technologies for massive parallel data generation in the life sciences: A review. Chemom Intell Lab Syst. 2011;108(1):64–75.

Tripathi S, Dehmer M, Emmert-Streib F. NetBioV: an R package for visualizing large network data in biology and medicine. Bioinformatics. 2014;30(19):2834–6. doi:10.1093/bioinformatics/btu384. Epub 2014 Jun 12.

Turing AM. Computing machinery and intelligence. Mind. 1950;49:433–60.

van Ommen GJ, Törnwall O, Bréchot C, Dagher G, Galli J, Hveem K, et al. BBMRI-ERIC as a resource for pharmaceutical and life science industries: the development of biobank-based Expert Centres. Eur J Hum Genet. 2015;23(7):893–900. doi:10.1038/ejhg.2014.235. Epub 2014 Nov 19.

Vivó-Truyols G, Torres-Lapasió JR, van Nederkassel AM, Vander Heyden Y, Massart DL. Automatic program for peak detection and deconvolution of multi-overlapped chromatographic signals part I: peak detection. J Chromatogr A. 2005;1096(1-2):133–45.

Vogeser M, Kirchhoff F. Progress in automation of LC-MS in laboratory medicine. Clin Biochem. 2011;44(1):4–13. doi:10.1016/j.clinbiochem.2010.06.005. Epub 2010 Jun 19. Review.

Voit EO. Computational analysis of biochemical systems – a practical guide for biochemists and molecular biologists. Cambridge: Cambridge University Press; 2000.

Voit E, Martens H, Omholt SW. 150 years of the mass action law. PLoS Comput Biol. 2015;11(1), e1004012.

Wang X, Liotta L. Clinical bioinformatics: a new emerging science. J Clin Bioinforma. 2011;1 (1):1. doi:10.1186/2043-9113-1-1.

Wang W, Zhou H, Lin H, Roy S, Shaler TA, Hill LR, et al. Quantification of proteins and metabolites by mass spectrometry without isotopic labeling or spiked standards. Anal Chem. 2003;75(18):4818–26.

Warrack BM, Hnatyshyn S, Ott KH, Reily MD, Sanders M, Zhang H, Drexler DM. Normalization strategies for metabonomic analysis of urine samples. J Chromatogr B Analyt Technol Biomed Life Sci. 2009;877:547–52.

Weckwerth W, Morgenthal K. Metabolomics: from pattern recognition to biological interpretation. Drug Discov Today. 2005;10(22):1551–8. Review.

Weinberger KM. [Metabolomics in diagnosing metabolic diseases]. Ther Umsch. 2008;65 (9):487–91. doi:10.1024/0040-5930.65.9.487. [Article in German]

Weinberger KM, Graber A. Using comprehensive metabolomics to identify novel biomarkers. Screen Trends Drug Discov. 2005;6:42–5.

Weinberger KM, Ramsay SL, Graber A. Towards the biochemical fingerprint. Biosyst Solut. 2005;12:36–7.

Wiener N. Cybernetics – control and communication in the animal and the machine. New York: Wiley; 1948.

Wishart DS. Current progress in computational metabolomics. Brief Bioinform. 2007;8 (5):279–93. Epub 2007 Jul 11. Review.

Wishart DS. Chapter 3: Small molecules and disease. PLoS Comput Biol. 2012;8(12), e1002805. doi:10.1371/journal.pcbi.1002805. Epub 2012 Dec 27.

Wishart DS, Knox C, Guo AC, Shrivastava S, Hassanali M, Stothard P, Chang Z, Woolsey J. DrugBank: a comprehensive resource for in silico drug discovery and exploration. Nucleic Acids Res. 2006;34(Database issue):D668–72.

Wishart DS, Knox C, Guo AC, Eisner R, Young N, Gautam B, et al. HMDB: a knowledgebase for the human metabolome. Nucleic Acids Res. 2009;37(Database issue):D603–10. doi:10.1093/nar/gkn810. Epub 2008 Oct 25.

Wishart DS, Jewison T, Guo AC, Wilson M, Knox C, Liu Y, et al. HMDB 3.0—the human metabolome database in 2013. Nucleic Acids Res. 2013;41(Database issue):D801–7. doi:10.1093/nar/gks1065. Epub 2012 Nov 17.

Wolkenhauer O. Systems biology: the reincarnation of systems theory applied in biology? Brief Bioinform. 2001;2:258–70.

Xia J, Wishart DS. MSEA: a web-based tool to identify biologically meaningful patterns in quantitative metabolomic data. Nucleic Acids Res. 2010;38(Web Server issue):W71–7.

Xia J, Psychogios N, Young N, Wishart DS. MetaboAnalyst: a web server for metabolomic data analysis and interpretation. Nucleic Acids Res. 2009;37(Web Server issue):W652–60. doi:10.1093/nar/gkp356. Epub 2009 May 8.

Xia D, Zheng H, Liu Z, Li G, Li J, Hong J, Zhao K. MRSD: a web server for metabolic route search and design. Bioinformatics. 2011;27(11):1581–2. doi:10.1093/bioinformatics/btr160. Epub 2011 Mar 30.

Xia J, Mandal R, Sinelnikov IV, Broadhurst D, WishartDS. MetaboAnalyst 2.0–a comprehensive server for metabolomic data analysis. Nucleic Acids Res. 2012;40(Web Server issue): W127–33.

Yuille M, van Ommen GJ, Bréchot C, Cambon-Thomsen A, Dagher G, Landegren U, et al. Biobanking for Europe. Brief Bioinform. 2008;9(1):14–24.

Zhao Q, Stoyanova R, Du S, Sajda P, Brown TR. HiRes—a tool for comprehensive assessment and interpretation of metabolomic data. Bioinformatics. 2006;22(20):2562–4.

Zhu W, Wang X, Ma Y, Rao M, Glimm J, Kovach JS. Detection of cancer-specific markers amid massive mass spectral data. Proc Natl Acad Sci U S A. 2003;100(25):14666–71.

Chapter 9
Metagenomic Profiling, Interaction of Genomics with Meta-genomics

Ruifeng Wang, Yu Zhou, Shaolong Cao, Yuping Wang, Jigang Zhang, and Hong-Wen Deng

Abstract Metagenomics is about the sequencing and characterization of genomic DNA of uncultured microbes sampled directly from their habitats. Next-generation sequencing (NGS) technologies and the ability of sequencing uncultured microbes have dramatically expanded and transformed our knowledge of the microbial world. In this chapter, we provide an introduction and flavor to metagenomic studies from sampling to data analysis. Also, workflow and several common methodologies are summarized for the sequence-driven metagenomic analysis to identify the composition of microbes, compare different microbial communities, characterize the functional potential of microbial communities and infer the microbes, which are involved in the metabolic pathways. Additionally, we describe some well-established platforms and software, and briefly review their utilities. Finally, an explication of interactions between genomics and meta-genomics gives a new view for host phenotype-genotype analysis.

Keywords Metagenomics • 16rRNA sequencing • Whole-genome shotgun sequencing • Analytical tools and databases for metagenomics

---

R. Wang (✉) • J. Zhang
Center for Bioinformatics and Genomics, Department of Biostatistics and Bioinformatics, Tulane University, New Orleans, LA 70112, USA
e-mail: rwang6@tulane.edu

Y. Zhou
Center for Bioinformatics and Genomics, Tulane University, New Orleans, LA, USA

Department of Cell and Moleculer Biology, Tulane University, New Orleans, LA, USA

S. Cao • Y. Wang
Center for Bioinformatics and Genomics, Tulane University, New Orleans, LA, USA

Department of Biomedical Engineering, Tulane University, New Orleans, LA, USA

H.-W. Deng
Center for Bioinformatics and Genomics, Department of Biostatistics and Bioinformatics, Tulane University School of Public Health and Tropical Medicine, New Orleans, LA, USA
e-mail: hdeng2@tulane.edu

X. Wang et al. (eds.), *Application of Clinical Bioinformatics*,
Translational Bioinformatics 11, DOI 10.1007/978-94-017-7543-4_9

Abbreviation

Egg	NOG Evolutionary genealogy of genes: non-supervised ortholog groups
EVA	Entity attribute value
FDR	False discovery rate
FP	Fecal protease
GWAS	Genome-wide association study
HMP	Human Microbiome Project
IBD	Inflammatory bowel disease
IBS	Irritable bowel syndrome
KEGG	Kyoto encyclopedia of genes and genomes
KR	Kantorovich-Rubinstein
MDA	Multiple displacement amplification
MGWAS	Metagenome-wide association study
MiRKAT	Microbiome regression based on kernel association test
MSA	Multiple sequence alignment
NGS	Next-generation sequencing
OTU	Operational taxonomic unit
PCR	Polymerase chain reaction
PCoA	Principal Coordinate Analysis
PERMANOVA	Permutational multivariate analysis of variance
QIIME	Quantitative insights into microbial ecology
RDP	Ribosomal Database Project
T2D	Type 2 diabetes

9.1 Introduction of Metagenomics

The human microbiome includes all microbes that inhabit in or on the human bodies. The human microorganisms exist by interacting with each other and the traditional pure-culture approaches (Widdel 1983) are not sufficient to fully understand the microbial communities. A new technology, metagenomics, is developed and applied to such kind of metagenomes and produces more accurate analysis of taxonomic diversity and functional diversity in a given microbial community.

9.1.1 The Human Microbiome

The human bodies are home to many microorganisms. Microbes are ubiquitous everywhere on the human bodies. A healthy human body carries more microbial cells than human cells (Savage 1977; Berg 1996). These microorganisms are vital to our health. For example, some vaccines derived from microorganisms provide active acquired immunity to a number of devastating diseases (Valdez et al. 2014). Better understanding of the microbiome can help human beings better understand their own body and health conditions (Collins and McKusick 2001; Kaput et al. 2009).

9.1.2 A Pure Culture Is not Enough for Microbiology Studies

Traditional culturing has provided us with a useful platform for building on a profundity and characteristic of modern microbiological knowledge. However, in nature, microorganisms exist by interacting with each other and cultured microorganisms do not represent much of the microbial world (Gupta and Vakhlu 2011). Several microbial interactions among microorganisms have been well recognized, including parasitic (i.e., one organism benefits at the cost of another), mutualistic (i.e., both organisms benefit from each other), and commensal (i.e., one organism benefits to the other one with no cost) (Phelan et al. 2012). The confluence of them shows that a pure laboratory culture only containing a single species of microorganism is not enough to realize a microbial world (Handelsman 2004).

9.1.3 Metagenomics

As a means of overcoming the limitations corresponding to studies using cultivation approach, a new biotechnology metagenomics is developed and growing rapidly. Metagenomics is a discipline that involves sequencing and analyzing genetic material across all microorganisms sampled directly from their inhabitants (Hugenholtz et al. 1998). Although, metagenomics is a relatively new discipline, it has produced a wealth of knowledge and revolutionized our understanding of microbial life and community, particularly the diversity, function, and evolution of the uncultivable majority of microbes (Rondon et al. 2000; Fuhrman 2012). Metagenomics can be used to characterize the composition of microbes both in identification and abundance, identify functions of a microbial community, which give us more valuable information than the analysis of singly isolated microbes (Lal 2011). The typical workflow for metagenomics includes filtering particles, lysis, DNA extraction, cloning and library construction (NGS could skip this step) (Fig. 9.1) (Thomas et al. 2012; Wooley et al. 2010). After that, two main approaches, sequence-driven analysis and function-driven analysis may be applied. (Thomas et al. 2012).

9.2 Study Design and Sampling

The goal of the clinical study design is to assess the efficacy or the mechanism of study actions. Designing a successful metagenomic project would help scientists to find out reliable results. In this chapter, we will introduce some aspects that researchers may consider when creating a study design and conducting sampling.

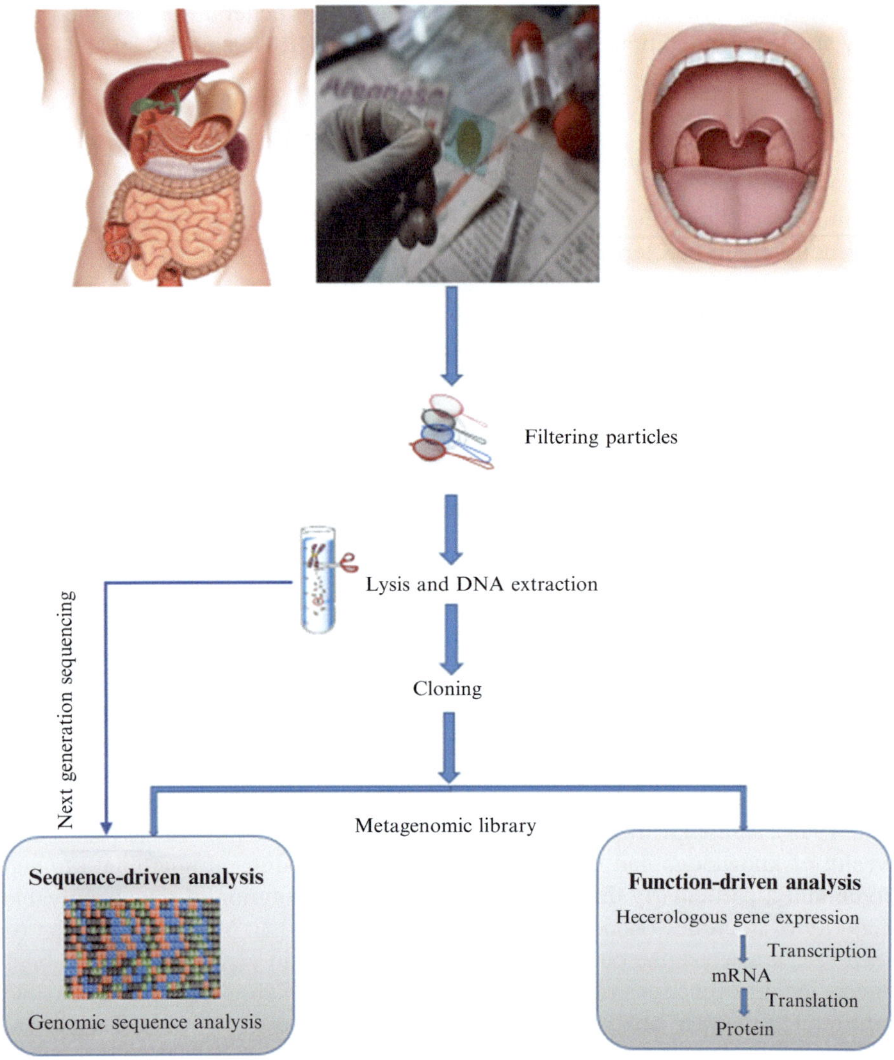

Fig. 9.1 Construction and screening of metagenomic libraries

9.2.1 Study Design

The first thing for metagenomic studies is to identify a specific primary objective. It is necessary to think about some questions, for example, what is the main scientific aim (e.g., identify taxonomic diversity in the community)? How to find the best microbial community sample that is related to a target disease?

Prior to the experiment, a specific pipeline should be developed. Several aspects should be taken into consideration: (1) the main characteristics of the study; (2) specific hypotheses to test; (3) sample size; (4) the potential statistical methods to be used; and (5) software can be used for processing data sets.

Normally, more than one hypothesis could be tested in a metagenomic project, which depend on the extent of the experiment and the generated data collected, that could be used for subsequent analysis. Most of the metagenomic datasets have high-dimensionality, meaning that the number of variables such as genetic features is much larger than the samples analyzed. The type of data collection will directly determine the type of analysis methodologies one may use.

9.2.2 Sampling

Sampling is a crucial step in metagenomic projects. An accurate representation of microbial sample will reduce the bias that can prejudice the evaluation results. Sample processing needs to consider sources of materials, sample physical size, scale, number of species, timing of sampling, storage and handling, and method choice for DNA extraction (Venter et al. 2004; Burke et al. 2009; Delmont et al. 2011). Samples should be representative of the microbiomes in their habitats. Efforts should be made to minimize DNA degradation before sequencing analysis. For example, in the metagenome-wide association study of gut microbiota in type 2 diabetes, the patients who were diagnosed with type 2 diabetes constituted the case group and those non-diabetic people were treated as the control group. For these two groups, their fresh fecal samples were obtained at home and immediately frozen in a home freezer for less than one day. Those frozen samples were transferred to BGI-Shenzhen and then stored at −80 °C until analysis (Qin et al. 2012).

9.2.2.1 Sample Collection for Metagenomes

Metagenome includes all genetic material present in a given sample and it consists of many individual organisms. High-quality genomic DNA extraction and the purification are critical and several robust DNA extraction methodologies are available (Venter et al. 2004; Burke et al. 2009; Delmont et al. 2011). As an example, processing of fecal samples is included in the Appendix.

The sufficient amount of extracted DNA from samples is important for the following experiment. Low biomass samples may not be sufficient for creating a complete metagenomic library (Thomas et al. 2012). If the target sample is associated with a host, both physical fractionation (like specific filtration or centrifugation process) and selective lysis might be two feasible ways to ensure obtaining minimal host DNA (Ballantyne et al. 2007; Arriola et al. 2007; Burke et al. 2009; Thomas et al. 2010). If the starting amount of the target material is small, the corresponding representation of DNAs may be weakened. The PCR can amplify and increase the amount of limited DNA samples. As an option different from the PCR method, a non-PCR based DNA amplification technique named multiple displacement amplification (MDA) uses random hexamers and phagephi29 polymerase to rapidly amplify minute amounts of DNA yields (Lizardi 2001). This method has been actively used for single-cell genome sequencing and a certain extent in metagenomics because it can amplify femtograms of DNA to produce micrograms of product (Zhang et al. 2015; Ishoey et al. 2008). In contrast to the PCR, MDA generates a larger amount of products with a lower mutation rate, minimal locus bias and ease of use (Shoaib et al. 2008).

However, depending on the amount of starting materials and requirement of DNA, no matter which of the amplification method is used, there exist potential problems (e.g., sequence errors, chimera formation, reagent contamination) associated with sequence bias (Lasken and Stockwell 2007). These issues may have a significant effect on metagenomic analysis. In addition, collecting extra sample material for complementary analyses is useful in metagenomic analysis.

9.2.2.2 Sample Metadata Collection

Metadata is broadly defined as the descriptive data that is used for describing sample information. It summarizes additional information to the microbial habitats and sample conditions which can make finding and working with a specific instance of data easier than no descriptive data and enhances the ability to interpret the sequence data, especially for a comparative analysis (DeLong et al. 2006). The metadata in clinical repositories, for instance, entity-attribute-value (EVA) data include study subject name, date of birth, sex, health status, race and other demographic variables in conventional tables. A good metadata makes it clear and easy for others to keep track of the resources. Different sample types may have different metadata. It is very important to identify some useful metadata when collecting the genomic sample. Recently, there exist several metagenomic databases with various degrees of metadata are available (Markowitz et al. 2006; Seshadri et al. 2007).

9.3 Approaches for Metagenomic Analysis

There are two approaches in metagenomic analysis. One is sequence-driven analysis and the other one is function-driven analysis. In this section, we will mainly introduce the sequence-driven analysis. Based on different types of sequencing, sequence-driven analysis can be categorized into two broad categories, including 16S rRNA sequencing analysis and whole-genome shotgun sequencing analysis. Both of these two analyses play an important role in the study of microorganisms and will be introduced here.

9.3.1 Sequence-Driven Analysis

Sequence-driven analysis aims to uncover the diversity of microorganisms, identify novel genes, compare different microbial communities and clarify the relationship between the microbial communities and host. There are two broad categories: 16S rRNA sequencing analysis and whole-genome shotgun sequencing analysis. 16S rRNA sequencing typically only sequence the informative marker (e.g., 16S rRNA gene) and can be used to identify and characterize microorganisms in a given sample. Whole-genome shotgun sequencing screens the entire metagenome that is used to characterize the diversity of microbes and their functions in a given sample (Kunin et al. 2008; Thomas et al. 2012). These two approaches provide some common and different information and offer unique advantages and trade-offs (Manichanh et al. 2008).

Here we give an overview of the sequence-driven analysis of metagenomics that involves significant steps in a typical sequence-based metagenomic project (Fig. 9.2), including sampling, genomic DNA extraction, DNA sequencing

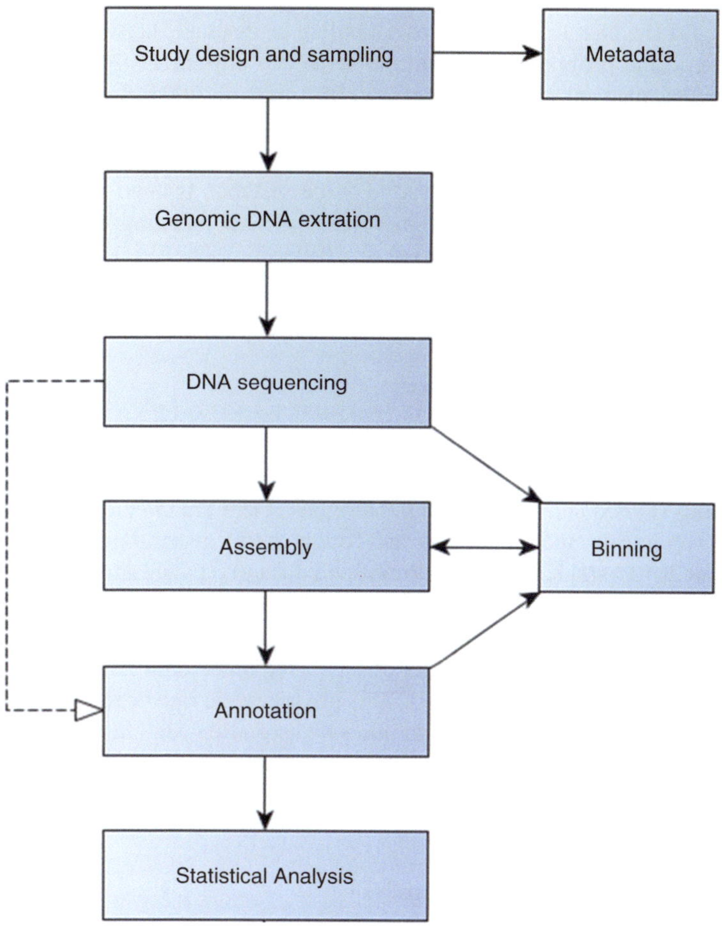

Fig. 9.2 Flow diagram of a typical metagenome projects (*Dashed arrows* indicate steps that can be omitted for 16S rRNA sequencing analysis)

technologies, assembly, binning, functional annotation, statistical analysis (Kunin et al. 2008). Some steps will be illustrated in the following. Information related to bioinformatic tools in metagenomic analysis will be expounded later.

First, we introduce some basic concepts used in sequence-driven metagenomic analysis.

9.3.1.1 Operational Taxonomic Unit (OTU)

In phylogeny, OTU is an operational definition of a species that can be used when DNA sequence data sets have been generated (Blaxter et al. 2005). In metagenomics, OTU is usually defined as a cluster of similar biomarker sequences (e.g., 16S rRNA genes), especially for those species that genic sequences' similarity over 97 % or more (Ellison et al. 2014). Generally, there are two mathematical steps to define OTU. The first step is to calculate a distance between each pair of sequences. Some programs use multiple sequence alignment (MSA) to align the sequence first, then calculate the distance (Blaxter et al. 2005; Schloss et al. 2009). The second step is to cluster the sequences based on the pairwise distance. Several programs use the maximum distance between two groups of sequences in hierarchical clustering while others use the average distance instead of maximum distance. The latter method may give more biologically meaningful OTUs (Blaxter et al. 2005; Schloss et al. 2009; Hao et al. 2011).

9.3.1.2 Distance-Based Analysis

After DNA sequencing, organism phylogenies can be derived from the sequenced data which can be used for creating a phylogenetic tree or taxonomic tree. Phylogenetic distance is used to identify the correlation and the difference between biological communities (Purvis and Hector 2000). Normally, there are three common distances, including UniFrac distance (Lozupone and Knight 2005), generalized UniFrac distance (Chen et al. 2012), and Kantorovich-Rubinstein distance (Evans and Matsen 2012).

Distance-based analysis is a widely used strategy for testing and evaluating the overall association between microbial diversity and the outcome of interest (Zhao et al. 2015). Distance-based analysis of DNA sequence data may incorporate the phylogenetic tree (or taxonomic tree). This phylogenetic tree or related information can be automatically created by available software after you cluster the OTUs.

Second, we will introduce 16S rRNA sequencing analysis and whole genome sequencing analysis respectively.

9.3.1.3 16S rRNA Sequencing Analysis

16S ribosomal gene is regarded as a taxonomically and phylogenetically informative biomarker for three main reasons: (1) almost all bacteria contain this gene;

(2) the function of the 16S rRNA gene will not change over time; (3) this gene (1,500 bp) is large enough for information purposes (Sharpton 2014; Chakravorty et al. 2007). Through analyzing different 16S ribosomal genes, we can identify the composition of the microbial community and discover some novel bacteria.

Normally, a complete 16S rRNA sequencing analysis of metagenomic project contains four steps, including genomic DNA extraction, PCR amplification, nucleotide sequencing and database homology search (Barry et al. 1991; Chen et al. 2004). We discussed sampling and genomic DNA extraction earlier. 16S rRNA genes of bacteria are amplified by using PCR-based primers and are subsequently cloned. These PCR amplicons can be sequenced directly after removal of some 'noise' sequences (e.g., sequencing errors) (Gilles et al. 2011), which will dramatically decrease the number of OTUs (Huse et al. 2010). During PCR amplification, an amount of chimeric sequences (artificial recombinants between two or more parental sequences) that can be generated. These artificial molecules make OTU clustering harder and we need an additional quality control process to remove these chimeras as well (Bradley and Hillis 1997). Several programs can process denoising and chimeras detection that we will review their utilities later. The next step is OTU clustering, which clusters the similar sequences into a taxon (Schloss et al. 2011). The OTU clusters may show us basic information of their microbial characteristics, such as epidemiology and physiology, and indirectly show the inference of their ecological roles (Kim et al. 2013). The OTUs constructed can be used for comparing homology with a public database, such as Ribosomal Database Project (RDP), in order to detect nearly identical sequences (same taxon) and characterize the diversity of microbial communities (Paulson et al. 2013).

There are some limitations and pitfalls when using 16S sequencing method such as

1. Some species within a taxon or even different taxon may have identical or more than 99 % similar 16S rRNA sequences, which will make sequencing-based analysis inaccurate. For example, in family *Enterobacteriaceae*, *E. coli* and *Shigella* have almost identical 16S rRNA and some *E. coli* strains can cause diarrhea similar to that caused by *Shigella* (Fukushima et al. 2002).
2. Some type of strains of sequence data do not have reference sequences in available database. Sometimes that is only a new isolate which does not have good matches in gene database and a false impression of a new taxa may ensue (Han et al. 2003).

Despite these limitations, 16S rRNA sequencing has been determined for an extremely large number of bacterial species and there is no other gene characterized as many and extensively as it (Clarridge 2004). More precise and logical bioinformatic tools are being developed for analyzing the 16S rRNA sequence data that will also help us get more robust results (Woo et al. 2008).

9.3.1.4 Whole-Genome Shotgun Sequencing

Whole-genome shotgun sequencing is increasingly widely used to detect the abundance of microbes, provide insight into community diversity and identify important metabolic pathways. In whole genome shotgun sequencing, all genomic DNA is broken up randomly into numerous small fragments, which are independently sequenced to obtain reads (Sharpton 2014). After these small DNA fragments are sequenced, software are available to assemble different reads into their original order based on overlaps, merge pairs of reads into longer contiguous sequences (i.e., contig), link contigs to form supercontigs and ultimately derive consensus sequence (Adams 2008). Several high-throughput sequencing technologies have been applied to process the metagenomic samples such as the 454/Roche and Illumina/Solex systems and they can sequence thousands of organisms in parallel (Liu et al. 2011b; Mardis 2008).

Here is the common metagenomic analytical strategies based on shotgun sequencing data (Fig. 9.3).

Two primary questions can be characterized by analyzing microbial community are 'who is there' and 'what are they doing'. These questions involve determining which microbes are present in a community and their abundance (e.g., taxonomic diversity), and what are their functions (e.g., protein family diversity) (Sharpton 2014).

9.3.1.5 Taxonomic Diversity Analysis

We will give an overview of taxonomic diversity analysis in metagenomics. Taxonomic diversity analysis, as a way of characterizing microbial community, involves analyzing what microbiomes present and their abundance in a given sample. Through this analysis, we can ascertain the similarity of two or more microbial communities and give an inference to distinguish their biological functions when containing members of functionally described taxa (Kim et al. 2013). In metagenomics, taxonomic diversity can be divided into three categories which include marker gene analysis, binning and assembly (assembling sequences into distinct genomes – genome diversity). These different approaches are not exclusive of each other, rather they may complement each other (Sharpton 2014).

1. Marker gene analysis

A marker gene, as the specific DNA sequence with known location on the chromosome, which can be used to characterize the taxonomic composition and phylogenetic diversity of a given sample. In metagenomics, ribosomal RNA genes (e.g., 16S rRNA) and single-copy protein coding genes are two major kinds of marker genes that have been used (Sharpton 2014). Marker gene analysis includes two steps: the first step is to compare sequencing reads with a database of taxonomical informative gene families database and the second step is to identify those reads

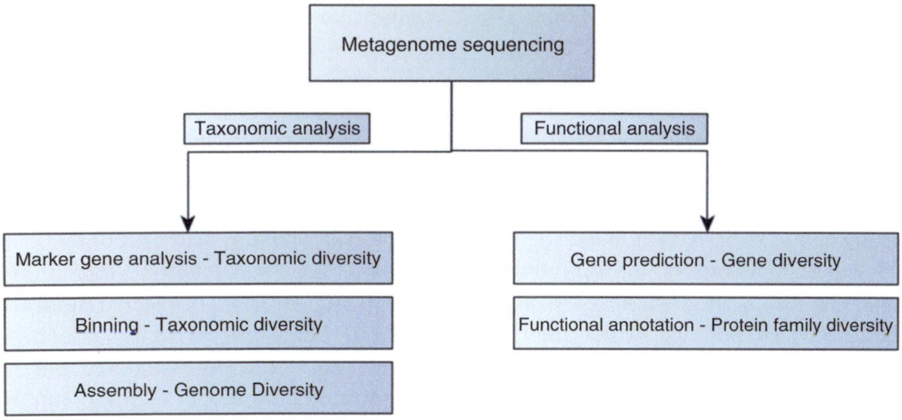

Fig. 9.3 Common shotgun sequencing metagenomic analytical strategies

that are marker gene homologs and taxonomically annotate each homolog (Sharpton 2014; Warnecke et al. 2007; Segata et al. 2012). Marker gene analysis is fast and relatively more accurate to estimate the taxonomic abundance when focusing on single-copy gene families. Additionally, this method can be applied to both assembled or unassembled reads that are more applicable than other methods (e.g., binning analysis) (Liu et al. 2011a).

2. Binning

Binning is a taxonomic classification method which aims to sort DNA sequences into groups that might represent genomes from closely related organisms (Thomas et al. 2012). Generally, there are two rules to classify each sequence into a taxonomic group. The first one, similarity-based algorithm, is to compare the unknown metagenome shotgun reads with a known referential data (e.g., small subunits rRNA for prokaryotes) and judge where the sequence should be classified (e.g., OTU genus). The second one, compositional-based algorithm, is to cluster those sequences that represent taxonomic groups based on conserved nucleotide composition of genomes (e.g., GC content) (Sharpton 2014; Thomas et al. 2012). The compositional-based algorithm generally does not require the alignment of reads to a reference sequence database. However, the similarity-based algorithm needs a search phase, such as BLAST (Wilkening et al. 2009), to identify an cluster those local similarity of a query sequence.

Binning provides a view of the presence of novel genomes; a way of reducing the complexity of sequencing data and a survey of taxa diversity in the community (Alneberg et al. 2014; Strous et al. 2012; Droge and McHardy 2012). These benefits will bring convenience to post-binning analysis.

3. Assembly

Assembly aims to obtain full length protein coding sequences or recover the genome of microorganisms. It merges overlapped metagenomic reads from the same genome into longer contigs. These contigs can dramatically simplify bioinformatic analysis (e.g., genome diversity and find novel genomes) comparing with unassembled sequencing reads, and especially make it easier to obtain accurate information for functional annotations (Wommack et al. 2008). Analysis of these longer sequences may provide insight into the genomic composition of uncultured organisms found in the given sample. Binning and classification of DNA fragments for taxonomic assignment can also receive benefits from them (Iverson et al. 2012; Ruby et al. 2013; Wrighton 2012). The main factors associated with the complexity of sequence assembly are the number of fragments and their lengths. Algorithms that can be applied to massive and long fragments are sophisticated (Boisvert et al. 2010). Chimeras, which may be generated by PCR, will affect the assembly results. Those chimeras, from two distinct genomes, could be assembled into a contig because of sharing similar sequence. The more complexity of the communities, the more chimeras could be generated (Luo et al. 2012).

9.3.1.6 Functional Diversity Analysis

To assess the functional diversity in a microbiome community, shotgun metagenomic reads or contigs are mapped to a known database of orthologous gene groups, i.e., KEGG (Kanehisa et al. 2012) to identify matches. By clarifying the common functions that encoded in the microorganism genomes, metagenomes provide insight into a community's physiology. This can be quantified through annotating metagenomic sequences (Lewis et al. 2012; Rup 2012). The functional diversity analysis usually contains two parts, including gene prediction and functional annotation (Looft et al. 2012; Morgan et al. 2012).

1. Gene prediction

Gene prediction is used to label sequences as genes or genomic elements. It is the fundamental step for annotation. Once coding sequences are identified, they can be functionally annotated. There are three major ways by which genes are predicted in metagenomics including gene fragment recruitment, protein family classification, and *de novo* gene prediction (Wu et al. 2009; Sharpton 2014). The gene fragment recruitment approach aligns metagenomic contigs or reads to the known gene sequence from database. If the contigs or reads are identical or almost identical to parts or full-length of a gene sequence, they could be considered to represent this gene (Qin et al. 2010). The principle of protein family classification method is very similar with gene fragment recruitment approach. The difference is that protein family classification method translates the metagenomic contigs or reads into possible peptide sequences and compares them with the known protein sequences from databases. The *de novo* gene prediction is used to identify novel genes without reference gene sequences. In this method, we can assess if the metagenomic contigs or reads belong to or contain a gene by analyzing them based on the characteristics of microbial genes, i.e., length, codon usage (Noguchi et al. 2006; Kelley et al. 2012).

2. Functional annotation

The most common method for sequence functional annotation is to classify the predicted metagenomic proteins into protein families. They are usually characterized by comparing full-length protein sequences with a genome sequencing program, such as the NHGRI genome sequencing program (http://www.genome.gov/10001691). If a metagenomic sequence is determined to be a homolog of one protein family, then it is inferred that the sequence encodes the family's function (Finn et al. 2014). There are many databases that can be used to functionally annotate metagenomic proteins and we will introduce them later.

As the NGS technique becomes less expensive, more researchers or bio-medical companies will adopt whole-genome shotgun sequencing instead of 16S rRNA sequencing. Characterizing microbial community diversity and functions through whole-genome shotgun sequencing is more precise than 16S rRNA sequencing.

9.3.2 Function-Driven Analysis

Function-driven analysis is invented to screen the metagenomic library followed by biochemical characterization and experimental methods. This could identify the genes expression of a desired or novel trait (Schloss and Handelsman 2003; Singh et al. 2008). This approach is mainly based on bench work with little bioinformatics involvement. Hence, we do not elaborate it here.

9.4 Analytical Tools and Databases for Metagenomics

Due to substantial cost reduction and massive data production by NGS, metagenomics studies increase rapidly in sheer amount and complexity. Meanwhile, metagenome databases and bioinformatic tools, which are used for handling and processing those datasets, become more and more crucial. Several recent associated bioinformatic analytical tools and databases that are widely used in metagenomics will be briefly reviewed here.

9.4.1 Analytical Tools for 16S rRNA Sequencing Data

16S rRNA genes are always treated as a taxonomic marker. This marker gene will be clustered and applied to OTU-based approaches, such as taxonomic analysis and phylogenetic analysis. These OTU-based approaches are common in many microbial community studies (Smith et al. 2015; Bik et al. 2010). Roche 454 Titanium, Ion Torrent PGM and Illumina MiSeq are three major NGS platforms to generate

Table 9.1 Bio-informatics resources for studying targeted metagenomics

Resources	Function	Website
PyroNoise	Denoising	http://code.google.com/p/ampliconnosie
DADA	Denoising	http://sites.google.com.site/dadadenoiser
Denoiser	Denoising	http://qiime.org
ChimeraSlayer	Cimera detection	http://microbiomeutil.sourceforge.net
DECIPHER	Chimera detection	http://decipher.cee.wisr.edu
UCHIME	Chimera detection	http://www.drive5.com/uchime
UCLUST	OUT clustering	http://www.drive5.com/usearch
TBC	OUT clustering	http://sw.ezbiocloud.net
CD-HIT-OTU	OUT clustering	http://weizhing-lab.ucsd.edu/cd-hit/otu
Mothur	All in one	http://mother.org
QIIME	All in one	http://qiime.org

metagenomic sequencing data. In this section, we will introduce several bioinformatic tools and their utilities in the analysis workflow (Tamaki et al. 2011; Seo et al. 2015; King et al. 2014) (Table 9.1).

The platforms that generate sequencing data may produce characteristic sequencing errors, especially imprecise signals for longer homopolymers runs. The overlapping pairwise alignments will be combined into a multiple sequence alignment and these alignments are often inaccurate near homopolymers (Hoberman et al. 2009). In addition, 16S rRNA gene sequencing requires the enrichment by PCR, which also can lead to artifacts and biases in coverage and allele representation (Acinas et al. 2005). If sequencing data contain enough errors, the data could be classified as additional rare OTU (Huse et al. 2010). Hence, a data pre-processing is necessary.

1. Denoising

Denoising removes 'noise' sequences from actual sequences. PyroNoise involves removal of noise from sequencing itself and PCR error points that produced by Roche 454 Titanium platform (Quince et al. 2009). DADA and Denoiser are two well-developed denoising resources, which use sequence abundance information in the denoising process (Erten et al. 2011; Reeder and Knight 2010).

2. Chimera detection

After denoising and additional quality control processes, e.g. remove low-quality reads and contaminating reads, artificial sequences should be removed from the dataset. It is very important to clean up those chimeric sequences because it is hard to differentiate the original sequence from combinants that will result in overestimating the microbial diversity if without removing artificial molecules. Several methods for chimera detection have already developed such as ChimeraSlayer (Haas et al. 2011), Decipher (Wright et al. 2012) and UCHIME (Edgar et al. 2011).

3. OTU clustering

The next step is OTU clustering. This method will cluster sequences with the closest matches into the same OTU as a taxa. Clustering algorithms, sequencing errors and artificial sequences have great influence on the quality of OTUs. Generally, there are two major clustering algorithms, one is alignment-based clustering and the other one is alignment-free clustering. Here we only list software platforms based on alignment-free clustering algorithms, including UCLUST (Edgar 2010), CD-HIT-OTU (Fu et al. 2012; Huang et al. 2010), TBC (Lee et al. 2012).

Several single software platforms that implement all the above three steps' algorithms. Two advanced all-in-one computational tools, Mothur (Yang et al. 2014) and QIIME (Kuczynski et al. 2012b; Navas-Molina et al. 2013), are more flexible and easily maintaining and also are able to address sophisticated targeted metagenomics studies.

9.4.1.1 Mothur

Mothur was initially developed by Dr. Patrick Schloss (http://www.mothur.org). It builds upon several previous tools including SONS (Schloss and Handelsman 2006), DOTUR (Schloss and Handelsman 2005), ARB (Ludwig et al. 2004) and UniFrac (Lozupone et al. 2006; Lozupone and Knight 2005) to provide a flexible software package, which is widely used for analyzing 16S rRNA gene sequences (Schloss et al. 2009). Mothur can be used to process data generated by the IonTorrent, the 454/Roche, and the Illumine (MiSeq/HiSeq). This software is free and available download from the project website (http://www.mothur.org).

9.4.1.2 QIIME

QIIME is an open-source software package, which is used to perform microbial community analysis of raw DNA sequencing data from sequencing technologies such as Illumina, 454/RocheSanger (Kuczynski et al. 2012a). There are several excellent functions that QIIME equipped (1) perform library de-multiplexing and quality filtering; (2) QIIME Denoiser; (3) OTU picking; (4) taxonomic assignment; (5) phylogenetic reconstruction (6) diversity analyses and visualizations (http://qiime.org/1.4.0/). It is available for download on its official website (http://qiime.org/).

9.4.2 Analytical Tools for Shotgun Sequencing Data

Assembly: There are several metagenome specialized assemblers such as Genovo (Laserson et al. 2011), Meta-IDBA (Peng et al. 2011), MetaVelvet (Namiki et al. 2012; Afiahayati and Sakakibara 2015), MAP (Lai et al. 2012) and Ray

Meta (Boisvert et al. 2012). These assemblers are designed to assemble single and clonal genomes. MetaVelet, Meta-IDBA and Ray Meta were developed to perform well on short reads (e.g., average 75~150 bp of Illumina sequencing) while Genovo and MAP were better for longer reads (e.g., average reads length between 600~800 bp of 454 sequencing).

Binning: Several popular tools have been developed based on binning algorithms as we mentioned before. The compositional-based binning algorithms include MEGAN (Huson et al. 2007; Huson and Weber 2013), MG-RAST (Glass et al. 2010a), and CARMA (Krause et al. 2008). The similarity-based binning algorithms include PCAHIER (Zheng and Wu 2010), PhyloPythiaS (Patil et al. 2012) and Phymm (Brady and Salzberg 2009). RITA (MacDonald et al. 2012). Some software combine both compositional-based algorithm and similarity-based algorithm such as MetaCluster (Leung et al. 2011).

Functional annotation: These are several online metagenome annotation services, such as MetaGene (Noguchi et al. 2006), MetaGeneAnnotator (Noguchi et al. 2008), METAREP (Goll et al. 2010), CAMERA (Seshadri et al. 2007), and MG-RAST (Meyer et al. 2008; Glass et al. 2010b), providing platforms for gene prediction, assignment of functional categories, protein families and gene ontologies, and inference of both protein interactions and metabolic pathways.

9.4.3 Databases

Taxonomic diversity analysis databases:

In order to identify the taxonomic groups in the microbial communities, we need to compare all the reads against a curated ribosomal RNA sequence database, such as RDP (Cole et al. 2009), SILVA (Quast et al. 2013), Greengenes (DeSantis et al. 2006), Ribosomal Differentiation of Medical Microorganisms (RIDOM). The database matches can be used to analyze the relative abundance of organisms in the community.

Functional annotation analysis databases:

Metagenomic annotation relies on classifying sequences to some known functions, which is based on comparing homology searches with available reference databases, such reference databases as COGs (Tatusov et al. 2003), eggNOGs (Powell et al. 2012), Pfam (Finn et al. 2014) and TIGRfam (Haft et al. 2013).

9.5 Clinical Example

Type 2 diabetes (T2D) is a prevalent endocrine disease that the body cannot use insulin properly. Several studies have shown that gut microbiota has a significant impact on this disease risk. In order to analyze gut microbial content in T2D patients, Qin et al. (2012) conducted a two-stage analysis of metagenome-wide association study (MGWAS) to identify metagenomic markers associated with T2D. In that study, stool samples from a total of 345 Chinese T2D patients and non-diabetic individuals were collected. DNA sequence data were generated through deep shotgun sequencing by Illumina GAIIx and HiSeq 2000.

To the study, first, developed a comprehensive metagenome reference gene set. The authors carried out whole-genome sequencing for 145 Chinese individuals (71 cases and 74 controls), and then performed *de novo* (Cao et al. 2015) assembly and metagenomic gene prediction for these samples. By integrating these data with the MetaHIT gene catalogue, a total of 1,090,889 genes were uniquely assembled from the Chinese samples that contributed ten more percent additional coverage of sequencing reads compared with MetaHIT gene catalogue. After that, taxonomic assignment and functional annotation were applied to the gene catalogue as a complete gene reference.

To identify T2D-associated metagenomic markers, the authors proposed a two-stage MGWAS strategy. In stage I, a sequence-based profiling method was used to quantify the gut microbiota in the 145 samples. On average, with the 90 % identity threshold, 77.40 % unique paired-end reads were mapped to the updated gene catalogue. However, sequence-based profiling method could reliably detect very low abundance genes. Hence, the author defined and prepared three types of profiles using the quantified gene results and applied PCA on these profiles. The results showed that several abundant genera including *Bacteroides, Prevotella, Bifidobacterium and Ruminococcus* composed three enterotypes but no significant relationship between enterotpes and T2D disease status. After examining several principal components, the study found that the first, the second and the fifth components were significantly correlated with T2D (p-value < 0.001). This result indicates that T2D was a determining factor in explaining gut microbial differences. To correct for population stratifications of metagenome-wide data, they employed a modified version of the EIGENSTRAT method (Price et al. 2006), which allow the use of covariance matrices instead of genotypes. The only difference between MGWAS and regular GWAS subpopulation correction is that they use microbial abundance rather than genotype. In addition, they modified the method further by replacing each PC axis with the residuals of this PC axis from a regression to T2D state. The number of PC axes of EIGENSTAT was determined by Tracy-Widom test at significance threshold of $P < 0.05$. With adjustment, the effects that correlated with non-T2D related factors disappeared. A Wilcoxon rank-sum test was performed on the adjusted gene profile to identify differential metagenomic gene between the T2D patients and controls. Substantial enrichment of a set of microbial genes had very small P values, indicating that these genes were highly likely to be T2D-associated gut microbial genes. To validate the significant associations identified in stage I, the author conducted the stage II analysis using additional 200 Chinese samples. By using whole genome sequencing and controlled false discovery rate (FDR) methods, it defined a total of 52,484 T2D associated genetic markers with 2.5 % FDR (stage II p-value < 0.01). In summary, the genes selected by both stage I and stage II tests are considered as T2D-associated gut microbial genes. The author also implemented the same two-stage analysis using the KEGG orthologous and eggNOG orthologous group profiles and identified a total of 1,345 KEGG orthologous markers (stage II $p\text{-value} < 0.05$ and 4.5 %FDR) and 5,612 eggNOG orthologous group markers (stage II $p\text{-value} < 0.05$ and 6.6 % FDR) that were associated with T2D.

In addition, the authors implemented PERMANOVA method to show that T2D was a significant factor for explaining the variation in the examined microbial samples.

9.6 Interaction of Genomics and Metagenomics

Several studies show that the composition of microbial communities change widely across different human individuals and different body sites because of life style, diet, antibiotic usage and other factors. In addition, the difference of structure and abundance of the microbiome are associated with multiple diseases. For example, a study of lean and obese germ-free mice trail showed that shifts in gut microbiome can influence host traits and cause obesity (Turnbaugh et al. 2008). Other studies show the similar results that gut microbiome has been proposed to contribute to a number of diseases, such as obesity and diabetes (Harley and Karp 2012; Burcelin et al. 2011). Recently, studies have indicated that specific gene variants of host can affect their composition of microbial communities and progress to increase the risk of developing many of the diseases. For example, MEFV gene, IBD-risk loci, has a strong association with gut microbiome composition (Khachatryan et al. 2008; Li et al. 2012). Understanding the impact of genomics difference is crucial to explaining the role of the microbial communities in disease.

Ran Blekhman et al. conducted a comprehensively study to profile the interactions between human genetic variation and the microbial communities' composition (Blekhman et al. 2015). The authors collected and analyzed 93 individuals' DNA reads from Human Microbiome Project (HMP) and their bacterial abundance data. They identified that there are significant associations between host genetic variation and microbiome composition in 10 of the 15 body sites. For example, they calculated the correlation between the host genetic variation principal component and alpha diversity in anterior nares was significant ($R^2 = 0.218$ with $p\text{-value} < 0.01$). In addition, they applied a mixed model to analyze this data which controlled population structure and other non-genetic factors that may cause correlations. Several genes of host were found that are correlated with microbiome composition. After that, they examined the correlations between genetic loci and microbiome composition that had been found to be associated with complex disease (e.g. IBD and other obesity-related disorders). These results highlighted the interaction between host genetic variation and their microbiome composition.

Another study, Daniel N. Frank et al. conducted a comprehensively study to determine whether human genetic variations underlie shifts in microbial populations (Frank et al. 2011). They focused on NOD2 and ATG16L1 risk alleles which are associated with abnormal Paneth cell function that can affect host ileum bacteria. A multivariate analyses was applied to analyze the effect of NOD2 and ATG16L1 risk alleles on the intestinal microbiota and confirmed that two taxa, *Clostridial XIVa* and *Proteobacteria* microbiota, have been affected differently, the frequencies of former taxa increased and later taxa decreased. After multiple comparisons, it showed that shifts in the relative frequencies of taxa have associated with NOD2 and ATG16LI genotype ($p < 0.024$ and $p < 0.011$ respectively). This study suggested that specific genetic loci which cause human disease can affect the microbiota composition.

This evidence shows that human genes can influence the microbiome's composition which contributes in many ways to shape the individual's phenotype. With increasing awareness of the impact of the microbes on different body sites, it might be a more accurate way to take into consideration of microbial affections when conducting host phenotype-genotype analysis.

Appendix

DNA isolation of fecal sample

- Collection and preparation of fecal samples

Fecal samples should be collected and processed timely, ideally within 4 h. After adding equal volume of sterile milli-q water (1:1 feces/water), fecal samples need to be homogenized thoroughly. Then about 1–10 ml slurries will be transferred to cryogenic tubes and frozen at −80 °C until DNA extraction.

- DNA extraction and purification

There are two common methods for fecal DNA purification. One is from the Metagenomics of the Human Intestinal Tract (MetaHIT) project in Europe and the other one is from the National Institutes of Health's Human Microbiome Project (HMP). The MetaHIT protocol mainly uses laboratory-made buffers and solutions, while HMP protocol is based on Mobio PowerLyzer™ PowerSoil® DNA isolation Kit (MO BIO Laboratories) (Wesolowska-Andersen et al. 2014). The MetaHIT method yields significantly higher DNA amount than HMP approach; however, yield and purity of DNA extracted with both protocols were sufficient for Illumina-based deep metagenome sequencing (Wesolowska-Andersen et al. 2014).

References

Acinas SG, Sarma-Rupavtarm R, Klepac-Ceraj V, Polz MF. PCR-induced sequence artifacts and bias: insights from comparison of two 16S rRNA clone libraries constructed from the same sample. Appl Environ Microbiol. 2005;71(12):8966–9. doi:10.1128/AEM.71.12.8966-8969.2005.

Adams J. Complex genomes: shotgun sequencing. Nat Educ. 2008;1(1):186.

Afiahayati SK, Sakakibara Y. MetaVelvet-SL: an extension of the Velvet assembler to a de novo metagenomic assembler utilizing supervised learning. DNA Res. 2015;22(1):69–77. doi:10.1093/dnares/dsu041.

Alneberg J, Bjarnason BS, de Bruijn I, Schirmer M, Quick J, Ijaz UZ, Lahti L, Loman NJ, Andersson AF, Quince C. Binning metagenomic contigs by coverage and composition. Nat Methods. 2014;11(11):1144–6. doi:10.1038/nmeth.3103.

Arriola E, Lambros MB, Jones C, Dexter T, Mackay A, Tan DS, Tamber N, Fenwick K, Ashworth A, Dowsett M, Reis-Filho JS. Evaluation of Phi29-based whole-genome amplification for microarray-based comparative genomic hybridisation. Lab Invest. 2007;87(1):75–83. doi:10.1038/labinvest.3700495.

Ballantyne KN, van Oorschot RA, Muharam I, van Daal A, John Mitchell R. Decreasing amplification bias associated with multiple displacement amplification and short tandem repeat genotyping. Anal Biochem. 2007;368(2):222–9. doi:10.1016/j.ab.2007.05.017.

Barry T, Glennon CM, Dunican LK, Gannon F. The 16s/23s ribosomal spacer region as a target for DNA probes to identify eubacteria. PCR Methods Appl. 1991;1(2):149.

Berg RD. The indigenous gastrointestinal microflora. Trends Microbiol. 1996;4(11):430–5.

Bik EM, Long CD, Armitage GC, Loomer P, Emerson J, Mongodin EF, Nelson KE, Gill SR, Fraser-Liggett CM, Relman DA. Bacterial diversity in the oral cavity of 10 healthy individuals. ISME J. 2010;4(8):962–74. doi:10.1038/ismej.2010.30.

Blaxter M, Mann J, Chapman T, Thomas F, Whitton C, Floyd R, Abebe E. Defining operational taxonomic units using DNA barcode data. Philos Trans R Soc Lond B Biol Sci. 2005;360 (1462):1935–43. doi:10.1098/rstb.2005.1725.

Blekhman R, Goodrich JK, Huang K, Sun Q, Bukowski R, Bell JT, Spector TD, Keinan A, Ley RE, Gevers D, Clark AG. Host genetic variation impacts microbiome composition across human body sites. Genome Biol. 2015;16:191. doi:10.1186/S13059-015-0759-1.

Boisvert S, Laviolette F, Corbeil J. Ray: simultaneous assembly of reads from a mix of high-throughput sequencing technologies. J Comput Biol: J Comput Mol Cell Biol. 2010;17 (11):1519–33. doi:10.1089/cmb.2009.0238.

Boisvert S, Raymond F, Godzaridis E, Laviolette F, Corbeil J. Ray meta: scalable de novo metagenome assembly and profiling. Genome Biol. 2012;13(12):R122. doi:10.1186/gb-2012-13-12-r122.

Bradley RD, Hillis DM. Recombinant DNA sequences generated by PCR amplification. Mol Biol Evol. 1997;14(5):592–3.

Brady A, Salzberg SL. Phymm and PhymmBL: metagenomic phylogenetic classification with interpolated Markov models. Nat Methods. 2009;6(9):673–6. doi:10.1038/nmeth.1358.

Burcelin R, Serino M, Chabo C, Blasco-Baque V, Amar J. Gut microbiota and diabetes: from pathogenesis to therapeutic perspective. Acta Diabetol. 2011;48(4):257–73. doi:10.1007/s00592-011-0333-6.

Burke C, Kjelleberg S, Thomas T. Selective extraction of bacterial DNA from the surfaces of macroalgae. Appl Environ Microbiol. 2009;75(1):252–6. doi:10.1128/AEM.01630-08.

Cao HZ, Wu HL, Luo RB, Huang SJ, Sun YH, Tong X, Xie YL, Liu BH, Yang HL, Zheng HC, Li J, Li B, Wang Y, Yang F, Sun P, Liu SY, Gao P, Huang HD, Sun J, Chen D, He GZ, Huang WH, Huang Z, Li Y, Tellier LCAM, Liu X, Feng Q, Xu X, Zhang XQ, Bolund L, Krogh A, Kristiansen K, Drmanac R, Drmanac S, Nielsen R, Li SG, Wang J, Yang HM, Li YR, Wong GKS, Wang J. De novo assembly of a haplotype-resolved human genome. Nat Biotechnol. 2015;33(6):617. doi:10.1038/nbt.3200.

Chakravorty S, Helb D, Burday M, Connell N, Alland D. A detailed analysis of 16S ribosomal RNA gene segments for the diagnosis of pathogenic bacteria. J Microbiol Methods. 2007;69 (2):330–9. doi:10.1016/j.mimet.2007.02.005.

Chen CC, Teng LJ, Chang TC. Identification of clinically relevant viridans group streptococci by sequence analysis of the 16S-23S ribosomal DNA spacer region. J Clin Microbiol. 2004;42 (6):2651–7. doi:10.1128/JCM.42.6.2651-2657.2004.

Chen J, Bittinger K, Charlson ES, Hoffmann C, Lewis J, Wu GD, Collman RG, Bushman FD, Li H. Associating microbiome composition with environmental covariates using generalized UniFrac distances. Bioinformatics. 2012;28(16):2106–13. doi:10.1093/bioinformatics/bts342.

Clarridge JE, 3rd. Impact of 16S rRNA gene sequence analysis for identification of bacteria on clinical microbiology and infectious diseases. Clin Microbiol Rev. 2004;17(4):840–62, table of contents. doi:10.1128/CMR.17.4.840-862.2004.

Cole JR, Wang Q, Cardenas E, Fish J, Chai B, Farris RJ, Kulam-Syed-Mohideen AS, McGarrell DM, Marsh T, Garrity GM, Tiedje JM. The Ribosomal Database Project: improved alignments and new tools for rRNA analysis. Nucleic Acids Res. 2009;37(Database issue):D141–5. doi:10.1093/nar/gkn879.

Collins FS, McKusick VA. Implications of the Human Genome Project for medical science. JAMA. 2001;285(5):540–4.

Delmont TO, Robe P, Clark I, Simonet P, Vogel TM. Metagenomic comparison of direct and indirect soil DNA extraction approaches. J Microbiol Methods. 2011;86(3):397–400. doi:10.1016/j.mimet.2011.06.013.

DeLong EF, Preston CM, Mincer T, Rich V, Hallam SJ, Frigaard NU, Martinez A, Sullivan MB, Edwards R, Brito BR, Chisholm SW, Karl DM. Community genomics among stratified microbial assemblages in the ocean's interior. Science. 2006;311(5760):496–503. doi:10.1126/science.1120250.

DeSantis TZ, Hugenholtz P, Larsen N, Rojas M, Brodie EL, Keller K, Huber T, Dalevi D, Hu P, Andersen GL. Greengenes, a chimera-checked 16S rRNA gene database and workbench compatible with ARB. Appl Environ Microbiol. 2006;72(7):5069–72. doi:10.1128/AEM.03006-05.

Droge J, McHardy AC. Taxonomic binning of metagenome samples generated by next-generation sequencing technologies. Brief Bioinform. 2012;13(6):646–55. doi:10.1093/bib/bbs031.

Edgar RC. Search and clustering orders of magnitude faster than BLAST. Bioinformatics. 2010;26(19):2460–1. doi:10.1093/bioinformatics/btq461.

Edgar RC, Haas BJ, Clemente JC, Quince C, Knight R. UCHIME improves sensitivity and speed of chimera detection. Bioinformatics. 2011;27(16):2194–200. doi:10.1093/bioinformatics/btr381.

Ellison MJ, Conant GC, Cockrum RR, Austin KJ, Truong H, Becchi M, Lamberson WR, Cammack KM. Diet alters both the structure and taxonomy of the ovine gut microbial ecosystem. DNA Res. 2014;21(2):115–25. doi:10.1093/dnares/dst044.

Erten S, Bebek G, Ewing RM, Koyuturk M. DADA: degree-aware algorithms for network-based disease gene prioritization. BioData Min. 2011;4:19. doi:10.1186/1756-0381-4-19.

Evans SN, Matsen FA. The phylogenetic Kantorovich-Rubinstein metric for environmental sequence samples. J R Stat Soc Series B Stat Methodol. 2012;74(3):569–92.

Finn RD, Bateman A, Clements J, Coggill P, Eberhardt RY, Eddy SR, Heger A, Hetherington K, Holm L, Mistry J, Sonnhammer EL, Tate J, Punta M. Pfam: the protein families database. Nucleic Acids Res. 2014;42(Database issue):D222–30. doi:10.1093/nar/gkt1223.

Frank DN, Robertson CE, Hamm CM, Kpadeh Z, Zhang TY, Chen HY, Zhu W, Sartor RB, Boedeker EC, Harpaz N, Pace NR, Li E. Disease phenotype and genotype are associated with shifts in intestinal-associated microbiota in inflammatory bowel diseases. Inflamm Bowel Dis. 2011;17(1):179–84. doi:10.1002/ibd.21339.

Fu L, Niu B, Zhu Z, Wu S, Li W. CD-HIT: accelerated for clustering the next-generation sequencing data. Bioinformatics. 2012;28(23):3150–2. doi:10.1093/bioinformatics/bts565.

Fuhrman JA. Metagenomics and its connection to microbial community organization. F1000 Biol Rep. 2012;4:15. doi:10.3410/B4-15.

Fukushima M, Kakinuma K, Kawaguchi R. Phylogenetic analysis of Salmonella, Shigella, and Escherichia coli strains on the basis of the gyrB gene sequence. J Clin Microbiol. 2002;40(8):2779–85.

Gilles A, Meglecz E, Pech N, Ferreira S, Malausa T, Martin JF. Accuracy and quality assessment of 454 GS-FLX Titanium pyrosequencing. BMC Genomics. 2011;12:245. doi:10.1186/1471-2164-12-245.

Glass EM, Wilkening J, Wilke A, Antonopoulos D, Meyer F. Using the metagenomics RAST server (MG-RAST) for analyzing shotgun metagenomes. Cold Spring Harb Protoc. 2010;(1): pdb. prot5368.

Glass EM, Wilkening J, Wilke A, Antonopoulos D, Meyer F. Using the metagenomics RAST server (MG-RAST) for analyzing shotgun metagenomes. Cold Spring Harbor Protoc. 2010 (1): pdb prot5368. doi:10.1101/pdb.prot5368.

Goll J, Rusch DB, Tanenbaum DM, Thiagarajan M, Li K, Methe BA, Yooseph S. METAREP: JCVI metagenomics reports—an open source tool for high-performance comparative metagenomics. Bioinformatics. 2010;26(20):2631–2. doi:10.1093/bioinformatics/btq455.

Gupta P, Vakhlu J. Metagenomics: a quantum jump from bacterial genomics. Indian J Microbiol. 2011;51(4):539–41. doi:10.1007/s12088-011-0231-1.

Haas BJ, Gevers D, Earl AM, Feldgarden M, Ward DV, Giannoukos G, Ciulla D, Tabbaa D, Highlander SK, Sodergren E, Methe B, DeSantis TZ, Human Microbiome C, Petrosino JF, Knight R, Birren BW. Chimeric 16S rRNA sequence formation and detection in Sanger and 454-pyrosequenced PCR amplicons. Genome Res. 2011;21(3):494–504. doi:10.1101/gr.112730.110.

Haft DH, Selengut JD, Richter RA, Harkins D, Basu MK, Beck E. TIGRFAMs and genome properties in 2013. Nucleic Acids Res. 2013;41(Database issue):D387–95. doi:10.1093/nar/gks1234.

Han XY, Pham AS, Tarrand JJ, Rolston KV, Helsel LO, Levett PN. Bacteriologic characterization of 36 strains of Roseomonas species and proposal of Roseomonas mucosa sp nov and Roseomonas gilardii subsp rosea subsp nov. Am J Clin Pathol. 2003;120(2):256–64. doi:10.1309/731V-VGVC-KK35-1Y4J.

Handelsman J. Metagenomics: application of genomics to uncultured microorganisms. Microbiol Mol Biol Rev: MMBR. 2004;68(4):669–85. doi:10.1128/MMBR.68.4.669-685.2004.

Hao X, Jiang R, Chen T. Clustering 16S rRNA for OTU prediction: a method of unsupervised Bayesian clustering. Bioinformatics. 2011;27(5):611–8. doi:10.1093/bioinformatics/btq725.

Harley IT, Karp CL. Obesity and the gut microbiome: striving for causality. Mol Metab. 2012;1(1-2):21–31. doi:10.1016/j.molmet.2012.07.002.

Hoberman R, Dias J, Ge B, Harmsen E, Mayhew M, Verlaan DJ, Kwan T, Dewar K, Blanchette M, Pastinen T. A probabilistic approach for SNP discovery in high-throughput human resequencing data. Genome Res. 2009;19(9):1542–52. doi:10.1101/gr.092072.109.

Huang Y, Niu B, Gao Y, Fu L, Li W. CD-HIT suite: a web server for clustering and comparing biological sequences. Bioinformatics. 2010;26(5):680–2. doi:10.1093/bioinformatics/btq003.

Hugenholtz P, Goebel BM, Pace NR. Impact of culture-independent studies on the emerging phylogenetic view of bacterial diversity. J Bacteriol. 1998;180(18):4765–74.

Huse SM, Welch DM, Morrison HG, Sogin ML. Ironing out the wrinkles in the rare biosphere through improved OTU clustering. Environ Microbiol. 2010;12(7):1889–98. doi:10.1111/j.1462-2920.2010.02193.x.

Huson DH, Weber N. Microbial community analysis using MEGAN. Methods Enzymol. 2013;531:465–85. doi:10.1016/B978-0-12-407863-5.00021-6.

Huson DH, Auch AF, Qi J, Schuster SC. MEGAN analysis of metagenomic data. Genome Res. 2007;17(3):377–86. doi:10.1101/gr.5969107.

Ishoey T, Woyke T, Stepanauskas R, Novotny M, Lasken RS. Genomic sequencing of single microbial cells from environmental samples. Curr Opin Microbiol. 2008;11(3):198–204. doi:10.1016/j.mib.2008.05.006.

Iverson V, Morris RM, Frazar CD, Berthiaume CT, Morales RL, Armbrust EV. Untangling genomes from metagenomes: revealing an uncultured class of marine Euryarchaeota. Science. 2012;335(6068):587–90. doi:10.1126/science.1212665.

Kanehisa M, Goto S, Sato Y, Furumichi M, Tanabe M. KEGG for integration and interpretation of large-scale molecular data sets. Nucleic Acids Res. 2012;40(Database issue):D109–14. doi:10.1093/nar/gkr988.

Kaput J, Cotton RG, Hardman L, Watson M, Al Aqeel AI, Al-Aama JY, Al-Mulla F, Alonso S, Aretz S, Auerbach AD, Bapat B, Bernstein IT, Bhak J, Bleoo SL, Blocker H, Brenner SE, Burn J, Bustamante M, Calzone R, Cambon-Thomsen A, Cargill M, Carrera P, Cavedon L, Cho YS, Chung YJ, Claustres M, Cutting G, Dalgleish R, den Dunnen JT, Diaz C,

Dobrowolski S, dos Santos MR, Ekong R, Flanagan SB, Flicek P, Furukawa Y, Genuardi M, Ghang H, Golubenko MV, Greenblatt MS, Hamosh A, Hancock JM, Hardison R, Harrison TM, Hoffmann R, Horaitis R, Howard HJ, Barash CI, Izagirre N, Jung J, Kojima T, Laradi S, Lee YS, Lee JY, Gil-da-Silva-Lopes VL, Macrae FA, Maglott D, Marafie MJ, Marsh SG, Matsubara Y, Messiaen LM, Moslein G, Netea MG, Norton ML, Oefner PJ, Oetting WS, O'Leary JC, de Ramirez AM, Paalman MH, Parboosingh J, Patrinos GP, Perozzi G, Phillips IR, Povey S, Prasad S, Qi M, Quin DJ, Ramesar RS, Richards CS, Savige J, Scheible DG, Scott RJ, Seminara D, Shephard EA, Sijmons RH, Smith TD, Sobrido MJ, Tanaka T, Tavtigian SV, Taylor GR, Teague J, Topel T, Ullman-Cullere M, Utsunomiya J, van Kranen HJ, Vihinen M, Webb E, Weber TK, Yeager M, Yeom YI, Yim SH, Yoo HS, Contributors to the Human Variome Project Planning M. Planning the human variome project: the Spain report. Hum Mutat. 2009;30(4):496–510. doi:10.1002/humu.20972.

Kelley DR, Liu B, Delcher AL, Pop M, Salzberg SL. Gene prediction with Glimmer for metagenomic sequences augmented by classification and clustering. Nucleic Acids Res. 2012;40(1):e9. doi:10.1093/nar/gkr1067.

Khachatryan ZA, Ktsoyan ZA, Manukyan GP, Kelly D, Ghazaryan KA, Aminov RI. Predominant role of host genetics in controlling the composition of gut microbiota. PLoS One. 2008;3(8): e3064. doi:10.1371/journal.pone.0003064.

Kim M, Lee KH, Yoon SW, Kim BS, Chun J, Yi H. Analytical tools and databases for metagenomics in the next-generation sequencing era. Genomics Inform. 2013;11(3):102–13. doi:10.5808/GI.2013.11.3.102.

King JL, LaRue BL, Novroski NM, Stoljarova M, Seo SB, Zeng X, Warshauer DH, Davis CP, Parson W, Sajantila A, Budowle B. High-quality and high-throughput massively parallel sequencing of the human mitochondrial genome using the Illumina MiSeq. Forensic Sci Int Genet. 2014;12:128–35. doi:10.1016/j.fsigen.2014.06.001.

Krause L, Diaz NN, Goesmann A, Kelley S, Nattkemper TW, Rohwer F, Edwards RA, Stoye J. Phylogenetic classification of short environmental DNA fragments. Nucleic Acids Res. 2008;36(7):2230–9. doi:10.1093/nar/gkn038.

Kuczynski J, Stombaugh J, Walters WA, González A, Caporaso JG, Knight R. Using QIIME to analyze 16S rRNA gene sequences from microbial communities. Current Protoc Microbiol. 2012a;1E. 5.1–1E. 5.20.

Kuczynski J, Stombaugh J, Walters WA, Gonzalez A, Caporaso JG, Knight R. Using QIIME to analyze 16S rRNA gene sequences from microbial communities. Curr Protoc Microbiol. 2012b; Chapter 1:Unit 1E 5. doi:10.1002/9780471729259.mc01e05s27.

Kunin V, Copeland A, Lapidus A, Mavromatis K, Hugenholtz P. A bioinformatician's guide to metagenomics. Microbiol Mol Biol Rev: MMBR. 2008;72(4):557–78, Table of Contents. doi:10.1128/MMBR.00009-08.

Lai B, Ding R, Li Y, Duan L, Zhu H. A de novo metagenomic assembly program for shotgun DNA reads. Bioinformatics. 2012;28(11):1455–62. doi:10.1093/bioinformatics/bts162.

Lal R. The new science of metagenomics: fourth domain of life. Indian J Microbiol. 2011;51 (3):245–6. doi:10.1007/s12088-011-0183-5.

Laserson J, Jojic V, Koller D. Genovo: de novo assembly for metagenomes. J Comput Biol. 2011;18(3):429–43. doi:10.1089/cmb.2010.0244.

Lasken RS, Stockwell TB. Mechanism of chimera formation during the Multiple Displacement Amplification reaction. BMC Biotechnol. 2007;7:19. doi:10.1186/1472-6750-7-19.

Lee JH, Yi H, Jeon YS, Won S, Chun J. TBC: a clustering algorithm based on prokaryotic taxonomy. J Microbiol. 2012;50(2):181–5. doi:10.1007/s12275-012-1214-6.

Leung HC, Yiu SM, Yang B, Peng Y, Wang Y, Liu Z, Chen J, Qin J, Li R, Chin FY. A robust and accurate binning algorithm for metagenomic sequences with arbitrary species abundance ratio. Bioinformatics. 2011;27(11):1489–95. doi:10.1093/bioinformatics/btr186.

Lewis Jr CM, Obregon-Tito A, Tito RY, Foster MW, Spicer PG. The Human Microbiome Project: lessons from human genomics. Trends Microbiol. 2012;20(1):1–4. doi:10.1016/j.tim.2011.10.004.

Li E, Hamm CM, Gulati AS, Sartor RB, Chen HY, Wu X, Zhang TY, Rohlf FJ, Zhu W, Gu C, Robertson CE, Pace NR, Boedeker EC, Harpaz N, Yuan J, Weinstock GM, Sodergren E, Frank DN. Inflammatory bowel diseases phenotype, C. difficile and NOD2 genotype are associated with shifts in human ileum associated microbial composition. PLoS One. 2012;7(6):e26284. doi:10.1371/journal.pone.0026284.

Liu B, Gibbons T, Ghodsi M, Treangen T, Pop M. Accurate and fast estimation of taxonomic profiles from metagenomic shotgun sequences. BMC Genomics. 2011a;12(Suppl 2):S4. doi:10.1186/1471-2164-12-S2-S4.

Liu S, Vijayendran D, Bonning BC. Next generation sequencing technologies for insect virus discovery. Viruses. 2011b;3(10):1849–69. doi:10.3390/v3101849.

Lizardi PM. Multiple displacement amplification. Google Patents. 2001.

Looft T, Johnson TA, Allen HK, Bayles DO, Alt DP, Stedtfeld RD, Sul WJ, Stedtfeld TM, Chai B, Cole JR, Hashsham SA, Tiedje JM, Stanton TB. In-feed antibiotic effects on the swine intestinal microbiome. Proc Natl Acad Sci U S A. 2012;109(5):1691–6. doi:10.1073/pnas.1120238109.

Lozupone C, Knight R. UniFrac: a new phylogenetic method for comparing microbial communities. Appl Environ Microbiol. 2005;71(12):8228–35. doi:10.1128/AEM.71.12.8228-8235.2005.

Lozupone C, Hamady M, Knight R. UniFrac--an online tool for comparing microbial community diversity in a phylogenetic context. BMC Bioinformatics. 2006;7:371. doi:10.1186/1471-2105-7-371.

Ludwig W, Strunk O, Westram R, Richter L, Meier H, Yadhukumar BA, Lai T, Steppi S, Jobb G, Forster W, Brettske I, Gerber S, Ginhart AW, Gross O, Grumann S, Hermann S, Jost R, Konig A, Liss T, Lussmann R, May M, Nonhoff B, Reichel B, Strehlow R, Stamatakis A, Stuckmann N, Vilbig A, Lenke M, Ludwig T, Bode A, Schleifer KH. ARB: a software environment for sequence data. Nucleic Acids Res. 2004;32(4):1363–71. doi:10.1093/nar/gkh293.

Luo C, Tsementzi D, Kyrpides NC, Konstantinidis KT. Individual genome assembly from complex community short-read metagenomic datasets. ISME J. 2012;6(4):898–901. doi:10.1038/ismej.2011.147.

MacDonald NJ, Parks DH, Beiko RG. Rapid identification of high-confidence taxonomic assignments for metagenomic data. Nucleic Acids Res. 2012;40(14):e111. doi:10.1093/nar/gks335.

Manichanh C, Chapple CE, Frangeul L, Gloux K, Guigo R, Dore J. A comparison of random sequence reads versus 16S rDNA sequences for estimating the biodiversity of a metagenomic library. Nucleic Acids Res. 2008;36(16):5180–8. doi:10.1093/nar/gkn496.

Mardis ER. The impact of next-generation sequencing technology on genetics. Trends Genet. 2008;24(3):133–41. doi:10.1016/j.tig.2007.12.007.

Markowitz VM, Ivanova N, Palaniappan K, Szeto E, Korzeniewski F, Lykidis A, Anderson I, Mavrommatis K, Kunin V, Martin HG, Dubchak I, Hugenholtz P, Kyrpides NC. An experimental metagenome data management and analysis system (vol 22, pg 359, 2006). Bioinformatics. 2006;22(20):e359–67. doi:10.1093/bioinformatics/btl436.

Meyer F, Paarmann D, D'Souza M, Olson R, Glass EM, Kubal M, Paczian T, Rodriguez A, Stevens R, Wilke A, Wilkening J, Edwards RA. The metagenomics RAST server – a public resource for the automatic phylogenetic and functional analysis of metagenomes. BMC Bioinformatics. 2008;9:386. doi:10.1186/1471-2105-9-386.

Morgan XC, Tickle TL, Sokol H, Gevers D, Devaney KL, Ward DV, Reyes JA, Shah SA, LeLeiko N, Snapper SB, Bousvaros A, Korzenik J, Sands BE, Xavier RJ, Huttenhower C. Dysfunction of the intestinal microbiome in inflammatory bowel disease and treatment. Genome Biol. 2012;13(9):R79. doi:10.1186/gb-2012-13-9-r79.

Namiki T, Hachiya T, Tanaka H, Sakakibara Y. MetaVelvet: an extension of Velvet assembler to de novo metagenome assembly from short sequence reads. Nucleic Acids Res. 2012;40(20):e155. doi:10.1093/nar/gks678.

Navas-Molina JA, Peralta-Sanchez JM, Gonzalez A, McMurdie PJ, Vazquez-Baeza Y, Xu Z, Ursell LK, Lauber C, Zhou H, Song SJ, Huntley J, Ackermann GL, Berg-Lyons D, Holmes S,

Caporaso JG, Knight R. Advancing our understanding of the human microbiome using QIIME. Methods Enzymol. 2013;531:371–444. doi:10.1016/B978-0-12-407863-5.00019-8.

Noguchi H, Park J, Takagi T. MetaGene: prokaryotic gene finding from environmental genome shotgun sequences. Nucleic Acids Res. 2006;34(19):5623–30. doi:10.1093/nar/gkl723.

Noguchi H, Taniguchi T, Itoh T. MetaGeneAnnotator: detecting species-specific patterns of ribosomal binding site for precise gene prediction in anonymous prokaryotic and phage genomes. DNA Res. 2008;15(6):387–96. doi:10.1093/dnares/dsn027.

Patil KR, Roune L, McHardy AC (2012) The PhyloPythiaS web server for taxonomic assignment of metagenome sequences. PLoS One. 2012;7(6):ARTN e38581. doi:10.1371/journal.pone.0038581.

Paulson JN, Stine OC, Bravo HC, Pop M. Differential abundance analysis for microbial marker-gene surveys. Nat Methods. 2013;10(12):1200–2. doi:10.1038/nmeth.2658.

Peng Y, Leung HC, Yiu SM, Chin FY. Meta-IDBA: a de Novo assembler for metagenomic data. Bioinformatics. 2011;27(13):i94–101. doi:10.1093/bioinformatics/btr216.

Phelan VV, Liu WT, Pogliano K, Dorrestein PC. Microbial metabolic exchange—the chemotype-to-phenotype link. Nat Chem Biol. 2012;8(1):26–35. doi:10.1038/nchembio.739.

Powell S, Szklarczyk D, Trachana K, Roth A, Kuhn M, Muller J, Arnold R, Rattei T, Letunic I, Doerks T, Jensen LJ, von Mering C, Bork P. eggNOG v3.0: orthologous groups covering 1133 organisms at 41 different taxonomic ranges. Nucleic Acids Res. 2012;40(Database issue):D284–9. doi:10.1093/nar/gkr1060.

Price AL, Patterson NJ, Plenge RM, Weinblatt ME, Shadick NA, Reich D. Principal components analysis corrects for stratification in genome-wide association studies. Nat Genet. 2006;38(8):904–9. doi:10.1038/ng1847.

Purvis A, Hector A. Getting the measure of biodiversity. Nature. 2000;405(6783):212–9. doi:10.1038/35012221.

Qin J, Li R, Raes J, Arumugam M, Burgdorf KS, Manichanh C, Nielsen T, Pons N, Levenez F, Yamada T, Mende DR, Li J, Xu J, Li S, Li D, Cao J, Wang B, Liang H, Zheng H, Xie Y, Tap J, Lepage P, Bertalan M, Batto JM, Hansen T, Le Paslier D, Linneberg A, Nielsen HB, Pelletier E, Renault P, Sicheritz-Ponten T, Turner K, Zhu H, Yu C, Li S, Jian M, Zhou Y, Li Y, Zhang X, Li S, Qin N, Yang H, Wang J, Brunak S, Dore J, Guarner F, Kristiansen K, Pedersen O, Parkhill J, Weissenbach J, Meta HITC, Bork P, Ehrlich SD, Wang J. A human gut microbial gene catalogue established by metagenomic sequencing. Nature. 2010;464(7285):59–65. doi:10.1038/nature08821.

Qin J, Li Y, Cai Z, Li S, Zhu J, Zhang F, Liang S, Zhang W, Guan Y, Shen D, Peng Y, Zhang D, Jie Z, Wu W, Qin Y, Xue W, Li J, Han L, Lu D, Wu P, Dai Y, Sun X, Li Z, Tang A, Zhong S, Li X, Chen W, Xu R, Wang M, Feng Q, Gong M, Yu J, Zhang Y, Zhang M, Hansen T, Sanchez G, Raes J, Falony G, Okuda S, Almeida M, LeChatelier E, Renault P, Pons N, Batto JM, Zhang Z, Chen H, Yang R, Zheng W, Li S, Yang H, Wang J, Ehrlich SD, Nielsen R, Pedersen O, Kristiansen K, Wang J. A metagenome-wide association study of gut microbiota in type 2 diabetes. Nature. 2012;490(7418):55–60. doi:10.1038/nature11450.

Quast C, Pruesse E, Yilmaz P, Gerken J, Schweer T, Yarza P, Peplies J, Glockner FO. The SILVA ribosomal RNA gene database project: improved data processing and web-based tools. Nucleic Acids Res. 2013;41(Database issue):D590–6. doi:10.1093/nar/gks1219.

Quince C, Lanzen A, Curtis TP, Davenport RJ, Hall N, Head IM, Read LF, Sloan WT. Accurate determination of microbial diversity from 454 pyrosequencing data. Nat Methods. 2009;6(9):639–U627. doi:10.1038/Nmeth.1361.

Reeder J, Knight R. Rapidly denoising pyrosequencing amplicon reads by exploiting rank-abundance distributions. Nat Methods. 2010;7(9):668–9. doi:10.1038/nmeth0910-668b.

Rondon MR, August PR, Bettermann AD, Brady SF, Grossman TH, Liles MR, Loiacono KA, Lynch BA, MacNeil IA, Minor C, Tiong CL, Gilman M, Osburne MS, Clardy J, Handelsman J, Goodman RM. Cloning the soil metagenome: a strategy for accessing the genetic and functional diversity of uncultured microorganisms. Appl Environ Microbiol. 2000;66(6):2541–7.

Ruby JG, Bellare P, Derisi JL. PRICE: software for the targeted assembly of components of (Meta) genomic sequence data. G3. 2013;3(5):865–80. doi:10.1534/g3.113.005967.
Rup L. The human microbiome project. Indian J Microbiol. 2012;52(3):315. doi:10.1007/s12088-012-0304-9.
Savage DC. Microbial ecology of the gastrointestinal tract. Annu Rev Microbiol. 1977;31:107–33. doi:10.1146/annurev.mi.31.100177.000543.
Schloss PD, Handelsman J. Biotechnological prospects from metagenomics. Curr Opin Biotechnol. 2003;14(3):303–10.
Schloss PD, Handelsman J. Introducing DOTUR, a computer program for defining operational taxonomic units and estimating species richness. Appl Environ Microbiol. 2005;71(3):1501–6. doi:10.1128/AEM.71.3.1501-1506.2005.
Schloss PD, Handelsman J. Introducing SONS, a tool for operational taxonomic unit-based comparisons of microbial community memberships and structures. Appl Environ Microbiol. 2006;72(10):6773–9. doi:10.1128/AEM.00474-06.
Schloss PD, Westcott SL, Ryabin T, Hall JR, Hartmann M, Hollister EB, Lesniewski RA, Oakley BB, Parks DH, Robinson CJ, Sahl JW, Stres B, Thallinger GG, Van Horn DJ, Weber CF. Introducing mothur: open-source, platform-independent, community-supported software for describing and comparing microbial communities. Appl Environ Microbiol. 2009;75 (23):7537–41. doi:10.1128/AEM.01541-09.
Schloss PD, Gevers D, Westcott SL. Reducing the effects of PCR amplification and sequencing artifacts on 16S rRNA-based studies. PLoS One. 2011;6(12):e27310. doi:10.1371/journal.pone.0027310.
Segata N, Waldron L, Ballarini A, Narasimhan V, Jousson O, Huttenhower C. Metagenomic microbial community profiling using unique clade-specific marker genes. Nat Methods. 2012;9 (8):811–4.
Seo SB, Zeng X, King JL, Larue BL, Assidi M, Al-Qahtani MH, Sajantila A, Budowle B. Underlying data for sequencing the mitochondrial genome with the massively parallel sequencing platform ion torrent PGM. BMC Genomics. 2015;6(Suppl 1):S4. doi:10.1186/1471-2164-16-S1-S4.
Seshadri R, Kravitz SA, Smarr L, Gilna P, Frazier M. CAMERA: a community resource for metagenomics. PLoS Biol. 2007;5(3):e75. doi:10.1371/journal.pbio.0050075.
Sharpton TJ. An introduction to the analysis of shotgun metagenomic data. Front Plant Sci. 2014;5:209. doi:10.3389/fpls.2014.00209.
Shoaib M, Baconnais S, Mechold U, Le Cam E, Lipinski M, Ogryzko V. Multiple displacement amplification for complex mixtures of DNA fragments. BMC Genomics. 2008;9:415. doi:10.1186/1471-2164-9-415.
Singh B, Gautam SK, Verma V, Kumar M, Singh B. Metagenomics in animal gastrointestinal ecosystem: potential biotechnological prospects. Anaerobe. 2008;14(3):138–44. doi:10.1016/j.anaerobe.2008.03.002.
Smith CC, Snowberg LK, Gregory Caporaso J, Knight R, Bolnick DI. Dietary input of microbes and host genetic variation shape among-population differences in stickleback gut microbiota. ISME J. 2015;9(11):2515–26. doi:10.1038/ismej.2015.64.
Strous M, Kraft B, Bisdorf R, Tegetmeyer HE. The binning of metagenomic contigs for microbial physiology of mixed cultures. Front Microbiol. 2012;3:410. doi:10.3389/fmicb.2012.00410.
Tamaki H, Wright CL, Li X, Lin Q, Hwang C, Wang S, Thimmapuram J, Kamagata Y, Liu WT. Analysis of 16S rRNA amplicon sequencing options on the Roche/454 next-generation titanium sequencing platform. PLoS One. 2011;6(9):e25263. doi:10.1371/journal.pone.0025263.
Tatusov RL, Fedorova ND, Jackson JD, Jacobs AR, Kiryutin B, Koonin EV, Krylov DM, Mazumder R, Mekhedov SL, Nikolskaya AN, Rao BS, Smirnov S, Sverdlov AV, Vasudevan S, Wolf YI, Yin JJ, Natale DA. The COG database: an updated version includes eukaryotes. BMC Bioinformatics. 2003;4:41. doi:10.1186/1471-2105-4-41.
Thomas T, Rusch D, DeMaere MZ, Yung PY, Lewis M, Halpern A, Heidelberg KB, Egan S, Steinberg PD, Kjelleberg S. Functional genomic signatures of sponge bacteria reveal unique and shared features of symbiosis. ISME J. 2010;4(12):1557–67. doi:10.1038/ismej.2010.74.

Thomas T, Gilbert J, Meyer F. Metagenomics – a guide from sampling to data analysis. Microb Inform Exp. 2012;2(1):3. doi:10.1186/2042-5783-2-3.

Turnbaugh PJ, Baeckhed F, Fulton L, Gordon JI. Diet-induced obesity is linked to marked but reversible alterations in the mouse distal gut microbiome. Cell Host Microbe. 2008;3 (4):213–23. doi:10.1016/j.chom.2008.02.015.

Valdez Y, Brown EM, Finlay BB. Influence of the microbiota on vaccine effectiveness. Trends Immunol. 2014;35(11):526–37. doi:10.1016/j.it.2014.07.003.

Venter JC, Remington K, Heidelberg JF, Halpern AL, Rusch D, Eisen JA, Wu D, Paulsen I, Nelson KE, Nelson W, Fouts DE, Levy S, Knap AH, Lomas MW, Nealson K, White O, Peterson J, Hoffman J, Parsons R, Baden-Tillson H, Pfannkoch C, Rogers YH, Smith HO. Environmental genome shotgun sequencing of the Sargasso Sea. Science. 2004;304(5667):66–74. doi:10.1126/science.1093857.

Warnecke F, Luginbühl P, Ivanova N, Ghassemian M, Richardson TH, Stege JT, Cayouette M, McHardy AC, Djordjevic G, Aboushadi N. Metagenomic and functional analysis of hindgut microbiota of a wood-feeding higher termite. Nature. 2007;450(7169):560–5.

Wesolowska-Andersen A, Bahl MI, Carvalho V, Kristiansen K, Sicheritz-Ponten T, Gupta R, Licht TR. Choice of bacterial DNA extraction method from fecal material influences community structure as evaluated by metagenomic analysis. Microbiome. 2014;2:19. doi:10.1186/2049-2618-2-19.

Widdel F. Methods for enrichment and pure culture isolation of filamentous gliding sulfate-reducing bacteria. Arch Microbiol. 1983;134(4):282–5. doi:10.1007/Bf00407803.

Wilkening J, Wilke A, Desai N, Meyer F. Using clouds for metagenomics: a case study. In: Cluster computing and workshops, 2009. CLUSTER'09. IEEE international conference on. IEEE, pp. 1–6; 2009.

Wommack KE, Bhavsar J, Ravel J. Metagenomics: read length matters. Appl Environ Microbiol. 2008;74(5):1453–63. doi:10.1128/AEM.02181-07.

Woo PC LS, Teng JL, Tse H, Yuen KY. Then and now: use of 16S rDNA gene sequencing for bacterial identification and discovery of novel bacteria in clinical microbiology laboratories. ELSEVIER. 2008. doi:10.1111/j.1469-0691.2008.02070.x.

Wooley JC, Godzik A, Friedberg I. A primer on metagenomics. PLoS Comput Biol. 2010;6(2): e1000667. doi:10.1371/journal.pcbi.1000667.

Wright ES, Yilmaz LS, Noguera DR. DECIPHER, a search-based approach to chimera identification for 16S rRNA sequences. Appl Environ Microbiol. 2012;78(3):717–25. doi:10.1128/Aem.06516-11.

Wrighton KC. Fermentation, hydrogen, and sulfur metabolism in multiple uncultivated bacterial phyla (vol 337, pg 1661, 2012). Science. 2012;338(6108):742–42

Wu D, Hugenholtz P, Mavromatis K, Pukall R, Dalin E, Ivanova NN, Kunin V, Goodwin L, Wu M, Tindall BJ, Hooper SD, Pati A, Lykidis A, Spring S, Anderson IJ, D'Haeseleer P, Zemla A, Singer M, Lapidus A, Nolan M, Copeland A, Han C, Chen F, Cheng JF, Lucas S, Kerfeld C, Lang E, Gronow S, Chain P, Bruce D, Rubin EM, Kyrpides NC, Klenk HP, Eisen JA. A phylogeny-driven genomic encyclopaedia of Bacteria and Archaea. Nature. 2009;462 (7276):1056–60. doi:10.1038/nature08656.

Yang S, Liebner S, Alawi M, Ebenhoh O, Wagner D. Taxonomic database and cut-off value for processing mcrA gene 454 pyrosequencing data by MOTHUR. J Microbiol Methods. 2014;103:3–5. doi:10.1016/j.mimet.2014.05.006.

Zhang R, Ma ZH, Wu BM. Multiple displacement amplification of whole genomic DNA from urediospores of Puccinia striiformis f. sp. tritici. Curr Genet. 2015;61(2):221–30. doi:10.1007/s00294-014-0470-x.

Zhao N, Chen J, Carroll IM, Ringel-Kulka T, Epstein MP, Zhou H, Zhou JJ, Ringel Y, Li H, Wu MC. Testing in microbiome-profiling studies with MiRKAT, the microbiome regression-based kernel association test. Am J Hum Genet. 2015;96(5):797–807. doi:10.1016/j.ajhg.2015.04.003.

Zheng H, Wu H. Short prokaryotic DNA fragment binning using a hierarchical classifier based on linear discriminant analysis and principal component analysis. J Bioinforma Comput Biol. 2010;8(6):995–1011.

Chapter 10
Clinical Epigenetics and Epigenomics

Chuan Qiu, Fangtang Yu, Hong-Wen Deng, and Hui Shen

Abstract Epigenetics is the study of somatically heritable changes in gene expression that occur without alterations in DNA sequence, mainly including DNA methylation and histone modification. Epigenomics refers to the complete study of these somatically heritable changes across the whole genome. Epigenetic mechanisms are critical components in the growth of cells and normal development. Aberrant epigenetic changes have been found to be causative factors in cancer, autoimmune diseases as well as others. Significant progress has been made towards epigenomic profiling by using molecular techniques. In this chapter, we introduce the basics of epigenetics and epigenomics; describe the remarkable advances in current epigenomic mapping and analysis technologies, especially microarray-based and next-generation sequencing-based applications. Then we focus on the recent studies of epigenetic changes in normal and diseased cells with the aim to translate basic epigenetic and epigenomics research into clinical applications. We also discuss some critical challenges ahead and provide a perspective on the progress of epigenomics field.

Keywords Epigenetics • Epigenomics • DNA methylation • Histone modification • Mapping and analyzing technologies • Human disease

Abbreviations

5-hmC	5-hydroxymethyl-cytosine
5-fC	5-formyl-cytosine
5-mC	5-methylcytosine
5-caC	5-carboxylcytosine

C. Qiu • F. Yu • H. Shen (✉)
Center for Bioinformatics & Genomics, Department of Biostatistics & Bioinformatics, School of Public Health & Tropical Medicine, Tulane University, New Orleans, LA 70112, USA
e-mail: hshen3@tulane.edu

H.-W. Deng
Center for Bioinformatics and Genomics, Department of Biostatistics and Bioinformatics, Tulane University School of Public Health and Tropical Medicine, New Orleans, LA, USA
e-mail: hdeng2@tulane.edu

X. Wang et al. (eds.), *Application of Clinical Bioinformatics*,
Translational Bioinformatics 11, DOI 10.1007/978-94-017-7543-4_10

CGI	CpG island
DMH	differential methylation hybridization
DMRs	differentially methylated regions
DNMTs	DNA methyltransferases
EWAS	epigenome-wide association studies
FDR	false discovery rate
HATs	histone acetyltransferases
HDACs	histone deacetylases
HMTs	histone methyltransferases
IMA	Illumina Methylation Analyzer
LOESS	locally weighted scatterplot smoothing
MBD	methyl-CpG binding domain
MeCPs	Methyl-CpG Binding Proteins
MeDIP	methylated DNA immunoprecipitation
PRC	polycomb repressive complexes
RRBS	reduced representation bisulfite sequencing
SNPs	Single Nucleotide Polymorphisms
TET	ten-eleven translocation
UHRF	ubiquitin plant homeodomain RING finger
WGBS	whole genome bisulfite sequencing

In 1959, Waddington first coined the concept "epigenetics" (Waddington 1959), which now refers to the mechanism for stable maintenance of gene expression changes that involves physically "marking" DNA or its associated proteins other than alterations in DNA sequence.

A variety of epigenetic factors have been identified, such as DNA methylation, histone modification, and non-coding RNAs (e.g., microRNAs and long non-coding RNAs) etc. These epigenetic factors coordinatively regulate gene expression and provide heritable epigenetic information that is not encoded in DNA sequence (Cedar and Bergman 2009; Esteller 2011). Epigenome refers to the entire constitution of epigenetic marks in a cell type at a given time point, it is cell-specific and tissue-specific (Varley et al. 2013). In each type of cells, epigenetic factors regulate gene expression in different ways, for example, facilitate or restrict transcription factor access to DNA sequence (Rivera and Ren 2013). Epigenome may change over the lifetime (Fraga et al. 2005a) and are prone to environmental influences, such as stress, social interactions, physical activity, exposure to toxins and diet (Alegria-Torres et al. 2011). Aberrant epigenomic alternations have been implicated in a wide variety of human disorders, such as cancer and autoimmune diseases etc. (Portela and Esteller 2010), and epigenetic drugs may revolutionize the treatment of many human diseases (Heerboth et al. 2014). In this chapter, we briefly reviewed the molecular basis for two major epigenetic factors, DNA methylation and histone modification, and discussed some commonly used epigenome-wide analytic approaches for these two factors, as well as their involvement in some human complex disorders.

10.1 Molecular Basis of DNA Methylation and Histone Modification

10.1.1 DNA Methylation

DNA methylation, commonly called the 'fifth base' in the genome, is one of the most extensively studied epigenetic mechanisms. It is a direct chemical modification of the fifth carbon of a cytosine that adds a methyl (-CH_3) group through a covalent bond resulting in 5-methylcytosine (5-mC).

In adult somatic tissues, DNA methylation typically occurs in a "CpG" (C-phosphate-G) dinucleotide context (Bird and Southern 1978; Cedar et al. 1979). An exception to this is seen in embryonic stem cells (Haines et al. 2001), where a substantial amount of 5-mC is also observed in non-CpG sites (mCHG, mCHH). In human genome, there are 28 million CpG sites, which are not evenly distributed throughout the genome (Lister et al. 2009) but tend to cluster in regions, known as "CpG islands" (CGIs). CGIs usually occur near gene transcription start sites (TSS) and ~60 % of human gene promoters are associated with CGIs (Bird 1986; Gardiner-Garden and Frommer 1987). DNA methylation is catalyzed by a family of enzymes termed DNA methyltransferases (DNMTs) (Okano et al. 1998), including DNMT1, DNMT3A, DNMT3B, DNMT2 and DNMT3L, which cooperate in establishing and maintaining DNA methylation patterns (Kulis and Esteller 2010; Okano et al. 1999). Equally important and opposite with DNA methylation is DNA demethylation. DNA demethylation can be either passive or active, or a combination of both. Passive DNA demethylation usually takes place when DNMT1 cannot effectively restore the DNA methylation patterns on newly synthesized DNA strands during replication rounds (Wu and Zhang 2010), whereas active demethylation is usually mediated by the ten-eleven translocation (TET) family enzymes (TET1, TET2 and TET3) and subsequent restoration of unmodified cytosine by the thymine DNA glycosylase (TDG)-mediated base excision repair (Kohli and Zhang 2013).

The importance of DNA methylation as a major epigenetic modification in gene expression has been widely recognized. Hypermethylation of CpGs in TSS proximal regions, particularly in promoter CGIs, is largely associated with repressed gene transcription (Wagner et al. 2014), whereas methylation of CpGs located within gene bodies is usually associated with an increase in transcriptional activity (Ramsahoye et al. 2000; Hellman and Chess 2007). However, several recent studies have revealed that there is no simple relationship between inter-individual DNA methylation and gene expression with respect to the location of the methylated CpGs and both negative and positive inter-individual methylation-expression correlations were detected for CpGs located in gene body and transcription start site proximal regions, as well as in intergenic regions (Wagner et al. 2014; Bell et al. 2011).

10.1.2 Histone Modification

The basic unit of chromatin is the nucleosome, which is composed of an octomer of histone proteins (containing two copies each of histones H2A, H2B, H3, and H4) around which is wrapped a length of 147 bp DNA. The degree to which chromatin are condensed or packed is a critical determinant of the transcriptional activity of the associated DNA and this is mediated in part by diverse post-translational covalent modifications of the N-terminal tails of histone proteins (Fig. 10.1).

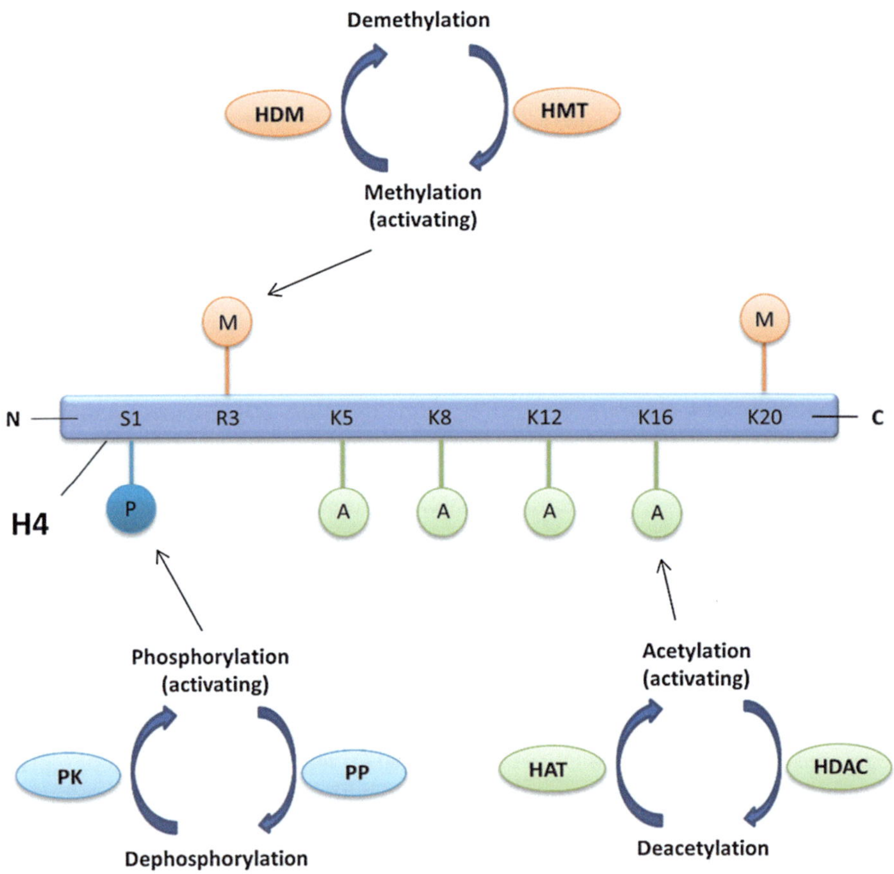

Fig. 10.1 Post-translational modifications of histones. The first 20 amino acids in the N-terminus of the human histone H4 are illustrated. Many sites in the N-terminus can be targets for epigenetic tagging such as acetylation (A), phosphorylation (P) and methylation (M). Acetylation is catalyzed by histone acetyltransferase (HAT) and removed by histone deacetylase (HDAC); Phosphorylation is catalyzed by protein kinases (PK) and removed by protein phosphatase (PP); Methylation is catalyzed by histone methyltransferases (HMT) and removed by histone demethylase (HDM). Some histone modification marks are associated with gene activation while others are associated with gene repression, and the integration of multiple marks leads to a finely tuned transcriptional response

At least eight different types of histone modification have been identified: acetylation, methylation, phosphorylation, ubiquitination, sumoylation, ADP ribosylation, deimination, and proline isomerization. All the modifications are reversible and dynamic, mediated by enzymes that add/remove modification.

Histone acetylation occurs via an enzymatic transferring of an acetyl group from acetyl-CoA to the ε-NH+ group of the lysine residues within a histone. This enzymatic activity is catalyzed by enzymes called histone acetyltransferases (HATs) and reversed by histone deacetylases (HDACs) (Hodawadekar and Marmorstein 2007). Histone acetylation is a hallmark of transcriptional activation (Sterner and Berger 2000) and the histone acetylation patterns are tightly associated with many cellular processes including chromatin dynamics and transcription, gene silencing, cell cycle progression, apoptosis, differentiation, DNA replication, DNA repair, nuclear import, and neuronal repression (Cohen et al. 2011).

Histone methylation is another extensively studies histone modification marks. It is defined as the transfer of one, two, or three methyl groups from S-adenosyl-L-methionine to lysine or arginine residues of histone proteins by histone methyltransferases (HMTs). In the cell nucleus, when histone methylation occurs, specific genes within the DNA complexed with the histone may be activated or silenced (Greer and Shi 2012). For instance, the tri-methylation of histone H3 at lysine 4 (H3K4me3) is positively correlated with gene transcription and commonly detected in a tight, localized area at 5′-ends/promoter regions of active genes (Barski et al. 2007). H3K36me3 is strongly enriched across the gene body and at the 3′-end of active genes and may link to transcriptional elongation (Barski et al. 2007). In contrast, H3K27me3 is the classic repressive histone modification mark, which shows a broad peak at promoters and throughout the gene body of the silent genes (Barski et al. 2007).

10.2 Epigenome-Wide Analyses of DNA Methylation and Histone Modification

10.2.1 Epigenome-Wide DNA Methylation Analysis

10.2.1.1 DNA Methylation Profiling Assays

DNA methylation analysis normally relies on three strategies (Fig. 10.2): (1) Digestion of genomic DNA with methylation-sensitive restriction enzymes; (2) Affinity-based enrichment of methylated DNA fragments; and (3) Bisulfite conversion. Each of the three strategies can be combined with either microarray or next-generation sequencing technique to interrogate epigenome-wide DNA methylation patterns, and each with unique advantages and drawbacks (Table 10.1).

- **Digestion of genomic DNA with methylation-sensitive restriction enzymes**: Some restriction enzymes (e.g., HpaII and SmaI) are methylation sensitive –

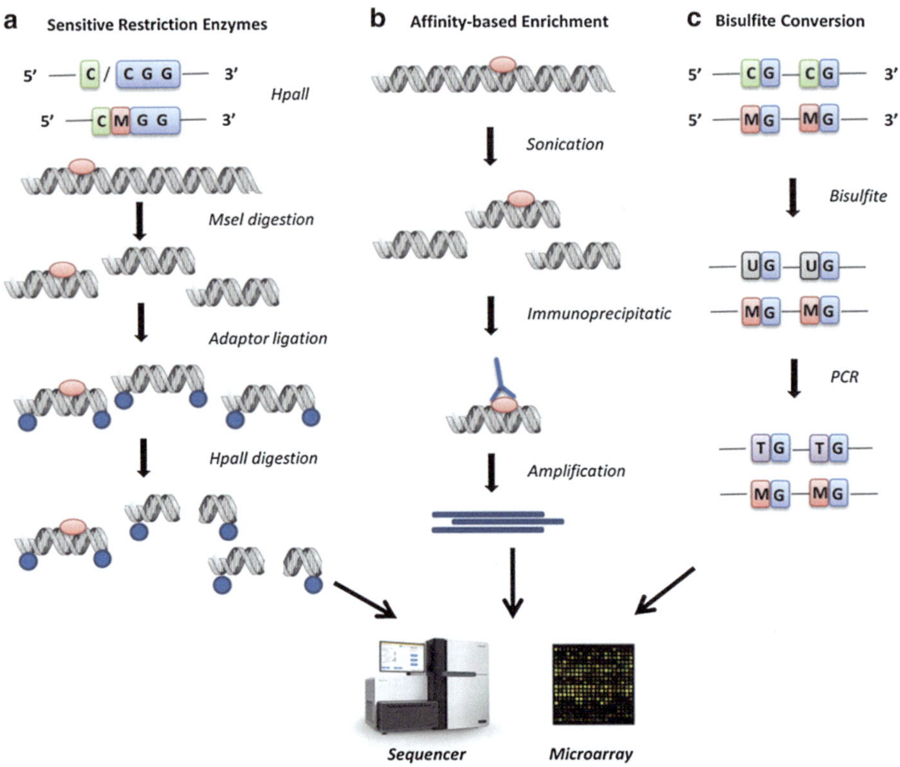

Fig. 10.2 Strategies for pretreatment of DNA sample. (**a**) Digestion of genomic DNA with methylation-sensitive restriction enzymes. (**b**) Affinity-based enrichment of methylated DNA fragments. (**c**) Chemical treatment of DNA with sodium bisulfite results in the conversion of unmethylated cytosines to uracils. In contrast, methylated cytosines are protected. Subsequently, microarray or next-generation sequencing of these libraries reveals the methylation status

their activity is affected by the presence of a methyl CpG within restriction sites. Therefore, when genomic DNA is digested with a methylation-sensitive restriction enzyme, difference in methylation status is converted into difference in sequence fragment size. For example, differential methylation hybridization (DMH) uses combinations of methylation-sensitive and methylation-insensitive restriction enzyme digestion, followed by ligation-mediated PCR to enrich for methylated or unmethylated fragments. PCR products are labeled and hybridized to arrays, or tested by next-generation sequencing (Fig. 10.2a). Methods based on this strategy were used in early epigenome-wide DNA methylation studies (Rakyan et al. 2011), but the genome-wide CpG coverage and resolution are limited by the cutting frequency and the fragment size of the chosen restriction enzymes.

- **Affinity-based enrichment of methylated DNA fragments**: Affinity-based enrichment assays capture methylated DNA fragments with a methyl-CpG

Table 10.1 Comparison of DNA methylation detection procedure

	Advantages	Disadvantages	Suitable application	Methods
Affinity-based enrichment	Rapid and efficient genome-wide assessment of DNA methylation	Possibility of antibody cross-reactivity	Rapid, large scale, low resolution study of DNA methylation	MeDIP-seq
	Powerful tool for comprehensive profiling of DNA methylation in complex genomes	Resolution depends on the fragment size of the enriched methylated DNA		MBD-seq
		Does not provide single base pair resolution		
Bisulfite conversion	Resolution at the nucleotide level	Often leads to damaged DNA	High resolution study of DNA methylation at small or large scale	WGBS
	Effectively converts an epigenetic difference into a genetic difference, easily detectable by sequencing	Potentially incomplete conversion of DNA		RRBS
		Cannot distinguish 5-mc and 5-hmc		Infinium
		Whole genome sequencing requires intensive downstream analysis		
Restriction enzyme-based digestion	Easy to use	Determination of methylation status is limited by the enzyme recognition site	Targeted, site specific study of DNA methylation	DMH
	High enzyme turnover			MSDK

Note: The table is modified from Ku et al. (2011)

binding protein (e.g., MBD2) or 5-mC specific antibody (Fig. 10.2b). For example, when performing methylated DNA immunoprecipitation (MeDIP)-chip/-seq, genomic DNAs are first fragmented by sonication and then incubated with anti-5mC antibody. The anti-5mC bound DNA fragments are isolated, deproteinized, and then hybridized onto microarrays (MeDIP-chip) or analyzed by next-generation sequencing (MeDIP-seq). The frequency of DNA fragments bound to specific probes or mapped to specific genomic regions provides the raw data from which DNA methylation levels can be inferred. The affinity-based DNA methylation assays allow for rapid and efficient genome-wide assessment of DNA methylation, however, as the affinity-captured DNA fragments are generally hundreds of nucleotides in size, the major limitation of these methods is its inability to pinpoint methylation changes at a single CpG resolution (Robinson et al. 2010).

- **Bisulfite Conversion**: Bisulfite conversion of DNA is the most commonly used method for DNA methylation studies. It uses bisulfite salt to deaminate cytosine residues on single-stranded DNA, converting them to uracil while leaving 5-methylcytosine intact. Once a difference of methylation status is converted into a difference of DNA sequence, it can be detected by various techniques (Fig. 10.2c). Bisulfite sequencing applies routine DNA sequencing methods on bisulfite-converted genomic DNA. It can provide quantitative methylation measurement at single nucleotide resolution and is widely accepted as a gold standard for DNA methylation analysis. Recent development of next-generation sequencing technology makes it feasible to perform whole genome bisulfite sequencing (WGBS) (Suzuki and Bird 2008). Though WGBS can provide a comprehensive coverage of almost all CpGs in the human genome, its usage is currently limited by its high cost. Thus, several more cost-effective bisulfite conversion-based approaches, such as reduced representation bisulfite sequencing (RRBS) and Illumina 450 k array, are widely employed in the current epigenomics research field. In RRBS, genomic DNA is digested by the methylation-insensitive restriction enzyme MspI (5′-C′CGG-3′) and separated by gel electrophoresis, and then size-selected DNA fragments are bisulfite converted and analyzed by next-generation sequencing platforms (Meissner et al. 2005). After the MspI digestion and size selection, CpG sites were enriched in sequencing library that reduce the amount of nucleotides needed to be sequenced.

10.2.1.2 Data Processing and Analysis

In this section, we will review the data processing approaches for three most popular epigenome-wide DNA methylation profiling methods, namely, Illumina 450k array, RRBS and MeDIP-seq. We also discussed the approaches for data visualization and identification of differential methylation in the following section.

Data Processing for Illumina 450k Array

The Illumina 450k array adapted BeadArray technology to recognize the bisulfite-converted DNA for interrogation of DNA methylation. It offers a unique combination of comprehensive, expert-selected coverage, high sample throughput and an affordable price, making it the most widely used method for current epigenome-wide association studies (EWAS). The Illumina 450k array tests more than 485,000 CpGs at single-nucleotide resolution, which covers 99 % of RefSeq genes and 96 % of CGIs (Bibikova et al. 2011). The data processing procedures for Illumina 450k array contain several main steps, including quality control (QC), normalization, adjustment of batch effect, and calculation of DNA methylation levels.

- **QC**: The aim of this step is to detect and filter out samples and probes that do not meet the experimental standard. The Illumina 450k arrays contain several control probes for determining the data quality. Diagnostic plots of control probes in Illumina Genome-Studio program can be used to detect poorly performed samples (Bibikova et al. 2011). Assessing for poor quality samples can also be carried out by functions embedded in several R-packages specifically for analyzing Illumina 450k arrays, such as HumMethQCReport (Mancuso et al. 2011), IMA (Wang et al. 2012), Minfi (Aryee et al. 2014) and MethyLumi (Davis et al. 2012).

  For QC of probes, some packages such as IMA (Wang et al. 2012) filter out probes for which a large proportion of samples (i.e., >25 %) have a detection P-value >0.05. LumiWCluster avoids to discard probes (Kuan et al. 2010), instead it incorporates all the data while accounting for the quality of individual observations. A particular issue for QC of 450k array is that certain probes contain single nucleotide polymorphisms (SNPs) within the targeted sequences and thus the methylation levels assessed by these probes may be influenced by the DNA genotype (Dedeurwaerder et al. 2011). Hence, several programs (e.g., IMA) have incorporated functions to filter out these SNP-associated CpG probes (Wang et al. 2012; Touleimat and Tost 2012).
- **Normalization**: Normalization step is used to remove technical and systematic variation which could mask true biological differences. There are two types of normalization approach: (1) between-array normalization: address the comparability of intensity distribution between multiple arrays; (2) within-array normalization: correction for dye, intensity and spatial dependent bias within individual arrays (Siegmund 2011). The Illumina GenomeStudio uses a basic normalization approach by treating the first sample in the array as the reference but allows the user to reselect the reference sample if the first sample shows poor quality. This approach is also implemented in R-package MethyLumi (Davis et al. 2012) and Minfi (Aryee et al. 2014). Locally weighted scatterplot smoothing (LOESS) and quantile normalization assume similar total methylation signals across samples and may potentially discard the true biological signals (Laird 2010). There also exist several other approaches for normalizing the probe intensities (Marabita et al. 2013), but currently a lack of consensus exists regarding to the optimal normalization algorithm.

• **Adjustment of Batch Effect**: Batch effects represent measurements that have different behavior across conditions but are not related to the biological or scientific questions in a study (i.e. experiment time, chip or instrument used and laboratory conditions.). Some of the factors can be corrected by careful study design, for example, equally splitting the cases and controls into different batches by random sampling (Johnson et al. 2007) Other potential confounders may be corrected by several computational methods. For example, R-packages ComBat is a widely used adjustment method. It is based on empirical Bayes procedure (Johnson et al. 2007) and is robust to outliers in small sample sizes (Sun et al. 2011).

- **Calculation of DNA Methylation Levels**: DNA methylation levels are determined based on the intensities of the fluorescence signals from probes. The main output is the β-value and M-value which are ready for downstream statistical analysis. The β-value is calculated with the intensity of signal from methylated alleles (Max(M,0)) and the intensity of signal from unmethylated alleles (Max (U,0)) by the following formula:

$$\beta = \frac{\mathrm{Max}(\mathrm{M}, 0)}{\mathrm{Max}(\mathrm{M}, 0) + \mathrm{Max}(\mathrm{U}, 0) + 100}$$

The obtained β-value denotes the average methylation level for each CpG site. It ranges from 0 (unmethylated) to 1 (fully methylated) on a continuous scale. Alternative, some researchers use M-value to indicate the methylation level, which is calculated as

$$\mathrm{M} = \log_2 \frac{\mathrm{Max}\,(\mathrm{M}, 0) + 1}{\mathrm{Max}(\mathrm{U}, 0) + 1}$$

The range of M-values is negative infinity to positive infinity, which is consistent with data from normal distribution. However, the interpretation is of M-values is not as intuitive as β-value. The relationship of M-values and β-value is:

$$\mathrm{M} = \log_2 \frac{\beta}{1 - \beta}$$

Thus, positive M-values correspond to a methylation rate greater than 50 %, while negative M-values indicate a methylation rate less than 50 %.

Data Processing for RRBS

Processing of RRBS data mainly involves two steps, QC and alignment of sequencing reads.

- **QC**: The raw sequencing reads are normally generated in the fastq format, which records the sequence of nucleotides and their base call confidence levels. In order to obtain high quality RRBS data, several technical details require careful attention. For example, the incomplete bisulfite conversion will lead to spuriously elevated DNA methylation levels. One should use spike-in control DNAs with known DNA methylation levels to monitor the sensitivity and specificity of bisulfite conversion. Alternatively, elevated levels of observed CpC methylation can also provide an indication of incomplete bisulfite conversion because CpC dinucleotides are rarely methylated in mammalian cells (Bock 2012). Some of the QC steps for the RRBS data can be performed by QC tools (e.g., NGS QC toolkit) (Patel and Jain 2012) that are generally applicable to the next-generation sequencing produced reads, while other QC criteria such as efficiency of bisulfite conversion require QC tools that are dedicated to bisulfite sequencing, such as BSeQC (Lin et al. 2013).

- **Alignment**: Because of the reduced sequence complexity of the bisulfite converted sequence reads, alignment of bisulfite converted sequence reads to the reference genome require specific alignment tools. Generally, the alignment tools can be categorized into two groups: three-letter aligners and wild-card aligners. Bismark (Krueger and Andrews 2011) and BS-Seeker (Chen et al. 2010) are examples of three-letter aligners, which convert C to T in both sequenced reads and reference sequences prior to alignment. In contrast, wild-card aligners like BSMAP/RRBSMAP (Xi and Li 2009; Xi et al. 2012) replace Cs in the sequenced reads with wild-card Y but do not need the reference genome conversion step. Compare with whole-genome bisulfite alignment tool, such as an extensively validated MAQ-based pipeline, these specific aligners (e.g. RRBSMAP) could maintain high mapping accuracy and consistency between replicates, and also significantly improve runtime performance and memory efficiency (Xi et al. 2012).
- **Calculation of DNA Methylation Signals**: As unmethylated cytosines will be converted to Ts by the bisulfite treatment and methylated cytosines will stay Cs, absolute DNA methylation level could be calculate by counting the number of Cs and Ts at each C and simply divide the number of Cs by the total number of Cs and Ts.

Data Processing for MeDIP-Seq

In MeDIP-seq, the information of enrichment or depletion of extended sequencing reads will be used to estimate the methylation level of specific regions in the genome, the reads sequence itself does not provide methylation information. As a result, specific data processing approaches are needed to estimate the DNA methylation levels from MeDIP-seq method.

- **QC and alignment**: similar to other sequencing-based methods, the first step in the analysis of MeDIP-/MBD-seq is QC and alignment of sequencing reads to the reference genome, which can be conducted by using a standard quality control program and aligner, such as Bowtie2 (http://bowtie-bio.sourceforge.net/bowtie2/index.shtml) and BWA (Li and Durbin 2009).
- **Estimation of DNA methylation levels**: after alignment, the unique mapped reads are then extended to MeDIP-enriched DNA fragment size, the DNA sequence of each chromosome is divided into a series of certain base pair intervals (e.g. 50bp), and the extended reads in each interval are counted as the methylation signal in this region. These estimated DNA methylation signal can be confounded by varying density of methylated CpG sites. That is, regions with high CpG densities can give rise to high enrichment scores even with low absolute DNA

methylation levels and low CpG density regions can produce low enrichment scores even with high levels of DNA methylation. Down et al. developed the tool BATMAN which applies a Bayesian method to estimate absolute methylation values from MeDIP-chip or MeDIP-seq data (Down et al. 2008). It provides accurate estimations of methylation value, however it is not especially user-friendly and is quite a computationally technical process. Another tool is R package MEDIPS, it is a comprehensive approach for normalizing and analyzing MeDIP-seq data (Chavez et al. 2010). This method is based on the valuable concept of coupling factors presented by BATMAN (Down et al. 2008). MEDIPS incorporates a statistical frame work developed for count data which models the read number by an overdispersed Poisson model. This method could significantly reduce run time for processing MeDIP-seq data and easy to use.

10.2.1.3 Identifying Differentially Methylated Regions (DMRs)

In clinical study (e.g. case and control study), it is crucial important to identify the DMRs between different experimental condition. There are several different types of DMR, such as tissue-specific DMR and aging-specific DMR (Rakyan et al. 2011). According to DNA methylation profiling methods we use, these DMR can be a single CpG site or a region of interest (e.g. promoters, CGIs). The Student's t-test and Wilcoxon rank sum test can be used to identify DMRs by using the normalized methylation signal between two groups. Bock, C (Bock 2012) well summarized several other advanced methods which aim to improve DMR detection (e.g. mixture models (Wang 2011), stratification of t-test (Chen et al. 2012) and point out it is difficult to predict which methods will work best for real-world DNA methylation data sets. There are several different tools used for identification of DMRs. For Illumina 450k array, most commonly used tools including R package IMA (Wang et al. 2012) and Minfi (Aryee et al. 2014) etc. The IMA (Wang et al. 2012) apply Student's t-test and empirical Bayes statistics, it allows identification of DMRs in both single CpG sites and regions of interest. For regions of interest differential methylation analysis, IMA will compute the mean, median or Tukey's biweight robust average for the loci within that region and create an index. limma uses an empirical Bayes moderated t-test to improve power in small sample sizes. M-values should be used in these cases as they will rely much more heavily on the assumption of normality. Minfi (Aryee et al. 2014) uses an F-test or linear regression to test each genomic position for association between methylation and categorical or continuous phenotype, respectively. R package methylKit (Akalin et al. 2012) is most commonly used tools for RRBS data analysis. It applies a t-test or logistic regression to calculate p-values which are adjusted to q-values for multiple test correction. For MeDIP-seq data, R package MEDIPS is sufficiently fast and could be practical for routine processing of MeDIP–seq (Bock 2012). Importantly, we need to concern the issue of correction multiple hypothesis testing since the tests for differential DNA methylation are performed simultaneously at a large number of genomic loci.

10.2.2 Epigenome-Wide Histone Modification Analysis

10.2.2.1 Histone Modification Profiling Assays

Methods for epigenome-wide analysis of histone modification marks rely heavily on a procedure called chromatin immunoprecipitation (ChIP). The basic steps of ChIP includes: (1) Crosslink DNA and associated proteins on chromatin in cells; (2) Sonicate the DNA-protein complexes into ~500 bp fragments; (3) Immunoprecipitate DNA fragments using specific antibody against the particular histone mark; (4) Purify the immunoprecipitated DNA fragments and subsequently analyze by microarrays (ChIP-chip) or sequencing (ChIP-seq) (Fig. 10.3). To control for the effects of non-specific bindings, nonspecific immunoglobulin G (IgG) antibodies and input chromatin have been commonly used as controls (Kidder et al. 2011). Regions showing enrichment of ChIP products over controls represent DNA sequences where the specific histone modification marks are associated with in vivo. In addition to histone modification marks, the ChIP-chip/-seq methods can also be used to map global binding sites for specific transcription factors, RNA polymerases, or in principle any DNA-associated proteins.

10.2.2.2 Data Processing and Analysis

Using standard QC and alignment programs, the high quality sequencing reads from ChIP-seq data can be selected and mapped to the reference genome. The aligned reads are then used to identify regions of increased read tag density relative to the background estimated from the IgG/input controls. One straightforward approach is simply to use a minimum fold enrichment threshold of ChIP tags over normalized control tags in candidate regions/tiling windows. However, any threshold is arbitrary and prone to error, this approach does little to assist the user in assessing the significance of peaks (Wilbanks and Facciotti 2010). More sophisticated statistical approaches have been incorporated to identify and assess the significance of putative peaks (Pepke et al. 2009). So far, over 40 different 'peak calling' programs have been developed under a variety of statistical models, such as Poisson, local Poisson, t-distribution, conditional binomial, and hidden Markov models. Though a few studies attempted to compare the performance of some of these peak calling programs (Wilbanks and Facciotti 2010; Micsinai et al. 2012), there does not appear to be a clear winner and many program have multiple parameters that can be adjusted by the user. As using different programs or different parameter settings can significantly affect the final peak lists, care must be taken that data sets that are to be compared must be analyzed using the same methods and settings.

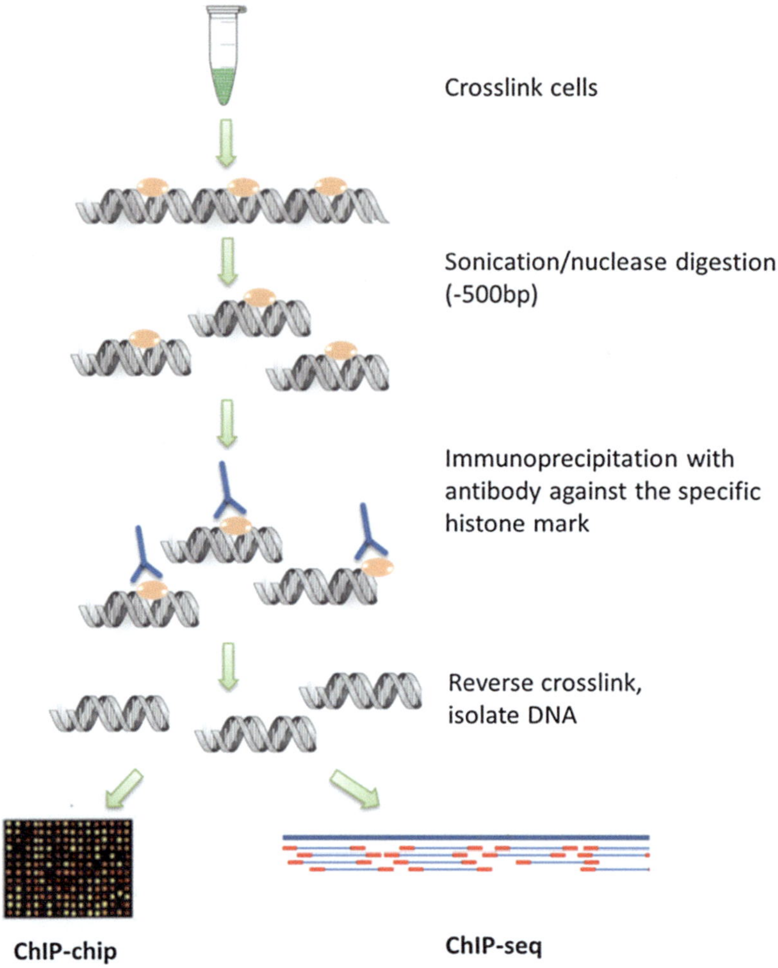

Fig. 10.3 Workflow for ChIP-chip and ChIP-seq. DNA and associated proteins are crosslinked and sheared into ~500 bp DNA fragments by sonication or nuclease digestion, DNA fragments associated with the histone mark of interest are selectively immunoprecipitated using an antibody specifically against the particular histone mark. Purified DNA can be analyzed by microarrays (ChIP-chip) or sequencing (ChIP-seq)

10.2.3 Epigenomic Data Visualization and Interpretation

Visualization of DNA methylation data (e.g. MeDIP-seq, RRBS) and histone modification (ChIP-seq) data is incredibly important. It enables you to investigate the data and may help you come up with new ideas about how to analyze the data. The ability to visualize these kinds of data is enabled through the use of some popular genome browsers, such as UCSC Genome Browser (Kent et al. 2002) and Integrative Genome Viewer (IGV) (Robinson et al. 2011). UCSC Genome Browser includes lots of published studies and ENCODE data, it is useful for data integration and visualization. However, the data files need to be uploaded to this web-based genome browser which makes a little more difficult to upload large custom data sets. IGV is Java based genome browser. It runs locally on your own computer. It does not have the same degree of public available data as UCSC genome browser, but tend to be somewhat faster for browsing across the genome. Also, it is better for looking at individual reads. There are several types of file format, such as BED, Wiggle and bedGraph format. BED files are very basic as they simply describe a region in the genome. They are usually used to describe MeDIP-seq and ChIP-Seq peaks. Nearly every genome browser supports visualization of BED files. Wiggle files are used to display quantitative information across genomic regions. Wiggle format is compact and displays data at regular intervals. Similar to Wiggle format, bedGraph use variable length intervals instead of constant intervals found in wiggle files, and are usually a little bigger in size. There are a bunch of specialized programs for creating genome browser files, such as bedToBigBed (https://www.encodeproject.org/software/bedToBigBed/) and igvtools (https://www.broadinstitute.org/igv/igvtools).

Several bioinformatics tools were used to interpret biological meaning from epigenomic data results. For example, EpiExplorer (Halachev et al. 2012) empowers biologists to explore large epigenome datasets in real time and over the Internet. It facilitates interactive hypothesis generation and identification of candidates for experimental follow-up. Cytoscape (Shannon et al. 2003) is an open source software platform for visualizing molecular interaction networks and biological pathways and integrating these networks with annotations, gene expression profiles and other state data. DAVID (da Huang et al. 2009) provides a comprehensive set of functional annotation tools for investigators to understand biological meaning behind large list of genes.

10.3 Epigenomics of Human Diseases

With the advent of new technologies we are starting to unravel the epigenomic mechanisms underlying a diverse range of human disorders, such as cancer and autoimmune diseases. A comprehensive understanding of epigenetic mechanisms, their interactions and alterations in human disease, has become a priority in clinical research (Portela and Esteller 2010).

10.3.1 Epigenomics of Cancer

Diverse altered DNA methylation patterns have been implicated in the pathogenesis and metastasis of various cancers. Genome-wide hypomethylation has been revealed in several common cancer types, such as stomach, liver and lung cancers

(Kulis and Esteller 2010) DNA hypomethylation occurs mostly at DNA-repetitive regions resulting in activation of genes with growth and tumor promoting functions and loss of genome stability and imprinting (Esteller 2008). A clear case is the long interspersed nuclear element (LINE) family member LINE1. Many studies have support correlations between LINE1 hypomethylation and increased risk of cancer (Barchitta et al. 2014). For example, hypomethylation of a specific LINE-1 promoter was found to induce an alternate transcript of the MET oncogene in bladder tumors and across the entire urothelium of tumor-bearing bladders (Wolff et al. 2010). A high degree of LINE-1 hypomethylation is a unique feature of early-onset colorectal cancer (Antelo et al. 2012; Ogino et al. 2013), and hypomethylation of LINE-1 in primary tumor has been associated with poor prognosis and survival in young breast cancer patients (van Hoesel et al. 2012) and prominent hypomethylation of Alu and LINE-1 in HER2 enriched subtype may be related to chromosomal instability (Park et al. 2014). In addition to the effects on repetitive elements, promoter hypomethylation can activate the aberrant expression of oncogenes and result in loss of imprinting in some loci (Portela and Esteller 2010). For instance, loss of imprinting of IGF2 gene has been associated with an increased risk of different types of cancer (Lim and Maher 2010). Recent study also shows that hypomethylation in TP73 and TERT gene body alter the transcriptional landscape of growth rate of glioblastoma through the activation of a limited number of normally silenced promoters within gene bodies, result in activating the aberrant expression of an oncogenic protein (Nagarajan et al. 2014). Hypermethylation at the CGIs of certain promoters causing transcriptional silencing of tumor suppressor gene were also observed. The transcriptional silencing caused by promoter hypermethylation affects genes involved in the multiple cellular pathways (Portela and Esteller 2010), such as DNA repair (e.g., MGMT, MLH1, MSH2, GSTP1), Ras signaling (e.g., DAPK, NOREIA, RASSFIA, RECK) etc. (Esteller 2007). For example, hypermethylation at CGI of MLH1 gene is reported in the majority of sporadic primary colorectal cancers with microsatellite instability, and that this methylation was often associated with loss of MLH1 protein expression (Herman et al. 1998).

Another epigenomic hallmark of cancer is the aberrant patterns of histone modifications. Epigenome-wide studies have characterized the overall profiles of various histone modification marks in cancer cells. For example, there is a global loss in H4K16ac in nearly all human cancer cell lines (Fraga et al. 2005b). Loss of acetylation is mediated by HDACs, which have been found to be overexpressed (Zhu et al. 2004) or mutated (Ropero et al. 2006) in different tumor types. Two different studies reported that global levels of H4K12ac and H3K18ac increased in adenocarcinomas in respect to normal tissue or adenoma (Ashktorab et al. 2009; Nakazawa et al. 2012). Cancer cells also bear global alterations of several histone methylation marks, such as a global loss of the active mark H3K4me3 (Hamamoto et al. 2004), and the repressive mark H4K20me3 (Fraga et al. 2005b), as well as a gain in the repressive marks H3K9me (Kondo et al. 2007) and H3K27me3 (Vire et al. 2006; Muller-Tidow et al. 2010).

Alterations of histone methylation marks in cancer cells are mainly due to the aberrant expression of both HMTs and histone demethylases (Chi et al. 2010). Gillian et al. reported the inactivating mutations in two genes encoding enzymes involved in histone modification: SETD2 gene (H3K36 methyltransferase) and JARID1C genes (H3K4 demethylase) in renal carcinomas (Dalgliesh et al. 2010). EZH2 gene (H3K27 methyltransferase) was reported overexpressed in several cancer types and enhances proliferation and neoplastic transformation (Kleer et al. 2003; Raman et al. 2005; Rhodes et al. 2004). NSD1, another HMT (H3K36 and H4K20), has been reported to undergo promoter DNA methylation-dependent silencing in neuroblastomas (Berdasco et al. 2009). H3K79 methyltransferase DOT1L is essential for development and maintenance the mixed lineage leukaemia. The presence of DOT1L results in H3K79 hypermethylation, which induces aberrant gene expression and contributes to leukemic transformation (Okada et al. 2006).

10.3.2 Epigenomics of Autoimmune Diseases

DNA methylation alteration has been increasingly associated with several autoimmune diseases in recent years; for which most studies focus on systemic autoimmune rheumatic diseases like systemic lupus erythematosus (SLE) and rheumatoid arthritis (RA). SLE is characterized by autoantibody response to nuclear and/or cytoplasmic antigens. Several studies have shown that there is a global hypomethylation of promoter regions, which drive the genes that are overexpressed in the disease such as PRF1, CD70, CD154, IFGNR2, MMP14, LCN2, CSF3R and AIM2 genes, and also in the ribosomal RNA gene promoter (18S and 28S) (Portela and Esteller 2010; Ballestar 2011). This global loss of methylation has been attributed to induce the activation of endogenous retroviruses such that they erase imprinting signals and deregulate gene expression and consequently break immune tolerance for active flaring of the disease (Okada et al. 2002). The hypomethylation in SLE may be partially mediated by miR-21 and miR-148a that directly and indirectly target DNMT1 (Pan et al. 2010; Zhu et al. 2011). RA is a chronic inflammatory disease that largely affects peripheral joints by invasive synovial fibroblasts. Global changes in DNA methylation measured in fibroblast like synoviocytes showed distinct methylation profiles of RA patients, particularly in genes with key roles in inflammation, immune responses and matrix deconvolution. Hypomethylated loci were identified in key genes relevant to RA, such as CHI3L1, CASP1, STAT3, MAP3K5, MEFV and WISP3. Hypermethylation was also observed at some RA related genes, including TGFBR2 and FOXO1. Differentially methylated genes could alter fibroblast like synoviocytes gene expression and contribute to the pathogenesis of RA. Histone modification studies in human autoimmune diseases have found that during apoptosis, histones can be modified to make them immunogenic. Hypoacetylated histones H3 and H4 and H3K9 hypomethylation in $CD4^+$ T cells were found to be a characteristic feature of SLE patients (Hu et al. 2008). In RA, the reduced activity of HDACs plays a key role in regulating NF-κB–mediated gene expression (Huber et al. 2007).

10.4 Discussion and Perspectives

Advances in technological development have enabled epigenomic analysis on a large scale. Remarkably, several international projects and consortia (Table 10.2) have been formed to comprehensively characterize epigenome-wide DNA methylation, histone modification, and other epigenetic profiles in healthy and disease tissues, such as the Encyclopedia of DNA Elements (ENCODE) Project (Consortium 2012), the Cancer Genome Atlas (TCGA) (TCGA. The Cancer Genome Atlas. http://cancergenome.nih.gov/) and the NIH Roadmap Epigenomics Project (The NIH Roadmap Epigenomics Project, http://www.epigenomebrowser.org/).

Although the number of epigenomic studies has grown exponentially in recent years, several issues need to be carefully considered when planning and interpretation of such studies. First, disease-associated epigenetic variation is likely to be cell-/tissue-specific. For studies using heterogeneous cell/tissue samples (e. g. blood, tumor), detection of differential DNA methylation or histone modification profiles is a problem of validity: molecular profile variation and changes in cell type proportions between tissue samples are confounded (Jacobsen et al. 2006; Jaffe and Irizarry 2014). If the disease-associated variation is restricted to a certain cell type that represents only a small proportion of the tissue sampled, then the

Table 10.2 Large-scale national and international epigenomic consortia

Project name	Start date	Affiliations	Data contributions	Access data
The Encyclopedia of DNA Elements (ENCODE) Project	2003	NIH	ChIP-seq, RNA-seq, DNase-seq, shRNA knockdown followed by RNA-seq, RRBS, shotgun bisulfite-seq assay, DNA methylation profiling by array assay etc.in more than 200 of primary human tissues and cell lines	http://encodeproject.org/
The Cancer Genome Atlas (TCGA)	2006	NIH	Matched tumor and normal tissues from 11,000 patients, allowing for the comprehensive characterization of 33 cancer types and subtypes, including 10 rare cancers	http://cancergenome.nih.gov/
Roadmap Epigenomics Project	2008	NIH	Bisulfite-seq, MeDIP-seq, MRE-seq, RRBS, DNasel, smRNA-seq, ChIP-seq etc. in more than 160s of normal primary cells, hESC, and hESC derived cells	http://www.epigenomebrowser.org/
International Cancer Genome Consortium (ICGC)	2008	17 countries, includes TCGA	DNA methylation profiles in thousands of patient samples from 31 tumor types	https://icgc.org/

variation may not be detected in the whole tissue (Jaffe and Irizarry 2014). Purified samples consisting only of a single cell type are preferable to mixed cell samples. Second, the complex system of the human body has many research areas, including genomics, epigenomics, transcriptomics, proteomics and metabolomics. Each research area provides insight into the system, but the entire complex of "omics" research offers more comprehensive insights. As costs of analysis of a human genome have dramatically plummeted, data integration is now a very commonly used notion. Integration between different epigenetic mechanisms and with other omics disciplines becomes easier and necessary for clinical research. For clinicians with access to omics data, being able to understand and appropriate interpret the data will become a key requirement for patient care. Along with the recent advancement in epigenetic drugs, there is a great potential for personalized epigenetic treatment of many human diseases in the near future.

References

Akalin A, Kormaksson M, Li S, Garrett-Bakelman FE, Figueroa ME, Melnick A, et al. MethylKit: a comprehensive R package for the analysis of genome-wide DNA methylation profiles. Genome Biol. 2012;13(10):R87. doi:10.1186/gb-2012-13-10-r87. PubMed PMID: 23034086; PubMed Central PMCID: PMC3491415.

Alegria-Torres JA, Baccarelli A, Bollati V. Epigenetics and lifestyle. Epigenomics. 2011;3 (3):267–77. doi:10.2217/epi.11.22. PubMed PMID: 22122337, PubMed Central PMCID: PMC3752894.

Antelo M, Balaguer F, Shia J, Shen Y, Hur K, Moreira L, et al. A high degree of LINE-1 hypomethylation is a unique feature of early-onset colorectal cancer. PLoS One. 2012;7(9): e45357. doi:10.1371/journal.pone.0045357. PubMed PMID: 23049789, PubMed Central PMCID: PMC3458035.

Aryee MJ, Jaffe AE, Corrada-Bravo H, Ladd-Acosta C, Feinberg AP, Hansen KD, et al. Minfi: a flexible and comprehensive Bioconductor package for the analysis of Infinium DNA methylation microarrays. Bioinformatics. 2014;30(10):1363–9. doi:10.1093/bioinformatics/btu049. PubMed PMID: 24478339, PubMed Central PMCID: PMC4016708.

Ashktorab H, Belgrave K, Hosseinkhah F, Brim H, Nouraie M, Takkikto M, et al. Global histone H4 acetylation and HDAC2 expression in colon adenoma and carcinoma. Dig Dis Sci. 2009;54 (10):2109–17. doi:10.1007/s10620-008-0601-7. PubMed PMID: 19057998, PubMed Central PMCID: PMC2737733.

Ballestar E. Epigenetic alterations in autoimmune rheumatic diseases. Nat Rev Rheumatol. 2011;7 (5):263–71. doi:10.1038/nrrheum.2011.16. PubMed.

Barchitta M, Quattrocchi A, Maugeri A, Vinciguerra M, Agodi A. LINE-1 hypomethylation in blood and tissue samples as an epigenetic marker for cancer risk: a systematic review and meta-analysis. PLoS One. 2014;9(10):e109478. doi:10.1371/journal.pone.0109478. PubMed PMID: 25275447, PubMed Central PMCID: PMC4183594.

Barski A, Cuddapah S, Cui K, Roh TY, Schones DE, Wang Z, et al. High-resolution profiling of histone methylations in the human genome. Cell. 2007;129(4):823–37. doi:10.1016/j.cell.2007.05.009. PubMed.

Bell JT, Pai AA, Pickrell JK, Gaffney DJ, Pique-Regi R, Degner JF, et al. DNA methylation patterns associate with genetic and gene expression variation in HapMap cell lines. Genome Biol. 2011;12(1):R10. doi:10.1186/gb-2011-12-1-r10. PubMed PMID: 21251332, PubMed Central PMCID: PMC3091299.

Berdasco M, Ropero S, Setien F, Fraga MF, Lapunzina P, Losson R, et al. Epigenetic inactivation of the Sotos overgrowth syndrome gene histone methyltransferase NSD1 in human neuroblastoma and glioma. Proc Natl Acad Sci U S A. 2009;106(51):21830–5. doi:10.1073/pnas.0906831106. PubMed PMID: 20018718, PubMed Central PMCID: PMC2793312.

Bibikova M, Barnes B, Tsan C, Ho V, Klotzle B, Le JM, et al. High density DNA methylation array with single CpG site resolution. Genomics. 2011;98(4):288–95. doi:10.1016/j.ygeno.2011.07.007. PubMed.

Bird AP. CpG-rich islands and the function of DNA methylation. Nature. 1986;321(6067):209–13. doi:10.1038/321209a0. PubMed PMID: WOS:A1986C330700035.

Bird AP, Southern EM. Use of restriction enzymes to study eukaryotic dna methylation .1. Methylation pattern in ribosomal dna from xenopus-laevis. J Mol Biol. 1978;118(1):27–47. doi:10.1016/0022-2836(78)90242-5. PubMed PMID: WOS:A1978EM86300002.

Bock C. Analysing and interpreting DNA methylation data. Nat Rev Genet. 2012;13(10):705–19. doi:10.1038/nrg3273. PubMed.

Cedar H, Bergman Y. Linking DNA methylation and histone modification: patterns and paradigms. Nat Rev Genet. 2009;10(5):295–304. doi:10.1038/nrg2540. PubMed.

Cedar H, Solage A, Glaser G, Razin A. Direct detection of methylated cytosine in DNA by use of the restriction enzyme MspI. Nucleic Acids Res. 1979;6(6):2125–32. doi:10.1093/nar/6.6.2125. PubMed PMID: WOS:A1979GX25500006.

Chavez L, Jozefczuk J, Grimm C, Dietrich J, Timmermann B, Lehrach H, et al. Computational analysis of genome-wide DNA methylation during the differentiation of human embryonic stem cells along the endodermal lineage. Genome Res. 2010;20(10):1441–50. doi:10.1101/gr.110114.110. PubMed PMID: 20802089, PubMed Central PMCID: PMC2945193.

Chen PY, Cokus SJ, Pellegrini M. BS Seeker: precise mapping for bisulfite sequencing. BMC Bioinf. 2010;11:203. doi:10.1186/1471-2105-11-203. PubMed PMID: 20416082, PubMed Central PMCID: PMC2871274.

Chen Z, Liu Q, Nadarajah S. A new statistical approach to detecting differentially methylated loci for case control Illumina array methylation data. Bioinformatics. 2012;28(8):1109–13. doi:10.1093/bioinformatics/bts093. PubMed PMID: 22368244, PubMed Central PMCID: PMC3324514.

Chi P, Allis CD, Wang GG. Covalent histone modifications–miswritten, misinterpreted and mis-erased in human cancers. Nat Rev Cancer. 2010;10(7):457–69. doi:10.1038/nrc2876. PubMed PMID: 20574448, PubMed Central PMCID: PMC3262678.

Cohen I, Poreba E, Kamieniarz K, Schneider R. Histone modifiers in cancer: friends or foes? Genes Cancer. 2011;2(6):631–47. doi:10.1177/1947601911417176. PubMed PMID: 21941619, PubMed Central PMCID: PMC3174261.

Consortium EP. An integrated encyclopedia of DNA elements in the human genome. Nature. 2012;489(7414):57–74. doi:10.1038/nature11247. PubMed PMID: 22955616, PubMed Central PMCID: PMC3439153.

da Huang W, Sherman BT, Lempicki RA. Bioinformatics enrichment tools: paths toward the comprehensive functional analysis of large gene lists. Nucleic Acids Res. 2009;37(1):1–13. doi:10.1093/nar/gkn923. PubMed PMID: 19033363, PubMed Central PMCID: PMC2615629.

Dalgliesh GL, Furge K, Greenman C, Chen L, Bignell G, Butler A, et al. Systematic sequencing of renal carcinoma reveals inactivation of histone modifying genes. Nature. 2010;463(7279):360–3. doi:10.1038/nature08672. PubMed PMID: 20054297, PubMed Central PMCID: PMC2820242.

Davis S DP, Bilke S, Triche JrT, Bootwalla M. Methylumi: handle illumina methylation data. R package version 220. 2012.

Dedeurwaerder S, Defrance M, Calonne E, Denis H, Sotiriou C, Fuks F. Evaluation of the infinium methylation 450K technology. Epigenomics. 2011;3(6):771–84. doi:10.2217/epi.11.105.

Down TA, Rakyan VK, Turner DJ, Flicek P, Li H, Kulesha E, et al. A Bayesian deconvolution strategy for immunoprecipitation-based DNA methylome analysis. Nat Biotechnol. 2008;26(7):779–85. doi:10.1038/nbt1414. PubMed PMID: 18612301, PubMed Central PMCID: PMC2644410.

Esteller M. Epigenetic gene silencing in cancer: the DNA hypermethylome. Hum Mol Genet. 2007;16(Spec No 1):R50–9. doi:10.1093/hmg/ddm018. PubMed.

Esteller M. Epigenetics in cancer. N Engl J Med. 2008;358(11):1148–59. doi:10.1056/NEJMra072067. PubMed.

Esteller M. Non-coding RNAs in human disease. Nat Rev Genet. 2011;12(12):861–74. doi:10.1038/nrg3074. PubMed.

Fraga MF, Ballestar E, Paz MF, Ropero S, Setien F, Ballestar ML, et al. Epigenetic differences arise during the lifetime of monozygotic twins. Proc Natl Acad Sci U S A. 2005a;102(30):10604–9. doi:10.1073/pnas.0500398102. PubMed PMID: 16009939, PubMed Central PMCID: PMC1174919.

Fraga MF, Ballestar E, Villar-Garea A, Boix-Chornet M, Espada J, Schotta G, et al. Loss of acetylation at Lys16 and trimethylation at Lys20 of histone H4 is a common hallmark of human cancer. Nat Genet. 2005b;37(4):391–400. doi:10.1038/ng1531. PubMed.

Gardiner-Garden M, Frommer M. CpG islands in vertebrate genomes. J Mol Biol. 1987;196(2):261–82. PubMed.

Greer EL, Shi Y. Histone methylation: a dynamic mark in health, disease and inheritance. Nat Rev Genet. 2012;13(5):343–57. doi:10.1038/nrg3173. PubMed PMID: WOS:000303045400011.

Haines TR, Rodenhiser DI, Ainsworth PJ. Allele-specific non-CpG methylation of the Nf1 gene during early mouse development. Dev Biol. 2001;240(2):585–98. doi:10.1006/dbio.2001.0504. PubMed PMID: WOS:000173216300021.

Halachev K, Bast H, Albrecht F, Lengauer T, Bock C. EpiExplorer: live exploration and global analysis of large epigenomic datasets. Genome Biol. 2012;13(10):R96. doi:10.1186/gb-2012-13-10-r96. PubMed PMID: 23034089, PubMed Central PMCID: PMC3491424.

Hamamoto R, Furukawa Y, Morita M, Iimura Y, Silva FP, Li M, et al. SMYD3 encodes a histone methyltransferase involved in the proliferation of cancer cells. Nat Cell Biol. 2004;6(8):731–40. doi:10.1038/ncb1151. PubMed.

Heerboth S, Lapinska K, Snyder N, Leary M, Rollinson S, Sarkar S. Use of epigenetic drugs in disease: an overview. Genet Epigenet. 2014;6:9–19. doi:10.4137/GEG.S12270. PubMed PMID: 25512710, PubMed Central PMCID: PMC4251063.

Hellman A, Chess A. Gene body-specific methylation on the active X chromosome. Science. 2007;315(5815):1141–3. doi:10.1126/science.1136352. PubMed PMID: WOS:000244387600041.

Herman JG, Umar A, Polyak K, Graff JR, Ahuja N, Issa JP, et al. Incidence and functional consequences of hMLH1 promoter hypermethylation in colorectal carcinoma. Proc Natl Acad Sci U S A. 1998;95(12):6870–5. PubMed PMID: 9618505, PubMed Central PMCID: PMC22665.

Hodawadekar SC, Marmorstein R. Chemistry of acetyl transfer by histone modifying enzymes: structure, mechanism and implications for effector design. Oncogene. 2007;26(37):5528–40. doi:10.1038/sj.onc.1210619. PubMed.

Hu N, Qiu X, Luo Y, Yuan J, Li Y, Lei W, et al. Abnormal histone modification patterns in lupus CD4+ T cells. J Rheumatol. 2008;35(5):804–10. PubMed.

Huber LC, Stanczyk J, Jungel A, Gay S. Epigenetics in inflammatory rheumatic diseases. Arthritis Rheum. 2007;56(11):3523–31. doi:10.1002/art.22948. PubMed.

Jacobsen M, Repsilber D, Gutschmidt A, Neher A, Feldmann K, Mollenkopf HJ, et al. Deconfounding microarray analysis – independent measurements of cell type proportions used in a regression model to resolve tissue heterogeneity bias. Methods Inf Med. 2006;45(5):557–63. PubMed.

Jaffe AE, Irizarry RA. Accounting for cellular heterogeneity is critical in epigenome-wide association studies. Genome Biol. 2014;15(2):R31. doi:10.1186/gb-2014-15-2-r31. PubMed PMID: 24495553, PubMed Central PMCID: PMC4053810.

Johnson WE, Li C, Rabinovic A. Adjusting batch effects in microarray expression data using empirical Bayes methods. Biostatistics. 2007;8(1):118–27. doi:10.1093/biostatistics/kxj037. PubMed.

Kent WJ, Sugnet CW, Furey TS, Roskin KM, Pringle TH, Zahler AM, et al. The human genome browser at UCSC. Genome Res. 2002;12(6):996–1006. doi:10.1101/gr.229102. Article published online before print in May 2002. PubMed PMID: 12045153; PubMed Central PMCID: PMC186604.

Kidder BL, Hu G, Zhao K. ChIP-Seq: technical considerations for obtaining high-quality data. Nat Immunol. 2011;12(10):918–22. doi:10.1038/ni.2117. PubMed PMID: 21934668, PubMed Central PMCID: PMC3541830.

Kleer CG, Cao Q, Varambally S, Shen R, Ota I, Tomlins SA, et al. EZH2 is a marker of aggressive breast cancer and promotes neoplastic transformation of breast epithelial cells. Proc Natl Acad Sci U S A. 2003;100(20):11606–11. doi:10.1073/pnas.1933744100. PubMed PMID: 14500907, PubMed Central PMCID: PMC208805.

Kohli RM, Zhang Y. TET enzymes, TDG and the dynamics of DNA demethylation. Nature. 2013;502(7472):472–9. doi:10.1038/nature12750. PubMed PMID: 24153300, PubMed Central PMCID: PMC4046508.

Kondo Y, Shen L, Suzuki S, Kurokawa T, Masuko K, Tanaka Y, et al. Alterations of DNA methylation and histone modifications contribute to gene silencing in hepatocellular carcinomas. Hepatol Res. 2007;37(11):974–83. doi:10.1111/j.1872-034X.2007.00141.x. PubMed.

Krueger F, Andrews SR. Bismark: a flexible aligner and methylation caller for Bisulfite-Seq applications. Bioinformatics. 2011;27(11):1571–2. doi:10.1093/bioinformatics/btr167. PubMed PMID: 21493656, PubMed Central PMCID: PMC3102221.

Ku CS, Naidoo N, Wu M, Soong R. Studying the epigenome using next generation sequencing. J Med Genet. 2011;48(11):721–30. doi:10.1136/jmedgenet-2011-100242. PubMed.

Kuan PF, Wang S, Zhou X, Chu H. A statistical framework for Illumina DNA methylation arrays. Bioinformatics. 2010;26(22):2849–55. doi:10.1093/bioinformatics/btq553. PubMed PMID: 20880956, PubMed Central PMCID: PMC3025715.

Kulis M, Esteller M. DNA methylation and cancer. Adv Genet. 2010;70:27–56. doi:10.1016/B978-0-12-380866-0.60002-2. PubMed.

Laird PW. Principles and challenges of genomewide DNA methylation analysis. Nat Rev Genet. 2010;11(3):191–203. doi:10.1038/nrg2732. PubMed.

Li H, Durbin R. Fast and accurate short read alignment with Burrows-Wheeler transform. Bioinformatics. 2009;25(14):1754–60. doi:10.1093/bioinformatics/btp324. PubMed PMID: 19451168, PubMed Central PMCID: PMC2705234.

Lim DH, Maher ER. Genomic imprinting syndromes and cancer. Adv Genet. 2010;70:145–75. doi:10.1016/B978-0-12-380866-0.60006-X. PubMed.

Lin X, Sun D, Rodriguez B, Zhao Q, Sun H, Zhang Y, et al. BSeQC: quality control of bisulfite sequencing experiments. Bioinformatics. 2013;29(24):3227–9. doi:10.1093/bioinformatics/btt548. PubMed PMID: 24064417, PubMed Central PMCID: PMC3842756.

Lister R, Pelizzola M, Dowen RH, Hawkins RD, Hon G, Tonti-Filippini J, et al. Human DNA methylomes at base resolution show widespread epigenomic differences. Nature. 2009;462(7271):315–22. doi:10.1038/nature08514. PubMed PMID: WOS:000271899300037.

Mancuso FM, Montfort M, Carreras A, Alibes A, Roma G. HumMeth27QCReport: an R package for quality control and primary analysis of Illumina Infinium methylation data. BMC Res Notes. 2011;4:546. doi:10.1186/1756-0500-4-546. PubMed PMID: 22182516, PubMed Central PMCID: PMC3285701.

Marabita F, Almgren M, Lindholm ME, Ruhrmann S, Fagerstrom-Billai F, Jagodic M, et al. An evaluation of analysis pipelines for DNA methylation profiling using the Illumina HumanMethylation450 BeadChip platform. Epigenetics. 2013;8(3):333–46. doi:10.4161/epi.24008. PubMed PMID: 23422812, PubMed Central PMCID: PMC3669124.

Meissner A, Gnirke A, Bell GW, Ramsahoye B, Lander ES, Jaenisch R. Reduced representation bisulfite sequencing for comparative high-resolution DNA methylation analysis. Nucleic Acids Res. 2005;33(18):5868–77. doi:10.1093/nar/gki901. PubMed PMID: 16224102, PubMed Central PMCID: PMC1258174.

Micsinai M, Parisi F, Strino F, Asp P, Dynlacht BD, Kluger Y. Picking ChIP-seq peak detectors for analyzing chromatin modification experiments. Nucleic Acids Res. 2012;40(9):e70. doi:10.1093/nar/gks048. PubMed PMID: 22307239, PubMed Central PMCID: PMC3351193.

Muller-Tidow C, Klein HU, Hascher A, Isken F, Tickenbrock L, Thoennissen N, et al. Profiling of histone H3 lysine 9 trimethylation levels predicts transcription factor activity and survival in acute myeloid leukemia. Blood. 2010;116(18):3564–71. doi:10.1182/blood-2009-09-240978. PubMed PMID: 20498303, PubMed Central PMCID: PMC2981478.

Nagarajan RP, Zhang B, Bell RJ, Johnson BE, Olshen AB, Sundaram V, et al. Recurrent epimutations activate gene body promoters in primary glioblastoma. Genome Res. 2014;24(5):761–74. doi:10.1101/gr.164707.113. PubMed PMID: 24709822, PubMed Central PMCID: PMC4009606.

Nakazawa T, Kondo T, Ma D, Niu D, Mochizuki K, Kawasaki T, et al. Global histone modification of histone H3 in colorectal cancer and its precursor lesions. Hum Pathol. 2012;43(6):834–42. doi:10.1016/j.humpath.2011.07.009. PubMed.

Ogino S, Nishihara R, Lochhead P, Imamura Y, Kuchiba A, Morikawa T, et al. Prospective study of family history and colorectal cancer risk by tumor LINE-1 methylation level. J Natl Cancer Inst. 2013;105(2):130–40. doi:10.1093/jnci/djs482. PubMed PMID: 23175808, PubMed Central PMCID: PMC3545905.

Okada M, Ogasawara H, Kaneko H, Hishikawa T, Sekigawa I, Hashimoto H, et al. Role of DNA methylation in transcription of human endogenous retrovirus in the pathogenesis of systemic lupus erythematosus. J Rheumatol. 2002;29(8):1678–82. PubMed.

Okada Y, Jiang Q, Lemieux M, Jeannotte L, Su L, Zhang Y. Leukaemic transformation by CALM-AF10 involves upregulation of Hoxa5 by hDOT1L. Nat Cell Biol. 2006;8(9):1017–24. doi:10.1038/ncb1464. PubMed PMID: 16921363, PubMed Central PMCID: PMC4425349.

Okano M, Xie SP, Li E. Cloning and characterization of a family of novel mammalian DNA (cytosine-5) methyltransferases. Nat Genet. 1998;19(3):219–20. PubMed PMID: WOS:000074565900012.

Okano M, Bell DW, Haber DA, Li E. DNA methyltransferases Dnmt3a and Dnmt3b are essential for de novo methylation and mammalian development. Cell. 1999;99(3):247–57. doi:10.1016/s0092-8674(00)81656-6. PubMed PMID: WOS:000083440600003.

Pan W, Zhu S, Yuan M, Cui H, Wang L, Luo X, et al. MicroRNA-21 and microRNA-148a contribute to DNA hypomethylation in lupus CD4+ T cells by directly and indirectly targeting DNA methyltransferase 1. J Immunol. 2010;184(12):6773–81. doi:10.4049/jimmunol.0904060. PubMed.

Park SY, Seo AN, Jung HY, Gwak JM, Jung N, Cho NY, et al. Alu and LINE-1 hypomethylation is associated with HER2 enriched subtype of breast cancer. PLoS One. 2014;9(6):e100429. doi:10.1371/journal.pone.0100429. PubMed PMID: 24971511, PubMed Central PMCID: PMC4074093.

Patel RK, Jain M. NGS QC Toolkit: a toolkit for quality control of next generation sequencing data. PLoS One. 2012;7(2):e30619. doi:10.1371/journal.pone.0030619. PubMed PMID: 22312429, PubMed Central PMCID: PMC3270013.

Pepke S, Wold B, Mortazavi A. Computation for ChIP-seq and RNA-seq studies. Nat Methods. 2009;6(11 Suppl):S22–32. doi:10.1038/nmeth.1371. PubMed PMID: 19844228, PubMed Central PMCID: PMC4121056.

Portela A, Esteller M. Epigenetic modifications and human disease. Nat Biotechnol. 2010;28(10):1057–68. doi:10.1038/nbt.1685. PubMed.

Rakyan VK, Down TA, Balding DJ, Beck S. Epigenome-wide association studies for common human diseases. Nat Rev Genet. 2011;12(8):529–41. doi:10.1038/nrg3000. PubMed PMID: 21747404, PubMed Central PMCID: PMC3508712.

Raman JD, Mongan NP, Tickoo SK, Boorjian SA, Scherr DS, Gudas LJ. Increased expression of the polycomb group gene, EZH2, in transitional cell carcinoma of the bladder. Clin Cancer Res. 2005;11(24 Pt 1):8570–6. doi:10.1158/1078-0432.CCR-05-1047. PubMed.

Ramsahoye BH, Biniszkiewicz D, Lyko F, Clark V, Bird AP, Jaenisch R. Non-CpG methylation is prevalent in embryonic stem cells and may be mediated by DNA methyltransferase 3a. Proc Natl Acad Sci U S A. 2000;97(10):5237–42. doi:10.1073/pnas.97.10.5237. PubMed PMID: WOS:000086998500044.

Rhodes DR, Yu J, Shanker K, Deshpande N, Varambally R, Ghosh D, et al. Large-scale meta-analysis of cancer microarray data identifies common transcriptional profiles of neoplastic transformation and progression. Proc Natl Acad Sci U S A. 2004;101(25):9309–14. doi:10.1073/pnas.0401994101. PubMed PMID: 15184677, PubMed Central PMCID: PMC438973.

Rivera CM, Ren B. Mapping human epigenomes. Cell. 2013;155(1):39–55. doi:10.1016/j.cell.2013.09.011. PubMed PMID: 24074860, PubMed Central PMCID: PMC3838898.

Robinson MD, Stirzaker C, Statham AL, Coolen MW, Song JZ, Nair SS, et al. Evaluation of affinity-based genome-wide DNA methylation data: effects of CpG density, amplification bias, and copy number variation. Genome Res. 2010;20(12):1719–29. doi:10.1101/gr.110601.110. PubMed PMID: 21045081, PubMed Central PMCID: PMC2989998.

Robinson JT, Thorvaldsdottir H, Winckler W, Guttman M, Lander ES, Getz G, et al. Integrative genomics viewer. Nat Biotechnol. 2011;29(1):24–6. doi:10.1038/nbt.1754. PubMed PMID: 21221095, PubMed Central PMCID: PMC3346182.

Ropero S, Fraga MF, Ballestar E, Hamelin R, Yamamoto H, Boix-Chornet M, et al. A truncating mutation of HDAC2 in human cancers confers resistance to histone deacetylase inhibition. Nat Genet. 2006;38(5):566–9. doi:10.1038/ng1773. PubMed.

Shannon P, Markiel A, Ozier O, Baliga NS, Wang JT, Ramage D, et al. Cytoscape: a software environment for integrated models of biomolecular interaction networks. Genome Res. 2003;13(11):2498–504. doi:10.1101/gr.1239303. PubMed PMID: 14597658, PubMed Central PMCID: PMC403769.

Siegmund KD. Statistical approaches for the analysis of DNA methylation microarray data. Hum Genet. 2011;129(6):585–95. doi:10.1007/s00439-011-0993-x. PubMed PMID: 21519831, PubMed Central PMCID: PMC3166559.

Sterner DE, Berger SL. Acetylation of histones and transcription-related factors. Microbiol Mol Biol Rev MMBR. 2000;64(2):435–59. PubMed PMID: 10839822, PubMed Central PMCID: PMC98999.

Sun Z, Chai HS, Wu Y, White WM, Donkena KV, Klein CJ, et al. Batch effect correction for genome-wide methylation data with Illumina Infinium platform. BMC Med Genomics. 2011;4:84. doi:10.1186/1755-8794-4-84. PubMed PMID: 22171553, PubMed Central PMCID: PMC3265417.

Suzuki MM, Bird A. DNA methylation landscapes: provocative insights from epigenomics. Nat Rev Genet. 2008;9(6):465–76. doi:10.1038/nrg2341. PubMed.

Touleimat N, Tost J. Complete pipeline for Infinium((R)) Human Methylation 450K BeadChip data processing using subset quantile normalization for accurate DNA methylation estimation. Epigenomics. 2012;4(3):325–41. doi:10.2217/epi.12.21. PubMed.

van Hoesel AQ, van de Velde CJ, Kuppen PJ, Liefers GJ, Putter H, Sato Y, et al. Hypomethylation of LINE-1 in primary tumor has poor prognosis in young breast cancer patients: a retrospective cohort study. Breast Cancer Res Treat. 2012;134(3):1103–14. doi:10.1007/s10549-012-2038-0. PubMed.

Varley KE, Gertz J, Bowling KM, Parker SL, Reddy TE, Pauli-Behn F, et al. Dynamic DNA methylation across diverse human cell lines and tissues. Genome Res. 2013;23(3):555–67. doi:10.1101/gr.147942.112. PubMed PMID: 23325432, PubMed Central PMCID: PMC3589544.

Vire E, Brenner C, Deplus R, Blanchon L, Fraga M, Didelot C, et al. The Polycomb group protein EZH2 directly controls DNA methylation. Nature. 2006;439(7078):871–4. doi:10.1038/nature04431. PubMed.

Waddington CH. Canalization of development and genetic assimilation of acquired characters. Nature. 1959;183(4676):1654–5. PubMed.

Wagner JR, Busche S, Ge B, Kwan T, Pastinen T, Blanchette M. The relationship between DNA methylation, genetic and expression inter-individual variation in untransformed human fibroblasts. Genome Biol. 2014;15(2):R37. doi:10.1186/gb-2014-15-2-r37. PubMed PMID: 24555846, PubMed Central PMCID: PMC4053980.

Wang S. Method to detect differentially methylated loci with case-control designs using Illumina arrays. Genet Epidemiol. 2011;35(7):686–94. doi:10.1002/gepi.20619. PubMed PMID: 21818777, PubMed Central PMCID: PMC3197755.

Wang D, Yan L, Hu Q, Sucheston LE, Higgins MJ, Ambrosone CB, et al. IMA: an R package for high-throughput analysis of Illumina's 450K Infinium methylation data. Bioinformatics. 2012;28(5):729–30. doi:10.1093/bioinformatics/bts013. PubMed PMID: 22253290, PubMed Central PMCID: PMC3289916.

Wilbanks EG, Facciotti MT. Evaluation of algorithm performance in ChIP-seq peak detection. PLoS One. 2010;5(7):e11471. doi:10.1371/journal.pone.0011471. PubMed PMID: 20628599, PubMed Central PMCID: PMC2900203.

Wolff EM, Byun HM, Han HF, Sharma S, Nichols PW, Siegmund KD, et al. Hypomethylation of a LINE-1 promoter activates an alternate transcript of the MET oncogene in bladders with cancer. PLoS Genet. 2010;6(4):e1000917. doi:10.1371/journal.pgen.1000917. PubMed PMID: 20421991, PubMed Central PMCID: PMC2858672.

Wu SC, Zhang Y. Active DNA demethylation: many roads lead to Rome. Nat Rev Mol Cell Biol. 2010;11(9):607–20. doi:10.1038/nrm2950. PubMed PMID: 20683471, PubMed Central PMCID: PMC3711520.

Xi Y, Li W. BSMAP: whole genome bisulfite sequence MAPping program. BMC Bioinf. 2009;10:232. doi:10.1186/1471-2105-10-232. PubMed PMID: 19635165, PubMed Central PMCID: PMC2724425.

Xi Y, Bock C, Muller F, Sun D, Meissner A, Li W. RRBSMAP: a fast, accurate and user-friendly alignment tool for reduced representation bisulfite sequencing. Bioinformatics. 2012;28 (3):430–2. doi:10.1093/bioinformatics/btr668. PubMed PMID: 22155871, PubMed Central PMCID: PMC3268241.

Zhu P, Martin E, Mengwasser J, Schlag P, Janssen KP, Gottlicher M. Induction of HDAC2 expression upon loss of APC in colorectal tumorigenesis. Cancer Cell. 2004;5(5):455–63. PubMed.

Zhu X, Liang J, Li F, Yang Y, Xiang L, Xu J. Analysis of associations between the patterns of global DNA hypomethylation and expression of DNA methyltransferase in patients with systemic lupus erythematosus. Int J Dermatol. 2011;50(6):697–704. doi:10.1111/j.1365-4632.2010.04804.x. PubMed.

Chapter 11
Integrative Biological Databases

Jinzeng Wang and Haiyun Wang

Abstract High throughput biotechnology brought an increasing number of the omics data across the species, making it possible to understand the genomic and genetic information in a systematic way. In this chapter, we introduce some valuable biological knowledgebase, covering the integrative aspects of the pathway structures, the molecular functions, macromolecular structures, molecular interactions, and so on. Moreover, a list of these databases is summarized in a table at the end of this chapter for a quick review.

Keywords Integrative databases • KEGG pathways • The gene ontology • Functional annotation • Molecular interactions

11.1 Introduction

The Human Genome Project (HGP) and the novel high throughput technology brought an increasing number of the omics data across the species, making it possible to understand the genomic and genetic information in a systematic way. As a new subject, Bioinformatics develops quickly in a few of decades along with the development of omics technologies. Bioinformatics develops algorithms, computational and statistical models, as well as some biological databases, to support the integrative research on biology and medicine. This chapter is aimed to introduce some integrative biological databases, which provides the detailed information of the pathway structures, the molecular functions, three-dimensional structures of macro-molecules, molecular interactions, and so on. These databases are listed in the following and take a comprehensive summary finally:

(1) Kyoto Encyclopedia of Genes and Genomes (KEGG)
(2) Gene ontology (GO)
(3) Search Tool for the Retrieval of Interacting Genes/Proteins (STRING)
(4) RCSB Protein Data Bank
(5) Other related databases for pathway and network

J. Wang • H. Wang (✉)
School of Life Science and Technology, Tongji University, 1239 Siping Road, 200092 Shanghai, People's Republic of China
e-mail: wanghaiyun@tongji.edu.cn

X. Wang et al. (eds.), *Application of Clinical Bioinformatics*, Translational Bioinformatics 11, DOI 10.1007/978-94-017-7543-4_11

11.2 Kyoto Encyclopedia of Genes and Genomes (KEGG)

KEGG is an integrative knowledgebase for systematic analysis of genomes, biological pathways, diseases, drugs and chemical substances. As part of Japanese Human Genome Program, it was developed in 1995 by Prof. Minoru Kanehisa at the Institute for Chemical Research, Kyoto University (Kanehisa and Coto 2000). KEGG (http://www.kegg.jp) currently consists of 17 main databases that cover four kinds of information including systems information, genomic information, chemical information and health information (Kanehisa et al. 2014) for understanding high-level functions and the biological system (Table 11.1).

KEGG PATHWAY is the most popular database in KEGG. The most unique feature in this database is the molecular networks—molecular interaction, reaction and relation networks interpreting systemic functions of the cell and the organism. Knowledge about the systemic mechanisms is manually collected and presented in the form called Pathway Map. Pathway Map is organized into a hierarchical structure from seven different aspects that cover the knowledge of metabolism, genetic information processing, environmental information processing, cellular processes, organismal systems, human diseases and drug development. Take a specific cancer type, non-small cell lung cancer (NSCLC) for example (Fig. 11.1).

Table 11.1 The overview of KEGG resource as of 2015/9/1

Category	Database	Content	Statistics
Systems information	KEGG PATHWAY	Pathway maps, reference (total)	477 (413,441)
	KEGG BRITE	Functional hierarchies, reference (total)	210 (141,124)
	KEGG MODULE	KEGG modules, reference (total)	710 (334,424)
Genomic information	KEGG ORTHOLOGY	KEGG Orthology (KO) groups	18,963
	KEGG GENOME	KEGG organisms with complete genomes	4029
	KEGG GENES	Genes catalogs of complete genomes	17,735,547
	KEGG SSDB	Sequence similarity database for GENES	104,942,041,227
Chemical information	KEGG COMPOUND	Metabolites and other small molecules	17,458
	KEGG GLYCAN	Glycans	10,989
	KEGG REACTION	Biochemical reactions	9925
	KEGG RPAIR	Reactant pair chemical transformations	15,074
	KEGG RCLASS	Reaction class	2980
	KEGG ENZYME	Enzyme nomenclature	6510
Health information	KEGG DISEASE	Human diseases	1430
	KEGG DRUG	Drugs	10,304
	KEGG DGROUP	Drug groups	1818
	KEGG ENVIRON	Crude drugs and health-related substances	850

http://www.kegg.jp/kegg/docs/statistics.html

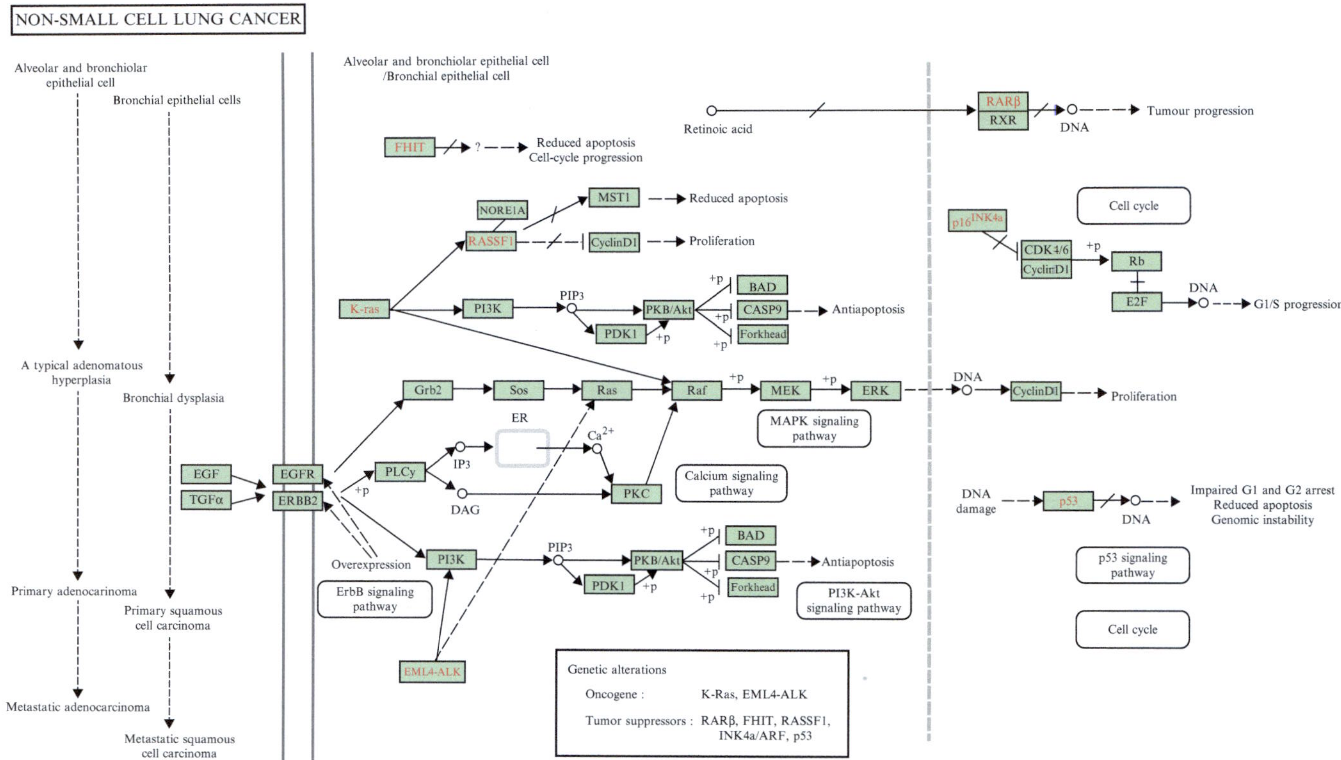

Fig. 11.1 KEGG pathway for Non-small cell lung cancer (NSCLC, http://www.genome.jp/kegg-bin/show_pathway?map=hsa05223&show_description=show)**:** Point mutations within the K-RAS gene inactivate GTPase activity and the p21-RAS protein transmits growth signals to the nucleus. Up-regulation of c-erbB-2 or EGFR leads to an advantageous proliferation. EML4-ALK fusion causes constitutive ALK activation, which leads to cell proliferation, invasion, and inhibition of apoptosis. Inactivation of p53 leads to rapid proliferation and reduced apoptosis. The protein coded by p16INK4a inhibits formation of CDK-cyclin-D complexes via competitive binding of CDK4 and CDK6. Loss of p16INK4a expression is a common characteristic of NSCLC. RAR-beta is a nuclear receptor with vitamin-A-dependent transcriptional activity. RASSF1A forms heterodimers with Nore-1—an RAS effector. Loss of RASSFIA may transform the balance of RAS activity towards continuously growing effect

NSCLC accounts for about 85 % of lung cancer, which is a leading cancer death among both men and women. Molecular mechanisms of alteration in NSCLC include activation of oncogenes K-RAS and EML4-ALK, and inactivation of tumor suppressor genes p53, RAR-beta, and RASSF1. A pathway basically consists of nodes (genes, proteins and other small molecules) and edges (relations, interactions and reactions). The notation of various kinds of nodes and edges is shown in Fig. 11.2.

In contrast to KEGG PATHWAY, which is limited to molecular interactions and reactions, KEGG BRITE provides many different types of relationships. It is a collection of hierarchical classifications representing knowledge on various aspects of biological systems. Thus, the genomic and molecular data of KEGG BRITE supplements the KEGG PATHWAY for inferring high-order functions.

The databases in the chemical information category, named KEGG LIGAND, are organized by capturing knowledge of the chemical network. Currently, KEGG LIGAND consists of six databases: KEGG COMPOUND for metabolites and other

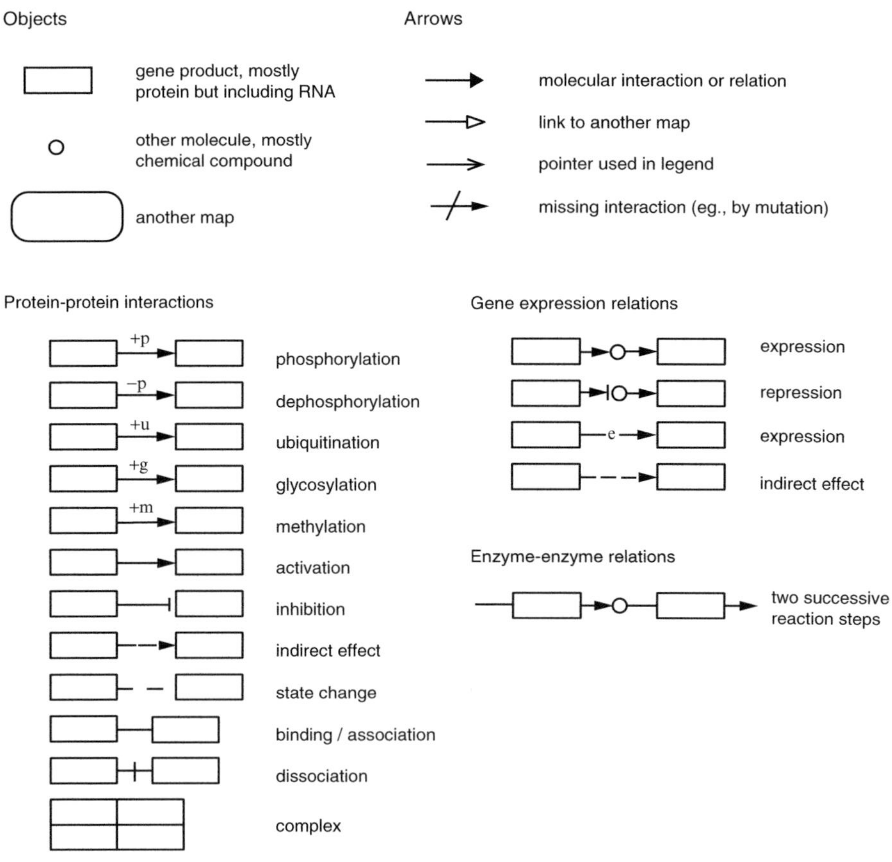

Fig. 11.2 The notation for the molecular network of KEGG pathway

small molecules, KEGG GLYCAN for glycans, KEGG REACTION for chemical reactions, KEGG RPAIR for reactant pair alignments, RCLASS for reaction classes defined by RPAIR and KEGG ENZYME for reactions in the enzyme nomenclature (Goto et al. 1999; Hashimoto et al. 2006; Muto et al. 2013).

In the health category of KEGG, there are two main kinds of information: disease and drug. Diseases are viewed as perturbed states of the biological system caused by perturbations of genetic or environmental factors while drugs are viewed as different types of perturbants. The KEGG DISEASE database incorporates known genetic and environmental factors of diseases and the KEGG DRUG database covers active ingredients of approved drugs in Japan, USA, and Europe based on the chemical structures and chemical components, and associated targets, metabolic enzymes, and other molecular interaction network information (Kanehisa et al. 2010).

11.3 Gene Ontology (GO)

The GO (The Gene Ontology Consortium 2008, 2015; Ashburner et al. 2000) project is aimed at providing detailed descriptions of gene and gene products in cells across different organisms using dynamic, controlled vocabularies. To achieve this goal, it has developed three basic ontologies that represent gene or protein attributes. They are categorized into cellular component, molecular function and biological process, respectively. To date, the GO (http://geneontology.org) project has developed formal ontologies that represent over 40,000 biological concepts, and are continually being revised to reflect new discoveries.

Cellular component is defined as the location in which proteins or other gene products perform their function in cell, such as mitochondria or nucleus. It also incorporates places like plasma membrane. Molecular function refers to the biological and chemical activities of gene products. It does not specify where, when or in what context the action occurs but describes accurately what is done. Generally, the molecular function term is defined as the activities that can be carried out by individual gene products, while they can also performed by assembled complex of gene products. The activities, like "binding activity" or "transporter activity", are broad functional terms and examples of more specifically molecular function terms are "Toll receptor binding" or "adenylate cyclase activity". Biological process is termed as a series of affairs or molecular functions executed by one or more organized gene products. Similarly, it contains broad biological terms like "cell growth" and narrower biological terms such as "alpha-glucoside transpot" or "cAMP biosynthesis".

The structure of GO can be described in a diagram, where each GO term is represented as a node while the relations between the terms are shown using edges between the nodes. It must be noted that GO is loosely hierarchical, since a child term may have more than one parent term. Terms are mainly linked by three relationships: "is-a", "part-of", and "regulates".

Accession	GO:0016049
Name	cell growth
Ontology	biological_process
Synonyms	**alt. id:** GO:0048591
	cellular growth
	growth of cell
	cell expansion
	metabolic process resulting in cell growth
	metabolism resulting in cell growth
	non-developmental cell growth
	non-developmental growth of a unicellular organism
Definition	The process in which a cell irreversibly increases in size over time by accretion and biosynthetic production of matter similar to that already present. *Source:* GOC:ai
Comment	None
History	See term history for GO:0016049 at QuickGO
Subset	goslim_metagenomics
	gosubset_prok
	goslim_plant
	goslim_pir
Community	**GN** Add usage comments for this term on the GONUTS wiki.
Related	Link to all **genes and gene products** associated to cell growth.
	Link to all direct and indirect **annotations** to cell growth.
	Link to all direct and indirect **annotations download** (limited to first 10,000) for cell growth.
Feedback	Contact the GO Helpdesk if you find mistakes or have concerns about the data you find here.

Fig. 11.3 A GO term named cell growth from Gene Ontology (http://amigo.geneontology.org/amigo/term/GO:0016049)

Typically, an ontology term has the essential elements (Fig. 11.3) with the "cell growth" as an example. Each term in the ontology has a term name and a unique seven digital identifier, which is called term accession number. The namespace indicates which of the three ontologies (cellular component, molecular function or biological) it belongs to. The definition describes what the term represents while the subset denotes the term belongs to an assigned subset of terms. Terms may also have synonyms, which are related to the term name in meaning. The scopes for GO synonyms are exact (an exact equivalent with the term name), broad (the synonym is broader than the term name), narrow (the synonym is narrower or more specific than the term name) and related (relevant to the term name in some cases). Each term in the ontology has defined relationships with one or more other terms, like is_a or part_of.

Along with the GO project, the GO Consortium developed two major tools: AmiGO and OBO-Edit. AmiGO (Carbon et al. 2008) is a web application that allows users to query, browse and visualize ontologies and gene or gene product annotation data. In addition, there are 4 other functionalities within AmiGO: BLAST, Term Enrichment, GO Slimmer and GO Online SQL Environment (GOOSE). AmiGO can be now freely used online at the Gene Ontology (GO) website to access the data provided by the GO Consortium; it can also be downloaded and installed to browse local ontologies and annotations. OBO-Edit (Day-Richter et al. 2007) is an open source, platform-independent ontology editor developed and maintained by the Gene Ontology Consortium. Based on Java, OBO-Edit uses a graph-oriented approach to display and edit ontologies. It incorporates a comprehensive search and filter interface for viewing and editing biochemical ontologies.

11.4 Search Tool for the Retrieval of Interacting Genes/ Proteins (STRING)

STRING (Snel et al. 2000; Jensen et al. 2009) is a web-based database that provides known and predicted protein interactions. The interactions include direct (physical) and indirect (functional) relationships and they are derived from four sources, including genomic context, high-throughput experiments, co-expression and public text mining. Currently, SRTING version 10 contains 9,643,763 proteins from 2031 organisms (Szklarczyk et al. 2015).

Protein-protein interaction network is of quite importance for better understanding the systemic cellular processes. STRING can be used for searching and viewing structural, functional and evolutionary properties of genes/proteins with intuitive platforms. Users can freely access the database with its online website (http://string-db.org) by inputting a protein name or ID. Alternatively, clicking on the other tabs, users can query by amino acid sequence, multiple names or multiple sequences. The organism can be specified by entering the name inside the relative input area. Take trpA in Escherichia coli K12 MG1655 for example (Fig. 11.4). It firstly provides the network view of the protein in which you are interested. The nodes in network denote proteins and the edges represent the predicted functional associations. Different colors indicate the evidence used to predict the interactions. The predicted associations for the input protein are also displayed in a summary view located below the view of the network. The input is shown at the top of the summary in red while the predicted interactions are shown just below, sorted by score. Users can view the data from different aspects by clicking the navigation buttons following the summary view. More detailed information about the protein or the prediction method scores can be attained via clicking on hyperlinks on the output web page.

The Info & Parameters panel at the bottom of the web page (Fig. 11.5) can be reset for preference, such as the prediction methods or confidence score.

11.5 RCSB Protein Data Bank

The Worldwide Protein Data Bank (wwPDB, http://www.wwpdb.org) is an organization that maintains the archive of macromolecular structure (Berman et al. 2003). As a member of wwPDB, RCSB Protein Data Bank (Berman et al. 2000; Rose et al. 2015) is a key database of structural biology that maintains the information for three-dimensional structural data of proteins, nucleic acids, and complex assemblies, which is for researchers to better understand structural, molecular and other aspects of biology. The data, mainly acquired by Nuclear magnetic resonance (NMR) Spectroscopy, X-ray crystallography or electron microscopy and submitted by biologists and biochemists from all over the world, are freely accessed via its online website (http://www.rcsb.org).

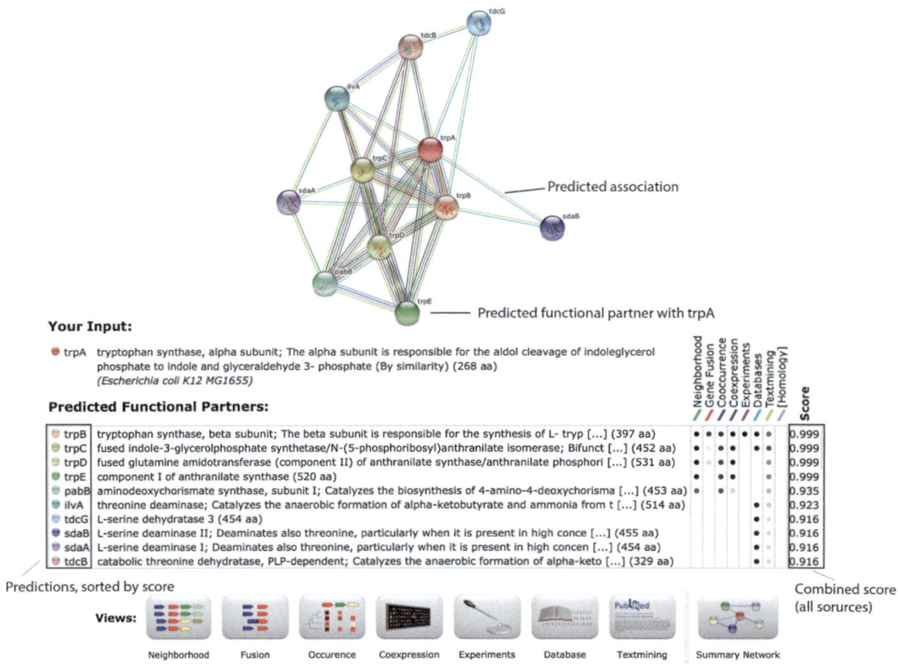

Fig. 11.4 The output web page of STRING (http://string-db.org/newstring_cgi/show_network_section.pl) by querying trpA in Escherichia coli K12 MG1655

At present, RCSB PDB covers 110,988 released entries across different species, such as Home sapiens, Escherichia coil, Mus musculus and so on. More detailed summary of RCSB PDB can be viewed through PDB Statistics Page (http://www.rcsb.org/pdb/static.do?p=general_information/pdb_statistics/index.html). Currently, users can access the RCSB PDB conveniently and search protein related information by various options (Fig. 11.6). Search categories can be PDB ID or name, sequence, structure annotation or features, chemical components, experimental methods, drug & drug targets and so on. Such queries can be also combined with AND or OR to perform complex searches.

Another important function of RCSB PDB is that users can visualize the structure, sequence and ligand binding of proteins or genes by different queries listed above. Meanwhile, RCSB PDB can do certain analysis for proteins, like structure quality, protein symmetry and sequence & structure alignment. Finally, researchers can download protein related data, which they are interested in.

In conclusion, RCSB PDB is a comprehensive protein resource, which holds detailed and various kinds of protein information.

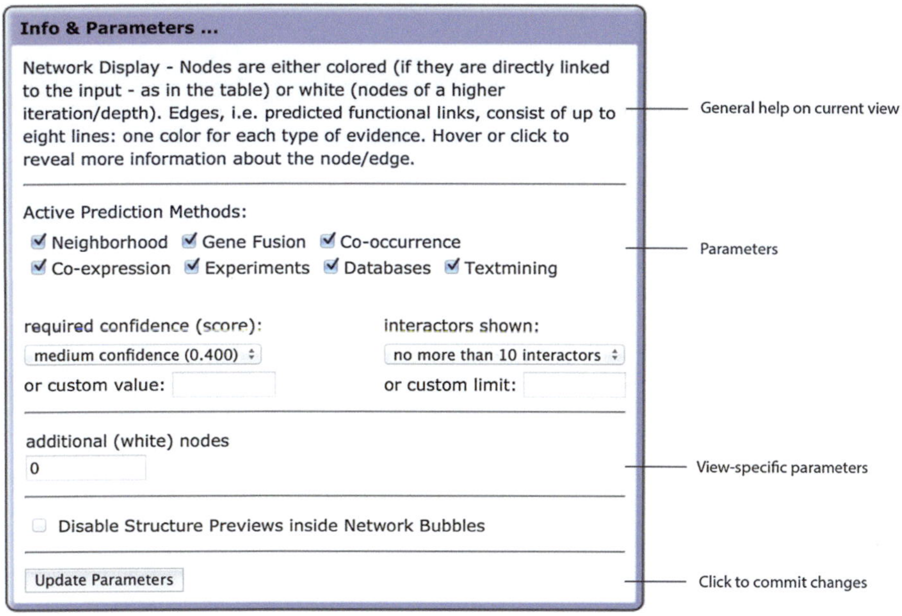

Fig. 11.5 The dialog box for parameters setting in STRING

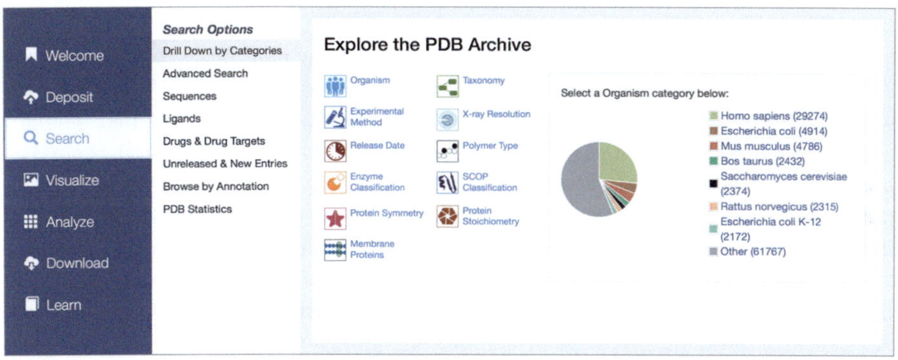

Fig. 11.6 The online search page of RCSB PDB (http://www.rcsb.org/pdb/home/home.do#Category-search)

11.6 Other Related Databases for Pathway and Networks

Other important tools or databases concerning pathways or networks are further summarized (Table 11.2).

The Database for Annotation, Visualization and Integrated Discovery (DAVID) (Huang et al. 2009a, b) now can provide a comprehensive set of functional annotation tools for understanding the biological function of a set of genes. For a given list, investigators can use DAVID (https://david.ncifcrf.gov) to identify enriched

Table 11.2 Other databases for pathway and networks

Name	Description	Url
DAVID	Database for annotation, visualization and integrated discovery	http://david.abcc.ncifcrf.gov
NCI/PID	National cancer institute/pathway interaction databases	http://pid.nci.nih.gov
REACTOME	An open, curated and peer-reviewed pathway database	http://www.reactome.org
PharmGKB	Pharmacogenomics knowledgebase	http://www.pharmgkb.org
Pathway Common	A network resource for biological pathway information	http://www.pathwaycommons.org
SIGNALING GATEWAY	Biological activity, regulation and localization for proteins	http://www.signalling-gateway.org

biological terms, discover enriched functional-related gene groups, cluster redundant annotation terms and more can be done with DAVID. National Cancer Institute/Pathway Interaction Database (NCI/PID, http://pid.nci.nih.gov) (Schaefer et al. 2009) is aimed at offering a highly structured, curated collection of information about known bio-molecular associations and key cellular processes. REACTOME (Joshi-Tope et al. 2004; Croft et al. 2010, 2014) provides an intuitive website (http://www.reactome.org) to navigate pathway knowledge and a series of data analysis tools to support the pathway-based analysis of complex experimental and computational data sets. The Pharmacogenomics Knowledgebase (PharmGKB) (Whirl-Carreillo et al. 2012) is a comprehensive database (https://www.pharmgkb.org) that covers knowledge about the impact of genetic variation on drug response and resistance for researchers and clinical therapists. It is a pharmacogenomics resource that includes clinical information, like drug labels, gene-drug involved pathways and genotype-phenotype associations. Pathway Commons (Cerami et al. 2011) is a network knowledge base (http://www.pathwaycommons.org) for biological pathway information collected from public pathway databases, which can be used to query, visualize and download. The Signaling Gateway (Dinasarapu et al. 2011) developed by UCSD (http://www.signalling-gateway.org) provides the essential information for more than thousands of proteins involved in cellular signaling, which depicts the biological activity, regulation and localization of proteins. Most importantly, all these data and software mentioned above are freely available for researchers.

11.7 Conclusion

Elucidating the complexities of biological pathways and protein-protein interaction is of immense importance to gain understanding the mechanism and what is the optimal treatment strategy of various diseases for investigators and clinicians. Undoubtedly, each existing database has its merits for particular objectives and the developed resources may overlap with each other in certain conditions for querying and analyzing biological data. Nevertheless, it is quite crucial for us to know when we should utilize one or more of these databases, and which is the best choice for biochemical research and clinical guidance.

To direct the database users, a good summary with detailed descriptions and spectrums of the existing pathway and interaction databases is definitely required. For systematically understanding the pathways involved in cellular processes or organismal systems, KEGG and REACTOME are strongly recommended. KEGG can be also used to view the pathways related to human diseases, like infectious illnesses, various cancer types. If you are only interested in signal pathway, the Signaling Gateway and NCI/PID are feasible too. To acquire a comprehensive knowledge about genes or gene products, Gene Ontology is absolutely preferred. Moreover, GO and DAVID databases can be also applied to analyze gene set enrichment. Whereas, for understanding the sequence and structure of proteins, like 3D shapes and protein symmetry, RCSB PDB is better than GO. STRING and Pathway Commons are of great importance to help us investigate the protein interactions and involved pathways related to our favorite proteins. To study the associations of drug and disease, PharmGKB and KEGG Drug database are primarily suggested.

It must be acknowledged that these resources listed above may have other distinct features and functions that are not described in detail. In addition, due to the space constrain in this chapter, we are not able to cover other different but also important types of databases. However, the contents of this chapter will surely give you a comprehensive and comparative review of existing biological databases, especially for systematic pathways and molecule interactions or associations.

References

Ashburner M, Ball CA, Blake JA, Botstein D, Butler H, Cherry JM, et al. Gene Ontology: tool for unification of biology. Nat Genet. 2000;25(1):25–9.

Berman HM, Westbrook J, Feng Z, Gilliland G, Bhat TN, Weissig H, et al. The protein data bank. Nucleic Acids Res. 2000;28:235–42.

Berman H, Henrich K, Nakamura H. Announcing the worldwide protein data bank. Nat Struct Biol. 2003;10(12):980.

Carbon S, Ireland A, Mungall CJ, Shu S, Marshall B, Lewis S, et al. AmiGO: online access to ontology and annotation data. Bioinformatics. 2008;25(2):288–9.

Cerami EG, Gross BE, Demir E, Rodchenkov I, Babur O, Anwar N, et al. Pathway commons, a web resource for biological pathway data. Nucleic Acids Res. 2011;39:685–90.

Croft D, O'Kelly G, Wu G, Haw R, Gillespie M, Matthews L, et al. Reactome: a database of reactions, pathways and biological processes. Nucleic Acids Res. 2010;39:691–7.

Croft D, Mundo AF, Haw R, Milacic M, Weiser J, Wu G, et al. The Reactome pathway knowledgebase. Nucleic Acids Res. 2014;42:472–7.

Day-Richter J, Harris MA, Haendel M. Gene Ontology OBO-Edit Working Group, Lewis S. - OBO-Edit: an ontology editor for biologists. Bioinformatics. 2007;23(16):2198–200.

Dinasarapu AR, Saunders B, Ozerlat I, Azam K, Subramaniam S. Signaling gateway molecule pages-a data model perspective. Bioinformatics. 2011;27(12):1736–8.

Goto S, Nishioka T, Kanehisa M. LIGAND database for enzymes, compounds and reactions. Nucleic Acids Res. 1999;27(1):377–9.

Hashimoto K, Goto S, Kawano S, Aoki-Kinoshita KF, Ueda N, Hamajima M, et al. KEGG as a glycome informatics resource. Glycobiology. 2006;16(5):63–70.

Huang DW, Sherman BT, Lempicki RA. Systematic and integrative analysis of large gene lists using DAVID Bioinformatics Resources. Nat Protoc. 2009a;4(1):44–57.

Huang DW, Sherman BT, Lempicki RA. Bioinformatics enrichment tools: paths toward the comprehensive functional analysis of large gene lists. Nucleic Acids Res. 2009b;37(1):1–13.

Jensen LJ, Kuhn M, Stark M, Chaffron S, Creevey C, Muller J, et al. STRING 8 – a global view on proteins and their functional interactions in 630 organisms. Nucleic Acids Res. 2009;37:412–16.

Joshi-Tope G, Gillespie M, Vastrik I, D'Eustachio P, Schmidt E, de Bono B, et al. Reactome: a knowledgebase of biological pathways. Nucleic Acids Res. 2004;33:428–32.

Kanehisa M, Coto S. KEGG: kyoto encyclopedia of genes and genomes. Nucleic Acids Res. 2000;28(1):27–30.

Kanehisa M, Goto S, Furumichi M, Tanabe M, Hirakawa M. KEGG for representation and analysis of molecular networks involving diseases and drugs. Nucleic Acids Res. 2010;38:355–60.

Kanehisa M, Goto S, Sato Y, Kawashima M, Furumichi M, Tanabe M. Data, information, knowledge and principle: back to metabolism in KEGG. Nucleic Acids Res. 2014;42:199–205.

Muto A, Kotera M, Tokimatsu T, Nakagawa Z, Goto S, Kanehisa M. Modular architecture of metabolic pathways revealed by conserved sequences of reations. J Chem Inf Model. 2013;53 (3):613–22.

Rose PW, Prlic A, Bi C, Bluhm WF, Christie CH, Dutta S, et al. The RCSB protein data bank: views of structural biology for basic and applied research and education. Nucleic Acids Res. 2015;43:345–56.

Schaefer CF, Anthony K, Krupa S, Buchoff J, Day M, Hannay T, et al. Buetow. PID: the pathway interaction database. Nucleic Acids Res. 2009;37:674–9.

Snel B, Lehmann G, Bork P, Huynen MA. STRING: a web-sever to retrieve and display the repeatedly occurring neighborhood of a gene. Nucleic Acids Res. 2000;28(18):3442–4.

Szklarczyk D, Franceschini A, Wyder S, Forslund K, Heller D, Huerta-Cepas J, et al. STRING v10: protein-protein interaction networks, integrated over the tree of life. Nucleic Acids Res. 2015;43:447–52.

The Gene Ontology Consortium. The Gene Ontology project in 2008. Nucleic Acids Res. 2008;36:440–4.

The Gene Ontology Consortium. Gene Ontology Consortium: going forward. Nucleic Acids Res. 2015;43:1049–56.

Whirl-Carreillo M, McDonagh EM, Heberg JM, Gong L, Sangkuhl K, Thorn CF, et al. Pharmacogenomics knowledge for personalized medicine. Clin Pharmacol Ther. 2012;92(4):414–17.

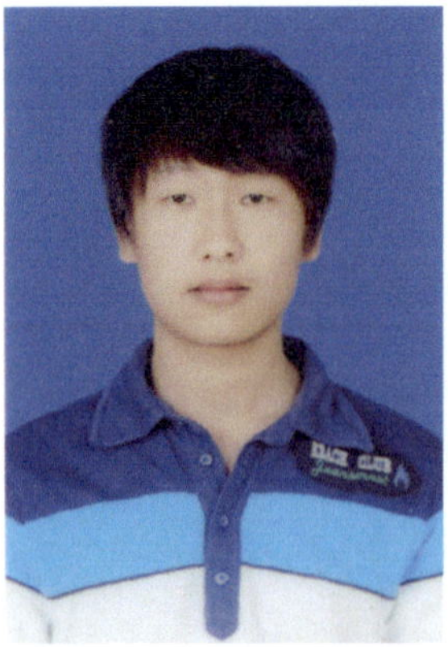

Jinzeng Wang is now a master student at School of Life Science and Technology, Tongji University in China. He is majoring in bioinformatics. Currently, he focuses on cancer immunology to reveal the mechanisms of tumor development, drug resistance and finally identify the optimal treatment strategies for cancer patients.

Haiyun Wang is an associate professor at School of Life Science and Technology, Tongji University in China. Her recent interests include integrative analysis of genomic data to explore the conservation and difference of genes relationship among different molecular levels, also computational and statistical methods used in analysis of epigenetics and genetics alterations, gene expression, and clinical phenotype with the goal of understanding complex human diseases or traits such as cancer and aging. Presently she is particularly working on identifying the novel biomarkers to develop new targeted treatment and combination therapies to overcome drug resistance and improve outcome in lung cancer.

Chapter 12
Standards and Regulations for (Bio)Medical Software

Jörg Schröttner, Robert Neubauer, and Christian Baumgartner

Abstract This chapter provides an introduction to the basic legal regulations and harmonized standards for the development and release of biomedical software, which is treated by law as a medical product. The authors will primarily focus on the regulations and recommendations within the European Union, and also delineate and discuss the legal situation exemplarily in selected countries such as the United States, Canada and Australia. In summary, this survey will provide a guideline for researchers and practitioners dealing with software development as a medical product according to the Medical Devices Act.

Keywords Medical software • Development • Risk management • Harmonized standards • Classification

12.1 Introduction

In the year 1993, when the Medical Device Directive (MDD 93/42 EWG) was published in the European Union, software played a minor role in (bio)medical devices and applications, and was most commonly used for driving display units or simple device functions. At that time it was the objective of the medical device directive to take care of medical device safety generally, based on hardware construction. How to deal with medical software, in particular stand-alone software, was not sufficiently considered in the directive. This situation has changed essentially. Today (bio)medical software is a central tool in biomedical engineering and computer science, and plays a major role in any aspects of healthcare and patient management. In 2007 the European Commission revised the medical device directive and stated that *"It is necessary to clarify that software in its own right, when specifically intended by the manufacturer to be used for one or more of the medical purposes set out in the definition of a medical device, is a medical device."* (Directive 2007/47/EC; 2007). Its importance is almost overtaking the status of hardware, even in special cases, when bio(medical) software is used for diagnosis,

J. Schröttner, Ph.D. (✉) • R. Neubauer • C. Baumgartner, Ph.D.
Institute of Health Care Engineering with European Notified Body of Medical Devices, Graz University of Technology, Stremayrgasse 16, A-8010 Graz, Austria
e-mail: schroettner@TUGraz.at

X. Wang et al. (eds.), *Application of Clinical Bioinformatics*,
Translational Bioinformatics 11, DOI 10.1007/978-94-017-7543-4_12

monitoring or safety-relevant purposes. As a consequence software is an important source of errors and adverse events in medical devices that should not be underestimated in the clinical application. Therefore an accurate procedure for the design and development of (bio)-medical software is crucial.

12.2 Legal Regulations and Harmonized Standards

The EC has published three directives that build the basis upon which medical devices provide patients, users and third parties with a high level of protection and attain the performance levels attributed to them by the manufacturer.

- Medical Device Directive (MDD 93/42/EEC; 1993)
- Directive on active implantable medical devices (AIMD 90/385/EEC; 1990)
- Directive on in vitro diagnostic medical devices (IVD 98/79/EC; 1998)

According to the definition of a medical device, as described in Sect. 12.3, (bio) medical software used for specified intentions is a medical device and is covered by one of these three directives. The implementation of these directives takes place under national laws of the European countries, like the Austrian Medical Devices Act (MPG BGBl. Nr. 657; 1996) and further regulations, such as the regulation for operators of medical devices (MPBV BGBl. II Nr. 70; 2007).

One key aspect for manufacturers of medical devices is the obligation to meet the *essential requirements*, which are specified in Annex I of each directive. One way to show compliance with these requirements is to develop and manufacture the medical device according to the relevant harmonized standards of the directive, which are published in the Official Journal of the European Union. In that case compliance with the essential requirements can be presumed, but has to be assessed by a notified body depending on the conformity class (see Sect. 12.3) of the product. An overview of legal regulations and harmonized standards for (bio)medical software within the European Union is shown in Fig. 12.1.

In the context of (bio) medical software the following harmonized standards may be applicable:

- Medical device software – software life cycle processes (IEC/EN 62304:2006)
- Medical devices-application of risk management to medical devices (EN ISO 14971:2012)
- Medical devices – Quality management systems (EN ISO 13485:2012)
- Application of usability engineering to medical devices (EN 62366:2008)
- Medical electrical equipment – Part 1: General requirements for basic safety and essential performance (IEC/EN 60601–1:2006)

This list comprises an overview of essential standards, but cannot be seen as a complete collection of all applicable standards for (bio)medical software, which of course mainly depends on the *intended use* defined by the manufacturer. Generally,

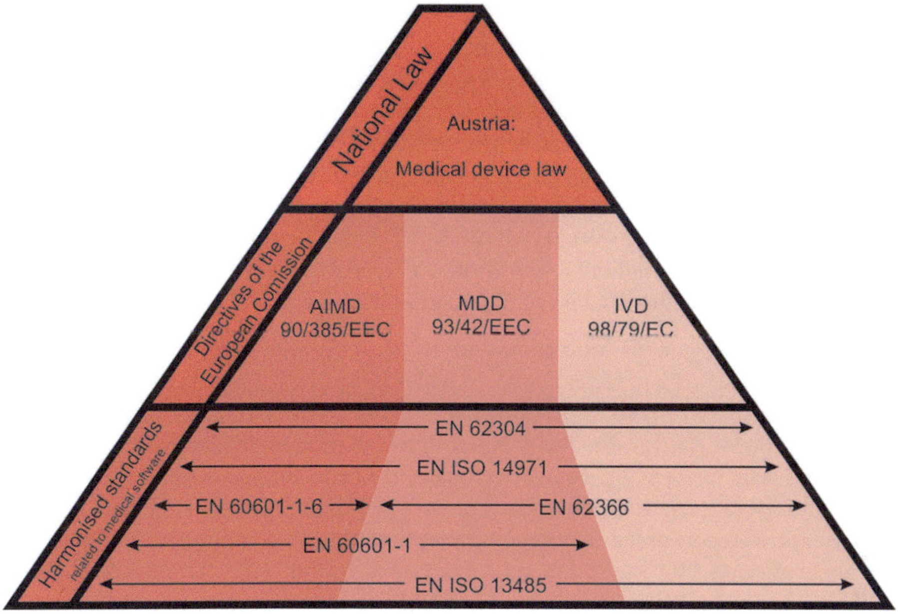

Fig. 12.1 Overview of legal regulations and harmonized standards for (bio)medical software within the European Union

a distinction is made between standards for products and standards for processes. Standards for products include specific recommendations for the design of a product, whereas process-orientated standards provide recommendations for the implementation of activities and procedures, such as software development processes or quality assurance activities. In terms of software development, the standard IEC/EN 62304 plays a major role. However it needs to be considered that relations to other appropriate standards exist and are combined when developing a medical device.

IEC/EN 62304

This international standard provides requirements for the development and maintenance of medical software and covers software embedded in medical devices as well as stand-alone software. In Europe, the EN 62304 version is a harmonized standard incorporated in all three medical device directives (see Fig. 12.1), that is technically equivalent to the IEC 62304. The purpose of this standard is to provide a development environment by means of implementing activities and tasks that permanently produce high quality and safe (bio)medical software. The requirements of the development life-cycle are explained in more detail in Sect. 12.4. In addition, the following fundamentals for developing medical software are assumed:

- an established *risk management process* according to EN ISO 14971
- development and maintenance within a *quality management system*

EN ISO 14971

This standard deals with the application of *risk management* to medical devices. It is a general standard for medical devices and does not solely focus on medical software. In any case, risk management is a very important issue for (bio)medical software. Several standards refer to the EN ISO 14971, for example the abovementioned EN 62304. The essential requirement of EN ISO 14971 is implementing and maintaining a process to discover and control risks arising from the medical device over its whole life cycle. This includes the elements of risk analysis, risk assessment, control of risk as well as information and feedback from the manufacturing process and market phase.

The risk management process has to be planned and documented in the *risk management plan* which, for example, includes allocation of responsibilities, criteria for the acceptance of risks or verification activities. All determinations and results from the process need to be documented in a *risk management dossier*. To obtain safety and effectiveness of medical software, it has to be proven that the software fulfills the specifications without causing unacceptable risks (EN ISO 14971:2012). The technical Report of the International Electrotechnical Commission "Guidance on the application of ISO 14971 to medical device software" (IEC/TR 80002–1; 2009) gives assistance for the application of the EN ISO 14971 for medical device software in respect of the requirements of the IEC/EN 62304. In Chap. 7 of the EN 62304 guidelines for a software risk management process are specified.

EN ISO 13485

This standard (EN ISO 13485:2012) contains requirements to implement and maintain a *quality management system* addressed to medical device developers and manufacturers. As mentioned above the IEC/EN 62304 demands a quality management system, however not a specific one. But as a benefit for the manufacturer, requirements of the IEC/EN 62304 are directly related to some of the requirements of the EN ISO 13485. Generally the manufacturer of medical software shall demonstrate the ability to develop software that consistently meets customer needs and applicable regulatory requirements. Therefore the EN ISO 13485 provides also a framework which enables a developer of medical products to meet some of the requirements of the medical device directives. Nevertheless it is important to note that compliance with EN ISO 13485 does not show conformity with all aspects of the quality systems of the medical device directives legally valid at this date.

IEC/EN 62366

The reduction of risks caused by usability problems is the main goal of this international standard, aspects like user satisfaction or efficiency of the product are not considered. It defines requirements for the analysis, specification, development, verification and validation processes, which should be implemented by the manufacturer. These processes are covered in a superior process termed *Usability Engineering Process*. The standard (EN 62366:2008) is heavily related to the risk management process according to EN ISO 14971. A *usability engineering file* has to be kept in evidence, which includes all activities set by the manufacturer (e.g. usability testing with a significant number of users) as well as the definition of the fundamental operating functions. The consideration of the EN 62366 is one of the essential requirements for proofing conformity of a medical device with the basic standard for safety aspects (EN 60601-1:2006+A1:2013).

IEC/EN 60601-1

This standard is fundamental for basic safety of medical electrical devices and includes also a special part for medical software, called PEMS. PEMS are Programmable Electrical Medical Systems, which consists of software and hardware parts.

The second edition of this standard relates to a special additional standard for programmable electrical systems (EN 60601-1-4:1996+A1/1999). The requirements are focused on prevention of hazardous situations and risk minimization. Compliance means that the inspected software is safe for the patient and user, not that it was properly developed. In the third edition of the EN 60601-1 the EN 60601-1-4 is already included under clause 14 (Programmable Electrical Medical Systems – PEMS). These requirements apply to all PEMS unless it does not provide functionality necessary for basic safety or essential performance or the risk management can demonstrate that the failure of the software does not lead to an unacceptable risk.

International/Global Approach

In the **United States of America** the Federal Food Drug & Cosmetic Act is the legal framework for medical products, hence for (bio)medical software. A specification of this law is given in a federal regulation (CFR-Code of Federal Regulations Title 21- Food and Drugs; 2006), which is composed and enforced by the Food and Drug Administration (FDA). Further the FDA provides the manufacturer with several guidelines that can be downloaded from the FDA homepage free of charge. These documents are not legally binding, but have a high relevance in practice. Software as a medical product is covered by a main guidance document (General Principles of Software Validation, U.S. Department of Health and Human Services (HHS) et al. 2002). This document recommends an implementation of software life cycle management and risk management. It further describes how requirements of the medical device quality system regulations (21 CFR 820) apply to medical software particularly with regard to software validation.

In addition to the FDA guidelines, the FDA considers so-called "Recognized Consensus Standards". In comparison to the European situation, for instance, the standards for risk management (ISO 14971), software life-cycle (IEC 62304), usability (IEC 62366) and the standard for basic safety of medical electrical devices (IEC 60601-1) are recognized as a whole or in parts. The American National Standard for software life cycle processes (ANSI/AMI/IEC 62304:2006) is equal to IEC/EN 62304 and it also addresses the minimum requirements for software of major levels of concern and requires a rating of each software component, assigning to one of three safety classes A, B or C (see Sect. 12.3).

A new guidance document from FDA aims at mobile apps. (Mobile Medical Applications, U.S. Department of Health and Human Services (HHS) et al. 2015). Its purpose is to apply regulations to only those mobile apps that are defined as a medical device and whose functionality could cause a risk for the patient. This guideline also refers to ANSI/AMI/IEC 62304:2006 and clause 14 of ANSI/AAMI ES 60601-1:2005 which is equal to clause 14 of IEC 60601-1:2005. Mobile apps with medical purposes are currently a widely discussed topic since the amount of medical apps will rise considerably within the next decade. Therefore this issue is also discussed in expert groups with regard to the upcoming standard IEC 82304. This standard focuses on "Health Software Products", which are intended to be used specifically for managing, maintaining or improving health of individual persons or the delivery of medical care.

In summary, the medical software standard situation in the United States is today similar to the European situation. However, one major exception is the implementation of a quality management system for medical devices, which does not refer to the international standard ISO 13485 in the US.

In **Canada** the definition of a medical device and therefore of medical software is regulated in the Food and Drugs Act (Food and Drugs Act, R.S.C.; 1985) and complies with the definition in the Medical Device Directive of the European Union. The recognized Standards are similar to harmonized standards in Europe and the United States, e.g.: IEC 60601-1, IEC 60601-1-4, CSA_ISO 14971, IEC 62304:2006, or ISO 14971. In contrast to the United States the standard of ISO 13485 for *quality management systems* of medical devices is recognized and approved in Canada.

In **Australia** a medical software is considered as a medical device if it fits in the definition of the Therapeutic Goods Act (TGA; 1990). The standards that may provide further guidance, but are not mandatory mentioned in the "Australian regulatory guidelines for medical devices" (TGA; 2011) are the IEC 60601- Family (including IEC 60601-1), the ISO 14971 as guidance for development of the risk management records and the IEC 62304 which is considered by the TGA, representing the state of the art for medical device software. Regarding quality management requirements, the manufacturer is responsible for implementing a quality management system according to ISO 13485 and has to meet the Australian regulatory requirements in addition, which is similar to the conformity assessment in the European Union.

This brief overview demonstrates that the IEC/EN 62304 is widely recognized for the development and maintenance of medical software. Apart from the examples above, it has also been translated into an identical Chinese standard (YY/T 0664; 2008).

12.3 Definition, Classification and Qualification

In principal, a medical device, including software as a medical device (devices incorporating software or stand-alone software) is defined by the directives (see Sect. 12.2) as follows:

Medical device means any instrument, apparatus, appliance, software, material or other article, whether used alone or in combination, including the software intended by its manufacturer to be used specifically for diagnostic and/or therapeutic purposes and necessary for its proper application, intended by the manufacturer to be used for human beings for the purpose of:

- *diagnosis, prevention, monitoring, treatment or alleviation of disease,*
- *diagnosis, monitoring, treatment, alleviation of or compensation for an injury or handicap,*
- *investigation, replacement or modification of the anatomy or of a physiological process,*
- *control of conception.*

According to this definition, which is very close to the definition of the document published by the International Medical Device Regulation Forum (IMDRF; 2013), software used for the intentions mentioned above is a medical device covered by one of these three directives.

Each medical device shall be classified into one of the four conformity classes (I, IIa, IIb or III) according to the classification rules of Annex IX in the EU medical device directive (MDD 93/42/EEC; 1993). These defined rules consider invasiveness, the duration of contact, part of body or body orifice in contact, rules for active and non-active products and others to determine the class of the medical device with respect to its potential risk. Appropriate to these rules and definitions medical software as stand-alone software is defined as an active medical device (see MDD 93/42/EEC; 1993), Annex IX, Chap. 1.4). This is an important decision for selecting the applicable classification rules. Stand-alone software is categorized according to its own risk level, whereas software, which drives a device or influences the use of a device, is automatically graded into the same class as the device itself.

Software – Medical Device or Not?

Although the definition above seems to be unambiguous, the question as to whether developed software, especially stand-alone software used in a medical surrounding, is a medical device or not is increasingly arising. In a lot of cases the definitions of the directives are not specific enough to gain a satisfying answer.

Therefore the European Commission provides manufacturers and notified bodies with guidance documents. One of these guidelines deals with the qualification and classification of stand-alone Software used in healthcare (MEDDEV 2.1/6; 2012) with the main purpose of solving the question whether the stand-alone software is a medical device or not. The decision can be based on the following steps, which are specified in the document. Figure 12.2 shows a simplified decision diagram based on the MEDDEV 2.1/6.

The decision making process (see Fig. 12.2) starts with the assumption that the product is a stand-alone software and not incorporated into a medical device, and the software is a computer program. If the software does not perform an action on data or performs an action limited to storage, archival, communication, simple search or lossless compression it is not a medical device (step 1). For example,

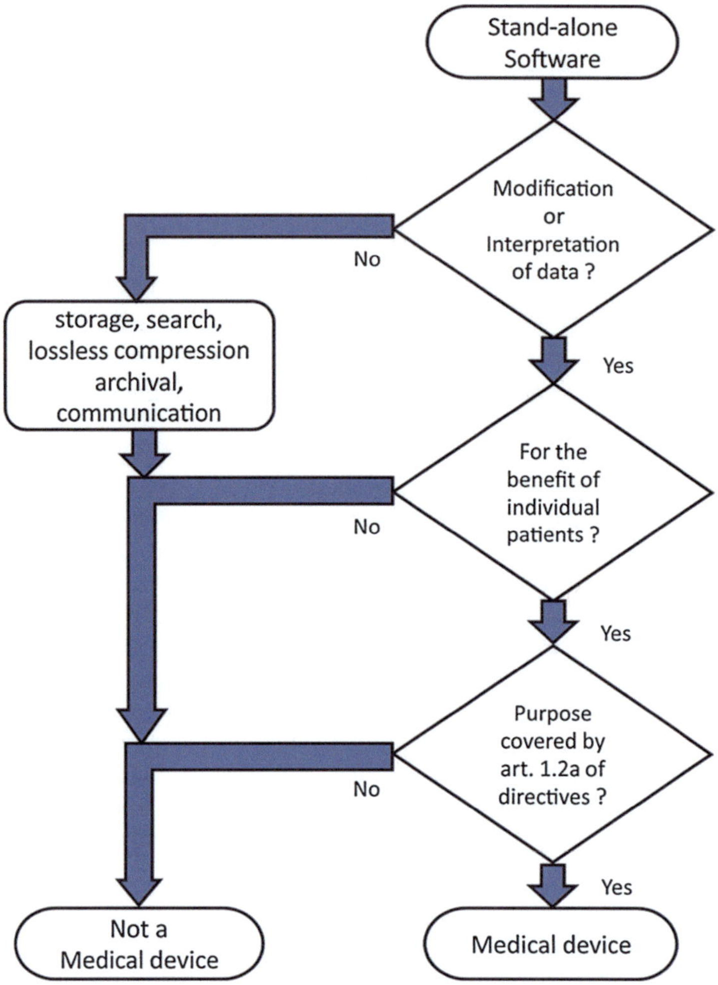

Fig. 12.2 Software as a medical device – simplified decision diagram based on the guideline on the classification of stand-alone software (MEDDEV 2.1/6; 2012)

image viewing programs that enable simple viewing features such as sharpening, zooming, contrast stretching etc. are not medical devices. Software that modifies medical raw data, creates new medical information or facilitates interpretation of data for the medical diagnosis might be classified as a medical device.

In the second step of decision the benefit of individual patients is in focus. This question should separate software intended to evaluate data of a single patient from software dealing with general data. Software for individual purposes, e.g. for diagnosis, prevention, monitoring and so on, is a medical device, while software intended to review common data, models for treatment pathways or software for epidemiological studies is not classified as a medical device.

Finally (step 3) the intended use of the software specified by the manufacturer has to be in accordance with one of the purposes listed in article 1.2a of the directives. For example, software is qualified as a medical device if the intended use is diagnosis, prevention, monitoring, treatment or alleviation of diseases.

Once the (bio)medical software meets the definition of a medical device, IEC/EN 62304 (see Sect. 12.2) is applicable and the software has to be categorized into one of the three software safety classes. The assignment to one safety class has to relate to the possible effects on the patient, operator or other people resulting from harm to which the software can contribute. Depending on the risk level and on the degree of hazard the following safety classes are defined:

Class A: No injury or health damage is possible
Class B: No serious injury is possible
Class C: Death or serious injury is possible

If it is possible to split the software into modules or sub systems, each module must be classified on its own. Finally the whole software is classified into the highest class, identified during the classification process for all sub systems. Depending on the software safety class, a different amount of requirements need to be fulfilled to proof the conformity with the standard IEC/EN 62304.

12.4 Software Development

In principle different models and strategies can be used for developing (bio)medical software. For example, the waterfall model, incremental models or the V-model, the latter being well known and very popular. Figure 12.3 shows the V-model including the development process steps of EN 62304. The illustration shows that the requirements are dependent on the software safety class of the product.

Software Development Process (IEC/EN 62304)
The main idea of this standard is to develop software according to a well-defined procedure. For this reason the standard requires to establish a software development plan, which should include all processes that are used, i.e. the deliverables, traceability between system requirements, software requirements, software system tests and risk control measures, as well as software configuration and modification management and software problem solving.

The next steps consist of the analysis of the software requirements, the software architectural design and the detailed design. At this point the design phase ends and the integration and testing-phase begins. These steps can be separated into software unit testing, software integration testing and software system testing (see Fig. 12.3). When these verification activities are completed and the results evaluated the software can be released.

It is important to mention that the EN 62304 does not cover software validation. Validation not only means the total amount of verification activities, but also that

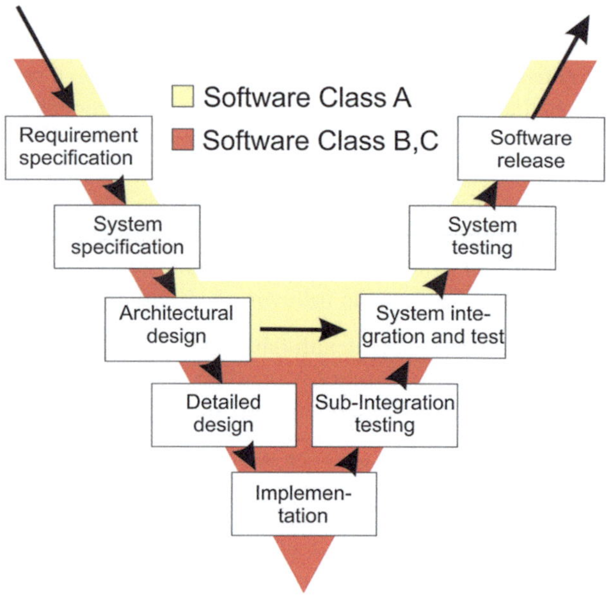

Fig. 12.3 Development of software according to the V-Model

the verified software satisfies its user needs and intended use. For embedded software, PEMS validation is a system level activity and is thus covered by EN 60601-1. For stand-alone software the link to validation can be made via the EN ISO 13485, which sets requirements for design and development validation. The future standard on health software (IEC 82304; 2015) will in any case cover validation of software-only products.

Furthermore a software maintenance plan has to be established, which specifies how to deal with feedbacks and in which case a feedback must be addressed as a problem. This procedure is connected to the software risk management process, the problem solving process as well as the software configuration management process, which is involved in the case of modifying the existing system.

As mentioned in Sect. 12.2 the risk management process of EN 62304 is based on the ISO 14971, but also covers special requirements for medical software. The first step is the detection of software or modules that could lead to hazardous situations. This analysis is based on the identification and documentation of possible reasons and events for these hazardous situations. The manufacturer has to develop risk control measures to keep the effects of identified reasons as low as possible. The documented verification of the implemented risk control measures is the final step in this risk management process, which also sets its focus on the prevention of risks that arise from the modification of software or software-updates. Another important process that needs to be considered is the problem-solving process. It is designed to analyse and solve problems that occur during the development, maintenance and application of the medical software. The aim is to provide the manufacturer with a tool that ensures documentation, analysis and solving of discovered problems in an efficient way.

Finally, another important process is the software configuration management process, which should provide a scheme for the unique identification of software items and software versioning with respect to the whole software life cycle. The software configuration management process is necessary for change control and the documentation of the modifications.

12.5 Conclusions

It is obvious that software as medical devices will play an important role on the healthcare market in the upcoming years. The overview of legal regulations and harmonized standards shows that the procedures of development and maintenance of medical software are widely harmonized and only a few differences exist according to the legal framework of individual countries. Concerning the safety aspect of software, the definition in the essential requirements of the European directive that devices which incorporate software or which are medical software themselves (stand-alone software) must be validated according to state of the art procedures, taking into account the principles of development lifecycle, risk management, validation and verification (see MDD 93/42/EEC, Annex I, 12.1a; 1993). Requirements regarding these aspects are defined in various standards (see Sect. 12.3), which should be considered to presume compliance with the essential requirements of legal regulations. However, development processes should not be focused on the admission requirements only, but should be based on well established procedures which are useful according to manufacturer environment with the aim to design and manufacture (bio)medical software in such a way that they will not compromise the clinical condition or the safety of patients, users or other persons. It is also pointed out in the proposal of the revision of the European legislation for medical products (COM(2012) 542; 2012) that the developers and manufacturers must take more responsibility regarding transparency and traceability of the medical devices they place on the European market.

Nowadays, great efforts are being made to minimise risks for patients and operators caused by software failures. However, efforts regarding software standardisation will become necessary in the near future, which has not been considered so far. Discussions on security of medical software are increasingly arising with respect to the problem of protecting individuals and health care providers against software attacks. The IEEE (Institute of Electrical and Electronics Engineers) has recently published a paper called "Building Code for Medical devices software security" (IEEE; 2015) which tries to demonstrate the direction for standardization concerning this issue. Thus, the area of standardization and regulations for (bio)medical software is an open, widely-discussed field that will not be exhausted in a short period of time.

References

62A/1013/CDV, IEC 82304-1: Health Software – Part 1: general requirements for product safety (working document). 2015.

Active Implantable Medical Devices 90/385/EEC. Off J Eur Union, L189. 1990;33.

ANSI/AAMI/IEC 62304:2006: Medical device software – software life cycle processes. 2006.

Australian Government, Therapeutic Goods Act 1989 – Sect 41BD, No. 21. 1990.

Bundesgesetz betreffend Medizinprodukte (Medizinproduktegesetz – MPG) StF: BGBl. Nr. 657/1996.

CFR-Code of Federal Regulations Title 21 – Food and Drugs. 2006.

Directive 98/79/EC on in vitro diagnostic medical devices. Off J Eur Union, L331. 1998; 41.

Directive 2007/47/EC. Off J Eur Union, L247. 2007;50.

EN 60601-1:2006+A1:2013 Medical electrical equipment – Part 1: general requirements for basic safety and essential performance.

EN 60601-1-4:1996+A1/1999 Medical electrical equipment – Part 1: general requirements for safety – 4. Collateral standard: Programmable electrical medical systems.

EN 62366:2008 Medical devices – application of usability engineering to medical devices.

EN ISO 13485:2012 Medical devices – quality management systems – requirements for regulatory purposes.

EN ISO 14971:2012 Medical devices – application of risk management to medical devices.

Food and Drugs Act, R.S.C., 1985, c. F-27.

IEC 62304:2006 Medical device software – software life cycle processes.

IEC/TR 80002-1:2009 Medical device software – Part 1: guidance on the application of ISO 14971 to medical device software.

IEEE cyber security, Building Code for Medical Device Software Security. 2015.

IMDRF SaMD Working Group, IMDRF/WG/N10FINAL:2013 – Software as a Medical Device (SaMD): Key Definitions, IMDRF (International Medical Device Regulators Forum). 2013.

MEDDEV 2.1/6. Guidelines on the qualification and classification of stand alone software used in healthcare within the regulatory framework of medical devices. 2012.

Medical Device Directive MDD 93/42/EEC. Off J Eur Union, L169. 1993;36.

Regulation of the European Parliament and of the Council on Medical Devices, COM(2012) 542 final. 2012.

State Food and Drug Administration of China (SFDA), YY/T 0664–2008 Medical device software – software life cycle processes. 2008.

Therapeutic Goods Administration (TGA), Australian regulatory guidelines for medical devices (ARGMD) Version 1.1. 2011.

U.S. Department of Health and Human Services (HHS), Food and Drug Administration (FDA), Center for Devices and Radiological Health, Center for Biologics Evaluation and Research: general Principles of Software Validation – Final Guidance for Industry and FDA Staff. 2002.

U.S. Department Of Health and Human Services (HHS), Food and Drug Administration (FDA), Center for Devices and Radiological Health, Center for Biologics Evaluation and Research: mobile Medical Applications, Guidance for Industry and Food and Drug Administration Staff. 2015.

Verordnung der Bundesministerin für Gesundheit, Familie und Jugend über das Errichten, Betreiben, Anwenden und Instandhalten von Medizinprodukten in Einrichtungen des Gesundheitswesens (Medizinproduktebetreiberverordnung – MPBV) StF: BGBl. II Nr. 70/2007.

Jörg Schröttner, PhD was born in Graz, Austria and received his Master (Dipl.-Ing.) and PhD in Electrical and Biomedical Engineering from Graz University of Technology in 2000 and 2003, respectively. Dr Schröttner is currently an Associate Professor with the Institute of Health Care Engineering, University of Technology Graz. His research interests comprise methodological, technical, operational, economic and quality assuring aspects of intra- and extramural health care. Since 2014, he has been the Head of the Testing- and Certifying Body for Medical Products Graz (PMG, European Notified Body Nr. 0636). Therefore an additional research focus lies on standards and regulations for medical products and biomedical software.

Robert Neubauer was born in Graz, Austria. He graduated in communications engineering from HTBLuVa Graz Gösting in 1994. He has worked at several medical device companies in the development and quality management department. In 2014 he was appointed Vice Head of the Testing- and Certifying Body for Medical Products Graz (PMG, European Notified Body Nr. 0636). He has more than 15 years' experience with standards and regulations for medical products including biomedical software.

Christian Baumgartner, PhD is Professor of Health Care Engineering at Graz University of Technology, Austria. He received his MSc (1994) and PhD degree (1998) in Electrical and Biomedical Engineering from Graz University of Technology, Austria, and his habilitation degree (Assoc.-Prof.) in Biomedical Engineering from the University for Health Sciences, Medical Informatics and Technology (UMIT), Hall in Tyrol, Austria (2006). From 1998–2002 Dr. Baumgartner held an R&D position at Tecan.com, where he developed confocal fluorescence laser scanning systems for micro array applications. From 2007–2008, he was part of the Barnett Institute of Chemical and Biological Analysis, Northeastern University and Harvard Medical School, Boston, MA, where Dr. Baumgartner worked in the field of computational biomarker discovery. In 2009, he was appointed Full Professor, Director of the Institute of Electrical and Biomedical Engineering, and Vice Chair of the Department of Biomedical Informatics and Mechatronics at UMIT. He has been a professor since 2015 and Head of the Institute of Health Care Engineering with European Notified Body of Medical Devices at Graz University of Technology in Austria since 2016. Dr. Baumgartner is the author of more than 150 publications in refereed journals, books and conference proceedings, and is a reviewer for more than 35 scientific journals. He served as a deputy editor of the "Journal of Clinical Bioinformatics", and is an editorial board member of "Clinical and Translational Medicine", "Methods of Information in Medicine" and "Cell Biology and Toxicology". His main research interests include cellular electrophysiology, biomedical sensors and signal processing, biomedical modeling & simulation, clinical bioinformatics and computational biology.

Chapter 13
Clinical Applications and Systems Biomedicine

Duojiao Wu, David E. Sanin, and Xiangdong Wang

Abstract A single disease can be caused by multiple mechanisms. System medicine is an emerging disciplinc that aims to address the problem that a disease is rarely caused by malfunction of one individual gene product, but instead depends on multiple gene products that interact in a complex network. Systems medicine integrates medicine, physics, and mathematical approaches with biologic and medical insights in an iterative process to visualize the interconnected events within a disease phenotype. System medicine comprises a series of concepts and approaches that have been used successfully both to delineate novel biological mechanisms and to drive translational advances in individualized healthcare. Here, we explain how and why systems medicine, and specifically network approaches, can be used to assist clinical decision making and to identify underlying disease mechanisms. We focus on describing how to use clinical bioinformatics to uncover pathogenic mechanisms in certain diseases. We finish by discussing the current problems and limitations of network and systems approaches and suggest possible solutions.

Keywords Systems medicine • Clinical bioinformatics • Clinical application • Network

---

D. Wu, M.D., Ph.D.
Biomedical Research Center, Zhongshan Hospital of Fudan University, Shanghai 200032, China

Shanghai Institute of Clinical Bioinformatics, Shanghai 200032, China

D.E. Sanin, Ph.D.
Max-Planck Institute of Immunobiology and Epigenetics, Stübeweg 51, 79108 Freiburg im Breisgau, Germany

X. Wang, M.D., Ph.D. (✉)
Zhongshan Hospital, Fudan University, Shanghai Institute of Clinical Bioinformatics, Shanghai, China
e-mail: xiangdong.wang@clintransmed.org

X. Wang et al. (eds.), *Application of Clinical Bioinformatics*,
Translational Bioinformatics 11, DOI 10.1007/978-94-017-7543-4_13

13.1 Introduction

Systems biomedicine is an interdisciplinary field of study that looks at the systems of the human body as part of an integrated whole, incorporating biochemical, physiological, and environment interactions. Systems biomedicine draws on systems science and systems biology, and considers complex interactions within the human body in light of a patient's genomics, behavior and environment (Federoff and Gostin 2009).

An important topic in systems biomedicine and systems medicine is the development of computational models that describe disease progression and the effect of therapeutic interventions. Recently, the National Institutes of Health (NIH) launched The Cancer Genome Atlas (TCGA) pilot project to integrate clinical data and high-throughput data for tumors (Weinstein et al. 2013; Akbani et al. 2014) (Piskorz et al. 2011). More and more evidence support that the incorporated evaluation of clinical and basic research could improve medical care, care provision data, and data exploitation methods in disease therapy and algorithms for the analysis of such heterogeneous data sets (Schwarz et al. 2009).

13.2 Clinical Application of Systems Biomedicine in Cancer Metastasis

Metastasis is a serious challenge for cancers. Genomic screening of cancers will continue to facilitate identification of molecular mechanisms of acquired resistance to targeted therapies. Ongoing translational and clinical research will facilitate a greater understanding of genomic alterations within cancer, with the aim of increasing benefit to wider population of cancer patients. At present, there are more urgent needs to develop systematic theory and methodology of systems biomedicine. It is a futuristic view of systems biomedicine shown in Fig. 13.1-the heterogeneous data sources from large-scale screening have to be integrated with clinical data and basic research.

Imaging-based biomarkers are used for staging, re-staging and monitoring the treatment of cancer patients. Several imaging modalities such as computed tomography (CT), magnetic resonance imaging (MRI), positron emission tomography (PET) and PET/CT play central roles in the detection and staging of the disease for cancer staging. Anatomical imaging that uses CT and uni-dimensional measurements of tumor size has been indispensable in establishing the Response Evaluation Criteria In Solid Tumors (RECIST) criteria as a standard for assessing response to therapy (Nishino et al. 2010; Lee et al. 2011). But for many reasons there has been great interest in carrying out translational research to improve these criteria. Such reasons include the complexity of the biology and heterogeneity of tumor masses, the precision of measurement that is needed, the possibility of misclassification (especially near the cut-off points) and the considerable interest in targeted cytostatic therapy, which requires improved measurements of stable disease (Imai et al. 2014). Therefore, there has been an increasing focus on developing novel techniques 'bioimage informatics' (Peng 2008) for the image data mining, database and visualization techniques to extract, compare, search and manage the biological knowledge in these data-intensive problems.

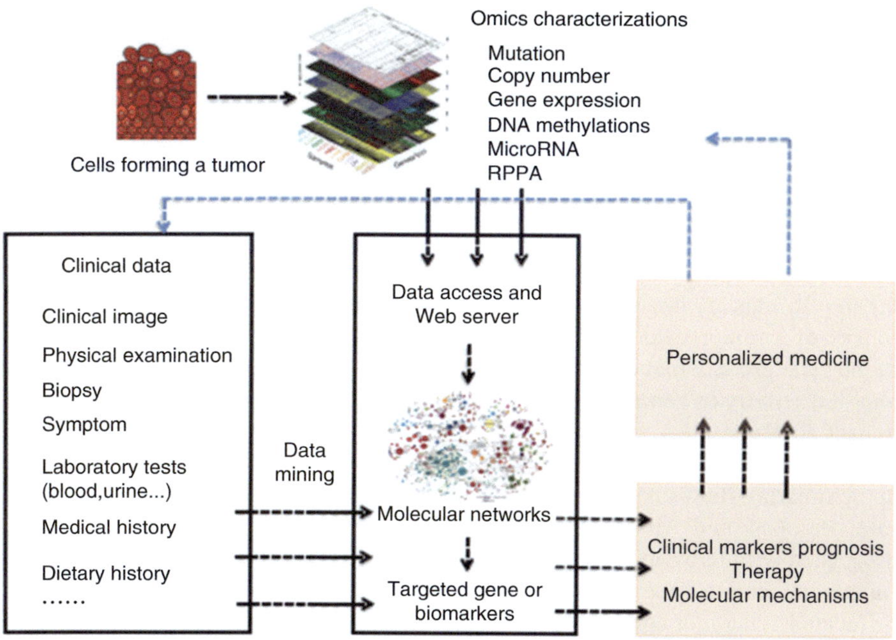

Fig. 13.1 Application of systems biomedicine in cancer metastasis research. The data flow shows heterogeneous data sources from large-scale screening have to be integrated with clinical data. Applying analytical methods on the integrated data can provide list(s) of target therapeutic genes and biomarkers, which can be used in personalized medicine

The essential techniques to the success of these applications are very complicated such as bioimage feature identification, segmentation and tracking, registration, annotation, mining, image data management and visualization. Image features are the fundamental description of pixels/voxels and all higher level objects. One way to extract features is based on domain knowledge; Another way for effective features extraction is to consider as many image transformations as possible, and thus generate a rich set of image features (Bauer et al. 2013). Image segmentation is one of the most basic processing steps in many bioimage informatics applications. While the goal is simply to segment out the meaningful objects of interest in the respective image, this task is non-trivial in many cases. Very complicated cases also exist due to problems such as a low signal–noise ratio and a big variability of image objects (Eklund et al. 2013; Caon et al. 2014). Image registration is essential in many applications that need to compare multiple image subjects of different conditions. Quantitative measurements and visualization of comparing patterns in the registered images can be done directly in a 'standard' space (Bauer et al. 2013; Kipli et al. 2013). Many applications such as phenotyping cells and determination of subcellular locations of proteins require the pattern clustering and classification techniques (Shi et al. 2014). Annotation of bioimage objects converts the image content information to concrete semantically meaningful information that is usually texts and can be conveniently organized and searched (Meyer et al. 2013; Yeh et al. 2014).

About 10 % of human genes have a known disease association (Amberger et al. 2009). The occurrence of cancer is a result of the genetic changes in tumor cells, the changes in the tumor microenvironment and the influences of the tumor microenvironment on the tumor (Junttila and de Sauvage 2013). Most cellular components exert their functions through interactions with other cellular components, which can be located either in the same cell or across cells, and even across organs. In cancer, the inter- and intracellular interconnectivity implies that the impact of a tumor-associated genetic abnormality is not restricted to the activity of the gene product that carries it, but can spread along the links of the network and alter the activity of gene products that otherwise carry no defects (Kipli et al. 2013; Schadt 2009).

In a random gene-gene or protein-protein interaction network, most nodes have approximately the same number of links, and highly connected nodes (hubs) are rare (Barabasi and Albert 1999). Essential genes show a strong tendency to be associated with hubs and are expressed in multiple tissues. They are located at the functional center of the disease interactome. Most tumor-associated genes tend to be located at the functional periphery of the interactome. Because if the mutation in essential genes will lead to spontaneous abortions, most tumor-associated genes in human are not essential genes (Barabasi et al. 2011) (Fig. 13.2).

Clinical cancer—the end result of this chaotic process—is characterized by unregulated cellular proliferation as well as cellular and clonal heterogeneity.

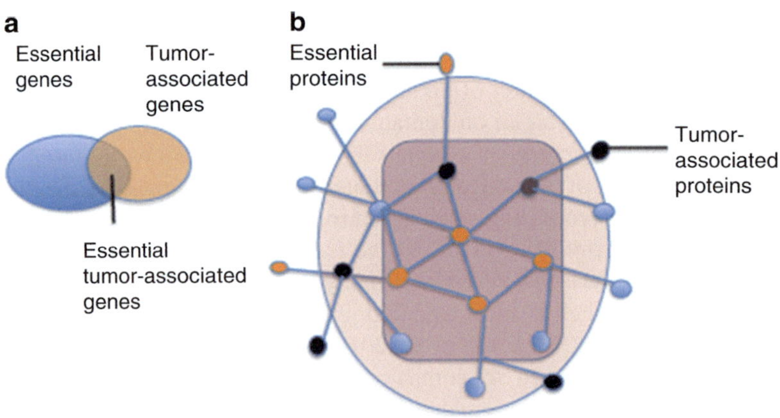

Fig. 13.2 Cancer and essential genes in the interactome. (**a**) shows the overlap between essential genes and tumor-associated genes. Most tumor-associated genes in human are not essential genes. (**b**) demonstrates the differences between essential proteins (shown as *orange nodes*) tend to be at the functional center; However, tumor-associated proteins(shown as *black nodes*) tend to be at the functional peripheral of the interactome

Consequently, an abnormality in a single gene or proteins could reflect lung cancer, but the perturbations of the complex intracellular and intercellular network that links tissue and organ systems will contribute to cancer biomarkers. With more dynamic information available, researchers' attention has recently shifted from static properties to dynamic properties of protein–protein interaction networks (Wang et al. 2014). Unlike traditional diagnosis of an existing disease state, detecting the pre-disease state just before the serious deterioration of a disease is a challenging task, because the state of the system may show little apparent change or symptoms before this critical transition during disease progression. Dynamic networks biomarker (DNB) can be constructed by involving proteomic, genomic, and transcriptome analyses (Wang et al. 2014; Wu et al. 2014); (Puig et al. 2012). By exploring the rich interaction information provided by high-throughput data, the dynamical network biomarker (DNB) can identify the pre-disease state have developed a novel computational approach based on the DNB theory and successful identified pre-disease samples from subjects or individuals before the emergence of disease symptoms for acute lung injury, influenza and breast cancer.

13.3 Understanding the Skin Microenvironment Using Bioinformatics

The skin is an interface between the human body and the environment. This tissue constitutes the first point of contact between the host's immune system and a plethora of infectious pathogens, including bacteria (Rudikoff and Lebwohl 1998), protozoa (Kedzierski and Evans 2014; Couper et al. 2008), filarial nematodes (Hoerauf et al. 2011), soil transmitted hookworms (Loukas and Prociv 2001), or the helminth *Schistosoma sp* (Paveley et al. 2009; Sanin et al. 2015). Likewise, the skin plays host to several populations of commensal microorganisms that it must learn to tolerate to avoid exacerbated inflammation (Heath and Carbone 2013; Pasparakis et al. 2014; Nestle et al. 2009). Previously, the diversity and role of commensal microorganism in the skin was greatly underrated. This notion was partly due to the limitations of conventional biochemical-based identification of microorganisms, especially in complex sites such as the skin.

With the advent of next generation sequencing, our understating of the diversity of commensal microorganisms in the skin has been greatly enhanced, due in particular to metagenomic studies that employ state of the art bioinformatics and 16S ribosomal RNA sequencing (Grice and Segre 2011). The four phyla of bacteria found on the skin (Actinobacteria, Firmicutes, Bacteroides and proteobacteria) was consistent with those found in gastrointestinal tract (Dewhirst et al. 2010; Eckburg et al. 2005), however the proportions varied vastly (Grice et al. 2009) with Actinobacteria being the most abundant phylum. Furthermore, commensal microorganism vary depending on the anatomical site with incredibly restrictive areas with one dominant population, such as *Corynebacteriaceae* in the umbilicus, or extremely diverse ecosystems, such as the at least nine different families present in the palm of the hand (Grice and Segre 2011).

The clinical implications of this newly found knowledge is beginning to unravel. Commensal microorganism play an important role in modulating T cell responses to *Leishmania major* cutaneous infection reducing lesion sizes (Naik et al. 2012), as well as limiting inflammation during *Schistosoma mansoni* percutaneous infection by inducing IL-10 (Sanin et al. 2015). Moreover, the number of commensal microorganisms with confirmed links to diseases, such as *Mealassezia sp.* in Seborrhoeic dermatitis (Gupta et al. 2004) or *Propionibacterium acnes* in acne (Dessinioti and Katsambas 2010), is likely to increase as metagenomic studies are applied systematically to different cases. For example a recent acne study in which bacteria isolated from hair follicles and adjacent skin were identified using 16S ribosomal RNA sequencing, found that hair follicles in healthy controls were colonised exclusively by *P. acnes*, whereas affected patients also had clones of *Staphylococcus epidermidis* (Bek-Thomsen et al. 2008).

These alterations in the composition of the skin microbiota are a result of many environmental factors, but the immune system plays a very important role in the process (Nestle et al. 2009; Grice et al. 2009). A clear example of this phenomenon was presented in a recent study which found that skin lacking T cells had different microbial composition compared to normal skin (Fig. 13.3) (Shen et al. 2014). Indeed, the researchers in that study were able to determine that T cells from normal mice (CD8+ T cells) located in the skin draining lymph nodes were able to respond to *Staphylococcus* (the most abundant commensal bacteria in murine skin). By contrast, T cells from animals bred under "germ-free" conditions, and thus having

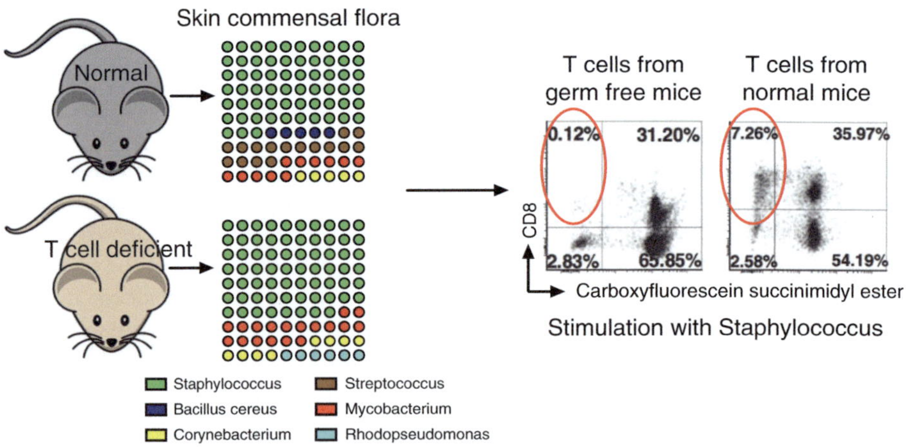

Fig. 13.3 Skin commensal microbiota. Microbial composition of murine skin in normal (*top*) or T cell deficient (*bottom*) animals. CD8+ T cells from germ free or normal mice proliferated differently to *Staphylococcus* antigens (Adapted from Shen et al. 2014)

no commensal bacteria, were unable to respond to *Staphylococcus* (Fig. 13.3). Together these results highlight the role of the immune system in modulating the composition of the commensal flora in the skin.

The clinical application of bioinformatics to the study of skin disorders extends beyond the study of microbiota in the surface of the skin. Differential gene expression coupled with pathway enrichment analysis provide invaluable tools for the study of human skin disorders without a robust animal model. Such is the case of psoriasis and atopic dermatitis, which were revealed as having opposing effects on pro-inflammatory and antimicrobial genes, with atopic dermatitis inducing a Th2 signature compared to psoriasis (Nomura et al. 2003). Additionally, skin from individuals suffering from psoriasis was shown to contain altered expression of genes involved with lipid metabolism, antimicrobial defenses, epidermal differentiation, and control of cutaneous vasculature when compared to skin from healthy individuals (Gudjonsson et al. 2009), highlighting the potential role of pathways that were not considered important before these bioinformatic tools were available.

13.4 Clinical Applications of Bioinformatics in Liver Disease

The liver plays a central role in homeostasis of glucose, fatty acids, amino acids and the synthesis of proteins. It is also involved in the detoxification of the body by removing noxious compounds from the blood. As such the liver is exposed to a variety of toxicants that invariably cause damage that this tissue must overcome. Consequently it is not surprising that the liver is involved in a vast number of diseases ranging from fatty liver disease, diabetes, hepatitis and liver cancer (Auger et al. 2015). In recent years a dramatic increase in the number of obese individuals, which affects 15–22 % of people in Western countries, has highlighted the threat posed by non-alcoholic fatty liver disease (NAFLD), which is by definition the deposition of fat in the liver in the absence of excessive alcohol intake, and is often accompanied by insulin resistance. As viral hepatitis is overcome by prevention and therapy, NAFLD could become the main cause for end stage liver disease, liver transplantation and hepatocellular carcinoma. These facts have triggered a wide search for biomarkers of clinical relevance to prevent and intervene in time before the liver is compromised (Dongiovanni et al. 2015).

Genome-wide association studies (GWAS), which can uncover links between single nucleotide polymorphisms (SNP) and particular phenotypes, have exposed a number of potential gene variants that could play a role both as biomarkers and in the progression of NAFLD, such as the *patatin-like phospholipase domain-containing 3 (PNPLA3)* gene (Romeo et al. 2008) or the *transmembrane 6 superfamily member 2 (TM6SF2)* gene (Zelber-Sagi et al. 2014). In particular *PNPLA3*, which is involved in the hepatocellular remodeling of lipid droplets as well as the secretion of very low density lipoprotein, has emerged as a dominant feature of NAFLD progression, although mechanistically the connection is yet unclear (Dongiovanni et al. 2015).

Obesity, cardiovascular disease and coronary artery disease are often linked in epidemiological studies, thus it is possible that these disorders share underlying genetic determinants. A recent study sought to address this question, and also using GWAS identified 56 pleiotropic genes in 87 autosomal regions with 181 SNPs, where most genes associated with plasma lipids and cardiovascular diseases and some with obesity and coronary artery disease (Rankinen et al. 2015). Forty-three of these genes could be associated in a network, connected by some 24 additional genes. Finally, within the 181 SNPs, regulatory elements such as enhancers and DNAase hypersensitive regions were enriched (Fig. 13.4). Thus the use of bioinformatics has permitted the identification of relevant biomarkers for NAFLD and obesity.

Another important set of tools to explore liver biology is modeling biological systems. As mentioned before, the liver is involved in the regulation of fatty acid homeostasis. In particular, cholesterol secretion and excretion is tightly controlled by this organ, yet the underpinning mechanism is unclear both at the cellular and organismal level. By employing mathematical models using differential equations it is possible to simulate the dynamics of cholesterol production and importantly the response to statins (Wattis et al. 2008), which are commonly used to treat high cholesterol. However, as of yet few models on cholesterol metabolism are able to pass functional tests (Paalvast et al. 1851), but they represent important initial attempts at tackling scientific questions that remain unresolved.

13.5 Conclusions

Systems biomedicine comprises a series of concepts and approaches that have been used successfully both to delineate novel biological mechanisms and to drive translational advances in individualized healthcare. In this article, we gave several examples of emerging systems biomedicine-based strategies as they apply to cancer, skin and liver diseases.

However, there is some challenges of systems biomedicine in clinical applications. For example, medical data entry, database management, and other processes vary among care providers, health insurers, and other members of the health sector. These differences impede the applications of systems biomedicine. The lack of strong interfaces and unified systems significantly complicates research and clinical decision support. Standards ensure consistency, integration, and accuracy. Greater application of standardized electronic record keeping appears to be a logical means to increase efficiency. It is therefore imperative to develop national or even universal standards (Gottlieb et al. 2014).

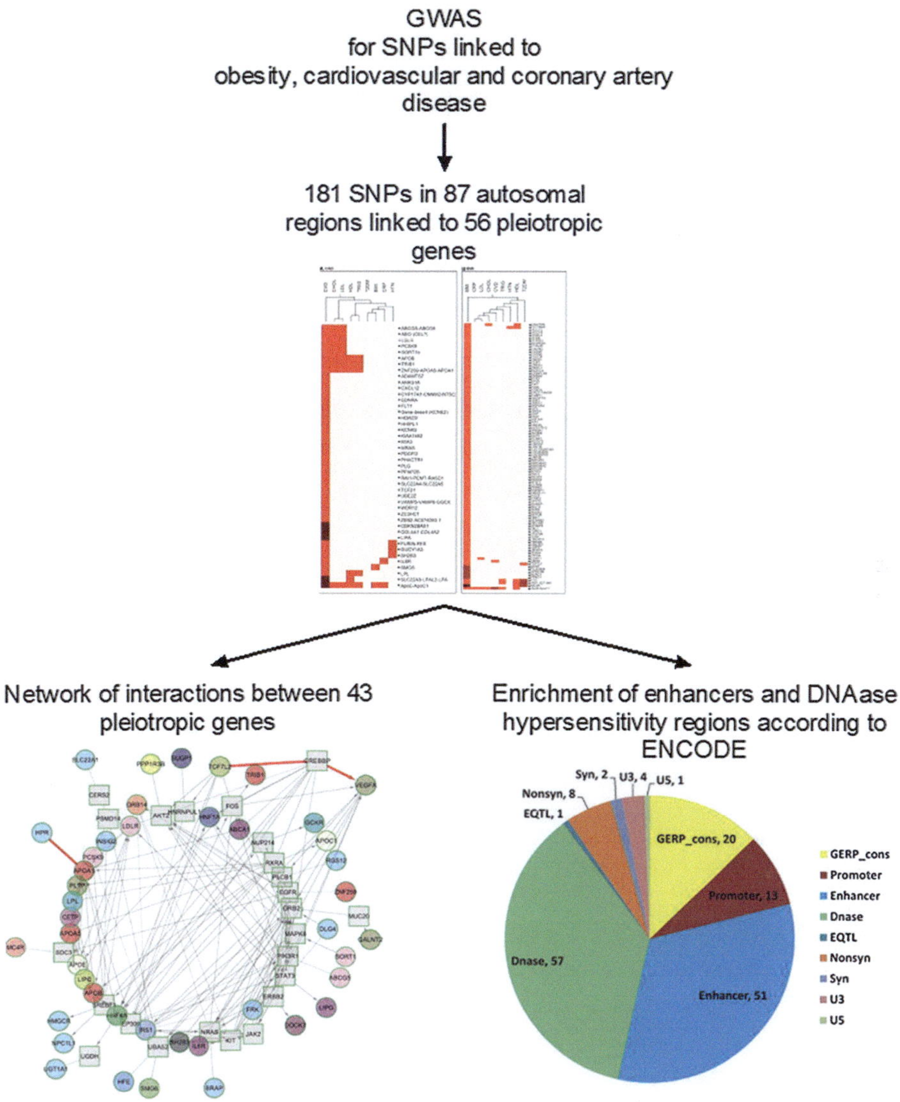

Fig. 13.4 Uncovering pleiotropic genes in obesity, cardiovascular and coronary artery disease. Based on GWAS results, 181 SNPs in 87 autosomal regions proximal to 56 genes were seen to correlate with plasma lipids and cardiovascular diseases or with obesity and coronary artery disease. Network analysis revealed further connections between these genes, whilst enrichment analysis suggested that the SNPs occurred predominantly within enhancer or DNAase hypersensitivity regions (Adapted from Rankinen et al. 2015)

References

Akbani R, Ng PK, Werner HM, Shahmoradgoli M, Zhang F, Ju Z, Liu W, Yang JY, Yoshihara K, Li J, et al. A pan-cancer proteomic perspective on The Cancer Genome Atlas. Nat Commun. 2014;5:3887.

Amberger J, Bocchini CA, Scott AF, Hamosh A. McKusick's Online Mendelian Inheritance in Man (OMIM). Nucleic Acids Res. 2009;37:D793–6.

Auger C, Alhasawi A, Contavadoo M, Appanna VD. Dysfunctional mitochondrial bioenergetics and the pathogenesis of hepatic disorders. Front Cell Dev Biol. 2015;3:40.

Barabasi AL, Albert R. Emergence of scaling in random networks. Science. 1999;286:509–12.

Barabasi AL, Gulbahce N, Loscalzo J. Network medicine: a network-based approach to human disease. Nat Rev Genet. 2011;12:56–68.

Bauer S, Wiest R, Nolte LP, Reyes M. A survey of MRI-based medical image analysis for brain tumor studies. Phys Med Biol. 2013;58:R97–129.

Bek-Thomsen M, Lomholt HB, Kilian M. Acne is not associated with yet-uncultured bacteria. J Clin Microbiol. 2008;46:3355–60.

Caon M, Sedlář J, Bajger M, Lee G. Computer-assisted segmentation of CT images by statistical region merging for the production of voxel models of anatomy for CT dosimetry. Australas Phys Eng Sci Med. 2014;37:393–403.

Couper KN, Blount DG, Wilson MS, Hafalla JC, Belkaid Y, Kamanaka M, Flavell RA, de Souza JB, Riley EM. IL-10 from CD4CD25Foxp3CD127 adaptive regulatory T cells modulates parasite clearance and pathology during malaria infection. PLoS Pathog. 2008;4:e1000004.

Dessinioti C, Katsambas AD. The role of Propionibacterium acnes in acne pathogenesis: facts and controversies. Clin Dermatol. 2010;28:2–7.

Dewhirst FE, Chen T, Izard J, Paster BJ, Tanner AC, Yu WH, Lakshmanan A, Wade WG. The human oral microbiome. J Bacteriol. 2010;192:5002–17.

Dongiovanni P, Romeo S, Valenti L. Genetic factors in the pathogenesis of nonalcoholic fatty liver and steatohepatitis. Biomed Res Int. 2015;2015:460190.

Eckburg PB, Bik EM, Bernstein CN, Purdom E, Dethlefsen L, Sargent M, Gill SR, Nelson KE, Relman DA. Diversity of the human intestinal microbial flora. Science. 2005;308:1635–8.

Eklund A, Dufort P, Forsberg D, LaConte SM. Medical image processing on the GPU – past, present and future. Med Image Anal. 2013;17:1073–94.

Federoff HJ, Gostin LO. Evolving from reductionism to holism: is there a future for systems medicine? JAMA. 2009;302:994–6.

Gottlieb LM, Tirozzi KJ, Manchanda R, Burns AR, Sandel MT. Moving electronic medical records upstream: incorporating social determinants of health. Am J Prev Med. 2014;48 (2):215–8.

Grice EA, Segre JA. The skin microbiome. Nat Rev Microbiol. 2011;9:244–53.

Grice EA, Kong HH, Conlan S, Deming CB, Davis J, Young AC, Program NCS, Bouffard GG, Blakesley RW, Murray PR, et al. Topographical and temporal diversity of the human skin microbiome. Science. 2009;324:1190–2.

Gudjonsson JE, Ding J, Li X, Nair RP, Tejasvi T, Qin ZS, Ghosh D, Aphale A, Gumucio DL, Voorhees JJ, et al. Global gene expression analysis reveals evidence for decreased lipid biosynthesis and increased innate immunity in uninvolved psoriatic skin. J Invest Dermatol. 2009;129:2795–804.

Gupta AK, Batra R, Bluhm R, Boekhout T, Dawson Jr TL. Skin diseases associated with Malassezia species. J Am Acad Dermatol. 2004;51:785–98.

Heath WR, Carbone FR. The skin-resident and migratory immune system in steady state and memory: innate lymphocytes, dendritic cells and T cells. Nat Immunol. 2013;14:978–85.

Hoerauf A, Pfarr K, Mand S, Debrah AY, Specht S. Filariasis in Africa–treatment challenges and prospects. Clin Microbiol Infect. 2011;17:977–85.

Imai K, Minamiya Y, Saito H, Motoyama S, Sato Y, Ito A, Yoshino K, Kudo S, Takashima S, Kawaharada Y, et al. Diagnostic imaging in the preoperative management of lung cancer. Surg Today. 2014;44:1197–206.

Junttila MR, de Sauvage FJ. Influence of tumour micro-environment heterogeneity on therapeutic response. Nature. 2013;501:346–54.

Kedzierski L, Evans KJ. Immune responses during cutaneous and visceral leishmaniasis. Parasitology. 2014;1–19.

Kipli K, Kouzani AZ, Williams LJ. Towards automated detection of depression from brain structural magnetic resonance images. Neuroradiology. 2013;55:567–84.

Lee HY, Lee KS, Ahn MJ, Hwang HS, Lee JW, Park K, Ahn JS, Kim TS, Yi CA, Chung MJ. New CT response criteria in non-small cell lung cancer: proposal and application in EGFR tyrosine kinase inhibitor therapy. Lung Cancer. 2011;73:63–9.

Loukas A, Prociv P. Immune responses in hookworm infections. Clin Microbiol Rev. 2001; 14:689–703, table of contents.

Meyer C, Ma B, Kunju LP, Davenport M, Piert M. Challenges in accurate registration of 3-D medical imaging and histopathology in primary prostate cancer. Eur J Nucl Med Mol Imaging. 2013;40 Suppl 1:S72–8.

Naik S, Bouladoux N, Wilhelm C, Molloy MJ, Salcedo R, Kastenmuller W, Deming C, Quinones M, Koo L, Conlan S, et al. Compartmentalized control of skin immunity by resident commensals. Science. 2012;337:1115–9.

Nestle FO, Di Meglio P, Qin JZ, Nickoloff BJ. Skin immune sentinels in health and disease. Nat Rev Immunol. 2009;9:679–91.

Nishino M, Jackman DM, Hatabu H, Yeap BY, Cioffredi LA, Yap JT, Jänne PA, Johnson BE, Van den Abbeele AD. New Response Evaluation Criteria in Solid Tumors (RECIST) guidelines for advanced non-small cell lung cancer: comparison with original RECIST and impact on assessment of tumor response to targeted therapy. AJR Am J Roentgenol. 2010;195:W221–8.

Nomura I, Goleva E, Howell MD, Hamid QA, Ong PY, Hall CF, Darst MA, Gao B, Boguniewicz M, Travers JB, Leung DY. Cytokine milieu of atopic dermatitis, as compared to psoriasis, skin prevents induction of innate immune response genes. J Immunol. 2003;171:3262–9.

Paalvast Y, Kuivenhoven JA, Groen AK. Evaluating computational models of cholesterol metabolism. Biochim Biophys Acta. 1851;2015:1360–76.

Pasparakis M, Haase I, Nestle FO. Mechanisms regulating skin immunity and inflammation. Nat Rev Immunol. 2014;14:289–301.

Paveley RA, Aynsley SA, Cook PC, Turner JD, Mountford AP. Fluorescent imaging of antigen released by a skin-invading helminth reveals differential uptake and activation profiles by antigen presenting cells. PLoS Negl Trop Dis. 2009;3:e528.

Peng H. Bioimage informatics: a new area of engineering biology. Bioinformatics. 2008;24:1827–36.

Piskorz L, Lesiak T, Brocki M, Klimek-Piskorz E, Smigielski J, Misiak P, Jablonski S. Biochemical and functional indices of malnutrition in patients with operable, non-microcelullar lung cancer. Nutr Hosp. 2011;26:1025–32.

Puig M, Tosh KW, Schramm LM, Grajkowska LT, Kirschman KD, Tami C, Beren J, Rabin RL, Verthelyi D. TLR9 and TLR7 agonists mediate distinct type I IFN responses in humans and nonhuman primates in vitro and in vivo. J Leukoc Biol. 2012;91:147–58.

Rankinen T, Sarzynski MA, Ghosh S, Bouchard C. Are there genetic paths common to obesity, cardiovascular disease outcomes, and cardiovascular risk factors? Circ Res. 2015;116:909–22.

Romeo S, Kozlitina J, Xing C, Pertsemlidis A, Cox D, Pennacchio LA, Boerwinkle E, Cohen JC, Hobbs HH. Genetic variation in PNPLA3 confers susceptibility to nonalcoholic fatty liver disease. Nat Genet. 2008;40:1461–5.

Rudikoff D, Lebwohl M. Atopic dermatitis. Lancet. 1998;351:1715–21.

Sanin DE, Prendergast CT, Bourke CD, Mountford AP. Helminth infection and commensal microbiota drive early IL-10 production in the skin by CD4+ T cells that are functionally suppressive. PLoS Pathog. 2015;11:e1004841.

Schadt EE. Molecular networks as sensors and drivers of common human diseases. Nature. 2009;461:218–23.

Schwarz E, Leweke FM, Bahn S, Liò P. Clinical bioinformatics for complex disorders: a schizophrenia case study. BMC Bioinformatics. 2009;10 Suppl 12:S6.

Shen W, Li W, Hixon JA, Bouladoux N, Belkaid Y, Dzutzev A, Durum SK. Adaptive immunity to murine skin commensals. Proc Natl Acad Sci U S A. 2014;111:E2977–86.

Shi P, Huang Y, Hong J. Automated three-dimensional reconstruction and morphological analysis of dendritic spines based on semi-supervised learning. Biomed Opt Express. 2014;5:1541–53.

Wang J, Peng X, Peng W, Wu FX. Dynamic protein interaction network construction and applications. Proteomics. 2014;14:338–52.

Wattis JA, O'Malley B, Blackburn H, Pickersgill L, Panovska J, Byrne HM, Jackson KG. Mathematical model for low density lipoprotein (LDL) endocytosis by hepatocytes. Bull Math Biol. 2008;70:2303–33.

Weinstein JN, Collisson EA, Mills GB, Shaw KR, Ozenberger BA, Ellrott K, Shmulevich I, Sander C, Stuart JM, Network CGAR. The Cancer Genome Atlas pan-cancer analysis project. Nat Genet. 2013;45:1113–20.

Wu X, Chen L, Wang X. Network biomarkers, interaction networks and dynamical network biomarkers in respiratory diseases. Clin Transl Med. 2014;3:16.

Yeh FC, Ye Q, Hitchens TK, Wu YL, Parwani AV, Ho C. Mapping stain distribution in pathology slides using whole slide imaging. J Pathol Inform. 2014;5:1.

Zelber-Sagi S, Salomone F, Yeshua H, Lotan R, Webb M, Halpern Z, Santo E, Oren R, Shibolet O. Non-high-density lipoprotein cholesterol independently predicts new onset of non-alcoholic fatty liver disease. Liver Int. 2014;34:e128–35.

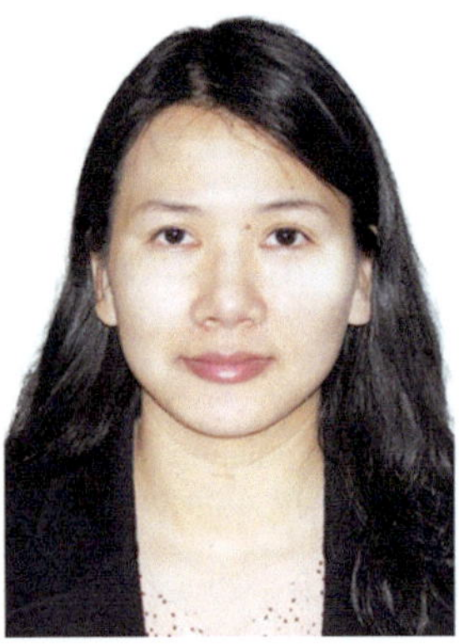

Duajiao Wu, MD, PhD Associate Professor of Systems Immunology in the Biomedical Research Center of Fudan University Zhongshan. Principle Investigator and Director of Systems Immunology Group/Platform in Fudan University Center for Clinical Bioinformatics. She was selected and honored as a Shanghai Rising-star Scientist of Science and Technology Commission of Shanghai Municipality, China. Her research focuses on clinical bioinformatics, transplantation immunology, and systems immunology.

Dr. Wu is funded by scientific grants including National Natural Science Foundation. She was invited to serve as reviewers of various international journals and as the faculty speaker in conferences. She has published more than 30 scientific papers in Journal of Translational Medicine, Journal of Proteome Research, and others.

David E. Sanin, PhD Postdoctoral Research Associate, Max-Planck Institute of Immunobiology and Epigenetics, Freiburg im Breisgau, Germany.

After receiving a scholarship granted by the National Institute of Science and Technology of Colombia (COLCIENCIAS), Dr Sanin did his PhD in the University of York, with Dr Adrian Mountford. His doctoral work focused on understanding the regulation and role of IL-10 production during the early stages of helminth skin infection. Dr Sanin has received prizes for the quality of his work awarded by the University of York and the American Society of Tropical Medicine and Hygiene. In addition to presenting his works in multiple international conferences, in some of which he has been a guest speaker, Dr Sanin has published in several international journals such as PLOS Pathogens and the Journal of Immunology. Currently Dr Sanin works on the metabolic regulation of antigen presenting cells with Dr Edward Pearce.

Xiangdong Wang, MD, PhD is a Distinguished Professor of Medicine at Fudan University, Director of Shanghai Institute of Clinical Bioinformatics, Director of Fudan University Center of Clinical Bioinformatics, Deputy Director of Shanghai Respiratory Research Institute, Adjunct Professor of Clinical Bioinformatics at Lund University, and visiting professor of King's College of London. He serves as a Director of Biomedical Research Center, Fudan University Zhongshan Hospital. His main research is focused on clinical bioinformatics, disease-specific biomarkers, lung chronic diseases, cancer immunology, and molecular & cellular therapies.

His group is working on precision medicine by integrating clinical informatics with omics science and bioinformatics to identify and validate disease-specific biomarkers and therapeutic targets in chronic lung diseases and lung cancer. His group initially developed the mirror-butterfly chemical structure of phosphoinositide 3-kinase inhibitor to prevent and treat chronic lung inflammation and injury, in combination of his pharmaceutical experience of drug discovery and development.

He serves as co-Editor-in-Chief of *Journal of Clinical Bioinformatics*, *Clinical & Translational Medicine*, and *Molecular & Cellular Therapies*; Editor of *Serial Book: Translational Bioinformatics*; Section Editor of Disease Biomarkers of *Journal of Translational Medicine* (IF = 3.99); Asian Editor of *Journal of Cellular Molecular Medicine* (IF = 3.67); and the editorial member of international journals, e.g. *American Journal of Pulmonary Critical Care Medicine* (IF = 11), *American Journal of Cellular & Molecular Biology* (IF = 5). He is the author of more than 200 scientific publications with the impact factor about 500, citation number about 2000, and cited journal impact factor about 5000.

Chapter 14
Key Law and Policy Considerations for Clinical Bioinformaticians

Mark Phillips

Abstract This chapter describes five key areas in which clinical bioinformatics activities are regulated by law and policy. These are, namely, open-data requirements, consent practices, anonymization strategies, restrictions on cross-border data transfer, and prohibitions on genetic discrimination. The discussion draws on examples of norms that are currently in effect in North America, Europe, Asia, and Oceania in order to illustrate the ways in which positions on various specific questions can either mutually converge or deviate from one another around the globe. The tension that animates virtually all of the debates throughout this area, whether explicitly or through proxy issues, this chapter argues, is between the promotion of the interests of research participants—particularly in ensuring data privacy—on the one hand, and in establishing a landscape optimized to best promote medical research discoveries, on the other.

Keywords ELSI • Privacy • Data protection • Open data • Funding agency policy

14.1 Introduction

Clinical bioinformaticians are governed by many of the same legal and policy norms as are researchers in genomics and the broader -omics fields. Physicians and health care practitioners who provide personalized medicine to their patients additionally remain bound by their professional ethics duties and laws applicable to the provision of health care.

Two new trends add complexity. First, a flurry of new rules have been adopted in recent years by law and policymakers who are scrambling to address ethical and privacy concerns that have emerged—and that often remain poorly understood—as a direct result of the rapid development of genomic technologies. Second, the increasing prevalence of data sharing, often across borders, as well as outsourcing to cloud service providers can mean that health projects must simultaneously contend with rule sets in multiple jurisdictions. Disparities between the rule sets

M. Phillips, B.Sc., LL.B., B.C.L. (✉)
Centre for Genomics and Policy, McGill University, Montreal, 740, avenue Dr. Penfield, suite 5200, Montreal, QC H3A 0G1, Canada
e-mail: mark.phillips2@mcgill.ca

X. Wang et al. (eds.), *Application of Clinical Bioinformatics*,
Translational Bioinformatics 11, DOI 10.1007/978-94-017-7543-4_14

can give rise to incompatibilities that may even needlessly make some medical projects impractical.

Despite this fragmentation, researchers have initiated significant efforts aimed at harmonizing these obligations at national, regional, and international levels. Although these efforts remain in the preliminary stages, their emphasis provides a helpful window through which to frame a general discussion of the law and policy as it currently affects clinical bioinformatics. Each of this chapter's following five sections describe one key area around which these discussions have centred, specifically (Sect. 14.2) open-data requirements; (Sect. 14.3) consent practices; (Sect. 14.4) anonymization strategies; (Sect. 14.5) restrictions on cross-border data transfer; and (Sect. 14.6) prohibitions on genetic discrimination. Each section draws on examples from existing laws and policies.

Before turning to these topics, the following subsections briefly distinguish and explain the two main sources of the duties that will be discussed: data-privacy law and funding agency policy.

14.1.1 Data-Privacy Law

Some countries use the term *privacy law* to describe law aimed at protection of personal data, while others speak of *data protection law*. This chapter follows the emerging trend of using the term *data-privacy law* to encompass both sets of laws.

Laws can guarantee the right to data privacy at the highest level of legal norms: in constitutions. The European Union (EU) has long been at the forefront of data-privacy law development, and its *Charter of Fundamental Rights of the European Union*—although not itself a formal constitution—conceives of personal data protection as a freestanding, fundamental right held by everyone (Fig. 14.1).

Across the Atlantic, the words 'data protection' or 'privacy' appear neither in the *United States Constitution* nor the *Canadian Charter of Rights and Freedoms*. But Supreme Court decisions have established that a degree of constitutional privacy protection is a necessary accessory to other explicit constitutional rights. The most prominent example is that the constitutional right to be secure against unreasonable searches and seizures has been found to necessarily flow from an underlying assumption that people enjoy a reasonable right to privacy (*Hunter v. Southam* 1984; *Katz v. United States* 1967).

The vast majority of data privacy law norms that are relevant in everyday practice are described in regular statutes. Data-privacy law varies significantly between countries: both in terms of the degree of protection provided, as well as the overall framework in which they are set out.

For decades, the European Union (EU) has encouraged its member countries to achieve a measure of harmonization. The preeminent EU data-privacy vehicle is currently the *Data Protection Directive 95/46/EC* (EU Directive), now 20 years old, which requires that each state subject to the *Directive* enact data-privacy legal protections that meet the minimum standards it describes. The EU Directive

1. Everyone has the right to the protection of personal data concerning him or her.
2. Such data must be processed fairly for specified purposes and on the basis of the consent of the person concerned or some other legitimate basis laid down by law. Everyone has the right of access to data which has been collected concerning him or her, and the right to have it rectified.
3. Compliance with these rules shall be subject to control by an independent authority.

Fig. 14.1 Article 8 of the *Charter of Fundamental Rights of the European Union*

maintains flexibility, however, by allowing each country a degree of leeway in their preferred implementation of the rules, to account for the legal traditions and social context in each country.

The EU Directive is now set to be superseded by a new *General Data Protection Regulation* (EU Regulation) in 2016. The new EU Regulation will further harmonize existing rules, as unlike the EU Directive, its rules will be directly enforceable throughout all of the European Economic Area and EU member states. At the time that this chapter was written, despite numerous draft iterations that had appeared, the text of the EU Regulation had not yet been finalized.

In contrast with this unified approach in the EU, the United States has relied on both a patchwork of highly specific laws and policies that address privacy, which exist alongside industry self-regulation mechanisms. Several federal laws include provisions that bear on privacy and that may be of particular interest to clinical bioinformaticians. These include the *Health Insurance Portability and Accountability Act of 1996* (HIPAA); the *Federal Policy for the Protection of Human Subjects* (Common Rule); and the *Genetic Information Nondiscrimination Act of 2008* (GINA).[1]

The siloed US approach has the advantage that each narrow topic was given lawmakers' undivided attention as the rules were drawn up, but it also comes with significant shortcomings. The first problem likely to be faced in practice is the difficulty in identifying which of the numerous federal and state laws do or do not regulate any given entity. HIPAA applies only to a list of covered entities including healthcare providers, health plans, and healthcare clearinghouses, as well as the covered entities' business associates and their subcontractors. The Common Rule applies to most organizations receiving federal funding for research. GINA applies to insurance companies and employers, and prohibits certain forms of discrimination based on genetic information.

But these laws can also apply to clinical bionformaticians in ways that may not be initially obvious. When a researcher receives genetic data from a HIPAA-covered entity for use in health research, for example, that researcher is bound to conform to the HIPAA Privacy Rule. As for the Common Rule, in one case in 2010, despite that it was unclear whether the Common Rule was legally enforceable against direct-to-consumer genetic testing company 23andMe, the company's non-compliance with the law's provisions was nonetheless argued to be a valid consideration in determining whether research based on the company's data was fit

[1] Each of these laws have been significantly amended since they were initially enacted.

for academic publication (Tobin et al. 2010). This last anecdote is one illustration of clinical bioinformaticians' interest in complying with generally accepted research and data-stewardship best practices even when the relevant law or policy sets a lower standard. This is especially true when the projects will persist over time.

A second shortcoming associated with the piecemeal US approach to data-privacy law is the risk of unintended gaps in protection. For example, although GINA provides legal protection against discrimination on the basis of asymptomatic genetic features, and the *Americans with Disabilities Act of 1990* protects against discrimination on the basis of disease that has manifested itself and imposes a substantial limitation, experts have cautioned that it remains unclear whether US law provides any protection against discrimination in situations between those two extremes, such as discrimination on the basis of mild symptoms (Rothstein 2008) or discrimination based on predictions made by machine learning techniques on genetic data (Horvitz and Mulligan 2015).

14.1.2 Funding Agency Policies

In addition to being subject to data-privacy law, clinical bioinformaticians must also often subject themselves to institutional policies, often as a condition of receiving funding. While data-privacy laws are enforceable either by national courts or by administrative entities to which the law delegates this task, the direct penalty for failing to adhere to funding policies is aimed at the professional medical activity itself. Funding may be withdrawn from a given project, and it may also be jeopardized for future projects. Policy breaches that become widely known may erode the confidence of both funders and participants so that continued research or practice is impossible.

But the duties described in policies adopted by funders can also become legally enforceable, either when they are explicitly made part of a contract (e.g. between researcher and funding agency), or when courts draw on the standards they establish to determine whether a defendant has met the standard that should be expected of a reasonable researcher or practitioner to determine liability in tort law or delict.[2]

Although certain laws, notably the US Common Rule, may require that medical research projects submit a detailed research proposal to a research ethics body and obtain prior approval, the requirement is commonplace in the realm of policy. Because the objectives of funding policies are not limited to protecting participants—the sole aim of medical and data-privacy law—and because they also seek to foster a context in which medical research will thrive, they contain some unique requirements not found in law. The following section discusses one such topic, open-data requirements.

---

[2] In the common law and civil law traditions, respectively.

14.2 Open Data

Data-sharing duties have rapidly proliferated in the policies adopted by funding agencies that they impose on grant recipients, although these duties have not found their way into data-privacy laws. This section explains the rationale behind open data policies, describes some of the obligations that apply to research data, notes the presence of sharing repositories, and finally discusses the extension of this current into open publishing.

14.2.1 Rationale

A recent report by the Expert Advisory Group on Data Access (EAGDA), a joint initiative of four of the largest UK research institutions, lists four factors that favour making research data openly available (see Fig. 14.2).

These factors weigh particularly heavily in bioinformatics and the -omics fields. The data in question is rich and multidimensional to the point that it is difficult to imagine ever exhausting its research potential. Research methodologies like genome-wide association studies (GWAS) rely on increasingly large sample sizes: the larger the better (McCarthy et al. 2008). The cost of collecting—let alone sequencing—this data directly from large numbers of people anew for each and every research initiative would be prohibitive.

Despite the strong trend toward data sharing, open-data requirements cannot be absolute. The exception that proves the rule is the Personal Genome Project (PGP), which makes the genetic data of 3500 volunteers freely available for download on its website (personalgenomes.org). But genetic research projects generally cannot meet their objectives without guaranteeing privacy protection to participants. This may also be true for the PGP, whose aim is to sequence and publicize the complete genomes and medical records of 100,000 volunteers.

1. The scale of datasets being collected has grown dramatically, and these datasets are assembled at significant cost.
2. It will usually not be possible for one group to analyse these data exhaustively, and there will often be significant potential for the data to be used to answer questions distinct from the original research questions of the data producers.
3. Developments in information technologies are transforming the ease with which large datasets can be shared, linked and analysed.
4. Both those who volunteer their data and samples for research, and those who pay for that research, hope for progress towards useful and eventually applicable results for human health and other societal benefits to be as rapid as possible. Indeed, there is a clear ethical requirement for efficient use of data from human research participants.

Fig. 14.2 'Drivers for data sharing' listed by the UK Expert Advisory Group on Data Access (Expert Advisory Group on Data Access 2015)

Genetic researchers have their own concerns regarding open data, primarily the fear that before they have the chance to publish their findings, their collected data will be used by rival researchers who will publish their results first. The most common mechanism to address this concern has been to mandate embargo periods, during which researchers temporarily holds exclusive publication rights over 'their' data. The embargo mechanism, however, has proven difficult to enforce in practice, and thus appears less often in recent data-sharing policies.

Enforcement difficulties were explicitly cited by the US National Institutes of Health (NIH), for example, as the reason abandoning embargo periods were abandoned in the 2014 *Genomic Data Sharing Policy* (GDS), which now applies to all large-scale, NIH-funded genomics research (National Institutes of Health 2014a).

14.2.2 Extent of the Duty

Open data obligations are about more than simply making research data available. Funding agencies commonly require that applicants include a data-sharing plan with their research funding proposal. Policies may also require that the data meet standards for quality and interoperability, and almost always encourage or even require that researchers release their data as rapidly as possible. The accepted delay for release of research data can be as little as 24 hours after they are generated, following the recommendation of the 'Bermuda Principles' put forward in 1996 by leaders in the Human Genome Project (Marshall 2001). A minor trend in the reverse direction has emerged, for example in the 2014 GDS policy, in which the NIH sought to account for its elimination of the embargo period by pushing some of its data-release deadlines back to the time of initial publication (National Institutes of Health 2014a).

The GDS policy provides detailed guidance regarding the NIH's expected deadlines. A slightly simplified version of the table it provides appears in Fig. 14.3, which is divided both between data submission and data publication deadlines, as well as into five distinct levels the NIH distinguishes based on the amount of processing and analysis that have been carried out on the data.

14.2.3 Repositories for Data-Sharing

To make compliance with their mandatory data-sharing requirements easier, funding agencies sometimes provide researchers with technological resources to assist in the process, and in particular have established repositories to which the data can be submitted for future access for secondary research. The NIH database of Genotypes and Phenotypes (dbGaP), a repository for individual-level data, is likely the most well-known of these. Data-sharing policies sometimes require that the data be submitted to a repository that has been specifically approved by the funding agency. The GDS policy is one such example, although it additionally allows

	General Description	Example Data Types	Data Submission Expected	Data Release Timeline
Level 0	Raw generated data	Instrument image data	Not expected	
Level 1	Initial sequence reads	DNA sequencing reads, ChIP-Seq reads	By publication time for non-human, de novo data Not expected for human data	
Level 2	After initial analysis or computation to clean data and assess quality	DNA sequence alignments to a reference sequence	Within 3 months of data generation, for human data By publication time, for non-human data	Within 6 months of acceptance for publication or data submission, whichever occurs first, for human data By publication time, for non-human data
Level 3	Analysis to identify genetic variants, gene expression patterns, or other features	SNP or structural variant calls, expression peaks, epigenomic features	Within 3 months of data generation, for human data By publication time, for non-human data	Within 6 months of acceptance for publication or data submission, whichever occurs first, for human data By publication time, for non-human data
Level 4	Final analysis relating genomic data to phenotype or other biological states	Genotype-phenotype relationships, relationships of epigenomic patterns to biological state	As analyses are completed, for human data By publication time, for non-human data	Data released with publication, for human data No later than the time of initial publication, for non-human data

Fig. 14.3 Data submission and release deadlines (Adapted and abridged from a supplement to the *NIH Genomic Data Sharing Policy* for the five general levels of data it specifies (National Institutes of Health 2014b))

researchers to submit their data to external repositories, so long as these include privacy and security features that meet the policy's requirements (National Institutes of Health 2014c).

14.2.4 Open Publishing

Funding agency policies also frequently apply the 'open' ethos more broadly, and require not only that researchers' data be made available, but also the academic analysis they ultimately publish. This trend has been made possible the proliferation of open-access academic journals. In Canada, for example, the *Tri-Agency Open Access Policy on Publications* not only requires the submission of bioinformatics data to a public database in certain circumstances, it also mandates that *any* funding from the country's three principal scientific research agencies[3] comes with the obligation that the funding recipient will 'ensure that any peer-reviewed journal

[3] Namely, the Canadian Institutes of Health Research (CIHR); the Natural Sciences and Engineering Research Council of Canada (NSERC), and the Social Sciences and Humanities Research Council of Canada (SSHRC).

publications arising from Agency-supported research are freely accessible within 12 months of publication' (Government of Canada 2015).

14.3 Consent

Informed consent has been a fundamental principle of healthcare and research for decades. Indeed, the topic dominated discussion in medical literature during the last half of the twentieth century (Manson and O'Neill 2007), and is seen as the essential mechanism for protecting patients and participants. Health care practitioners who provide personalized medicine to patients will be held to informed consent requirements, just as are other providers of health care.

But informed consent duties are usually less strict for secondary use of data or materials, for example, when research teams carry out studies on materials in biobanks or information in genomic data sharing repositories. This section focuses on consent to secondary research, and then also discusses dynamic consent, which has been proposed and has recently begun to be adopted in attempt to breathe new life into consent practices.

14.3.1 Secondary Use

Secondary research is difficult to reconcile with the normal approach to informed consent, which would require the initial participants to re-consent, which 'is costly and time-consuming, and difficulty in locating people can result in high drop-out rates' (Kaye et al. 2015). A variety of policy and legislative responses have emerged to decrease the intensity of specific consent requirements in the context of secondary use.

One approach is to simply abandon the consent requirement where it proves too onerous. Singapore's *Personal Data Protection Act 2012* (PDPA), for example, allows the use of personal data without consent for research when re-consent would be 'impracticable'. (To drop informed consent, the PDPA additionally requires that the research could not be accomplished without the data, it imposes limits on linkage with other data, and requires that the data subjects not be contacted.) In a guidance document, Singapore's Personal Data Protection Commission explained that the impracticability condition should be considered to be satisfied, for example, where the data was 'collected many years ago', because in that case the data subjects may have died or moved to another country in the intervening time (Personal Data Protection Commission (Singapore) 2014).

A second strategy is to eliminate the need for re-consent by seeking consent from participants at the moment when they initially consent to participate that is broad enough to also allow it to satisfy the consent needed for their participation in potential future research. Volunteers in the PGP, for example, whose data is available to any researcher or hobbyist who cares to make any conceivable use of them, must consent to all potential future uses of their data before they are included in the open, online PGP repository. Somewhat similarly, the NIH's GDS policy 'expects investigators generating genomic data to seek consent from participants for future research uses and the broadest possible sharing' (National Institutes of Health 2014a).

But consent can easily become too broad. Blanket consent to any possible future research is inconsistent with many law and policy protections. Even when these protections allow departures from full informed consent, the tradeoff is usually an increase of other obligations, such as external monitoring and governance requirements.

One of the fundamental principles of the 1980 *OECD Privacy Guidelines*—which have influenced nearly every other data-privacy law that currently exists—requires that whenever personal data is collected, the purposes of collection are specified, and that any subsequent use of the data must be limited to those purposes. If the purpose of collection is stated too broadly, for example if the purpose is simply to allow participation in future research, this may prove to be insufficient to satisfy the specification requirement. Under the HIPAA Privacy Rule, for example, unless a research team member 'anticipated and adequately described the purposes of the secondary research in the initial authorization received from a patient, that initial authorization may not constitute authorization for the use of identifiable registry data for secondary research purposes' (United States Agency for Healthcare Research and Quality 2007).

Similarly, Article 6 of the current draft of the forthcoming EU Regulation prescribes a general prohibition on secondary use '[w]here the purpose of further processing is incompatible with the one for which the personal data have been collected', with only few exceptions, although this provision has been among those in the EU Regulation that have been most actively contested.

A third approach is to allow secondary use without fresh consent when privacy guarantees in place are likely to prevent harm to the participant that might flow from data use. Canada's *Tri-Council Policy Statement* (TCPS) adopts an approach similar to that of Singapore's PDPA, but additionally requires that appropriate privacy safeguards are in place. Rather than stating what, precisely, those privacy safeguards must be, the TCPS leaves the competent research ethics body the discretion to consider the question according to each particular set of circumstances. The US Common Rule implements the privacy guarantee approach in a more rigid manner. It simply exempts data that have been anonymized from having to conform to its requirements, by deeming research on anonymized data not to involve human subjects. In a somewhat similar way, the HIPAA Privacy Rule also allows secondary use of anonymized data.

14.3.2 Dynamic Consent

One additional strategy that might be leveraged to address the difficulties with secondary consent is the adoption of dynamic consent mechanisms, although no prominent laws or policies currently explicitly require that they be used. The strategy is, however, being discussed with enthusiasm as a means to begin to address shortcomings associated with the traditional approach to informed consent more broadly (Erlich et al. 2014). The existing, standard notice-and-consent practice is characterized by lengthy consent forms that are presented to participants at the outset of their involvement research. The forms tend to leave the participant little meaningful choice, beyond the initial decision between whether to accept the conditions it describes, or to opt out of the research altogether. Critics liken this process to the lengthy terms and conditions often found in online contractual agreements, which invariably end with a single button marked 'I agree', which a person can choose to either click, or not.

If we are indeed entering into an era of *personalized medicine*, advocates of dynamic consent ask, why not also one of *personalized consent*? 'If biobank research is open-ended and ongoing then information technologies offer the possibility for participant involvement similarly to extend through time' (Kaye et al. 2015). The approach is most compelling where the participants' and patients' internet access is not overly hindered, either by technological, cultural, or educational barriers.

Research participants each have unique desires and expectations related to their research. Some may be comfortable with their data being shared for research into a specific disease, but feel that participation in unrelated research is not worth the privacy risks. Others may want their data to be available for a wide variety of medical research, but be opposed to their data being acquired or used by pharmaceutical corporations. Still others may want to prevent their personal health information from being used in studies that open it to a greater risk of government or law enforcement surveillance programs.

Dynamic consent strategies can allow not only for these decisions to be made by the participant and respected by researchers, especially in the clinical setting, but also allow for evolution over time of both the available options and preferences themselves. They can also allow participants' preferences to travel with their data samples. The approach seems to more fully embody and give meaning to the longstanding expectation that 'researchers will comply with any known preferences previously expressed by individuals about any use of their information' (TCPS).

Whether or not dynamic consent ultimately continues to expand in practice, consent will continue to retain its central role in medical practice despite undergoing significant changes in evolving contexts (Expert Advisory Group on Data Access 2015).

14.4 Anonymization

Until relatively recently, privacy experts invested a significant portion of their efforts into techniques to achieve data *anonymization* (or *de-identification*, which is sometimes used as a synonym, and other times, as a broader concept also encompassing pseudonymization). But a series of published re-identification attacks has led to vigorous debate and reappraisal of the merits of anonymization, and health professionals and privacy experts have now increasingly been driven toward alternative strategies.

The basic practice of anonymization can be illustrated by considering aggregate statistics. Even if thousands of people's personal data must be mobilized to determine that Berlin has a population of 3½ million, that statistic itself reveals effectively nothing about any of the city's specific residents. Even if the statistic relies on a great amount of personal information, it is itself an anonymized datum.

Anonymization, however, is commonly carried out without aggregation. The paradigmatic example is an operation on set of records, each of which relates to a single person, which excizes or obfuscates enough information in each record to make it becomes impossible to use the resulting data set to identify any of the people initially connected to the data.

Data-privacy legislation usually addresses anonymization only implicitly: The laws usually restrict their scope so that they have no application to information in general, but only to *personal information*,[4] defined as information about an identifiable individual. Information that cannot identify an individual—such as the statistics mentioned above, or data that has otherwise been anonymized—falls outside of this scope and is therefore not subject in any way to data-privacy legal or policy protections. Some specific data-privacy protections also explicitly state that they do not apply to data that has been anonymized.

This section first describes different legal standards that data has to meet in order to be considered properly anonymized. It then explains why anonymization as a technique has fallen into disfavour among privacy exports and health professionals alike. The section finally discusses the legal implications of some of the new approaches that are beginning to occupy the place formerly held by traditional anonymization techniques.

14.4.1 Thresholds

Perfect anonymization is impossible: 'Data Cannot be Fully Anonymized and Remain Useful' (Dwork and Roth 2014). Perhaps unsurprisingly then, the legal and policy requirements for data to be considered properly anonymized vary. Different thresholds are sometimes deliberately specified depending on the use that will be made of the data, on its sensitivity, or on a combination of both factors. This follows the principle that the degree of anonymization should be proportionate to the intensity of potential harms that might result from misuse of the data, and the likelihood that those harms will, in fact, materialize.

Coding or *pseudonymization* is a strategy related to, but distinct from, anonymization. The practice allows data sets to retain an identifier whose purpose is to allow the re-identification of data which have otherwise been anonymized, but

[4] Or any of various synonyms used, such as 'identifying data'.

to allow this only by people with access to a separate, private data set that links the identifiers back to individually identifying information.

While the HIPAA Privacy Rule's conception of 'de-identified' data, as discussed below, encompasses coded data, the EU Data Protection Directive excludes it by defining personal data as data relating to a person 'who can be identified, directly or indirectly, in particular by reference to an identification number'.

The various legal and policy definitions of personal data almost invariably remain at a highly abstracted level. In the broadest terms, personal data can be cast either *narrowly*, as occurs in definitions that include only data in which a person's identity is 'readily ascertainable', or *broadly*, as occurs in definitions that include any data for which it is reasonably foreseeable that the data (either alone or in combination with other data sets) will allow an individual to be identified.

One notable exception to the trend of defining anonymization in general terms is the HIPAA Privacy Rule, whose definition delves into an unusual level of technical detail. The Privacy Rule provides two alternative procedures, either of which allows data to be considered de-identified for HIPAA's purposes. The first option is to obtain a detailed written opinion from a statistician assuring that the re-identification risk that can reasonably be anticipated is 'very small'. The second option, sometimes referred to as the HIPAA 'Safe Harbor', requires that each of seventeen specified fields be removed from every record in the data set (see Fig. 14.4).

(A) Names;

(B) All geographic subdivisions smaller than a State ... except for the initial three digits of a zip code if ...

(1) The geographic unit formed ... contains more than 20,000 people; and

(2) The initial three digits of a zip code for all such geographic units containing 20,000 or fewer people is changed to 000.

(C) All elements of dates (except year) ... directly related to an individual ... ; and all ages over 89 and all elements of dates (including year) indicative of such age, except that such ages and elements may be aggregated into a single category of age 90 or older;

(D) Telephone numbers;

(E) Fax numbers;

(F) Electronic mail addresses;

(G) Social security numbers;

(H) Medical record numbers;

(I) Health plan beneficiary numbers;

(J) Account numbers;

(K) Certificate/license numbers;

(L) Vehicle identifiers and serial numbers, including license plate numbers;

(M) Device identifiers and serial numbers;

(N) Web Universal Resource Locators (URLs);

(O) Internet Protocol (IP) address numbers;

(P) Biometric identifiers, including finger and voice prints;

(Q) Full face photographic images and any comparable images;

Fig. 14.4 The seventeen HIPAA privacy rule de-identification fields

(i) Names;
(ii) Postal address information, other than town or city, State, and zip code;
(iii) Telephone numbers;
(iv) Fax numbers;
(v) Electronic mail addresses;
(vi) Social security numbers;
(vii) Medical record numbers;
(viii) Health plan beneficiary numbers;
(ix) Account numbers;
(x) Certificate/license numbers;
(xi) Vehicle identifiers and serial numbers, including license plate numbers;
(xii) Device identifiers and serial numbers;
(xiii) Web Universal Resource Locators (URLs);
(xiv) Internet Protocol (IP) address numbers;
(xv) Biometric identifiers, including finger and voice prints; and
(xvi) Full face photographic images and any comparable images.

Fig. 14.5 Protected health information that excludes these direct identifiers of the person to whom the data relates and of their relatives, employers, or household members qualifies as a Limited Dataset under the HIPAA Privacy Rule

HIPAA Safe Harbor anonymization additionally requires the removal of any other unique identifying number, with the exception of an optional re-identification number (which would thus result in a coded, rather than an anonymized, data set). The Safe Harbor's final requirement is that the person carrying out the anonymization must not have 'actual knowledge that the information could be used alone or in combination with other information to identify an individual'.

A less strict variation on Safe Harbor anonymization, called a limited data set, is also described by HIPAA. The use of limited data sets, as a tradeoff, requires that researchers sign a data-use agreement subjecting them to additional restrictions on how the data may be used and to whom it may be disclosed. The limited data set anonymization fields appear in Fig. 14.5. Limited data sets are most commonly used by researchers who want to analyze the additional data fields that can legitimately be retained, such as dates and five-digit zip codes.

14.4.2 Anonymization's Fall into Disfavour

The Safe Harbor's straightforward anonymization instructions may seem appealing when compared with a duty to anonymize data according to the vague standard requiring that it is no longer reasonably foreseeable that they will one day allow re-identification. But the apparent simplicity of the Safe Harbor is a false promise, especially where genetic data is concerned, which is notably absent from the anonymization fields listed in Figs. 14.4 and 14.5.

Because all but the shortest genetic sequences are high dimensional, these data are generally thought to be impractical to anonymize (El Emam and Arbuckle 2013). Some have argued that to comply with the Safe Harbor rules genetic information must be removed because it itself constitutes a unique identifying number under the eighteenth HIPAA Safe Harbor identifier. Guidance issued in the intervening years has, however, failed to address the issue (Office for Civil Rights 2012). The Safe Harbor's 'actual knowledge' requirement should also generally prevent genetic data from being included in a data set, though that test is drafted to depend on the mindset of the person doing the anonymization rather than on reasonable expectations of re-identifiability. The requirement is thus often ignored in practice (El Emam and Arbuckle 2013).

Beyond the challenges posed by genetic data, the HIPAA Safe Harbor is an illustration of the broader problems with attempts to set out a detailed anonymization procedures in law that do not take into account specific contexts. A 2009 report by the US Institute of Medicine found a number of failings with the HIPAA Privacy Rule, and emphasized that HIPAA's rigid procedure is simultaneously too strict and not strict enough. It is indeed trivial to construct an example data set for which the Safe Harbor both allows re-identification and also requires that data be unnecessarily removed.

Not only is HIPAA's approach to anonymization less than optimal, but the broader practice of anonymization itself has now increasingly fallen out of favour as an effective means of privacy protection, particularly when it comes to high dimensional data such as genomic sequences. Existing techniques are able to re-identify an individual given as few as thirty independent single nucleotide polymorphisms (SNPs) (El Emam and Arbuckle 2013), and so to anonymize any genetic sequence with confidence would often require obliterating most of the data, along with its research value. In the same vein, anonymization is coming to be seen as unhelpful to translational medicine, which relies on linkage between different data sets, and is impossible once anonymization effectively sterilizes them.

If debate about the continued relevance of anonymization has not been completely settled, perhaps it is because its remaining defenders have already conceded so much. It is increasingly rare for anonymization to be used as a privacy safeguard in practice on its own, without being supported by other mechanisms. After researchers showed that data in dbGaP could be re-identified despite having been anonymized according to HIPAA, the NIH converted dbGaP into a controlled- rather than open-access repository (Homer et al. 2008; National Institutes of Health 2008). In the UK, EAGDA now similarly recommends alternative protections such as access controls (Expert Advisory Group on Data Access 2013), although new re-identification attacks continue to be described in the literature (Cai et al. 2015). In 2014, the New Zealand Privacy Commissioner suggested addressing anonymization's weaknesses by going so far as to make it illegal to attempt to re-identify data (Edwards 2014).

14.4.3 Successors

Data-privacy experts are now turning away from anonymization and have begun to explore emerging alternative approaches to privacy protection, which sometimes include the potential to achieve provable security. The goal is no longer to anonymize datasets as much as possible so that they can be shared as widely as possible. Instead, many of the new strategies are based on cryptographic methods which aim to allow genomic research studies to be carried out without the need for any of the raw research data itself to ever need to be disclosed. At the forefront of these techniques are homomorphic encryption, secure multiparty computation, and differential privacy.

Homomorphic encryption is an attractive approach in cases where a third party is made responsible for storage and computation whose access to the data would itself be a privacy risk, such as in the context of the increasingly prevalent practice of genomic research using cloud computing services (Lauter et al. 2014). Genomic data is uploaded to the cloud in an encrypted form, and homomorphic encryption then allows researchers to submit calculations to have the cloud perform on the encrypted data and to ultimately receive the encrypted result, all the while maintaining the data in its encrypted form so that it remains unreadable to other parties, including the cloud service provider itself.

Secure multiparty computation is a related strategy. In this case the data set is split between multiple parties so that each one holds only a fraction of the overall data to be analyzed. Cryptographic methods then allow researchers the parties to collectively carry out calculations on the full data set without any individual party having to reveal any of their own raw data (Kamm et al. 2013). Similarly, techniques such as DataSHIELD allow researchers to perform aggregate calculations and studies on data sets held by third parties without the need to reveal any raw data to the researcher (Wolfson et al. 2010).

Differential privacy offers perhaps the most promise of all of these new methods, and is used in contexts of statistical aggregation. This method aims to mathematically determine whether an individual's decision to participate in a given study will have any effect on their privacy. Dwork and Roth describe differential privacy as a 'promise' that those holding data make to a data subject: 'You will not be affected, adversely or otherwise, by allowing your data to be used in any study or analysis, no matter what other studies, data sets, or information sources, are available' (Dwork and Roth 2014).

These techniques have not yet made their way into laws and policies, and even though they are largely departures from anonymization, the legal analysis of their use must consider the rubric of personal information. This is so because despite the mathematical proofs that have been published demonstrating some of the methods' abilities to securely protect privacy, none has yet resulted in a generalizable method to ensure privacy protection in practice. Current practical methods of homomorphic encryption, for example, still require an 'assum[ption] that all [collaborating] entities behave *semi-honestly*' (Lu et al. 2015). Because calculations in secure multiparty computation are always based on the raw data, the results they produce must reveal a degree of private information. For reasons such as these, it will remain necessary to ask whether it is reasonably foreseeable that the information revealed by these techniques allows individuals to be identified are likely to remain relevant. If the answer is yes, the associated data-privacy law restrictions will continue to apply.

14.5 Cross-Border Transfer

Concerns about cross-border transfer of data have grown considerably following Edward Snowden's revelations about the existence of widespread electronic surveillance programs. Both legal and policy restrictions now exist on cross-border transfer of personal information have increased in intensity and expanded in number. The rapidly expanding use of cloud computing in bioinformatics fields has also added to these concerns, given what is seen as the inherent borderlessness of cloud technologies.

The overarching concern with cross-border transfers and outsourcing of personal data is that these can place the data in contexts where they may be exposed to more serious privacy risks, and in particular, risks that the data holder is required not to expose them to. Because Canada's *Tri-Council Policy Statement*, for example, requires that researchers 'avoid being put in a position of becoming informants for authorities' (Canadian Institutes of Health Research, Natural Sciences and Engineering Research Council of Canada, Social Sciences and Humanities Research Council of Canada 2014), this requires that researchers seek to avoid cross-border transfer to a jurisdiction known to engage in such surveillance programs.

In the broadest terms, most laws and policies aim to allow cross-border transfer and outsourcing when this will not significantly undermine data privacy. Two general approaches have emerged in data-privacy law with the objective of achieving this aim in the context of cross-border personal data transfer.

The *accountability* approach requires the entity transferring data to ensure that it will enjoy a similar or greater degree of protection in the hands of the specific entity to which the data is transferred in another jurisdiction. Canada has adopted this approach in its *Personal Information Protection and Electronic Documents Act* (Canada 2000).

The *adequacy* approach, in contrast, allows cross-border transfer only if the target jurisdiction has previously been deemed adequate by the data-privacy authority tasked with making such determinations. Adequacy determinations are required for cross-border transfer by the EU *Directive*, and the same approach will be retained in the new EU Regulation, and existing adequacy determinations will remain in force. Data-privacy laws in thirteen different jurisdictions have currently been approved by the European Commission as providing adequate protection.

Other laws, particularly since the Snowden revelations, have imposed blanket prohibitions on transfer or storage outside the jurisdiction, which may be subject to exceptions. The law of the Canadian province of British Columbia, for example, has long included such a blanket provision, which prohibits public bodies from storing personal information outside of Canada (British Columbia 1996). In 2014, however, the Information & Privacy Commissioner of that province published updated guidance stating that it is possible for public bodies to store personal data outside of the country without violating the law if the data are protected by a data security technique called tokenization (Office of the Information & Privacy Commissioner for British Columbia 2014). Tokenization is somewhat similar to coding. In this case, it would allow data sets to be stored outside of Canada so long as any personal information has been replaced by a 'token'. The token allows the personal information it represents to then be retrieved using a separate data set known as a 'crosswalk table', which must be stored in Canada.

14.6 Genetic Nondiscrimination

Laws and guidelines have recently proliferated that prohibit certain forms of discrimination on the basis of genetic information. The US *Genetic Information Nondiscrimination Act of 2008*, for example, was discussed in the Introduction to this chapter, especially with respect to gaps in the protection it provides against certain 'milder' forms of genetic discrimination.

But like many other genetic nondiscrimination laws, GINA is also limited in terms of who it prevents from engaging in discrimination. GINA applies only to insurance and employment sectors. But this does not mean that researchers in clinical bioinformatics are free to disregard these laws, which often contribute to determining the risks of discrimination to which research participants are exposed. The Philippine *National Ethics Guidelines for Health Research 2011*, for example, explicitly require that research projects involving genetic data contend with the issue (Philippine National Health Research System 2011):

> There is potential harm to research participants arising from the use of genetic information, including stigmatization or discrimination. Researchers should take special care to protect the privacy and confidentiality of this information.

Beyond providing these privacy and confidentiality protections, clinical bioinformaticians must be aware of any participant or patient interaction that could reasonably increase the risk of becoming subject to genetic discrimination. For example, although Australia's *Insurance Contracts Act 1984* prohibits insurance companies from requiring that a customer undergo genetic testing, if the customer has already had a genetic test, and even if they simply know the results a family member's test, the results must be declared before entering into a new insurance contract (Liddell 2002).

Article 11 – Non-discrimination
Any form of discrimination against a person on grounds of his or her genetic heritage is prohibited.
Article 12 – Predictive genetic tests
Tests which are predictive of genetic diseases or which serve either to identify the subject as a carrier of a gene responsible for a disease or to detect a genetic predisposition or susceptibility to a disease may be performed only for health purposes or for scientific research linked to health purposes, and subject to appropriate genetic counselling.
Article 13 – Interventions on the human genome
An intervention seeking to modify the human genome may only be undertaken for preventive, diagnostic or therapeutic purposes and only if its aim is not to introduce any modification in the genome of any descendants.
Article 14 – Non-selection of sex
The use of techniques of medically assisted procreation shall not be allowed for the purpose of choosing a future child's sex, except where serious hereditary sex-related disease is to be avoided.

Fig. 14.6 Chapter IV of the Council of Europe's 1997 *Oviedo Convention*, which sets out basic protections with respect to the human genome (Council of Europe 1997)

In some jurisdictions, clinical bioinformaticians themselves are additionally directly subject to prohibitions on genetic discrimination. One of the earliest genetic nondiscrimination laws, for example, the Council of Europe's *Oviedo Convention*, does not limit its scope to any particular categories of potential discriminators (Fig. 14.6).

14.7 Conclusion

Although legal and policy duties regulate clinical bioinformaticians in areas beyond those discussed in here—including transaction logging, data-privacy breach notification, risk assessment, and reporting of incidental findings, among others—this chapter provided an introduction to five areas of key importance. Some were chosen because they hold a fundamentally important place in the legal and policy frameworks, while others have been the subject of extensive expert debate and discussion. In either case, familiarity with these concepts is helpful in contending with the broader issues. The discussion focused on law and policy from several world regions, to illustrate the ways in which positions on each question can either converge or deviate around the globe.

The fundamental tension in the field, which presents itself at every turn, remains finding the optimal balance between privacy protection and the facilitation of medical research and care. What often appear to be new areas of debate—such as the question of open data, or even of the continued relevance of anonymization techniques—each soon reveal themselves to be manifestations of that same initial underlying tension. If the issue of open data in genomics were entirely independent of this tension, the most vocal advocates of open bioinformatics research data might be expected to be seen applying the idea of a genomic commons to the issue of genetic patents, arguing to limit these or eliminate them altogether, but efforts in this direction are, if anything, currently declining (Contreras 2015). Thankfully, robust promotion of both health research and of privacy protections do not always have to be played off against one another in a zero-sum game. Many of the techniques described in this chapter that have only recently begun to be explored and that have yet to be internalized by law and policy at all, such as homomorphic encryption and dynamic consent, appear to have the potential to promote both.

Disclosure Statement Funding for this work was provided by the Canada Research Chair in Law and Medicine.

References

British Columbia. Freedom of information and protection of privacy act. Revised Statutes of British Columbia, chapter 165. Queen's Printer BC; 1996.

Cai R, Hao Z, Winslett M, Xiao X, Yang Y, Zhang Z, Zhou S. Deterministic identification of specific individuals from GWAS results. Bioinformatics. 2015;31(11):1701.

Canada. Personal information protection and electronic documents act. Statutes of Canada, chapter 5. Queen's Printer for Canada; 2000.

Canadian Institutes of Health Research, Natural Sciences and Engineering Research Council of Canada, Social Sciences and Humanities Research Council of Canada. Tri-council policy statement: ethical conduct for research involving humans. Ottawa: Secretariat on Responsible Conduct of Research; 2014.

Contreras JL. NIH's genomic data sharing policy: timing and tradeoffs. Trends Genet. 2015;31(2):55.

Council of Europe. Convention for the protection of human rights and dignity of the human being with regard to the application of biology and medicine: convention on human rights and biomedicine. Strasbourg: Council of Europe; 1997.

Dwork C, Roth A. The algorithmic foundations of differential privacy. Found Trends Theor Comp Sci. 2014;9(3–4):211.

Edwards J. Privacy and big data: presentation by privacy commissioner John Edwards. New Zealand Privacy Commissioner; 2014.

El Emam K, Arbuckle L. Anonymizing health data: case studies and methods to get you started. Beijing: O'Reilly; 2013.

Erlich Y, Williams JB, Glazer D, Yocum K, Farahany N, Olson M, et al. Redefining genomic privacy: trust and empowerment. PLoS Biol. 2014;12(11):e1001983. doi:10.1371/journal.pbio.1001983.

Expert Advisory Group on Data Access. Governance of data access. 2015. http://www.wellcome.ac.uk/stellent/groups/corporatesite/@msh_peda/documents/web_document/wtp059343.pdf

Expert Advisory Group on Data Access. Statement for EAGDA funders on re-identification. 2013. http://www.wellcome.ac.uk/stellent/groups/corporatesite/@policy_communications/documents/web_document/wtp055972.pdf

Government of Canada. Tri-agency open access policy on publications. 2015. http://www.science.gc.ca/default.asp?lang=En&n=F6765465-1

Homer N, Szelinger S, Redman M, Duggan D, Tembe W, Muehling J, et al. Resolving individuals contributing trace amounts of DNA to highly complex mixtures using high-density SNP genotyping microarrays. PLoS Genet. 2008;4(8):1000167.

Horvitz E, Mulligan D. Data, privacy, and the greater good. Science. 2015;349(6245):253.

Hunter v. Southam. 11 Dominion Law Reports, 4th Series, 641. 1984.
Kamm L, Bogdanov D, Laur S, Vilo J. A new way to protect privacy in large-scale genome-wide association studies. Bioinformatics. 2013;29(7):886.
Katz v. United States. 389 US 347. 1967.
Kaye J, Whitley EA, Lund D, Morrison M, Teare H, Melham K. Dynamic consent: a patient interface for twenty-first century research networks. Eu J Hum Gen. 2015;23:141.
Lauter K, López-Alt A, Naehrig M. Private computation on encrypted genomic data. 2014. http://research.microsoft.com/pubs/219979/genomics.pdf
Liddell K. Just genetic discrimination? The ethics of Australian law reform proposals. Univ NSW Law J. 2002;25(1):160.
Lu W, Yamada Y, Sakuma J. Efficient secure outsourcing of genome-wide association studies. IEEE CS Security and Privacy Workshops. 2015.
Manson NC, O'Neill O. Rethinking informed consent in bioethics. Cambridge: Cambridge University Press; 2007.
Marshall E. Bermuda rules: community spirit, with teeth. Science. 2001;291(5507):1192.
McCarthy MI, Abecasis GR, Cardon LR, Goldstein DB, Little J, Ioannidis JP, et al. Genome-wide association studies for complex traits: consensus, uncertainty and challenges. Nat Rev Genet. 2008;9(5):356.
National Institutes of Health. Modifications to genome-wide association studies (GWAS) data access. 2008. http://gds.nih.gov/pdf/Data%20Sharing%20Policy%20Modifications.pdf
National Institutes of Health. NIH genomic data sharing policy. 2014c. http://gds.nih.gov/PDF/NIH_GDS_Policy.pdf
National Institutes of Health. Notice number: NOT-OD-14-124. 2014a, 27 August. http://grants.nih.gov/grants/guide/notice-files/NOT-OD-14-124.html
National Institutes of Health. Supplemental information to the National Institutes of Health genomic data sharing policy. 2014b. http://gds.nih.gov/PDF/Supplemental_Info_GDS_Policy.pdf
Office for Civil Rights. Guidance regarding methods for de-identification of protected health information in accordance with the Health Insurance Portability and Accountability Act (HIPAA) privacy rule. 2012. http://www.hhs.gov/ocr/privacy/hipaa/understanding/coveredentities/De-identification/guidance.html
Office of the Information & Privacy Commissioner for British Columbia. Updated guidance on the storage of information outside of Canada by public bodies. 2014, 16 June. https://www.oipc.bc.ca/public-comments/1649
Personal Data Protection Commission (Singapore). Advisory guidelines for the healthcare sector. 2014, 11 September.
Philippine National Health Research System. National ethical guidelines for health research. 2011. http://www.ethics.healthresearch.ph/index.php/phoca-downloads/category/4-neg?download=9:pub-ethics-guidelines-2011
Rothstein MA. Currents in contemporary ethics: GINA, the ADA, and genetic discrimination in employment. J Law Med Ethics. 2008;36(4):837.
Tobin SL, Lee SSJ, Greely HT, Ormond KE, Cho MK. Not a loophole: commercial exploitation of an IRB error. Comment on: Gibson G, Copenhaver GP. Consent and internet-enabled human genomics. PLoS Genet. 2010;6(6):e1000965.
United States Agency for Healthcare Research and Quality. Registries for evaluating patient outcomes: a user's guide. Rockville: US Department of Health and Human Services; 2007.
Wolfson M, Wallace SE, Masca N, Rowe G, Sheehan NA, Ferretti V, et al. DataSHIELD: resolving a conflict in contemporary bioscience—performing a pooled analysis of individual-level data without sharing the data. Int J Epidemiol. 2010;39(5):1372.

Mark Phillips holds an LL.B. and a B.C.L. from McGill University's Faculty of Law, as well as a B.Sc. (Honours) in Computer Science from the University of Manitoba. His work at the Centre of Genomics and Policy is focused on comparative analyses of data protection, privacy, and cloud computing laws and policies, particularly as they relate to bioinformatics. He is a former editor of both the *McGill Journal of Law and Health* and of the *McGill Law Journal* and has published several peer-reviewed articles and book chapters. His research interests also include computer-assisted legal research methodologies, mental health and disability, and law and social movements.

Chapter 15
Challenges and Opportunities in Clinical Bioinformatics

Denis C. Shields

Abstract Clinically applied bioinformatics faces many specific challenges. Many of these are related to challenges faced by discovery bioinformatics, which is confronted by similar issues of data complexity, data scale, and the lack of statistical power to address key problems. These include issues of the scale and variety of data being handled and annotated, and the associated proliferation of errors of data annotation, and errors of statistical inference. Modelling issues include the choice of methods, and the flaws associated with overly simplistic and overly complex approaches.

Data aggregation among researchers and clinicians and patients is likely to represent a key step forward, but the main clinical gains are likely to emerge from the aggregation of relatively homogeneous data types, associated with clear prior hypotheses. It may be less useful to integrate analytical and predictive approaches across many different complex data types in smaller groups of subjects.

Clinical bioinformatics needs to be integrated into new regulatory paradigms for incorporation of knowledge into healthcare. We need to explore possibilities of one-person trials integrated with genotypic data but the theoretical and practical frameworks for such approaches are not worked out.

Keywords Clinical bioinformatics • Big data • Statistical power • Data integration • Data aggregation

Clinical bioinformatics presents considerable opportunities for advancing human health. In line with genomics advances (Stephens et al. 2015; Schatz and Langmead 2013), it faces considerable challenges. One clear example is the case of stratified analysis of patient data, where the disadvantages of smaller sample sizes that result from dealing with a subset are in a few cases outweighed by the very clearly defined improvement in risk definition.

An example of the opportunities from discovery bioinformatics is the identification of the BRCA1 gene for breast cancer, by focusing on early-onset patients

D.C. Shields (✉)
School of Medicine, University College Dublin, Dublin 4, Ireland
e-mail: denis.shields@ucd.ie

X. Wang et al. (eds.), *Application of Clinical Bioinformatics*,
Translational Bioinformatics 11, DOI 10.1007/978-94-017-7543-4_15

(Hall et al. 1990). An example of an opportunity from clinical bioinformatics is the treatment with antibodies versus the HER2 receptor to treat those breast cancer patients whose tumours test positive for this receptor (Vogel et al. 2002).

One of the challenges is the assumption that more knowledge, and further stratification based on this knowledge will inevitably improve health outcomes. To counter this, the medical literature is littered with examples of subset analyses or stratified analyses have been pursued doggedly in spite of a clear lack of evidence that the particular stratum focused on did indeed have an identifiable and different risk, whose treatment as a separate clinical entity clearly justified the distinctive treatment of those patients in a separate manner. Another related challenge is the spectre of "incidental" findings from high throughput molecular analysis (Berg et al. 2011) confusing and stifling our health systems with spurious and inappropriate follow up studies.

Bioinformatics in general, and translational and clinical bioinformatics are not truly a single field (Stein 2008), but are a collection of activities joined by a set of overlapping skillsets. Accordingly, there is no "single" challenge or opportunity. While there are features that separate discovery and applied clinical bioinformatics, there are also features that unite them (See Chap. 2). Tables 15.1 and 15.2 list challenges facing bioinformatics in general, and highlighted where they may be of particular relevance to clinical bioinformatics. One general problem is that we lack statistical power to address all the hypotheses we would love to investigate. This is at the heart of many challenges, as it may indeed reflect a central challenge in medicine. Many of these challenges represent traps that I have stumbled headlong into in the course of my own research, and have taken quite some time to emerge from: in particular the trap of wanting findings to be more significant than they are (Ioannidis 2005). It is very difficult for any human scientist to avoid this, and as long as we have humans doing science we have to read each-others' work in a spirit of extensive and healthy scepticism, which I hope you will apply in reading this review.

15.1 So Much Data...

While computational systems can handle ever increasing volumes and complexity of data in terms of storage and retrieval, the ability of humans to make sense of them require complex systems that create a layer of mystery over the data, such that the person interpreting the data may have little understanding of the underlying data, and is at the mercies of the assumptions of the models used to generate interpretations. This is a serious issue, but one that continuing serious scholarship can tackle over time, by addressing the fundamental flaws in modelling processes and providing clearer guidance to users on the limitations or limited assumptions of the models that are built into the associated systems.

The most serious consequence of data overload is simply the reduction in statistical power to formally test hypotheses. While some guidance through the

Table 15.1 Statistical and modelling challenges facing bioinformatics

Thirteen statistical and modelling challenges	Research bioinformatics	Clinical bioinformatics	Possible opportunities or solutions
So many hypotheses to test, so little time or money to test them.	Many genes have unknown functions	Cannot test every drug in every genetic sub-group	See directly below.
Too many tests, sample sizes too small	GWAS only discovers a small proportion of variants that modestly influence population attributable risk	"Incidental findings" of rare variants with possible effects from large-scale molecular analysis of individual patients	Worldwide aggregation of routine clinical data for research; one-person trial aggregation (Schork 2015; Nikles et al. 2011; Smith and Kingsmore 2014; Chen et al. 2012; van Ommen 2013; Kaput et al. 2012); associated tools to efficiently capture phenotypic data (Rehm et al. 2015; Stumbo et al. 2010)
Number of tests for synergy analyses greatly inflated over main effect tests	Hard to discover synergies among polymorphisms, even though we know they exist (Fitzpatrick et al. 2015)	Combination drug effects should in principal be mined from large patient datasets but relatively few applications emerging	New methods to focus on smaller subsets of interactions for testing, to increase power by reducing the search space.
"Most" published research findings are false (Ioannidis 2005) and databases built on them will carry forward those errors	Polymorphism disease association literature is littered with false positives	Erroneous application of "discovered" associations in clinical practice without any clear clinical validation of their predictive validity	Solution : an immediate ban on use of the word "significant" to mean a threshold of evidence separating truth from error
"Winner's curse" phenomenon (Costa-Font et al. 2013), inflating effect sizes.	GWAS studies: initial SNP-disease associations confer higher risks than follow up studies (Xiao and Boehnke 2009).	Published trials of successful drugs also exaggerate effect sizes (Dickersin et al. 1987; Young et al. 2008)	Adequate replication of findings.
Confused lines of evidence in presentation and analysis of complex multivariate datasets	Hyping of data integration in complex analyses	MALDI serum cancer prognostic failed to analyse data properly (Ransohoff 2005).	Call for rigorous evidence-based claims on complex analyses (Ransohoff 2005).
Excess reliance on simplistic reductionism	Hyped reports of "gene for disease X" blur boundaries between partial association and major causation	–	Greater emphasis on quantitative rather than qualitative research; clarify difference between association and causation

(continued)

Table 15.1 (continued)

Thirteen statistical and modelling challenges	Research bioinformatics	Clinical bioinformatics	Possible opportunities or solutions
Excess reliance on overly complex analyses	Many projects assume that complex analyses can overcome fundamental flaws in study design (Fregnac and Laurent 2014)	–	Require comparison with simpler analytical approaches, alongside complex proposals or analyses.
Biological evolution is complex and messy and does not follow clean engineering principles	Flawed assumption that features of biological networks reflect selection for function, rather than non-adaptive processes (Lynch 2007)	Flawed assumption that most non-conservative amino acid changes are deleterious.	
Clinical evolution of therapeutic solutions selects among complex alternatives without a full understanding of why the best solutions work		Promiscuous drugs with multiple paths of action may be buffered against inter-individual variation, making them better than "single target" drugs (Kell 2013).	
Needless re-invention of wheels, sometimes even square or rhomboidal versions of them.			Try to ensure that those proposing new methods read more widely the previous literature, e.g. thirty or more years old.
Multiple competing prediction methods unclear which ones are best.	In many cases, could the "best" predictors simply represent some kind of subtly uncontrolled over-fitting		Use of "consensus" predictors that combine results of multiple predictors
Users confused: so much data, so many analyses			Undergrad, postgrad and post-qualification bioinformatics training with statistical emphasis

Table 15.2 Data challenges facing bioinformatics

Ten data challenges	Research bioinformatics	Clinical bioinformatics	Possible opportunities or solutions
Poor traceability of evidence behind annotations	Gene/protein annotation (Mons et al. 2008)	Variant function annotation	Improved representation of evidence chain in databases
Error propagation	annotating genes based on homology (Gilks et al. 2002, 2005)	Linking rare diseases to chromosomal changes or single base variants (Quintans et al. 2014)	
Imperfect data standard formulation and compliance	Many standards have slow uptake in community, and are in parallel made obsolete by technology change		Enforcement of standards by journal editors; simplify standards
Ill-defined data entities	Overlapping, polymorphic duplicate, or chimeric genes cannot be defined as single object	Patient status and data changing over time.	
Too much data to physically handle or present	Challenge of finding resources to keep centralized databases going	Routine clinical imaging data: volume of data issues, and limited computer tools for their automated analysis.	Good informatics strategies led by those with close knowledge of the data and its useful meanings, and track record dealing with its existing issues
Increasing diversity of data	e.g. human tissue imaging, proteomic, antibody, RNA, disease state, integrated in protein atlas	See Chaps. 3, 4, 5, 6, 7, 8.	
Commercial interests, scientific ownership interests and concerns about human data privacy restrict data access and limit research	Slowing of disease SNP discovery by the restrictions freely and rapidly sharing anonymised datasets, even with relatively limited phenotypic data.	Limited sharing of patient genomic and phenotypic data; clinically important allele risk information hidden in commercial silos (Angrist and Cook-Deegan 2014)	Need to develop encrypted analysis approaches that maximise research benefits but minimise privacy breaches (Erlich and Narayanan 2014); need regulatory frameworks to avoid incentivising commercial and academic data hoarding (Quackenbush 2014), and systems that enable data sharing (Field et al. 2009).

(continued)

Table 15.2 (continued)

Ten data challenges	Research bioinformatics	Clinical bioinformatics	Possible opportunities or solutions
Cost and feasibility of distributing data	Shift of sequence databases from release to users to availability via web and via remote analysis tools (API, REST interfaces)		Evolution of common data and ethics practices and standards for access to a limited number of shared repositories of human genetic and phenotypic data
Computational cost of processing data		Variant calling in human genome (Langmead et al. 2009)	Hardware development focusing on string manipulation rather than floating point speed (Stephens et al. 2015); algorithms for more efficient genome comparisons (Stephens et al. 2015)
Struggle to maintain and develop controlled vocabularies/ ontologies/ definitions	Gene ontology only covers a minority of genes with some functional annotation	Proliferation of new definitions of clinical entities or pathogens, based on molecular analysis and sub-analysis	

data can help reduce somewhat the search space of hypotheses, and some conditioning on prior information can again assist in reducing the search space and the number of competing hypotheses, the consequences for the translation of research observations to clinical practice are serious. This problem is not new, but the volume and variety of data that may be collected in a clinical setting is likely to mushroom over the next few years, making this the single most challenging issue of modern clinical bioinformatics: how to prioritise what matters, and whether the existing system of moving from research observations to clinical practice needs to be overhauled, or at least supplemented, with alternative approaches that are mindful of this problem.

A very simple example is as follows. Clinical trials are deliberately powered to have enough patients to be likely to address the central issue: is the new drug effective versus an alternative? If half of a study size of 1,300 patients are on drug and half on placebo, the study has 95 % power to detect a drop of blood pressure from 84 to 82, assuming a standard deviation of 10. This is to say that, if you were to carry out this same study 100 times, 90 of the repeated studies would show a significant association (at a p-value in each of ≤ 0.05), while ten would fail to observe such an association. Suppose you want to ask a more complex question:

does the drug have a differential effect on those patients who carry a particular common polymorphism within the target protein of the drug? The study size is effectively halved, and there is only 72 % power to detect such an effect in this subgroup. If the polymorphism is only found in 10 % of patients, the power drops to 21 %, so the chances are greater of not observing the true effect, with such a small sample. If you extend this to a genome wide study of half a million polymorphisms (even assuming all are carried at a frequency of 0.5) there is no longer sufficient power to detect the association of genotype with drug (only 4 %), because a much stricter p-value must be employed. If you want to look and see what pairwise combinations of polymorphisms distinguish between drugs you have a total of over 100,000,000,000 tests, and thc power drops towards zero. Yet, responses in individual patients may be actually influenced by combinations of genetic factors, so we are in principal interested in such studies, yet the trial would require a study size of at least 12,000 subjects in order to address this. This is at least ten times the original trial size, and in practise much greater, when rarer allele frequencies are taken into account.

15.2 When We Have 'all' the Data, Will We Understand Everything?

Most likely not. There is substantial complexity in the relationships between data and outcomes. There is also problem of low predictive power for many datasets across a variety of applications from discovery to clinical application. So much so, that when investigators fail to find neat emerging patterns from a given strand of evidence, they are tempted to combine it with another strand of evidence, in the hope that data integration within a unifying model will have greater predictive power than just taking one strand of evidence alone. This aspiration is based on the observation that multiple factors influence outcomes, which leads to the spurious conclusion, namely that the more factors that are modelled, the better will be our understanding of the outcomes. This fails to recognize that when we combine two very poorly powered data sources, each which has a low predictive power, it may only in exceptional cases overcome the noise in both datasets to pinpoint a causal factor or really quantify its effect. More typically, it will simply inflate the noise of the source datasets, so that any source signal in a single data source may be effectively drowned. Since there have been no formal examinations of this general effect in the combination of noisy datasets, the scale of the problem is difficult to judge in different circumstances. Typically, authors who present complex data integration solutions lavish considerable effort on building an impressive model, but little time on checking the sensitivity of the model to removing components. So often an outcome that appears to be the result of a complex integration of multiple datasets is merely dependent primarily on a single data source, with some over-fitting effects of the other contributors adding a little apparent, but spurious, signal. The higher profile journals require that such predictions are followed up with experimental validations, but it is quite difficult to distinguish among three kinds of papers: (i) those which genuinely advance biological understanding (ii) those which have been deliberately retro-fitted to make an experimental validation match a supposed complex informatics procedure that identified the gene of interest, and (iii) those which were simply lucky by mistake. All three kinds of papers read remarkably similarly, with extensive analytical detail buried in many relatively arbitrary analysis decisions described in extensive supplementary material. In many cases, it is likely that the authors themselves are often fairly unclear which category their paper belongs in.

Bioinformatics has been around for a while now, and when we look at where it has really transformed and accelerated understanding, leading towards useful therapeutics and diagnostics, it has usually been in making simple links and finding simple associations, rather than in providing convoluted models that try to integrate everything. There is a place for integrative analysis, which may perform useful roles equivalent to the kinds of data visualisations that a statistician performs prior to carrying out a formal analysis. These analyses can guide the researcher's thinking about how complex processes may operate, and highlight potential biases and artefacts, but they should not be typically placed centre stage as the primary result. It should not be assumed that such integrative analyses are a panacea for all the things about biology and medicine that we do not understand. In contrast with the general field of bioinformatics, the field of clinical bioinformatics is better protected against the tendency to place too much faith in complex integrative analyses, since natural clinical caution avoids complexity in medical processes: each extra step or dependency in a medical procedure is another thing that can go wrong in the measurement or the application, reducing benefits to patients. Thus, the apparent initial success of MALDI proteomics in predicting ovarian cancer was relatively quickly made the focus of intensive investigation that revealed that the models were poorly fitted (Ransohoff 2005).

While increasing complexity of data is likely to present substantial problems, increasing the sample sizes of given classes of data (Risch and Merikangas 1996) is a key factor in improving discovery of associations that can have clinical application or be useful in defining new targets for therapeutics. Data accumulation of clinical instances of rare variants, and their likely functional effects, is likely to greatly improve the prognostic ability of clinicians to provide useful information to patients regarding their genetic risks. But the aggregation needs to impose some kinds of standards of consistency of functional annotation, or the aggregation is much less worthwhile (Rehm et al. 2015). More organised data sharing is needed, including domain-specific expert panels, so that clinical bioinformatics can benefit from the data sharing across large research consortia that has benefited research bioinformatics (Rehm et al. 2015). Thus, more organised data accumulation to create larger relatively homogeneous datasets will help partially overcome some of the issues surrounding weak predictive power.

15.3 Data Presentation and Error Propagation

While integrative complex analyses of biological data should be treated with more caution than is currently the case, integrative databases that allow researchers to navigate among data types are as important as ever, but growing harder to manage, in the face of the ever increasing complexity. Alternative integration schemes are available, such as data warehousing and federated database solutions (Zhang). However, the more intractable challenges lie within the data itself and the behaviour of scientists generating data. One key problem that will survive beyond the teething problems seen with every new data type, is the problem of automated curation that derives from analyses of other entries, using features such as homology to infer annotations. This has long been recognized as a major issue in degrading data quality (Table 15.2) and the evidence chain is difficult to work back through. This leads to so-called propagation or percolation of error through databases. We need better ways of representing more dynamically the annotation basis of data so that a user can more readily evaluate either computationally or manually the nature of the annotation quality and confidence.

15.4 Personalized or Precision Medicine

In the context of the above caveats, what is the prospect then for personalized medicine and big data on everyone, how can we make sense of things? Drug therapy is a good example of where personalized medicine is argued to be the area where there are the greatest gains to be made. This is on the basis that adverse drug events, many of which arise from measurable prior risk factors such as genetic variants, are one of the leading causes of death. If we can incorporate data on those adverse events, then we should be able to dose existing drugs more successfully (Meyer 2000). Given that the older drugs were developed to work on a population, often very ineffectively in individual cases, should newer drugs be developed from the start assuming that different drugs will be suitable for different people? The main issue is that the cost of developing personalized drugs for sub-populations is bound to be more expensive than population drugs, and is only justified for society if the clinical benefits outweigh that additional cost very substantially, in a significant fraction of patients. There are two very different perspectives from the modelling community that throw light on this topic from very different directions.

Douglas Kell (2013) pointed out that one of the features of many large blockbuster drugs that treat the general population is that they frequently have pleiotropic effects. If a drug lowers blood pressure by targeting not just one, but three different, pathways, then it may well be much more resistant, in terms of its clinical effect on blood pressure, to genetic variations. After all, people are unlikely to have drug-sensitivity factor in all three targets. The overall result is a smoother effect on blood pressure, and thus greater efficacy and fewer side-effects. In the competition among drugs, those drugs with such effects have been selected because their very pleiotropic nature confers advantages to the population by reducing interpersonal variability in responses, compared to a single-target drug. A related argument has been made, that drug combinations in lower doses may be beneficial (Lehár et al. 2009). From this perspective, knowledge about personalised differences should be used in order to develop medicines that are more robust against inter-personal differences, and personalised medicine research supports the design of better population-wide therapeutics.

A proposed alternative to population-targeted therapeutics is the proposal of 'one-person' trials in individual patients, which can then be aggregated in new ways to give new insights into treatment. This is not entirely novel for drugs that have a rapidly and easily quantified response, such as blood pressure lowering medications: patients are tried on a drug, and moved to the next if there is no response, or moved to drug combinations, and taken off drugs if there is an adverse event. But these are not trials. The value anticipated is that aggregation of many one-person trials across different groupings (any treatment versus a particular disease in individuals with a certain genetic background; or all treatments against a particular disease for example) may allow insights over time as the data accumulates. However, as he has pointed out, there are very substantial barriers to making this happen, and it is most likely to be driven first by patients demanding that their data may be useful to future patients, and pushing doctors to enrol them in such trials. While in principle one-person trial aggregation could be massively powerful, in practice the diversity of data collection and the complexity of integration, represent substantial hurdles in the management of these trials that may only be overcome in a minority of settings, where the initial data collection is likely to be partly motivated by some direct benefit of the individual trial to the patient. Costs of performing these trials and aggregating their data would need to be kept low to ensure substantial adoption.

One person trial design presupposes a very strong heterogeneity in responses, to justify the considerable complexity of implementing and analysing such aggregated trials. In many cases, even though there are clearly personalised responses, they may not be strong enough to justify the complex investment of resources and the increased uncertainties surrounding study control. It is possible that some preliminary trials with an alternative approach may be a good precursor to n-of-1 type trials, by estimating heterogeneity of response (Loop et al. 2012), but there has been little work in this area.

15.5 Conclusions

We are standing at the crossroads in a data swamp. The areas where clinical bioinformatics has the greatest opportunities to transform medical practise are those where the following conditions are met:

1. Certain types of relatively homogeneous risk factor and outcome data may be collected inexpensively from many patients and aggregated efficiently.
2. Patients can be randomised in a cost efficient manner within blinded one-person crossover trials.
3. Regulatory authorities may be willing to re-draw the boundaries of pre-approval and post-approval clinical trials.

The area where this may be first met most easily is in the area of nutritional supplements, since the safety concerns are relatively limited. Even in this area there are substantial challenges, where it is difficult to effectively blind many randomised intakes of interest. However, the greatest clinical benefit may come in the area of rare disease treatment, where the safety issues of drug interventions and small sample sizes both limit prospects for advancement. Clinical bioinformatics needs to be integrated into new regulatory paradigms for incorporation of knowledge into healthcare. We need to explore possibilities of one-person trials integrated with genotypic data but the theoretical and practical frameworks for such approaches are not worked out.

References

Angrist M, Cook-Deegan R. Distributing the future: the weak justifications for keeping human genomic databases secret and the challenges and opportunities in reverse engineering them. Appl Transl Genom. 2014;3:124–7.

Berg JS, Khoury MJ, Evans JP. Deploying whole genome sequencing in clinical practice and public health: meeting the challenge one bin at a time. Genet Med. 2011;13:499–504.

Chen R, et al. Personal omics profiling reveals dynamic molecular and medical phenotypes. Cell. 2012;148:1293–307.

Costa-Font J, McGuire A, Stanley T. Publication selection in health policy research: the winner's curse hypothesis. Health Policy. 2013;109:78–87.

Dickersin K, Chan S, Chalmers TC, Sacks HS, Smith Jr H. Publication bias and clinical trials. Control Clin Trials. 1987;8:343–53.

Erlich Y, Narayanan A. Routes for breaching and protecting genetic privacy. Nat Rev Genet. 2014;15:409–21.

Field D, et al. Megascience. 'Omics data sharing. Science. 2009;326:234–6.

Fitzpatrick DJ, et al. Genome-wide epistatic expression quantitative trait loci discovery in four human tissues reveals the importance of local chromosomal interactions governing gene expression. BMC Genomics. 2015;16:109.

Fregnac Y, Laurent G. Neuroscience: where is the brain in the human brain project? Nature. 2014;513:27–9.

Gilks WR, Audit B, De Angelis D, Tsoka S, Ouzounis CA. Modeling the percolation of annotation errors in a database of protein sequences. Bioinformatics. 2002;18:1641–9.

Gilks WR, Audit B, de Angelis D, Tsoka S, Ouzounis CA. Percolation of annotation errors through hierarchically structured protein sequence databases. Math Biosci. 2005;193:223–34.

Hall JM, et al. Linkage of early-onset familial breast cancer to chromosome 17q21. Science. 1990;250:1684–9.

Ioannidis JP. Why most published research findings are false. PLoS Med. 2005;2:e124.

Kaput J, Morine M. Discovery-based nutritional systems biology: developing N-of-1 nutrigenomic research. Int J Vitam Nutr Res. 2012;82:333–41.

Kell DB. Finding novel pharmaceuticals in the systems biology era using multiple effective drug targets, phenotypic screening and knowledge of transporters: where drug discovery went wrong and how to fix it. FEBS J. 2013;280:5957–80.

Langmead B, Schatz MC, Lin J, Pop M, Salzberg SL. Searching for SNPs with cloud computing. Genome Biol. 2009;10:R134.

Lehár J, Krueger AS, Avery W, Heilbut AM, Johansen LM, Price ER, Rickles RJ, Short GF 3rd, Staunton JE, Jin X, Lee MS, Zimmermann GR, Borisy AA. Synergistic drug combinations tend to improve therapeutically relevant selectivity. Nat Biotechnol. 2009;27:659–66. doi: 10.1038/nbt.1549 PMID: 19581876.

Loop MS, et al. Submitted for your consideration: potential advantages of a novel clinical trial design and initial patient reaction. Front Genet. 2012;3:145.

Lynch M. The evolution of genetic networks by non-adaptive processes. Nat Rev Genet. 2007;8:803–13.

Meyer UA. Pharmacogenetics and adverse drug reactions. Lancet. 2000;356:1667–71.

Mons B, et al. Calling on a million minds for community annotation in WikiProteins. Genome Biol. 2008;9:R89.

Nikles J, et al. Aggregating single patient (n-of-1) trials in populations where recruitment and retention was difficult: the case of palliative care. J Clin Epidemiol. 2011;64:471–80.

Quackenbush J. Learning to share. Sci Am. 2014;311:S22.

Quintans B, Ordonez-Ugalde A, Cacheiro P, Carracedo A, Sobrido MJ. Medical genomics: the intricate path from genetic variant identification to clinical interpretation. Appl Transl Genom. 2014;3:60–7.

Ransohoff DF. Lessons from controversy: ovarian cancer screening and serum proteomics. J Natl Cancer Inst. 2005;97:315–19.

Rehm HL, et al. ClinGen–the Clinical Genome Resource. N Engl J Med. 2015;372:2235–42.

Risch N, Merikangas K. The future of genetic studies of complex human diseases. Science. 1996;273:1516–17.

Schatz MC, Langmead B. The DNA data deluge: fast, efficient genome sequencing machines are spewing out more data than geneticists can analyze. IEEE Spectr. 2013;50:26–33.

Schork NJ. Personalized medicine: time for one-person trials. Nature. 2015;520:609–11.

Smith LD, Kingsmore SF. N-of-1 genomic medicine for the rare pediatric genetic diseases. Expert Opin Orphan Drugs. 2014;2:1279–90.

Stein LD. Bioinformatics: alive and kicking. Genome Biol. 2008;9:114.

Stephens ZD, et al. Big data: astronomical or genomical? PLoS Biol. 2015;13:e1002195.

Stumbo PJ, et al. Web-enabled and improved software tools and data are needed to measure nutrient intakes and physical activity for personalized health research. J Nutr. 2010;140:2104–15.

van Ommen B. The nutrition researcher cohort: toward a new generation of nutrition research and health optimization. Genes Nutr. 2013;8:343–4.

Vogel CL, et al. Efficacy and safety of trastuzumab as a single agent in first-line treatment of HER2-overexpressing metastatic breast cancer. J Clin Oncol. 2002;20:719–26.

Xiao R, Boehnke M. Quantifying and correcting for the winner's curse in genetic association studies. Genet Epidemiol. 2009;33:453–62.

Young NS, Ioannidis JP, Al-Ubaydli O. Why current publication practices may distort science. PLoS Med. 2008;5, e201.

Zhang Z, Yu J, Cheung, K.-H. Data integration in bioinformatics: current efforts and challenges. In: Mahdavi MA, editor. Bioinformatics – trends and methodologies. INTECH Open Access Publisher; 2011.

Chapter 16
Heterogeneity of Hepatocellular Carcinoma

Tingting Fang, Li Feng, and Jinglin Xia

Abstract Liver cancer is the third leading cause of cancer-related death worldwide. And Hepatocellular carcinoma (HCC) is the most common form of liver cancer. The extreme variability of the clinical outcome caused a major challenge of HCC, which makes it difficult to properly stage the disease and thereby estimate the prognosis. That's because the rapidly growing tumor displays heterogeneity of genetic and histopathologic characteristics. The risk of HCC may be affected by several known environmental factors such as hepatitis viruses, alcohol, cigarette smoking, and others. The aetiological factors associated with HCC have been well characterized; however, their effects on the accumulation of genomes changes and the influence of ethnic variation in risk factors still remain unclear. Advances in sequencing technologies have enabled the examination of liver cancer genomes at high resolution; somatic mutations, structural alterations, HBV integration, RNA editing and retrotransposon changes have been comprehensively identified. In addition, integrated analyses of trans-omics data have identified diverse critical genes and signaling pathways implicated in hepatocarcinogenesis. These analyses have revealed potential therapeutic targets, and prepared the way for new molecular classifications for clinical application. Therefore, the international collaborations of cancer genome sequencing projects are expected to contribute to an improved understanding of risk assessment, diagnosis and therapy for HCC. This review discusses the contribution of heterogeneity such as aetiological factors, tumor microenvironment, genetic variations, epigenetic changes and signaling pathways in HCC progression.

Keywords Hepatocellular carcinoma • Heterogeneity

T. Fang
Zhongshan Hospital, Fudan University, Shanghai, China
e-mail: fangttfddx@sina.com

L. Feng
Minhang Hospital, Fudan University, Shanghai, China
e-mail: fengweihong66666@sina.cn

J. Xia (✉)
Medical College, Fudan University, Shanghai, China
e-mail: xia.jinglin@zs-hospital.sh.cn

X. Wang et al. (eds.), *Application of Clinical Bioinformatics*,
Translational Bioinformatics 11, DOI 10.1007/978-94-017-7543-4_16

16.1 Introduction

HCC is a leading malignancy worldwide (Torre et al. 2015). Chronic liver damage which may result from chronic hepatitis, liver cirrhosis and fatty liver disease, is closely associated with HCC. Hepatitis virus infection, alcohol intake, aflatoxin B exposure, and some metabolic diseases such as obesity, haemochromatosis and diabetes mellitus are well-known risk factors for HCC (El-Serag 2012; Forner et al. 2012; Yu et al. 2013). The incidence of HCC is high in East Asian and African countries (Torre et al. 2015; El-Serag 2012; Forner et al. 2012; Shaib and El-Serag 2004). Africa and Asian countries (except Japan) have the highest rate of HBV infection in the world (El-Serag 2012). However, the number of patients infected with HCV has been rapidly increasing in Japan and Western countries, especially in the USA where viral hepatitis infection is partly mediated through drug abuse (El-Serag 2012; Forner et al. 2012). With the exception of environmental risk factors, individual genetic predisposition may be linked to the risk of HCC as suggested by the fact that in a relevant percentage of HCC cases, i.e., about 20 % of cases diagnosed in the United States, without known predisposing risk factors, including alcohol use or viral hepatitis, can be identified (El-Serag and Mason 2000). The role of genetic factors in the risk of HCC is supported by strong evidence from animal models, which have enabled the identification of the number and chromosomal location of loci affecting genetic susceptibility to chemically induced hepatocarcinogenesis in both mice and rats (Dragani et al. 1996; Feo et al. 2006). In this Review, we mainly focus on HCC, as HCC showed distinctive genomic alterations at present, which includes estimated risk of HCC according to particular genetic factors.

16.2 Aetiological Factors for HCC

The risk of HCC may be affected by several known environmental factors such as hepatitis viruses, alcohol, cigarette smoking and so on (IARC 2004; Bosch et al. 2004; Kuper et al. 2000; Llovet et al. 2003), among which the prevalence of chronic hepatitis B (HBV) or C (HCV) virus infections plays an identified role in the incidence of HCC. HCC is more prevalent in Southeast Asia and sub-Saharan Africa, where HBV infection is endemic, but HBV-related liver cancer cases also occur in western countries (Bosch et al. 2004; Llovet et al. 2003). Chronic carriers of HBV have up to a 30-fold increased risk of HCC (IARC 1994; Evans et al. 2002; Franceschi et al. 2006). In western countries, HCV infection plays a major role in the pathogenesis of HCC, and it has become more prevalent over the past decades, accompanied by a higher incidence and mortality from HCC (El-Serag and Mason 2000; IARC 1994). The fact that alcohol consumption causes liver cirrhosis and is an independent risk factor for primary liver cancer has been disclosed by a large number of cohort and case–control studies (Kuper et al. 2000; Baan et al. 2007;

Ogimoto et al. 2004). And epidemiological studies showed that increasing HCC risks associated with exposure to aflatoxins after adjustment for HBV exposure (IARC 2002). What's more, cigarette smoking has been causally associated with the risk of HCC (IARC 2004; Kuper et al. 2000), and heavy smoking and heavy drinking was reported to have a multiplicative effect in HCC development (Kuper et al. 2000).

In addition to environmental risk factors, individual genetic predisposition may also play a role in the risk of HCC with the current evidence from epidemiological/genetic studies in human populations, which argues for the important role of monogenic and polygenic factors in determining the risk of HCC development. Rare monogenic syndromes such as alpha1-antitrypsin deficiency, hemochromatosis, acute intermittent, cutanea tarda porphyria, and glycogen storage disease type I as well as hereditary tyrosinemia type I are associated with a high risk of HCC (Andant et al. 2000; Elmberg et al. 2003; Elzouki and Eriksson 1996; Fracanzani et al. 2001; Haddow et al. 2003; Janecke et al. 2001; Ostrowski et al. 1983; Scott 2006; Weinberg et al. 1976). Several common conditions or diseases inherited as polygenic traits e.g. autoimmune hepatitis, type 2 diabetes, non-alcoholic steatohepatitis, hypothyroidism, and a family history of HCC also show an increased risk of HCC compared to the normal population (El-Serag et al. 2006; Hashimoto et al. 2009; Hassan et al. 2009; Werner et al. 2009; Hemminki and Li 2003). Therefore, the increased risk of HCC may not be directly linked to genetic disorders, but instead single germ-line mutations or conditions regulated by complex genetics may cause chronic damage such as liver cirrhosis of the target organ, in turn causing the oncogenic mutations and/or promoting preexisting endogenous or virus- or chemical-induced mutations which lead to HCC. Indeed, similar to those occurring in human liver cirrhosis, conditions of hepatic necrosis and regeneration may promote carcinogen-induced hepatocarcinogenesis, as suggested by the experiments with rodent models (Dragani et al. 1986). Thus, cirrhosis from any cause appears to be the common signaling pathway by which some risk factors exert their hepatocarcinogenesis (Fig. 16.1). Overall, the genetic susceptibility to HCC is characterized by a genetic heterogeneity; With the fact that, a high individual risk of HCC may thus be caused by several unlinked single gene defects, whose carriers are rare in the general population, or by more common conditions inherited by complex genetics.

16.3 Heterogeneity of Tumor Microenvironment in HCC

As a highly heterogeneous disease, HCC displays differences in angiogenesis, extracellular matrix proteins and the immune microenvironment, which contribute to HCC progression. Therefore, a better understanding of its heterogeneity will greatly contribute to the development of strategies for the HCC treatment.

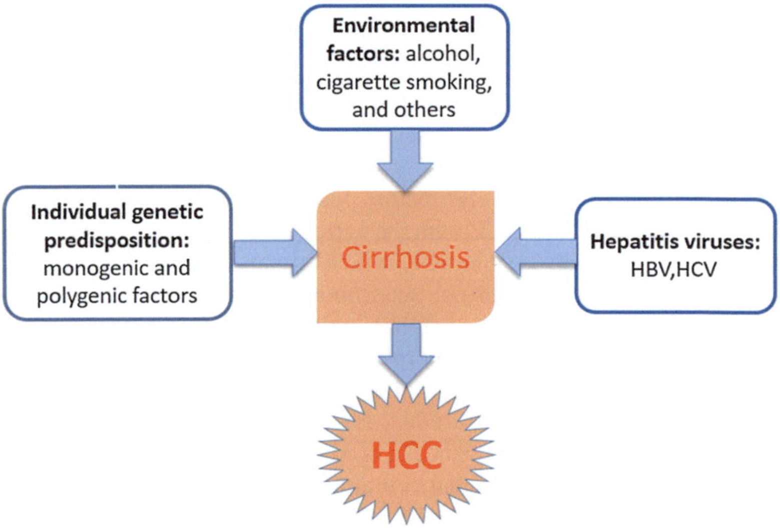

Fig. 16.1 Aetiological factors for HCC. The risk of HCC may be mainly affected by several known environmental factors, including hepatitis viruses, alcohol, cigarette smoking, and others. In addition to environmental risk factors, individual genetic predisposition may play a role in the risk of HCC as suggested by the fact that in a relevant percentage of HCC cases, i.e.

16.3.1 Angiogenic Heterogeneity

HCC has wide variations in vascularity that are dependent upon tumor size (T stage) and histological grade, and angiogenic switch depends on the balance between pro- and antiangiogenic factors at different stages of tumor progression (Baeriswyl and Christofori 2009). Pro-angiogenic factors include VEGF, fibroblast growth factor (FGF), platelet-derived growth factor (PDGF), angiopoietin-1 and angiopoietin-2. And anti-angiogenic factors include thrombospondin-1 (TSP1), endostatin, interferon-α, interferon-β and angiostatin. VEGF expression is up-regulated by hypoxia-induced factor-1α (HIF-1α) to switch angiogenic phenotype (Fang et al. 2001). Therefore, HCC is a hypervascularized tumor because of increased angiogenic phenotype (Muto et al. 2015), which is not only required for tumor growth supplied with oxygen and essential nutrients but also facilitates metastasis. A higher level of VEGF mRNA in tumor tissue correlates with increased post-resection recurrences, suggesting that an altered balance between angiogenic stimulators and inhibitors contributes to cancer progression. Therefore, angiogenic heterogeneity is associated with angiogenic molecules such as VEGF, PEDF and HIF-1α (Fig. 16.2a) that could be different among various tumor sizes and time intervals during hepatocarcinogenesis, which needs to be taken into the consideration when we decide to carry out an anti-angiogenic therapy to prevent recurrence in HCC patients (Wu et al. 2007).

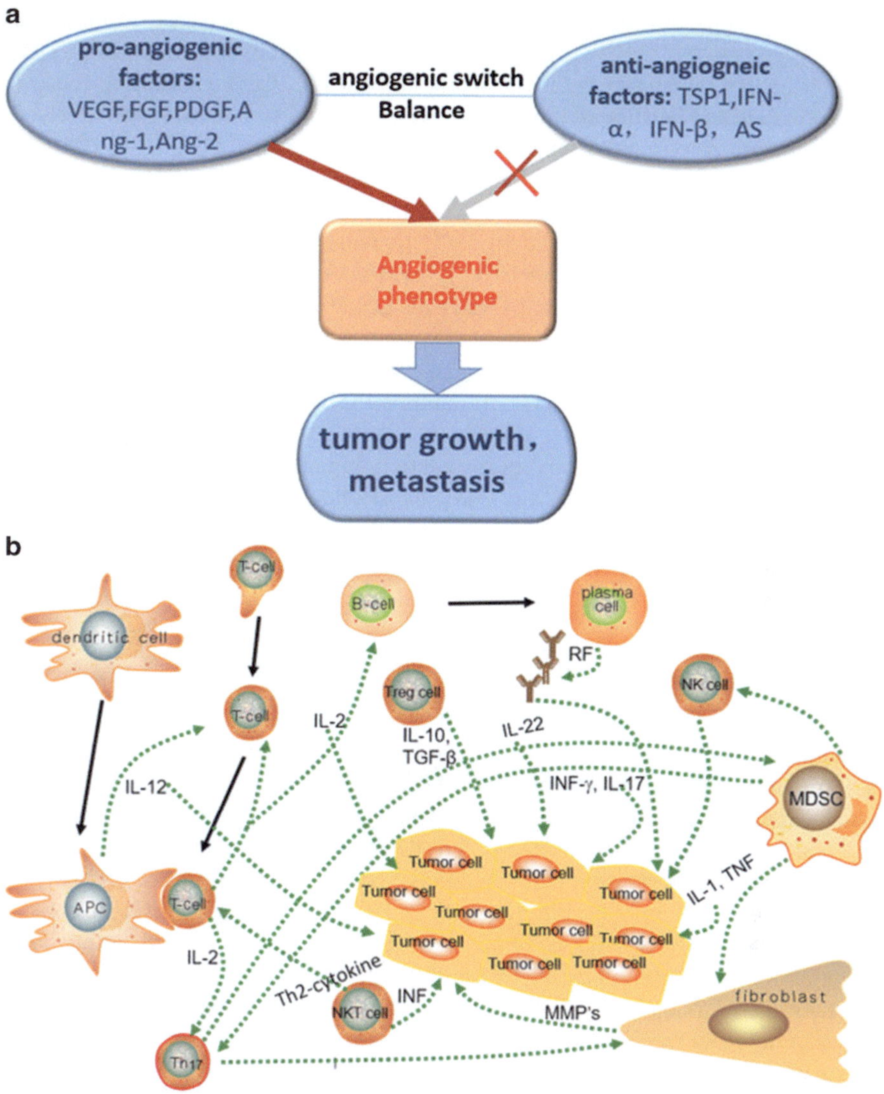

Fig. 16.2 Heterogeneity of hepatocellular carcinoma. (**a**) Angiogenesis (**b**) immune microenvironment

16.3.2 Heterogeneity of Extracellular Matrix

Extracellular matrix (ECM) components mainly consist of collagen, laminin, fibronectin, glycosaminoglycan and proteoglycan. Because of continuous repatterning of the ECM, HCC tumor cells can invade via direct or indirect interactions among ECM, stroma cells and HCC (Carloni et al. 2014). The major tumor ECM concerned in this process are collagen type IV, lysyl oxidase (LOX) and matricellular proteins (MCPs), whereas MCPs is prime metastatic niches in HCC (Chew et al. 2012; Fang et al. 2013; Wong and Rustgi 2013). Overall, a dynamic ECM contributes to hepatocarcinogenesis. Matrix metalloproteinases (MMPs) were associated with tumor invasion and migration, particularly MMP2, MMP9 and MT1-MMP, which play a pivotal role in the degradation of ECM to facilitate HCC metastasis (Ogasawara et al. 2005). Furthermore, connective tissue growth factor (CTGF) was overexpressed in HCC patients whereas downregulating the expression of CTGF could inhibit HCC growth which could be a potential therapeutic strategy for HCC treatment (Jia et al. 2011). As we all known, epithelial mesenchymal transition (EMT) is an very important step in hepatocarcinogenesis, which involves the interactions between HCC cells and ECM mediated by transforming growth factor-β1(TGF-β1) and/or PDGFR signaling pathway (Dorn et al. 2010).

The heterogeneity of ECM makes it a challenging topic to inhibit ECM proteins due to the various ECM proteins and complex mechanisms. However, it still needs to be considered for the target therapy in which the proteins required to maintain or degrade ECM-related proteins could be used.

16.3.3 Heterogeneity of the Immune Microenvironment

The immune microenvironment in HCC is also found to be heterogeneous. Cell types within or around tumors include cytotoxic T cells (CD8+), regulatory T cells (Treg), natural killer (NK), natural killer T cells (NKT), myeloid-derived suppressor cells (MDSCs) and so on (Fig. 16.2b). These cells can play an important role in promoting or inhibiting HCC progression (Junttila and de Sauvage 2013) (Fig. 16.3).

CD8+ T cells are found infiltrating among HCC tumor cells, whereas CD4+ T cells are found mainly around the tumor or liver interface (Kasper et al. 2009). Treg cells promote immune suppression by secreting IL-10 and TGF-β and direct contact with tumor cells (Wang et al. 2012). On the other hand, Tregs could inhibit CD8+ T cells responses and would enhance immune responses when the Treg number is low (Huang et al. 2012a). Cytotoxic T cells (CTLs) have the cytotoxicity to kill tumor cells which lead to less immune response against HCC (Gao et al. 2007). Therefore, low number of Tregs and increased number of activated CTLs are associated with a favorable prognosis. A higher frequency of Th17 cells which secret IL-22 are found in advanced HCC patients with poor survival (Zhang et al. 2009; Liao et al. 2013). And a higher expression of IL-22 can activate Stat-3 signaling and promote tumor growth (Jiang et al. 2011).

Some studies have reported that the frequency and cytotoxic function of NK cells to be reduced both in the liver and peripheral blood of HCC patients (Cai et al. 2008; Gao et al. 2009; Hoechst et al. 2009). The reduced NK cell function was associated with lower expression of NK cell receptor, NKG2D (Sha et al. 2014). Invariant natural killer T (iNKT) cells was also shown to be increased in patients

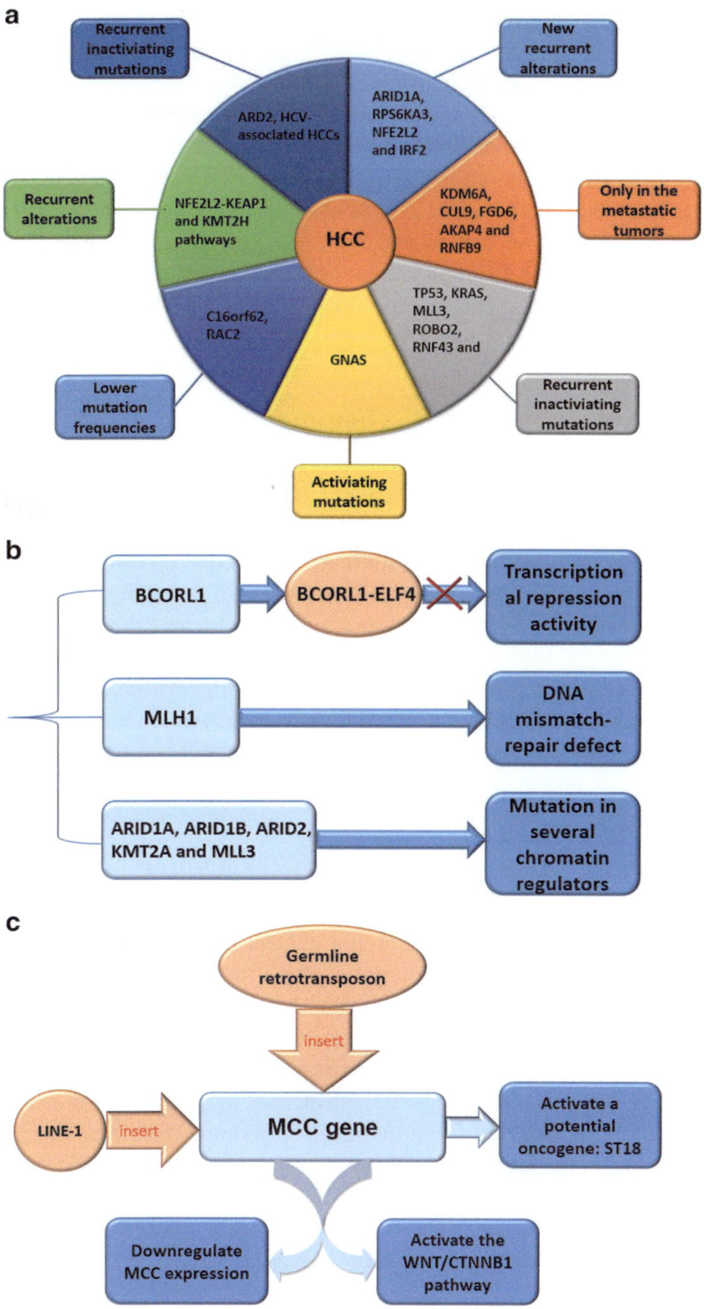

Fig. 16.3 Somatic alterations in the HCC genome. (**a**) Some representative somatic mutations in the whole exon domain (exome), which is determined by massively parallel sequencing (**b**) Some representative somatic mutations in the whole-genome domain, which is performed by whole-genome sequencing (**c**) representative somatic change of retrotransposons in HCC

produced Interferon-gamma (IFN) to inhibit tumor growth (Crowe et al. 2005). In addition, CD4+ NKT cells produced Th2-cytokine which could also inhibit CD8 T cell expansion and function (Bricard et al. 2009).

Myeloid-derived suppressor cells (MDSCs) are heterogeneous in HCC patients that includes macrophages, dendritic cells, immature granulocytes and early myeloid progenitors. MDSCs could inhibit T cell responses as well as natural killer cell function via the NKp30 receptor (Hoechst et al. 2009). Overall, the development of immunotherapy requires an understanding of the heterogeneous microenvironment, regulation of cytokines at different stages of HCC, and the functional activity of T cells, CTLs, NK cells and MDSCs etc.

16.4 Heterogeneity of HCC Genomes

Progress in sequencing technologies have made it possible to examine liver cancer genomes at high resolution. Somatic mutations, structural alterations, HBV integration, RNA editing, retrotransposon changes and so on have been comprehensively identified. In addition, integrated analyses of genome, transcriptome and methylome data have identified various critical genes and pathways involved in hepatocarcinogenesis, and paved the way for new molecular classifications for clinical application. Furthermore, the international collaborations of cancer genome sequencing projects are expected to contribute to an improved understanding of risk evaluation, diagnosis and therapy strategy for this cancer.

16.4.1 Somatic Alterations in the HCC Genome

Whole-genome and whole-exome sequencing have provided a comprehensive and high-resolution view of somatic genomic alterations in HCC. The liver cancer genome contains multiple types of somatic alterations, including mutations such as single nucleotide substitutions, small insertions and deletions, changes of gene copy numbers, intra-chromosomal rearrangements and inter-chromosomal rearrangements. For the past few years, an increasing appreciation and identification of somatic mutations that drive human tumors have enable us within reach of personalized cancer medicine.

16.4.1.1 Genome-Wide Copy Number Analysis

Somatic DNA copy number changes in human cancers genomes have been detected mainly by array-based comparative genome hybridization methods (CGH). That's because array-based CGH can enable high-throughput and high-resolution screening of genome-wide DNA copy number changes (Pollack et al. 1999). In addition to

Table 16.1 Amplified and deleted genes in HCC (Wang et al. 2012)

Gene name	Locus	Function
Recurrently amplified genes in HCC		
MDM4	1q32.1	P53 pathway
BCL9	1q21.1	WNT pathway
ARNT	1q21.2	Xenobiotics metabolism
ABL2	1q25.2	Proliferation
MET	7q31.2	Proliferation
COPS5	8q13.1	Proteolysis
MTDH	8q22.1	Metastasis
COX6C	8q22.2	Mitochondria
MYC	8q24.21	Proliferation
CCND1	11q13.2	Proliferation
FGF19	11q13.2	WNT pathway
RPS6KB1	11q23.1	Proliferation
EEF1A2	20q13.33	Translation
Recurrently amplified genes in HCC		
TNFRSF14	1p36.33	Immune response
CDKN2C	1p36.11	Cell cycle
ARID1A	1p36.11	Chromatin remodelling
TNFAIP3	6q26	NF-κB pathway
CSMD1	8p23.2	Immune response
DLC1	8P22	Small GTpase
SORBS3	8p21.3	Migration
WRN	8p21.3	DNA repair
SH2D4A	8p21.2	Proliferation
PROSC	8p11.2	Unknown
CDKN2A	9p21.3	Cell cycle
CDKN2B	9p21.3	Cell cycle
PTEN	10q23.31	Proliferation
SPRY2	13q31.1	Proliferation
BRCA2	13q13.1	DNA repair
RB1	13q14.3	Cell cycle
XPO4	13q11	Nuclear export
SMAD4	18q21.3	TGF-β signalling

well-known oncogenes e.g. MYC and CCND1, and tumour suppressor genes, such as TP53 and RB, liver cancers harbour multiple chromosomal amplifications and deletions, and Shibata et al. have summarized these recurrent copy number alterations on the Table 16.1 (Shibata and Aburatani 2014).

In recent years, several studies reported chromosomal alterations in HCC using array CGH (Chochi et al. 2009; Kakar et al. 2009; Patil et al. 2005; Schlaeger et al. 2008). Guo X et al. discovered significant gains in 5p15.33 and 9q34.2–34.3 and losses in 6q, 9p and 14q in addition to the regions that were previously identified by conventional CGH analyses by a meta-analysis of 159 HCC array

CGHs (Chochi et al. 2009; Kakar et al. 2009; Patil et al. 2005; Schlaeger et al. 2008). In a study by Patil et al. (2005), the correlation between DNA copy numbers and gene expression pattern at the 8q region was demonstrated, which was frequently amplified in 49 HCC samples. A study of Roessler et al. (2012) identified ten driver genes that were associated with poor survival by integrating high-resolution array CGH data and gene expression profiles of 256 HCC cases to gain the genes which have the significant correlation between somatic copy number alterations and the whole genome expression patterns. In order to identify potential cancer driver genes, Woo et al. (2009) integrated whole genome copy number profiles of 15 HCC cases with gene expression profiles of 139 HCC cases. They analyzed genes that have a correlation between expression levels and copy number changes, finally discovered 50 potential driver genes that are linked to HCC prognosis.

16.4.1.2 Whole-Exome Sequencing

Advance in sequencing technologies have enabled researchers to explore the liver cancer genome more deeply. Whole exome sequencing (WES) can efficiently identify mutations in protein-coding exons, which are much more easily identifiable than the mutations or variants in non-coding regions. This approach concerns target-enrichment of whole protein-coding exons across the human genome (30–40 Mb, approximately 1 % of the whole human genome) adopting in-solution RNA or oligonucleotide DNA probe hybridization technologies (Gnirke et al. 2009; Hodges et al. 2007) which enable the comprehensive detection of somatic alterations in the protein-coding exons, and have discovered many novel genes involved in liver cancer. In the research of Li M et al., the recurrent inactivating mutations of the ARID2 gene in 18.2 % of HCV-associated HCCs were identified by exomic sequencing of 10 HCV-positive HCCs and analysis of an additional tumour cohort of various aetiological backgrounds (Li et al. 2011). Huang et al. (2012b) sequenced nine pairs of HCCs and their intrahepatic metastases across whole exome to come out with the result that although about 94.2 % substitutions were common in both primary and metastatic tumours, a fraction of mutations were detected in 1.1 % primary or 4.7 % metastatic tumours. Among these mutations, KDM6A, CUL9, RNF139, AKAP4 and FGD6 were only identified in the metastatic tumors of three individuals. Using whole-exome sequencing of 87 HCC cases, Cleary et al. (2013) found recurrent alterations in the NFE2L2–KEAP1 and MLL pathways, while C16orf62 and RAC2 with lower mutation frequencies. According to copy number analysis of 125 HCC cases and whole exome sequencing of 24 of these cases, Guichard et al. (2012) detected novel recurrent mutations in the ARID1A, RPS6KA3, NFE2L2 and IRF2 genes. Interestingly, inactivation of the IRF2 gene was exclusively observed in HBV-related HCC, which led to disruption of TP53 function. In addition, alterations in chromatin remodelers were found in association with alcohol-related HCC.

16.4.1.3 Whole-Genome Sequencing

Many research groups have sequenced the whole liver cancer genome in further attempts to detect all somatic driver events involved in hepatocarcinogenesis. Whole genome sequencing (WGS) can cover almost all the genome sequences in human and detect variants in non-coding regions, copy number alterations, genomic rearrangements, and virus genome integrations except single nucleotide changes (Nakagawa and Shibata 2013). By sequencing HCV-related HCC cases, >16,000 somatic mutations and 26 intra-chromosomal and interchromosomal rearrangements inducing four fusion transcripts were identified, including the TP53, AXIN1, ADAM22, JAK2, KHDRBS2, NEK8, TRRAP and BCORL1 genes, as well as a large number of somatic mutations in genes encoding phosphoproteins and those with bipartite nuclear signals. Through high-resolution analysis, the authors also identified intratumor heterogeneity of the mutations, including inactivation of the TSC complex in a subpopulation of HCV-related HCCs (Totoki et al. 2011). By performed whole-genome sequencing of 27 HCCs and matched normal genomes, Fujimoto et al. showed that 25 of which were associated with HBV or HCV infection. The average number of somatic point mutations at the whole-genome level was 4.2 per Mb. Moreover, several chromatin regulators mutations, including ARID1A, ARID1B, ARID2, MLL, MLL3, BAZ2B, BRD8, BPTF, BRE and HIST1H4B, were identified in 50 % tumors. These mutations were marginally linked to the stage of liver fibrosis and hepatic invasion (Fujimoto et al. 2012). By a whole-genome sequencing study of 88 matched HCC tumor/normal pairs, 81 of which are Hepatitis B virus (HBV) positive, Kan et al. (2013) seeked to identify genetically altered genes and pathways implicated in HBV-associated HCC cases. They found the most frequently mutated oncogene (15.9 %) and the most frequently mutated tumor suppressor (35.2 %) are beta-catenin and TP53, respectively. The Wnt/beta-catenin and JAK/STAT pathways, mutated in 62.5 % and 45.5 % of cases, respectively, are possible to be two major oncogenic drivers in HCC. This research also identified several prevalent and potentially actionable mutations, such as activating mutations of Janus kinase 1 (JAK1) in 9.1 % of patients, suggesting that these genes or pathways could be new therapeutic targets in HCC (Kan et al. 2013).

16.4.2 Somatic Change of Retrotransposons in HCC

The human genome contains a variety of repetitive genome sequences, including tandem repeats and retrotransposons e.g. short interspersed nuclear elements (SINEs) and long interspersed nuclear elements (LINEs). In the human genome, Alu and LINE-1 are major forms of SINEs and LINEs, respectively (Treangen and Salzberg 2012). LINE-1 retrotransposons are a major source of endogenous mutagenesis in humans (Burns and Boeke 2012; Levin and Moran 2011).

Retrotransposon insertions can deeply alter gene structure and expression (Levin and Moran 2011; Cordaux and Batzer 2009; Han et al. 2004; Faulkner et al. 2009) and have been identified in nearly 100 cases of diseases (Faulkner 2011; Hancks and Kazazian 2012). LINE-1 activity is consequently suppressed in most somatic cells by methylation of a CpG island in the internal LINE-1 promoter (Coufal et al. 2009; Swergold 1990). By contrast, LINE-1 is often hypomethylated in tumor cells, removing a key obstacle to retrotransposition (Levin and Moran 2011).

Shukla et al. (2013) used enhanced retrotransposon capture sequencing (RC-seq) to detect 19 HCC tumors and matched adjacent liver tissue that were confirmed positive for HBV or HCV infection and elucidated endogenous LINE-1-mediated retrotransposition in the germline and somatic cells of HCC patients. The authors reported two archetypal mechanisms revealing MCC and ST18 as HCC candidate genes. MCC is a highly plausible liver tumor suppressor. However distinct germline LINE-1 or Alu insertions contribute to MCC suppression in tumor and nontumor liver tissue and then activate the WNT/CTNNB1 pathway. The other event is a tumor-specific LINE-1 insertion which activates a potential oncogene, Suppression of tumorigenicity 18 (ST18), in liver tumors (Shukla et al. 2013).

16.4.3 HBV Genome Integrations in the Host Genome

Chronic HBV infection is a major risk factor for HCC, and more than half of HCC cases in the world are attributed to HBV infection. HBV is a DNA virus whose genome can be integrated into the host genome. The integration of the viral DNA sequences affect host gene expression near the integration site and its effect on the integrity of the host genome is associated with virus-mediated hapatocarcinogenesis (Neuveut et al. 2010). By Southern blot analysis or inverse PCR, previous studies identified the integration of HBV DNA sequences into host genomes both in HCC samples and non-tumorous tissues from patients with chronic HBV hepatitis (Brechot et al. 1980). Advanced current genome sequencing technology have enabled researchers to detect virus integration events more comprehensively and at higher resolution than previously.

HBV integration was reportedly observed within or upstream of the TERT gene in tumor tissues in HCC cases with HBV infection (Fujimoto et al. 2012). Furthermore, Sung et al. (2012) reported integration events at the MLL4, CCNE1, SENP5, FN1 and ROCK1 genes. They conducted whole-genome sequencing of 81 - HBV-positive and seven HBV-negative HCC samples and found that most HBV breakpoints in HCC were close to active coding genes, which potentially enabled HBV to integrate into the open chromatin region more effectively (Sung et al. 2012). Jiang et al. (2012a) also made comprehensive analyses of HBV-related HCC and their corresponding normal tissues. They found clonal and high-abundance viral integrations in tumor tissue, while many viral integration sites randomly scattered throughout the genome in nontumor liver tissues (Jiang et al. 2012b). These research indicated that a heterogeneous and widespread viral integration landscape in HCC as well as nontumor liver tissue and integration events may cause aberrant expression of genes near the integration sites, alterations of DNA copy number and emergence of fusion genes (Sung et al. 2012; Jiang et al. 2012b). Moreover, recurrent integration of HBV was also detected in the FAR2, ITPR1, MAPK1, IRAK2 and MLL genes (Sung et al. 2012; Gozuacik et al. 2001; Paterlini-Brechot et al. 2003; Murakami et al. 2005; Saigo et al. 2008).

16.4.4 DNA Methylation in HCC

(update) DNA methylation and demethylation is an important mechanism of regulating gene expression and chromatin structure in normal cells. DNA methylase contribute to the methylation of cytosine at CpG islands at the gene promoter region. Aberrant DNA methylation at the gene promoter region is an important mechanism in inactivation of tumor suppressor gene (Nagae et al. 2011; Hendrich and Bird 1998).

Altered DNA methylation is an early event in HCC development. Global hypomethylation has a critical role in increasing chromosomal instability and mainly affected intergenic regions of the genome (Eden et al. 2003). DNA hypermethylation is the state where the methylation of "normally" undermethylated DNA domains, which predominantly consist of CpG islands (Rollins et al. 2006), increases. Abnormal gains of DNA methylation (hypermethylation) of typically unmethylated CpG island-containing promoters can lead to transcriptional repression and loss of gene function. In addition, non-CpG island-containing promoter coding region hypermethylation contribute to genes inactivation (Pogribny and James 2002; Tomasi et al. 2012).

The study of Udali et al. (2015) used array-based DNA methylation and gene expression data of all annotated genes from eight HCC patients undergoing curative surgery to analyze by comparing HCC tissue and homologous cancer-free liver tissue. They identified 159 hypermethylated-repressed, 56 hypomethylated-repressed, 49 hypermethylated-induced, and 30 hypomethylated-induced genes. Notably, promoter DNA methylation proved to be a novel regulatory mechanism for the transcriptional repression of genes e.g. involving the retinol metabolism (ADH1A, ADH1B, ADH6, CYP3A43, CYP4A22, RDH16), one-carbon metabolism (SHMT1), iron homeostasis (HAMP), and potential tumor suppressors (FAM107A, IGFALS, MT1G, MT1H, RNF180).

Nishida et al. (2014) applied Infinium Human Methylation 450 Bead Chip array which can analyze >485,000 CpG sites distributed throughout the genome to analyze comprehensive methylation from 117 liver tissues consisting of 59 HCC and 58 noncancerous livers. They identified 38,330 CpG sites with significant differences in methylation levels between HCCs and nontumors livers (DMCpGs). Among the DMCpGs, 92 % were hypomethylated and only 3051 CpGs (8 %) were hypermethylated in HCC. The DMCpGs were more common within intergenic regions with isolated CpGs. However, DMCpGs that were hypermethylated in HCC were predominantly located within promoter regions and CpG islands.

Shen et al. (2012) analyzed tumor and adjacent nontumor tissues from 62 Taiwanese HCC cases using Illumina methylation arrays which can screen 26,486 autosomal CpG sites. They found that a total of 2324 CpG sites significantly differed in methylation level. Among these CpG sites, 684 CpG sites significantly hypermethylated and 1640 hypomethylated in tumor compared to nontumor tissues. The 684 hypermethylation markers could be utilized for plasma DNA diagnostics. In addition, They identified the top 500 significant CpG sites using a 450 K array from 66 HCC cases. These differential methylations were able to distinguish HCC from adjacent nontumor tissues (Shen et al. 2013).

Previous study (Nishida et al. 2007) reported that extensive methylation is involved in CTNNB1 mutations, while TP53 mutation in HCC is often characterized by chromosomal instability. CpG islands promoter of the tumour suppressor genes CDKN2A and CDKN2B are frequently hypermethylated, leading to inactivation of the RB pathway (Zang et al. 2011). Methylation of the CDKN2A gene promoter occurs in 73 % of HCC tissues (Wong et al. 1999), 56 % of HBV-related HCC, and 84 % of HCV-related HCC (Narimatsu et al. 2004). Moreover, RASSF1A is methylated in up to 85 % of HCCs (Zhang et al. 2002), GSTP1 in 50–90 % (Yang et al. 2003; Zhong et al. 2002), and MGMT in 40 % (Zhang et al. 2003).

16.5 Heterogeneity of Signaling Pathways Affects the Progression of HCC

16.5.1 p53 Gene Pathway

As a tumor suppressor, p53 can initiate cell-cycle arrest, apoptosis, and senescence in response to cellular stress to maintain the integrity of the genome. About 50 % human tumors carry mutant p53, and many p53 mutants facilitate oncogenic functions such as increased proliferation, viability, and invasion or dominant-negative regulate the remaining wild-type p53 (Muller and Vousden 2013).

p53 plays important and unique roles in HCC. A study indicated that ablation of the p53-mediated senescence program in hepatic stellate cells under chronic liver damage promotes liver fibrosis and cirrhosis, which are associated with reduced survival; in addition, loss of p53 enhances the transformation of adjacent epithelial cells into HCC (Lujambio et al. 2013). p53 is mainly regulated by the E3 ubiquitin ligase MDM2. MDM2 binds p53 blocking p53-mediated transcriptional regulation, while simultaneously promoting its degradation (Brown et al. 2011). In addition, the MDM2–p53 binding can be disrupted by a small inhibitor Nutlin-3, which thereby activates p53 dependent apoptosis in different HCC cell lines (Zheng et al. 2010a). Therefore, Inhibition of MDM2–p53 binding could reactivate p53 in cancer cells with wild-type p53 and may offer an effective therapeutic approach for millions of cancer patients (Brown et al. 2011).

16.5.2 Hedgehog Pathway

Hedgehog signaling contributes to many aspects of cell differentiation, organ formation, cancergenesis and cancer metastasis. It is widely accepted that Hedgehog activity plays an important role in the progression of HCC. Many studies report that aberrant activation of Hedgehog signaling promote proliferation, viability, migration and invasion of HCC cells with complex underlying mechanisms (Zheng et al. 2010b, 2012; Lu et al. 2012).

Gli, Smo and PTCH were found to be overexpressed in HCC patients (Che et al. 2012; Jeng et al. 2013; Zhang et al. 2013). Lu et al. (2012) reported that Shh treatment can stimulate Hedgehog signaling to promote HCC cell invasion and migration by increasing GLI1 expression. Sicklick et al. (2006) found overexpression of Smo and an increase in the stoichiometric ratio of Smo to PTCH mRNA levels in HCC, this effect is associated with tumor size and Smo and PTCH may be prognostic marker of HCC. Downstream transcription factors, Gli, affect the proliferation, migration, invasion, angiogenesis, aberrant autophagy and stem cell regeneration in HCC (Zheng et al. 2013). Previous studies have found that GLI1 contributes to the EMT phenotype and intrahepatic metastasis and portal venous invasion of human HCCs (Zheng et al. 2010b). Other studies reported that GLI1 expression in HCC tissues is associated with disease-free survival, overall survival and rapid recurrence (Zheng et al. 2012). *In vitro* experiments indicated that GLI1 promotes proliferation, viability, colony formation, migration and invasion of Huh7 cells. In addition, inhibition of Hedgeho signaling by GANT61, which is a small-molecule inhibitor of GLI1, led to autophagy. The result demonstrate that Hedgehog signaling is involved in aberrant autophagy of HCC cells (Wang et al. 2013). Furthermore, Several Gli target genes have been identified such as cMyc, Cyclin D1 and FOXM1 (cell proliferation) and Bcl-2 (survival) (Lin et al. 2010). For example, the down-regulation Hedgehog signaling pathways could induce cell arrest at G1 and cause apoptosis by downregulation of Bcl-2 (Chen et al. 2008; Cheng et al. 2009; Kim et al. 2007; Zhang et al. 2011).

16.5.3 Wnt/β-Catenin Signaling

The Wnt/β-catenin signaling pathway is mainly composed of the Wnt protein, Wnt protein ligand frizzled protein, and related regulator proteins such as GSK-3β and β-catenin. Previous study indicated that aberrant activation of WNT signalling is a driving molecular event in many types of tumors, including liver cancers (Polakis 2012). The aberrant Wnt/β-catenin signaling pathway plays an important role in liver physiology and pathology. Various molecular and genetic factors such as CTNNB1, AXIN1 and AXIN2 participate to the aberrant activation of the Wnt/-catenin pathway. Gain-of-function mutations of CTNNB1 which encode for β-catenin are occurred in about 90 % HCCs (Bruix and Sherman 2011). In contrast, loss-of-function mutations of negative regulators such as *AXIN1, AXIN2* and APC genes are also observed in such aberrant pathway (Laurent-Puig and Zucman-Rossi 2006). When upstream stimulation activate the pathway, the Wnt protein binds to its ligand and β-catenin accumulates in cells, where β-catenin is activated and transferred into nucleus. In the nucleus, β-catenin dimerizes with the downstream specific transcription factor LEF/TCF, which regulates the transcription of key genes such as cyclin D (Thompson and Monga 2007; Langeswaran et al. 2013).

16.5.4 PI3K/AKT/mTOR Signaling Pathway

The PI3K/AKT/mTOR signaling pathway is a central regulator of various oncogenic processes including cell growth, proliferation, metabolism, survival regulation, antiapoptosis and angiogenesis. It also plays significant function in HCC and is activated in 30–50 % of HCC. There is growing evidence to suggest that activation of the PI3K/AKT/mTOR pathway is associated with less differentiated tumors, earlier tumor recurrence, and worse survival outcomes (Zhou et al. 2010). In normal tissue, this pathway is negatively regulated by the tumor suppressor phosphatase and tensin homolog (PTEN) on chromosome 10. Abnormal PTEN function and expression may lead to excessively activation of the PI3K/AKT/mTOR pathway in HCC (Zhou et al. 2011). Previous study has found that the loss of PTEN and overexpression of pAkt and p-mTOR were linked to the tumor differentiation, TNM stage, intrahepatic metastasis, vascular invasion Ki-67 labeling index, and MMP-2 and MMP-9 upregulation of human HCCs (Chen et al. 2009; Grabinski et al. 2012). Furthermore, Mcl-1, an anti-apoptotic molecule transcribed via a PI3K/Akt dependent pathway, was associated with HCC poor survival (Personeni et al. 2013).

16.5.5 Ras/Raf/MAPK Signaling Pathway

The MAPK intracellular signaling network is often activated in cancer cells. Recent researches show that HCC cells activation and proliferation is known to involve various different signaling pathways as previously mentioned (Laurent-Puig and Zucman-Rossi 2006). Among them, the Ras/Raf/MAPK signaling pathways is one of the most critical pathways in pathogenesis, development and proliferation of HCC and have been extensively investigated (Llovet and Bruix 2008).

The intracellular part of Ras/Raf/MAPK pathway is downstream of several receptor tyrosine kinases such as the EGFR, PDGFR and VEGFR which transmit growth factor signals from the cell membrane to the nucleus regulating multiple cellular functions including cell growth and survival, and differentiation. However, multiple upstream receptors including other receptor tyrosine kinases, integrins, serpentine receptors, heterotrimeric G-proteins, and cytokine receptors are able to activate Ras (Cantrell 2003).

Mechanisms for the increased activity of the Ras/Raf/MAPK signaling pathway in HCC include aberrant upstream signals, inactivation of Raf kinase inhibitor protein and induction by hepatitis viral proteins (Galuppo et al. 2014).

Several components of this pathway are mutated in HCC. Bos (1989) found that about 30 % of HCC bear Ras mutations. The Raf family consists of three isoforms, A-Raf, B-Raf and C-Raf. Overexpression of wild-type *C-Raf-1* proto-oncogene has been reported in liver cirrhosis and HCC (Jenke et al. 1994; Huang and Sinicrope 2010). Sorafenib has activity inhibiting B-Raf (Tannapfel et al. 2003). Huynh et al. (2003) found overexpression of MEK1/2 and ERK1/2, and phosphorylation of ERK1/2 in 100 % (46/46), 91 % (42/46) and 69 % (32/46) HCC, respectively.

16.5.6 Notch Signaling

Notch signalling is an evolutionarily conserved pathway that involves in a variety of fundamental cellular processes such as cell fate and differentiation (Artavanis-Tsakonas et al. 1999; Lai 2004). The effects of Notch signaling seem heterogeneous in HCC progression (Strazzabosco and Fabris 2012). Activation of Notch signaling could lead to reduced cell proliferation and tumor growth in HCC (Viatour et al. 2011). And in addition, it also participates in invasion and migration of HCC cells (Zhou et al. 2013). Several researches indicated that NOTCH is activated in mice and human HCC samples (Tschaharganeh et al. 2013; Villanueva et al. 2012). However, other reports found the activation of NOTCH signalling as a suppressor feedback mechanism during HCC progression (Viatour et al. 2011; Qi et al. 2003). These contradictions suggest that biological activities of NOTCH signaling during hepatocarcinogenesis mainly depend on the cellular environment, which is also reported in other tumor types (Radtke and Raj 2003).

16.5.7 KEAP1-NFE2L2 Pathway

A sequence-specific transcriptional factor, encoded by the NFE2L2 gene, upregulates genes associated with oxidative stress and other metabolic pathways (Taguchi et al. 2011). And the level of the NFE2L2 protein is regulated by the ubiquitin proteasome pathway, and KEAP1 functions as an E3 ubiquitin ligase. A study found that NFE2L2 coding for NRF2 a transcription factor crucial for cellular redox homeostasis, was mutated in 6.4 % of HCC (Shibata et al. 2008). The mutation disrupts direct NFE2L2–KEAP1 interaction, or inactivating mutations of the KEAP1 gene are recurrently reported in HCC (Guichard et al. 2012).

16.6 Conclusion

Heterogeneities in aetiological factors, tumor microenvironment, genetic variations and signaling pathways contribute to HCC progression, which makes it difficult to properly stage the disease and thereby estimate the prognosis. Besides the established main role of hepatitis virus infections and of alcohol use in the risk of HCC, multiple genetic factors also play an significant role. Advances in sequencing technologies have guided the examination of HCC genomes into a new view. In addition to copy number changes and mutations, analyses have identified additional genome alterations, including DNA methylation, HBV integration, retrotransposon changes an so on. The integration of data from different levels of global analyses have identified various critical genes and pathways involved in hepatocarcinogenesis. The heterogeneity of HCC makes it difficult to clarify the mechanism of cancer development and to develop effective therapeutics. For future clinical research design, it is essential to take into account how to eliminate the confounding effects from interpatients and intratumor heterogeneity of genome, aetiological factors and tumor microenvironment. Precision medicine based on global genetic analysis will become more and more important to overcome the heterogeneity of HCC. While some genetic profiles or signaling pathways may prove to be potential targets for clinical application. Therefore, targeting these heterogeneity in HCC patients will definitely create a new field for developing personal treatment options.

References

Andant C, Puy H, Bogard C, Faivre J, Soule JC, Nordmann Y, Deybach JC. Hepatocellular carcinoma in patients with acute hepatic porphyria: frequency of occurrence and related factors. J Hepatol. 2000;32:933–9.

Artavanis-Tsakonas S, Rand MD, Lake RJ. Notch signaling: cell fate control and signal integration in development. Science. 1999;284:770–6.

Baan R, Straif K, Grosse Y, Secretan B, El GF, Bouvard V, Altieri A, Cogliano V. Carcinogenicity of alcoholic beverages. Lancet Oncol. 2007;8:292–3.

Baeriswyl V, Christofori G. The angiogenic switch in carcinogenesis. Semin Cancer Biol. 2009;19:329–37.

Bos JL. Ras oncogenes in human cancer: a review. Cancer Res. 1989;49:4682–9.

Bosch FX, Ribes J, Diaz M, Cleries R. Primary liver cancer: worldwide incidence and trends. Gastroenterology. 2004;127:S5–16.

Brechot C, Pourcel C, Louise A, Rain B, Tiollais P. Presence of integrated hepatitis b virus DNA sequences in cellular DNA of human hepatocellular carcinoma. Nature. 1980;286:533–5.

Bricard G, Cesson V, Devevre E, Bouzourene H, Barbey C, Rufer N, Im JS, Alves PM, Martinet O, Halkic N, Cerottini JC, Romero P, Porcelli SA, Macdonald HR, Speiser DE. Enrichment of human CD4+ V(alpha)24/Vbeta11 invariant NKT cells in intrahepatic malignant tumors. J Immunol. 2009;182:5140–51.

Brown CJ, Cheok CF, Verma CS, Lane DP. Reactivation of p53: from peptides to small molecules. Trends Pharmacol Sci. 2011;32:53–62.

Bruix J, Sherman M. Management of hepatocellular carcinoma: an update. Hepatology. 2011;53:1020–2.

Burns KH, Boeke JD. Human transposon tectonics. Cell. 2012;149:740–52.

Cai L, Zhang Z, Zhou L, Wang H, Fu J, Zhang S, Shi M, Zhang H, Yang Y, Wu H, Tien P, Wang FS. Functional impairment in circulating and intrahepatic NK cells and relative mechanism in hepatocellular carcinoma patients. Clin Immunol. 2008;129:428–37.

Cantrell DA. GTPases and T cell activation. Immunol Rev. 2003;192:122–30.

Carloni V, Luong TV, Rombouts K. Hepatic stellate cells and extracellular matrix in hepatocellular carcinoma: more complicated than ever. Liver Int. 2014;34:834–43.

Che L, Yuan YH, Jia J, Ren J. Activation of sonic hedgehog signaling pathway is an independent potential prognosis predictor in human hepatocellular carcinoma patients. Chin J Cancer Res. 2012;24:323–31.

Chen XL, Cao LQ, She MR, Wang Q, Huang XH, Fu XH. Gli-1 siRNA induced apoptosis in Huh7 cells. World J Gastroenterol. 2008;14:582–9.

Chen JS, Wang Q, Fu XH, Huang XH, Chen XL, Cao LQ, Chen LZ, Tan HX, Li W, Bi J, Zhang LJ. Involvement of PI3K/PTEN/AKT/mTOR pathway in invasion and metastasis in hepatocellular carcinoma: association with MMP-9. Hepatol Res. 2009;39:177–86.

Cheng WT, Xu K, Tian DY, Zhang ZG, Liu LJ, Chen Y. Role of hedgehog signaling pathway in proliferation and invasiveness of hepatocellular carcinoma cells. Int J Oncol. 2009;34:829–36.

Chew V, Tow C, Huang C, Bard-Chapeau E, Copeland NG, Jenkins NA, Weber A, Lim KH, Toh HC, Heikenwalder M, Ng IO, Nardin A, Abastado JP. Toll-like receptor 3 expressing tumor parenchyma and infiltrating natural killer cells in hepatocellular carcinoma patients. J Natl Cancer Inst. 2012;104:1796–807.

Chochi Y, Kawauchi S, Nakao M, Furuya T, Hashimoto K, Oga A, Oka M, Sasaki K. A copy number gain of the 6p arm is linked with advanced hepatocellular carcinoma: an array-based comparative genomic hybridization study. J Pathol. 2009;217:677–84.

Cleary SP, Jeck WR, Zhao X, Chen K, Selitsky SR, Savich GL, Tan TX, Wu MC, Getz G, Lawrence MS, Parker JS, Li J, Powers S, Kim H, Fischer S, Guindi M, Ghanekar A, Chiang DY. Identification of driver genes in hepatocellular carcinoma by exome sequencing. Hepatology. 2013;58:1693–702.

Cordaux R, Batzer MA. The impact of retrotransposons on human genome evolution. Nat Rev Genet. 2009;10:691–703.

Coufal NG, Garcia-Perez JL, Peng GE, Yeo GW, Mu Y, Lovci MT, Morell M, O'Shea KS, Moran JV, Gage FH. L1 retrotransposition in human neural progenitor cells. Nature. 2009;460:1127–31.

Crowe NY, Coquet JM, Berzins SP, Kyparissoudis K, Keating R, Pellicci DG, Hayakawa Y, Godfrey DI, Smyth MJ. Differential antitumor immunity mediated by NKT cell subsets in vivo. J Exp Med. 2005;202:1279–88.

Dorn C, Riener MO, Kirovski G, Saugspier M, Steib K, Weiss TS, Gabele E, Kristiansen G, Hartmann A, Hellerbrand C. Expression of fatty acid synthase in nonalcoholic fatty liver disease. Int J Clin Exp Pathol. 2010;3:505–14.

Dragani TA, Manenti G, Della PG. Enhancing effects of carbon tetrachloride in mouse hepatocarcinogenesis. Cancer Lett. 1986;31:171–9.

Dragani TA, Canzian F, Manenti G, Pierotti MA. Hepatocarcinogenesis: a polygenic model of inherited predisposition to cancer. Tumori. 1996;82:1–5.

Eden A, Gaudet F, Waghmare A, Jaenisch R. Chromosomal instability and tumors promoted by DNA hypomethylation. Science. 2003;300:455.

Elmberg M, Hultcrantz R, Ekbom A, Brandt L, Olsson S, Olsson R, Lindgren S, Loof L, Stal P, Wallerstedt S, Almer S, Sandberg-Gertzen H, Askling J. Cancer risk in patients with hereditary hemochromatosis and in their first-degree relatives. Gastroenterology. 2003;125:1733–41.

El-Serag HB. Epidemiology of viral hepatitis and hepatocellular carcinoma. Gastroenterology. 2012;142:1264–73.

El-Serag HB, Mason AC. Risk factors for the rising rates of primary liver cancer in the United States. Arch Intern Med. 2000;160:3227–30.

El-Serag HB, Hampel H, Javadi F. The association between diabetes and hepatocellular carcinoma: a systematic review of epidemiologic evidence. Clin Gastroenterol Hepatol. 2006;4:369–80.

Elzouki AN, Eriksson S. Risk of hepatobiliary disease in adults with severe alpha 1-antitrypsin deficiency (Pizz): is chronic viral hepatitis B or C an additional risk factor for cirrhosis and hepatocellular carcinoma? Eur J Gastroenterol Hepatol. 1996;8:989–94.

Evans AA, Chen G, Ross EA, Shen FM, Lin WY, London WT. Eight-year follow-up of the 90,000-person Haimen City cohort: I. Hepatocellular carcinoma mortality, risk factors, and gender differences. Cancer Epidemiol Biomarkers Prev. 2002;11:369–76.

Fang J, Yan L, Shing Y, Moses MA. Hif-1alpha-mediated up-regulation of vascular endothelial growth factor, independent of basic fibroblast growth factor, is important in the switch to the angiogenic phenotype during early tumorigenesis. Cancer Res. 2001;61:5731–5.

Fang M, Yuan JP, Peng CW, Pang DW, Li Y. Quantum dots-based in situ molecular imaging of dynamic changes of collagen IV during cancer invasion. Biomaterials. 2013;34:8708–17.

Faulkner GJ. Retrotransposons: mobile and mutagenic from conception to death. Febs Lett. 2011;585:1589–94.

Faulkner GJ, Kimura Y, Daub CO, Wani S, Plessy C, Irvine KM, Schroder K, Cloonan N, Steptoe AL, Lassmann T, Waki K, Hornig N, Arakawa T, Takahashi H, Kawai J, Forrest AR, Suzuki H, Hayashizaki Y, Hume DA, Orlando V, Grimmond SM, Carninci P. The regulated retrotransposon transcriptome of mammalian cells. Nat Genet. 2009;41:563–71.

Feo F, De Miglio MR, Simile MM, Muroni MR, Calvisi DF, Frau M, Pascale RM. Hepatocellular carcinoma as a complex polygenic disease. Interpretive analysis of recent developments on genetic predisposition. Biochim Biophys Acta. 2006;1765:126–47.

Forner A, Llovet JM, Bruix J. Hepatocellular carcinoma. Lancet. 2012;379:1245–55.

Fracanzani AL, Taioli E, Sampietro M, Fatta E, Bertelli C, Fiorelli G, Fargion S. Liver cancer risk is increased in patients with porphyria cutanea tarda in comparison to matched control patients with chronic liver disease. J Hepatol. 2001;35:498–503.

Franceschi S, Montella M, Polesel J, La Vecchia C, Crispo A, Dal Maso L, Casarin P, Izzo F, Tommasi LG, Chemin I, Trepo C, Crovatto M, Talamini R. Hepatitis viruses, alcohol, and tobacco in the etiology of hepatocellular carcinoma in Italy. Cancer Epidemiol Biomarkers Prev. 2006;15:683–9.

Fujimoto A, Totoki Y, Abe T, Boroevich KA, Hosoda F, Nguyen HH, Aoki M, Hosono N, Kubo M, Miya F, Arai Y, Takahashi H, Shirakihara T, Nagasaki M, Shibuya T, Nakano K, Watanabe-Makino K, Tanaka H, Nakamura H, Kusuda J, Ojima H, Shimada K, Okusaka T, Ueno M, Shigekawa Y, Kawakami Y, Arihiro K, Ohdan H, Gotoh K, Ishikawa O, Ariizumi S, Yamamoto M, Yamada T, Chayama K, Kosuge T, Yamaue H, Kamatani N, Miyano S, Nakagama H, Nakamura Y, Tsunoda T, Shibata T, Nakagawa H. Whole-genome sequencing of liver cancers identifies etiological influences on mutation patterns and recurrent mutations in chromatin regulators. Nat Genet. 2012;44:760–4.

Galuppo R, Maynard E, Shah M, Daily MF, Chen C, Spear BT, Gedaly R. Synergistic inhibition of HCC and liver cancer stem cell proliferation by targeting RAS/RAF/MAPK and WNT/beta-catenin pathways. Anticancer Res. 2014;34:1709–13.

Gao Q, Qiu SJ, Fan J, Zhou J, Wang XY, Xiao YS, Xu Y, Li YW, Tang ZY. Intratumoral balance of regulatory and cytotoxic T cells is associated with prognosis of hepatocellular carcinoma after resection. J Clin Oncol. 2007;25:2586–93.

Gao B, Radaeva S, Park O. Liver natural killer and natural killer T cells: immunobiology and emerging roles in liver diseases. J Leukoc Biol. 2009;86:513–28.

Gnirke A, Melnikov A, Maguire J, Rogov P, LeProust EM, Brockman W, Fennell T, Giannoukos G, Fisher S, Russ C, Gabriel S, Jaffe DB, Lander ES, Nusbaum C. Solution hybrid selection with ultra-long oligonucleotides for massively parallel targeted sequencing. Nat Biotechnol. 2009;27:182–9.

Gozuacik D, Murakami Y, Saigo K, Chami M, Mugnier C, Lagorce D, Okanoue T, Urashima T, Brechot C, Paterlini-Brechot P. Identification of human cancer-related genes by naturally occurring hepatitis b virus DNA tagging. Oncogene. 2001;20:6233–40.

Grabinski N, Ewald F, Hofmann BT, Staufer K, Schumacher U, Nashan B, Jucker M. Combined targeting of AKT and mTOR synergistically inhibits proliferation of hepatocellular carcinoma cells. Mol Cancer. 2012;11:85.

Guichard C, Amaddeo G, Imbeaud S, Ladeiro Y, Pelletier L, Maad IB, Calderaro J, Bioulac-Sage P, Letexier M, Degos F, Clement B, Balabaud C, Chevet E, Laurent A, Couchy G, Letouze E, Calvo F, Zucman-Rossi J. Integrated analysis of somatic mutations and focal copy-number changes identifies key genes and pathways in hepatocellular carcinoma. Nat Genet. 2012;44:694–8.

Haddow JE, Palomaki GE, McClain M, Craig W. Hereditary haemochromatosis and hepatocellular carcinoma in males: a strategy for estimating the potential for primary prevention. J Med Screen. 2003;10:11–3.

Han JS, Szak ST, Boeke JD. Transcriptional disruption by the L1 retrotransposon and implications for mammalian transcriptomes. Nature. 2004;429:268–74.

Hancks DC, Kazazian HJ. Active human retrotransposons: variation and disease. Curr Opin Genet Dev. 2012;22:191–203.

Hashimoto E, Yatsuji S, Tobari M, Taniai M, Torii N, Tokushige K, Shiratori K. Hepatocellular carcinoma in patients with nonalcoholic steatohepatitis. J Gastroenterol. 2009;44 Suppl 19:89–95.

Hassan MM, Kaseb A, Li D, Patt YZ, Vauthey JN, Thomas MB, Curley SA, Spitz MR, Sherman SI, Abdalla EK, Davila M, Lozano RD, Hassan DM, Chan W, Brown TD, Abbruzzese JL. Association between hypothyroidism and hepatocellular carcinoma: a case-control study in the United States. Hepatology. 2009;49:1563–70.

Hemminki K, Li X. Familial liver and gall bladder cancer: a nationwide epidemiological study from Sweden. Gut. 2003;52:592–6.

Hendrich B, Bird A. Identification and characterization of a family of mammalian methyl-CpG binding proteins. Mol Cell Biol. 1998;18:6538–47.

Hodges E, Xuan Z, Balija V, Kramer M, Molla MN, Smith SW, Middle CM, Rodesch MJ, Albert TJ, Hannon GJ, McCombie WR. Genome-wide in situ exon capture for selective resequencing. Nat Genet. 2007;39:1522–7.

Hoechst B, Voigtlaender T, Ormandy L, Gamrekelashvili J, Zhao F, Wedemeyer H, Lehner F, Manns MP, Greten TF, Korangy F. Myeloid derived suppressor cells inhibit natural killer cells in patients with hepatocellular carcinoma via the NKp30 receptor. Hepatology. 2009;50:799–807.

Huang S, Sinicrope FA. Sorafenib inhibits STAT3 activation to enhance trail-mediated apoptosis in human pancreatic cancer cells. Mol Cancer Ther. 2010;9:742–50.

Huang Y, Wang FM, Wang T, Wang YJ, Zhu ZY, Gao YT, Du Z. Tumor-infiltrating FoxP3+ Tregs and CD8+ T cells affect the prognosis of hepatocellular carcinoma patients. Digestion. 2012a;86:329–37.

Huang J, Deng Q, Wang Q, Li KY, Dai JH, Li N, Zhu ZD, Zhou B, Liu XY, Liu RF, Fei QL, Chen H, Cai B, Zhou B, Xiao HS, Qin LX, Han ZG. Exome sequencing of hepatitis B virus-associated hepatocellular carcinoma. Nat Genet. 2012b;44:1117–21.

Huynh H, Nguyen TT, Chow KH, Tan PH, Soo KC, Tran E. Over-expression of the mitogen-activated protein kinase (MAPK) kinase (MEK)-MAPK in hepatocellular carcinoma: its role in tumor progression and apoptosis. BMC Gastroenterol. 2003;3:19.

IARC Working Group on the Evaluation of Carcinogenic Risks to Humans. Hepatitis viruses. IARC Monogr Eval Carcinog Risks Hum. 1994;59:1–255.

IARC Working Group on the Evaluation of Carcinogenic Risks to Humans. Some traditional herbal medicines, some mycotoxins, naphthalene and styrene. IARC Monogr Eval Carcinog Risks Hum. 2002;82:1–556.

IARC Working Group on the Evaluation of Carcinogenic Risks to Humans. Tobacco smoke and involuntary smoking. IARC Monogr Eval Carcinog Risks Hum. 2004;83:1–1438.

Janecke AR, Mayatepek E, Utermann G. Molecular genetics of type 1 glycogen storage disease. Mol Genet Metab. 2001;73:117–25.

Jeng KS, Sheen IS, Jeng WJ, Yu MC, Hsiau HI, Chang FY. High expression of sonic hedgehog signaling pathway genes indicates a risk of recurrence of breast carcinoma. Onco Targets Ther. 2013;7:79–86.

Jenke HS, Deml E, Oesterle D. C-raf expression in early rat liver tumorigenesis after promotion with polychlorinated biphenyls or phenobarbital. Xenobiotica. 1994;24:569–80.

Jia XQ, Cheng HQ, Li H, Zhu Y, Li YH, Feng ZQ, Zhang JP. Inhibition of connective tissue growth factor overexpression decreases growth of hepatocellular carcinoma cells in vitro and in vivo. Chin Med J (Engl). 2011;124:3794–9.

Jiang R, Tan Z, Deng L, Chen Y, Xia Y, Gao Y, Wang X, Sun B. Interleukin-22 promotes human hepatocellular carcinoma by activation of STAT3. Hepatology. 2011;54:900–9.

Jiang Z, Jhunjhunwala S, Liu J, Haverty PM, Kennemer MI, Guan Y, Lee W, Carnevali P, Stinson J, Johnson S, Diao J, Yeung S, Jubb A, Ye W, Wu TD, Kapadia SB, de Sauvage FJ, Gentleman RC, Stern HM, Seshagiri S, Pant KP, Modrusan Z, Ballinger DG, Zhang Z. The effects of hepatitis B virus integration into the genomes of hepatocellular carcinoma patients. Genome Res. 2012a;22:593–601.

Jiang S, Yang Z, Li W, Li X, Wang Y, Zhang J, Xu C, Chen PJ, Hou J, McCrae MA, Chen X, Zhuang H, Lu F. Re-evaluation of the carcinogenic significance of hepatitis B virus integration in hepatocarcinogenesis. PLoS One. 2012b;7:e40363.

Junttila MR, de Sauvage FJ. Influence of tumour micro-environment heterogeneity on therapeutic response. Nature. 2013;501:346–54.

Kakar S, Chen X, Ho C, Burgart LJ, Adeyi O, Jain D, Sahai V, Ferrell LD. Chromosomal abnormalities determined by comparative genomic hybridization are helpful in the diagnosis of atypical hepatocellular neoplasms. Histopathology. 2009;55:197–205.

Kan Z, Zheng H, Liu X, Li S, Barber TD, Gong Z, Gao H, Hao K, Willard MD, Xu J, Hauptschein R, Rejto PA, Fernandez J, Wang G, Zhang Q, Wang B, Chen R, Wang J, Lee NP, Zhou W, Lin Z, Peng Z, Yi K, Chen S, Li L, Fan X, Yang J, Ye R, Ju J, Wang K, Estrella H, Deng S, Wei P, Qiu M, Wulur IH, Liu J, Ehsani ME, Zhang C, Loboda A, Sung WK, Aggarwal A, Poon RT, Fan ST, Wang J, Hardwick J, Reinhard C, Dai H, Li Y, Luk JM, Mao M. Whole-genome sequencing identifies recurrent mutations in hepatocellular carcinoma. Genome Res. 2013;23:1422–33.

Kasper HU, Drebber U, Stippel DL, Dienes HP, Gillessen A. Liver tumor infiltrating lymphocytes: comparison of hepatocellular and cholangiolar carcinoma. World J Gastroenterol. 2009;15:5053–7.

Kim Y, Yoon JW, Xiao X, Dean NM, Monia BP, Marcusson EG. Selective down-regulation of glioma-associated oncogene 2 inhibits the proliferation of hepatocellular carcinoma cells. Cancer Res. 2007;67:3583–93.

Kuper H, Tzonou A, Kaklamani E, Hsieh CC, Lagiou P, Adami HO, Trichopoulos D, Stuver SO. Tobacco smoking, alcohol consumption and their interaction in the causation of hepatocellular carcinoma. Int J Cancer. 2000;85:498–502.

Lai EC. Notch signaling: control of cell communication and cell fate. Development. 2004;131:965–73.

Langeswaran K, Gowthamkumar S, Vijayaprakash S, Revathy R, Balasubramanian MP. Influence of limonin on Wnt signalling molecule in HepG2 cell lines. J Nat Sci Biol Med. 2013;4:126–33.

Laurent-Puig P, Zucman-Rossi J. Genetics of hepatocellular tumors. Oncogene. 2006;25:3778–86.

Levin HL, Moran JV. Dynamic interactions between transposable elements and their hosts. Nat Rev Genet. 2011;12:615–27.

Li M, Zhao H, Zhang X, Wood LD, Anders RA, Choti MA, Pawlik TM, Daniel HD, Kannangai R, Offerhaus GJ, Velculescu VE, Wang L, Zhou S, Vogelstein B, Hruban RH, Papadopoulos N, Cai J, Torbenson MS, Kinzler KW. Inactivating mutations of the chromatin remodeling gene ARID2 in hepatocellular carcinoma. Nat Genet. 2011;43:828–9.

Liao Y, Wang B, Huang ZL, Shi M, Yu XJ, Zheng L, Li S, Li L. Increased circulating Th17 cells after transarterial chemoembolization correlate with improved survival in stage III hepatocellular carcinoma: a prospective study. PLoS One. 2013;8:e60444.

Lin M, Guo LM, Liu H, Du J, Yang J, Zhang LJ, Zhang B. Nuclear accumulation of glioma-associated oncogene 2 protein and enhanced expression of forkhead-box transcription factor M1 protein in human hepatocellular carcinoma. Histol Histopathol. 2010;25:1269–75.

Llovet JM, Bruix J. Molecular targeted therapies in hepatocellular carcinoma. Hepatology. 2008;48:1312–27.

Llovet JM, Burroughs A, Bruix J. Hepatocellular carcinoma. Lancet. 2003;362:1907–17.

Lu JT, Zhao WD, He W, Wei W. Hedgehog signaling pathway mediates invasion and metastasis of hepatocellular carcinoma via ERK pathway. Acta Pharmacol Sin. 2012;33:691–700.

Lujambio A, Akkari L, Simon J, Grace D, Tschaharganeh DF, Bolden JE, Zhao Z, Thapar V, Joyce JA, Krizhanovsky V, Lowe SW. Non-cell-autonomous tumor suppression by p53. Cell. 2013;153:449–60.

Muller PA, Vousden KH. P53 mutations in cancer. Nat Cell Biol. 2013;15:2–8.

Murakami Y, Saigo K, Takashima H, Minami M, Okanoue T, Brechot C, Paterlini-Brechot P. Large scaled analysis of hepatitis b virus (HBV) DNA integration in HBV related hepatocellular carcinomas. Gut. 2005;54:1162–8.

Muto J, Shirabe K, Sugimachi K, Maehara Y. Review of angiogenesis in hepatocellular carcinoma. Hepatol Res. 2015;45:1–9.

Nagae G, Isagawa T, Shiraki N, Fujita T, Yamamoto S, Tsutsumi S, Nonaka A, Yoshiba S, Matsusaka K, Midorikawa Y, Ishikawa S, Soejima H, Fukayama M, Suemori H, Nakatsuji N, Kume S, Aburatani H. Tissue-specific demethylation in CpG-poor promoters during cellular differentiation. Hum Mol Genet. 2011;20:2710–21.

Nakagawa H, Shibata T. Comprehensive genome sequencing of the liver cancer genome. Cancer Lett. 2013;340:234–40.

Narimatsu T, Tamori A, Koh N, Kubo S, Hirohashi K, Yano Y, Arakawa T, Otani S, Nishiguchi S. P16 promoter hypermethylation in human hepatocellular carcinoma with or without hepatitis virus infection. Intervirology. 2004;47:26–31.

Neuveut C, Wei Y, Buendia MA. Mechanisms of HBV-related hepatocarcinogenesis. J Hepatol. 2010;52:594–604.

Nishida N, Nishimura T, Nagasaka T, Ikai I, Goel A, Boland CR. Extensive methylation is associated with beta-catenin mutations in hepatocellular carcinoma: evidence for two distinct pathways of human hepatocarcinogenesis. Cancer Res. 2007;67:4586–94.

Nishida N, Nishimura T, Nakai T, Chishina H, Arizumi T, Takita M, Kitai S, Yada N, Hagiwara S, Inoue T, Minami Y, Ueshima K, Sakurai T, Kudo M. Genome-wide profiling of DNA methylation and tumor progression in human hepatocellular carcinoma. Dig Dis. 2014;32:658–63.

Ogasawara S, Yano H, Momosaki S, Nishida N, Takemoto Y, Kojiro S, Kojiro M. Expression of matrix metalloproteinases (MMPs) in cultured hepatocellular carcinoma (HCC) cells and surgically resected HCC tissues. Oncol Rep. 2005;13:1043–8.

Ogimoto I, Shibata A, Kurozawa Y, Nose T, Yoshimura T, Suzuki H, Iwai N, Sakata R, Fujita Y, Ichikawa S, Fukuda K, Tamakoshi A. Risk of death due to hepatocellular carcinoma among smokers and ex-smokers, Univariate analysis of JACC study data. Kurume Med J. 2004;51:71–81.

Ostrowski J, Kostrzewska E, Michalak T, Zawirska B, Medrzejewski W, Gregor A. Abnormalities in liver function and morphology and impaired aminopyrine metabolism in hereditary hepatic porphyrias. Gastroenterology. 1983;85:1131–7.

Paterlini-Brechot P, Saigo K, Murakami Y, Chami M, Gozuacik D, Mugnier C, Lagorce D, Brechot C. Hepatitis B virus-related insertional mutagenesis occurs frequently in human liver cancers and recurrently targets human telomerase gene. Oncogene. 2003;22:3911–6.

Patil MA, Gutgemann I, Zhang J, Ho C, Cheung ST, Ginzinger D, Li R, Dykema KJ, So S, Fan ST, Kakar S, Furge KA, Buttner R, Chen X. Array-based comparative genomic hybridization reveals recurrent chromosomal aberrations and Jab1 as a potential target for 8q gain in hepatocellular carcinoma. Carcinogenesis. 2005;26:2050–7.

Personeni N, Rimassa L, Pressiani T, Destro A, Ligorio C, Tronconi MC, Bozzarelli S, Carnaghi C, Di Tommaso L, Giordano L, Roncalli M, Santoro A. Molecular determinants of outcome in sorafenib-treated patients with hepatocellular carcinoma. J Cancer Res Clin Oncol. 2013;139:1179–87.

Pogribny IP, James SJ. Reduction of p53 gene expression in human primary hepatocellular carcinoma is associated with promoter region methylation without coding region mutation. Cancer Lett. 2002;176:169–74.

Polakis P. Wnt signaling in cancer. Cold Spring Harb Perspect Biol. 2012;4:a0080052.

Pollack JR, Perou CM, Alizadeh AA, Eisen MB, Pergamenschikov A, Williams CF, Jeffrey SS, Botstein D, Brown PO. Genome-wide analysis of DNA copy-number changes using cDNA microarrays. Nat Genet. 1999;23:41–6.

Qi R, An H, Yu Y, Zhang M, Liu S, Xu H, Guo Z, Cheng T, Cao X. Notch1 signaling inhibits growth of human hepatocellular carcinoma through induction of cell cycle arrest and apoptosis. Cancer Res. 2003;63:8323–9.

Radtke F, Raj K. The role of notch in tumorigenesis: oncogene or tumour suppressor? Nat Rev Cancer. 2003;3:756–67.

Roessler S, Long EL, Budhu A, Chen Y, Zhao X, Ji J, Walker R, Jia HL, Ye QH, Qin LX, Tang ZY, He P, Hunter KW, Thorgeirsson SS, Meltzer PS, Wang XW. Integrative genomic identification of genes on 8p associated with hepatocellular carcinoma progression and patient survival. Gastroenterology. 2012;142:957–66.

Rollins RA, Haghighi F, Edwards JR, Das R, Zhang MQ, Ju J, Bestor TH. Large-scale structure of genomic methylation patterns. Genome Res. 2006;16:157–63.

Saigo K, Yoshida K, Ikeda R, Sakamoto Y, Murakami Y, Urashima T, Asano T, Kenmochi T, Inoue I. Integration of hepatitis B virus DNA into the myeloid/lymphoid or mixed-lineage leukemia (MLL4) gene and rearrangements of MLL4 in human hepatocellular carcinoma. Hum Mutat. 2008;29:703–8.

Schlaeger C, Longerich T, Schiller C, Bewerunge P, Mehrabi A, Toedt G, Kleeff J, Ehemann V, Eils R, Lichter P, Schirmacher P, Radlwimmer B. Etiology-dependent molecular mechanisms in human hepatocarcinogenesis. Hepatology. 2008;47:511–20.

Scott CR. The genetic tyrosinemias. Am J Med Genet C: Semin Med Genet. 2006;142C:121–6.

Sha WH, Zeng XH, Min L. The correlation between NK cell and liver function in patients with primary hepatocellular carcinoma. Gut Liver. 2014;8:298–305.

Shaib Y, El-Serag HB. The epidemiology of cholangiocarcinoma. Semin Liver Dis. 2004;24:115–25.

Shen J, Wang S, Zhang YJ, Kappil M, Wu HC, Kibriya MG, Wang Q, Jasmine F, Ahsan H, Lee PH, Yu MW, Chen CJ, Santella RM. Genome-wide DNA methylation profiles in hepatocellular carcinoma. Hepatology. 2012;55:1799–808.

Shen J, Wang S, Zhang YJ, Wu HC, Kibriya MG, Jasmine F, Ahsan H, Wu DP, Siegel AB, Remotti H, Santella RM. Exploring genome-wide DNA methylation profiles altered in hepatocellular carcinoma using Infinium Humanmethylation 450 Beadchips. Epigenetics-Us. 2013;8:34–43.

Shibata T, Aburatani H. Exploration of liver cancer genomes. Nat Rev Gastroenterol Hepatol. 2014;11:340–9.

Shibata T, Ohta T, Tong KI, Kokubu A, Odogawa R, Tsuta K, Asamura H, Yamamoto M, Hirohashi S. Cancer related mutations in NRF2 impair its recognition by Keap1-Cul3 E3 ligase and promote malignancy. Proc Natl Acad Sci U S A. 2008;105:13568–73.

Shukla R, Upton KR, Munoz-Lopez M, Gerhardt DJ, Fisher ME, Nguyen T, Brennan PM, Baillie JK, Collino A, Ghisletti S, Sinha S, Iannelli F, Radaelli E, Dos SA, Rapoud D, Guettier C, Samuel D, Natoli G, Carninci P, Ciccarelli FD, Garcia-Perez JL, Faivre J, Faulkner GJ. Endogenous retrotransposition activates oncogenic pathways in hepatocellular carcinoma. Cell. 2013;153:101–11.

Sicklick JK, Li YX, Jayaraman A, Kannangai R, Qi Y, Vivekanandan P, Ludlow JW, Owzar K, Chen W, Torbenson MS, Diehl AM. Dysregulation of the hedgehog pathway in human hepatocarcinogenesis. Carcinogenesis. 2006;27:748–57.

Strazzabosco M, Fabris L. Notch signaling in hepatocellular carcinoma: guilty in association! Gastroenterology. 2012;143:1430–4.

Sung WK, Zheng H, Li S, Chen R, Liu X, Li Y, Lee NP, Lee WH, Ariyaratne PN, Tennakoon C, Mulawadi FH, Wong KF, Liu AM, Poon RT, Fan ST, Chan KL, Gong Z, Hu Y, Lin Z, Wang G, Zhang Q, Barber TD, Chou WC, Aggarwal A, Hao K, Zhou W, Zhang C, Hardwick J, Buser C, Xu J, Kan Z, Dai H, Mao M, Reinhard C, Wang J, Luk JM. Genome-wide survey of recurrent HBV integration in hepatocellular carcinoma. Nat Genet. 2012;44:765–9.

Swergold GD. Identification, characterization, and cell specificity of a human line-1 promoter. Mol Cell Biol. 1990;10:6718–29.

Taguchi K, Motohashi H, Yamamoto M. Molecular mechanisms of the Keap1-Nrf2 pathway in stress response and cancer evolution. Genes Cells. 2011;16:123–40.

Tannapfel A, Sommerer F, Benicke M, Katalinic A, Uhlmann D, Witzigmann H, Hauss J, Wittekind C. Mutations of the BRAF gene in cholangiocarcinoma but not in hepatocellular carcinoma. Gut. 2003;52:706–12.

Thompson MD, Monga SP. Wnt/beta-catenin signaling in liver health and disease. Hepatology. 2007;45:1298–305.

Tomasi ML, Li TW, Li M, Mato JM, Lu SC. Inhibition of human methionine adenosyltransferase 1A transcription by coding region methylation. J Cell Physiol. 2012;227:1583–91.

Torre LA, Bray F, Siegel RL, Ferlay J, Lortet-Tieulent J, Jemal A. Global cancer statistics, 2012. CA Cancer J Clin. 2015;65:87–108.

Totoki Y, Tatsuno K, Yamamoto S, Arai Y, Hosoda F, Ishikawa S, Tsutsumi S, Sonoda K, Totsuka H, Shirakihara T, Sakamoto H, Wang L, Ojima H, Shimada K, Kosuge T, Okusaka T, Kato K, Kusuda J, Yoshida T, Aburatani H, Shibata T. High-resolution characterization of a hepatocellular carcinoma genome. Nat Genet. 2011;43:464–9.

Treangen TJ, Salzberg SL. Repetitive DNA and next-generation sequencing: computational challenges and solutions. Nat Rev Genet. 2012;13:36–46.

Tschaharganeh DF, Chen X, Latzko P, Malz M, Gaida MM, Felix K, Ladu S, Singer S, Pinna F, Gretz N, Sticht C, Tomasi ML, Delogu S, Evert M, Fan B, Ribback S, Jiang L, Brozzetti S, Bergmann F, Dombrowski F, Schirmacher P, Calvisi DF, Breuhahn K. Yes-associated protein up-regulates Jagged-1 and activates the Notch pathway in human hepatocellular carcinoma. Gastroenterology. 2013;144:1530–42.

Udali S, Guarini P, Ruzzenente A, Ferrarini A, Guglielmi A, Lotto V, Tononi P, Pattini P, Moruzzi S, Campagnaro T, Conci S, Olivieri O, Corrocher R, Delledonne M, Choi SW, Friso S. DNA methylation and gene expression profiles show novel regulatory pathways in hepatocellular carcinoma. Clin Epigenetics. 2015;7:43.

Viatour P, Ehmer U, Saddic LA, Dorrell C, Andersen JB, Lin C, Zmoos AF, Mazur PK, Schaffer BE, Ostermeier A, Vogel H, Sylvester KG, Thorgeirsson SS, Grompe M, Sage J. Notch signaling inhibits hepatocellular carcinoma following inactivation of the RB pathway. J Exp Med. 2011;208:1963–76.

Villanueva A, Alsinet C, Yanger K, Hoshida Y, Zong Y, Toffanin S, Rodriguez-Carunchio L, Sole M, Thung S, Stanger BZ, Llovet JM. Notch signaling is activated in human hepatocellular carcinoma and induces tumor formation in mice. Gastroenterology. 2012;143:1660–9.

Wang F, Jing X, Li G, Wang T, Yang B, Zhu Z, Gao Y, Zhang Q, Yang Y, Wang Y, Wang P, Du Z. Foxp3+ regulatory T cells are associated with the natural history of chronic hepatitis B and poor prognosis of hepatocellular carcinoma. Liver Int. 2012;32:644–55.

Wang Y, Han C, Lu L, Magliato S, Wu T. Hedgehog signaling pathway regulates autophagy in human hepatocellular carcinoma cells. Hepatology. 2013;58:995–1010.

Weinberg AG, Mize CE, Worthen HG. The occurrence of hepatoma in the chronic form of hereditary tyrosinemia. J Pediatr. 1976;88:434–8.

Werner M, Almer S, Prytz H, Lindgren S, Wallerstedt S, Bjornsson E, Bergquist A, Sandberg-Gertzen H, Hultcrantz R, Sangfelt P, Weiland O, Danielsson A. Hepatic and extrahepatic malignancies in autoimmune hepatitis. A long-term follow-up in 473 Swedish patients. J Hepatol. 2009;50:388–93.

Wong GS, Rustgi AK. Matricellular proteins: priming the tumour microenvironment for cancer development and metastasis. Br J Cancer. 2013;108:755–61.

Wong IH, Lo YM, Zhang J, Liew CT, Ng MH, Wong N, Lai PB, Lau WY, Hjelm NM, Johnson PJ. Detection of aberrant p16 methylation in the plasma and serum of liver cancer patients. Cancer Res. 1999;59:71–3.

Woo HG, Park ES, Lee JS, Lee YH, Ishikawa T, Kim YJ, Thorgeirsson SS. Identification of potential driver genes in human liver carcinoma by genomewide screening. Cancer Res. 2009;69:4059–66.

Wu XZ, Xie GR, Chen D. Hypoxia and hepatocellular carcinoma: the therapeutic target for hepatocellular carcinoma. J Gastroenterol Hepatol. 2007;22:1178–82.

Yang B, Guo M, Herman JG, Clark DP. Aberrant promoter methylation profiles of tumor suppressor genes in hepatocellular carcinoma. Am J Pathol. 2003;163:1101–7.

Yu J, Shen J, Sun TT, Zhang X, Wong N. Obesity, insulin resistance, NASH and hepatocellular carcinoma. Semin Cancer Biol. 2013;23:483–91.

Zang JJ, Xie F, Xu JF, Qin YY, Shen RX, Yang JM, He J. P16 gene hypermethylation and hepatocellular carcinoma: a systematic review and meta-analysis. World J Gastroenterol. 2011;17:3043–8.

Zhang YJ, Ahsan H, Chen Y, Lunn RM, Wang LY, Chen SY, Lee PH, Chen CJ, Santella RM. High frequency of promoter hypermethylation of RASSF1A and p16 and its relationship to aflatoxin B1-DNA adduct levels in human hepatocellular carcinoma. Mol Carcinog. 2002;35:85–92.

Zhang YJ, Chen Y, Ahsan H, Lunn RM, Lee PH, Chen CJ, Santella RM. Inactivation of the DNA repair gene O6-methylguanine-DNA methyltransferase by promoter hypermethylation and its relationship to aflatoxin B1-DNA adducts and p53 mutation in hepatocellular carcinoma. Int J Cancer. 2003;103:440–4.

Zhang JP, Yan J, Xu J, Pang XH, Chen MS, Li L, Wu C, Li SP, Zheng L. Increased intratumoral IL-17-producing cells correlate with poor survival in hepatocellular carcinoma patients. J Hepatol. 2009;50:980–9.

Zhang D, Liu J, Wang Y, Chen J, Chen T. shRNA-mediated silencing of Gli2 gene inhibits proliferation and sensitizes human hepatocellular carcinoma cells towards trail-induced apoptosis. J Cell Biochem. 2011;112:3140–50.

Zhang D, Cao L, Li Y, Lu H, Yang X, Xue P. Expression of glioma-associated oncogene 2 (Gli 2) is correlated with poor prognosis in patients with hepatocellular carcinoma undergoing hepatectomy. World J Surg Oncol. 2013;11:25.

Zheng T, Wang J, Song X, Meng X, Pan S, Jiang H, Liu L. Nutlin-3 cooperates with doxorubicin to induce apoptosis of human hepatocellular carcinoma cells through p53 or p73 signaling pathways. J Cancer Res Clin Oncol. 2010a;136:1597–604.

Zheng X, Yao Y, Xu Q, Tu K, Liu Q. Evaluation of glioma-associated oncogene 1 expression and its correlation with the expression of sonic hedgehog, E-cadherin and S100a4 in human hepatocellular carcinoma. Mol Med Rep. 2010b;3:965–70.

Zheng X, Vittar NB, Gai X, Fernandez-Barrena MG, Moser CD, Hu C, Almada LL, McCleary-Wheeler AL, Elsawa SF, Vrabel AM, Shire AM, Comba A, Thorgeirsson SS, Kim Y, Liu Q, Fernandez-Zapico ME, Roberts LR. The transcription factor GLI1 mediates TGFBETA1 driven EMT in hepatocellular carcinoma via a SNAI1-dependent mechanism. PLoS One. 2012;7:e49581.

Zheng X, Zeng W, Gai X, Xu Q, Li C, Liang Z, Tuo H, Liu Q. Role of the hedgehog pathway in hepatocellular carcinoma (review). Oncol Rep. 2013;30:2020–6.

Zhong S, Tang MW, Yeo W, Liu C, Lo YM, Johnson PJ. Silencing of GSTP1 gene by CpG island DNA hypermethylation in HBV-associated hepatocellular carcinomas. Clin Cancer Res. 2002;8:1087–92.

Zhou L, Huang Y, Li J, Wang Z. The mTOR pathway is associated with the poor prognosis of human hepatocellular carcinoma. Med Oncol. 2010;27:255–61.

Zhou Q, Lui VW, Yeo W. Targeting the PI3K/AKT/mTOR pathway in hepatocellular carcinoma. Future Oncol. 2011;7:1149–67.

Zhou L, Wang DS, Li QJ, Sun W, Zhang Y, Dou KF. The down-regulation of Notch1 inhibits the invasion and migration of hepatocellular carcinoma cells by inactivating the Cyclooxygenase-2/Snail/E-cadherin pathway in vitro. Dig Dis Sci. 2013;58:1016–25.

Dr Tingting Fang got her master's degree of medicine at Fudan University in June 2015, with the thesis on metastasis of hepatocellular carcinoma. She is now studying in Zhongshan hospital affiliated to Fudan University for Doctor's degree of medicine.

Dr Li Feng is a Professor and Director of Department of Science and Research, and Chief-in-Physician of Department of Internal Medicine, Minhang Hospital of Fudan University, Shanghai, China. She also serves as the General Secretary and Committee Member of Shanghai Medical Association of Digestive Endoscopy Committee, Digestive Endoscopy Esophageal Division and, Shanghai Traditional Chinese and Western Medical Association Digestion Professional Committee, Dr Feng won the "Shanghai Medical Science and Technology Award", "Minhang District Science and Technology Achievement Award" and others. She published more than 50 articles, acted as an Editor-in-Chief of Medical Monographs 2, and authored 8 books. Dr Feng also won the honorable title of "Shanghai Leading Talents", "Shanghai Workers of Outstanding Innovative Technology Achievement Award", "Shanghai Medical Women Association Outstanding Female Physician Award".

Dr. Jinglin Xia is the vice president of Medical College Fudan University Shanghai China. He finished postdoctoral research at the Medical Center of Pittsburgh University in 2002–2003. He serves as the editor of a number of international medical journals executive member of the International Association of translational medicine vice chairman of the Shanghai Medical Association. Dr. Xia's main research focuses on tumor angiogenesis and tumor inflammatory microenvironment. Dr. Xia has published over 60 papers and plenty of books covering tumor angiogenesis, microRNAs, and inflammatory factors and so on.

MIX
Papier aus verantwortungsvollen Quellen
Paper from responsible sources
FSC® C105338

If you have any concerns about our products,
you can contact us on
ProductSafety@springernature.com

In case Publisher is established outside the EU,
the EU authorized representative is:
Springer Nature Customer Service Center GmbH
Europaplatz 3, 69115 Heidelberg, Germany

Printed by Libri Plureos GmbH
in Hamburg, Germany